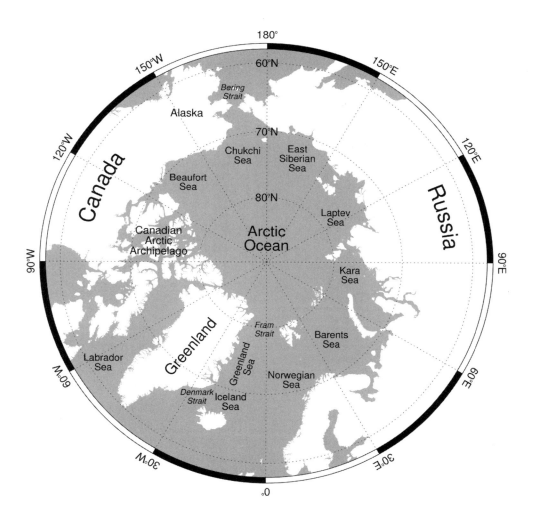

CLIMATE CHANGE IN THE POLAR REGIONS

The polar regions have experienced some remarkable environmental changes in recent decades, such as the Antarctic ozone hole, the loss of large amounts of sea ice from the Arctic Ocean and major warming on the Antarctic Peninsula. The polar regions are also predicted to warm more than any other region on Earth over the next century if greenhouse gas concentrations continue to rise. Yet trying to separate natural climate variability from anthropogenic forcing still presents many problems. This book presents a thorough review of how the polar climates have changed over the last million years and sets recent changes within the long-term perspective, as determined from ice and ocean sediment cores. The approach taken is highly cross-disciplinary and the close links between the atmosphere, ocean and ice at high latitudes are stressed. The volume will be invaluable for researchers and advanced students in polar science, climatology, global change, meteorology, oceanography and glaciology.

JOHN TURNER is a research scientist at the British Antarctic Survey in Cambridge, UK, where he leads a project investigating recent Antarctic climate change and how it may change over the next century. He has had a long involvement with the Scientific Committee on Antarctic Research (SCAR): he was the Chief Officer of the Physical Sciences Standing Scientific Group from 2002 to 2006 and chaired the steering committee of the SCAR programme on Antarctica and the Global Climate System from 2005 to 2008. He is the co-author of *Antarctic Meteorology and Climatology* (1997) and *Polar Lows: Mesoscale Weather Systems in the Polar Regions* (2003), both of which are published by Cambridge University Press. He was awarded the International Journal of Climatology Prize of the Royal Meteorological Society in 2005.

GARETH J. MARSHALL is a climatologist at the British Antarctic Survey where he is the Climate programme coordinator. He has worked at BAS since 1995 after completing his PhD at the University of Cambridge. His research has included field work in both polar regions and he has contributed to more than 50 scientific papers. Recently, he was a corresponding author on the Intergovernmental Panel on Climate Change Fourth Assessment Report. He is also a member of the World Climate Research Programme CLIVAR Southern Ocean panel, which addresses climate variability and predictability in this region.

CLIMATE CHANGE IN THE POLAR REGIONS

JOHN TURNER
British Antarctic Survey

AND

GARETH J. MARSHALL
British Antarctic Survey

CAMBRIDGE UNIVERSITY PRESS
Cambridge, New York, Melbourne, Madrid, Cape Town,
Singapore, São Paulo, Delhi, Tokyo, Mexico City

Cambridge University Press
The Edinburgh Building, Cambridge CB2 8RU, UK

Published in the United States of America by Cambridge University Press, New York

www.cambridge.org
Information on this title: www.cambridge.org/9780521850100

© John Turner and Gareth J. Marshall 2011

This publication is in copyright. Subject to statutory exception
and to the provisions of relevant collective licensing agreements,
no reproduction of any part may take place without the written
permission of Cambridge University Press.

First published 2011

Printed in the United Kingdom at the University Press, Cambridge

A catalogue record for this publication is available from the British Library

Library of Congress Cataloguing in Publication data
Turner, J. (John)
Climate change in the polar regions / John Turner and Gareth J. Marshall.
p.   cm.
Includes bibliographical references and index.
ISBN 978-0-521-85010-0 (hardback)
1. Climatic changes – z Polar regions – History.   2. Meteorology – Polar regions – Observations.
3. Climatic changes – Environmental aspects – Polar regions.   4. Sea ice – Polar regions.
5. Global warming – Polar regions.   I. Marshall, Gareth.   II. Title.
QC903.2.P73T87   2011
551.6911–dc22
2010052384

ISBN 978-0-521-85010-0 Hardback

Cambridge University Press has no responsibility for the persistence or
accuracy of URLs for external or third-party internet websites referred to
in this publication, and does not guarantee that any content on such
websites is, or will remain, accurate or appropriate.

Contents

Preface		*page* ix
1	**Introduction**	1
	1.1 The environment of the polar regions	1
	1.2 The role of the polar regions in the global climate system	9
	1.3 Possible implications of high latitude climate change	12
2	**Polar climate data and models**	16
	2.1 Introduction	16
	2.2 Instrumental observations	17
	2.3 Meteorological analysis fields	28
	2.4 Remotely sensed data	36
	2.5 Proxy climate data	42
	2.6 Models	51
3	**The high latitude climates and mechanisms of change**	61
	3.1 Introduction	61
	3.2 Factors influencing the broadscale climates of the polar regions	63
	3.3 Processes of the high latitude climates	71
	3.4 The mechanisms of high latitude climate change	80
	3.5 Atmospheric circulation	114
	3.6 Temperature	126
	3.7 Cloud and precipitation	132
	3.8 Sea ice	142
	3.9 The ocean circulation	154
	3.10 Concluding remarks	161
4	**The last million years**	162
	4.1 Introduction	162
	4.2 The Arctic	164
	4.3 The Antarctic	181
	4.4 Linking high latitude climate change in the two hemispheres	190

vi   *Contents*

5	The Holocene	194
	5.1 Introduction	194
	5.2 Forcing of the climate system during the Holocene	197
	5.3 Atmospheric circulation	202
	5.4 Temperature	207
	5.5 The ocean circulation	234
	5.6 Sea ice and sea surface temperatures	238
	5.7 Atmospheric gases and aerosols	245
	5.8 The cryosphere, precipitation and sea level	247
	5.9 Concluding remarks	254
6	The instrumental period	256
	6.1 Introduction	256
	6.2 The main meteorological elements	257
	6.3 Changes in the atmospheric circulation	270
	6.4 The ocean environment	280
	6.5 Sea ice	289
	6.6 Snow cover	299
	6.7 Permafrost	307
	6.8 Atmospheric gases and aerosols	315
	6.9 Terrestrial ice and sea level	320
	6.10 Attribution of recent changes	331
	6.11 Concluding remarks	334
7	Predictions for the next 100 years	336
	7.1 Introduction	336
	7.2 Possible future greenhouse gas emission scenarios and the IPCC models	337
	7.3 Changes in the atmospheric circulation and the modes of climate variability	340
	7.4 The main meteorological elements	344
	7.5 The ocean circulation and water masses	353
	7.6 Sea ice	356
	7.7 Seasonal snow cover and the terrestrial environment	361
	7.8 Permafrost	362
	7.9 Atmospheric gases and aerosols	364
	7.10 Terrestrial ice, the ice shelves and sea level	366
	7.11 Concluding remarks	372
8	Summary and future research needs	374
	8.1 Introduction	374
	8.2 Gaining improved understanding of past climate change	374

Contents vii

8.3	Modelling the high latitude climate system	381
8.4	Data required	382
8.5	Concluding remarks	385

References	386
Index	428

Preface

The last few years have seen an unprecedented level of interest in the climate of the polar regions. The discovery of the Antarctic ozone hole, the reduction in extent of Arctic sea ice, the disintegration of floating ice shelves around the Antarctic and the high levels of aerosols reaching the Arctic have all been reported extensively in the media. This has been coupled with climate model predictions showing that the high latitude areas will warm more than any other region on Earth over the next century if 'greenhouse gas' concentrations continue to rise. Yet some have pointed to rapid climatic fluctuations that have taken place in the polar regions over the last few centuries and millennia and questioned whether the recent changes that we have seen are not simply a result of natural climate variability. Hence the time is right for a reappraisal of our understanding of recent high latitude climate change in the context of increasing anthropogenic influence on the Earth and our greater understanding of the reasons for past climate variability.

This book seeks to assess the climatic and environmental changes that have taken place over the last century and set these in the context of our understanding of natural climate variability in the pre-industrial period. We will draw on many of the new climate data sets that have become available in recent years and also make use of the results of modelling experiments. The last few years have seen great advances in our ability to observe, monitor and model the present and past polar climates. In particular, the International Polar Year of 2007–08 gave us an unprecedented amount of data from the two polar regions and increased our understanding of the mechanisms responsible for climate variability and change at high latitudes.

The record of in-situ meteorological measurements from observatories and research stations extends back about a century in many parts of the Arctic and about 50 years in the Antarctic. These observations provide us with the most accurate measurements of atmospheric conditions, yet the early observations are widely separated in many areas, with few observations from the ocean areas. However, from the mid 1970s an increasing number of observations became available from polar-orbiting satellites, which allowed the production of increasingly reliable atmospheric analyses of the high latitude areas. Over the past few decades the historical archive of in-situ and satellite observations have been reprocessed using current data assimilation schemes to produce so-called 'reanalysis' data sets. These provide a particularly valuable source for investigation of climate variability and change in the polar regions over approximately the last 30 years.

In the pre-instrumental period the most valuable data on climate variability comes from analysis of chemical species and accumulation in ice cores drilled on high latitude ice sheets. Annual layers and/or specific events, such as volcanic eruptions, can often be identified in these cores, allowing the dating of the core. Many short cores covering a few years to a few decades have been collected in the Arctic and Antarctic but there are far fewer longer cores extending back several centuries or more. However, the Vostok, Dome C and Dome F cores from East Antarctica and several long cores collected from the plateau of Greenland provide extremely valuable records of climatic conditions extending back over one or more ice ages.

The proxy data described above provide us with a reasonable picture of changes in atmospheric conditions in the past, although with decreasing resolution in the earlier part of the record. However, even today we do not have a synoptic picture of the distribution of water masses across the world's ocean and of ocean circulation, so understanding oceanic conditions in the past presents a number of difficulties. Nevertheless, analysis of ocean sediment cores has provided key information on past oceanic conditions to complement the palaeoclimatic atmospheric data.

Complementary to the climate observations from the polar regions that are used in investigating high latitude climate change are atmospheric and coupled atmosphere–ocean models. These have developed rapidly over the last couple of decades from relatively coarse resolution atmosphere-only models, which were only able to simulate the broadest features of the polar climates, to complex, high-resolution models capable of simulating the non-linear interactions between the atmosphere, ocean and cryosphere. Moreover, they have been applied to the present-day climate, and conditions during previous climatic regimes. They are the only tool that we have for predicting how the Earth's climate will evolve in the future.

In this volume we have used the data discussed above to describe past and possible future climate scenarios for the polar regions. The emphasis is on explaining the forcing mechanisms behind the observed changes and the difficulties in differentiating natural climate variability from anthropogenic effects. A priority is to integrate our understanding of the atmospheric, oceanic and cryospheric changes and to present the polar regions within the context of Earth System Science studies.

The geographical focus is obviously the Arctic and Antarctic, but research has shown that there is a close, but non-linear, coupling between the climates of the polar regions and lower latitudes. For example, recent analysis of chemical species in Antarctic ice cores has shown that signals of the El Niño–Southern Oscillation (ENSO) are present in temperature and precipitation data, but that the high latitude response varies between different ENSO events. So, where appropriate, we will not limit our coverage to just the polar regions, as defined as the areas poleward of the Arctic and Antarctic circles, but set the climatic changes of the polar regions in a global context.

Regarding the time period that the book should cover in terms of the past, we will obviously deal with the period when in-situ meteorological observations are available, approximately the last 100 years. The Holocene, which covers approximately the last 11.7 kyr, had roughly the same solar forcing as today and the Mid Holocene warm period was when some of the ice shelves around the Antarctic Peninsula disintegrated in a similar fashion to the way they have collapsed in recent decades. However, the most dramatic climatic changes at high latitudes have been the ice ages, and the latest Antarctic ice cores provide a unique record of such events through a significant part of the Pleistocene. We therefore felt that it would be logical to cover the period covered by the Dome C ice core and consider the last million years. In terms of future changes, we will deal with the Intergovernmental Panel on Climate Change (IPCC) scenarios considering the next 100 years.

As a practical note on timing within this book, for periods of more than 2000 years ago we will use the 'before present (BP)' notation, where the present is taken as AD 1950; thus the start of the Holocene will be indicated as 11.7 kyr BP.

In the first chapter we provide an introduction to the environments of the Arctic and Antarctic and consider the role of the polar regions in the global climate system. Although the book does not generally deal with the societal consequences of climate change, here we provide a brief account of the possible implications of major changes to the high latitude icecaps.

Chapter 2 is concerned with the data and the models that we have available to investigate the past and present polar climates and how they will evolve in the future. We review the availability and quality of the instrumental observations and assess the climatic information that can be derived from ice cores and ocean sediment cores. We also review the reliability of the meteorological reanalysis fields.

In Chapter 3 we consider the mechanisms that are responsible for variability and change in the high latitude climates on a range of timescales and explain why the climates of the two polar regions have their particular form. We deal with the radiation regime, ice/atmosphere feedbacks, the impact of the different land–sea distributions and orography in the two hemispheres and, in particular, the role of the Arctic Ocean compared with the Antarctic continent. We also present some mean meteorological/cryospheric fields for the two areas based on data from recent decades. These provide reference fields for the discussion on past conditions and future predictions.

Chapter 4 discusses our understanding of climate change over the last one million years, which is the most recent half of the Pleistocene. We describe the broadscale climate changes associated with the different ice ages and examine some of the more recent episodes of major change in detail, including the peak of the most recent ice age, the Last Glacial Maximum.

The Holocene is dealt with in Chapter 5. For this period we have much more detailed climatic information as there is greater temporal resolution in the available ice and ocean sediment cores. We consider our current knowledge of atmospheric and oceanic circulation, along with changes in temperature, precipitation, sea ice and the ice sheets.

The instrumental period of the last 50–100 years is covered in Chapter 6. Over this period increasingly sophisticated observations have been obtained of many aspects of the polar environment and these are used to examine variability and change. We consider the main meteorological elements, as well as the ocean environment, sea ice and the icecaps.

xii *Preface*

Although a great deal of atmospheric and cryospheric data are now available, a major gap in our knowledge is still the oceans, where many records are short.

Chapter 7 examines the prospects for the evolution of the polar climates over the next 100 years. The atmospheric and oceanic predictions come from state-of-the-art climate models, many of which were used in the production of the IPCC Fourth Assessment. Such models can also provide information on future changes in sea ice extent; however, prediction of changes in the major ice sheets is still very difficult to quantify.

In Chapter 8 we summarise our current understanding of high latitude climate change and consider future research and data collection needs.

Many people have provided assistance in the production of this book. We would particularly like to thank Drs Dominic Hodgson, Liz Thomas, Hugh Venables and Eric Wolff for reviewing sections of the book. The figures were prepared by Phoebe Allan, Emma Critchey, Katherine Dolan, Tony Phillips and Eleanor Tomlinson. Penny Goodearl obtained permission to use selected figures. Peter Fretwell provided statistics on the area and elevation of the Antarctic continent.

1
Introduction

1.1 The environment of the polar regions

The polar regions can be defined in a number of ways, based on geographical, topographic and even political factors. However, geometrically, the Arctic and Antarctic are considered as the areas of the Earth poleward of the Arctic and Antarctic Circles, which are located at latitudes of 66° 33′ 39″ north and south of the Equator (see maps on the end papers). These areas experience at least one day each year when the Sun does not set, and one day when the Sun is always below the horizon. At the poles themselves there is only one sunrise and one sunset each year, which occurs on the equinoxes of 21 September and 21 March. Together the Arctic and Antarctic comprise about 8% of the surface area of the Earth.

The regions of perpetual summer sunlight and winter darkness are present because the Earth is tilted away from the plane of its orbit around the Sun by 23° 27′, resulting in the high latitude areas having periods when they are orientated away from or towards the Sun. The tilt of the Earth's axis changes over long periods of time (millennia), resulting in variations in the latitude of the Arctic and Antarctic Circles. The change in the tilt, along with slow variations in the Earth's orbit about the Sun, alter the amount of solar radiation arriving at different parts of the Earth, which is a major factor in long-term, millennial-scale climate variability. This subject is covered in depth in Section 3.4.

The Arctic Circle runs through the North Atlantic, southern Greenland, northern Canada, central Alaska, northern Russia and northern Fennoscandia. The Antarctic Circle by contrast is primarily located over the Southern Ocean, only crossing the Antarctic continent on the Antarctic Peninsula, and in the northern edges of Enderby Land, Wilkes Land and George V Land.

In the polar regions the Sun always has a relatively low elevation during the summer, never reaching more than about 23° above the horizon. The instantaneous amount of solar radiation received at the top of the atmosphere is therefore quite small, and much less than occurs in tropical latitudes where the Sun is nearly overhead.

A number of factors influence the amount of solar radiation that is absorbed by the surface and is therefore available to heat the ground or ocean, or melt snow and ice. Although the Sun is low in the sky in summer, the high latitude atmosphere is relatively clear (especially in the Antarctic) with few suspended particles (aerosols) that can scatter or reflect the incoming radiation. The amount of water vapour in the atmosphere is also quite low, especially on the Antarctic Plateau, so the amount of radiation absorbed by this gas is very limited. However, a major factor in influencing the amount of radiation arriving at the surface is the length of the period of sunlight. On the Antarctic Plateau the long period of continuous daylight, coupled with the clear, frequently cloud-free atmosphere means that this region receives more solar radiation than anywhere else on Earth. Nevertheless, the lowest temperature ever recorded at the surface of the Earth was at Vostok Station (78.5° S, 106.9° E, 3488 m) high on the plateau.

Although large amounts of solar radiation can be received at the surface in summer, much of this insolation is reflected back to space because of the high albedo (reflectivity) of the snow and ice surfaces. Freshly fallen snow can have a very high albedo of 90%, but this drops as the snow pack ages, typically reaching values of 80%. Exposed glacier ice (known as blue ice) typically has an albedo of around 70%. Snow with dust particles on the surface will have a lower albedo, and as snow melts, gradually exposing the rock or soil surface below, the albedo will gradually drop to that of bare ground, which is typically 15–20%. The Antarctic Plateau is therefore a unique location, receiving large amounts of solar radiation in summer, most of which is reflected back to space, resulting in the lowest summer temperatures on Earth.

In the sea ice zone the surface consists of a mix of ice floes that typically have an albedo of 70–80% and open water with an albedo of 10–15%. So the fractional ice cover is critical in determining the amount of solar radiation that is absorbed by the surface. A major difference between the Arctic and Antarctic is the amount of multi-year sea ice (sea ice that has survived one summer) that is present. In the Antarctic most sea ice melts by the late summer, with only small amounts of ice persisting into the following winter along the coast of East Antarctic and over the western Weddell Sea. In the Arctic, however, there is a higher proportion of multi-year ice present.

During periods when solar radiation is received, a number of different 'polar feedback' mechanisms can come into play, which can amplify small environmental changes. For example, in the case where the ocean is partially covered by sea ice, once the fractional ice cover has dropped to a certain level, enough solar radiation may be received to warm the upper layers of the ocean, resulting in the rapid melting of the remaining sea ice in a region. This is often the case along the coast of Dronning Maud Land, on the eastern side of the Weddell Sea, Antarctica, where there can be a rapid expansion of the coastal polynya (the ice-free region next to the coast) during December. A similar positive feedback is found over snow-covered land areas during the high latitude spring, where the snow can rapidly retreat once enough bare ground is exposed and sufficient heat has been absorbed by the surface. This takes place across northern Eurasia and results in extensive river runoff into the Arctic Ocean.

As discussed in later chapters, many projections of future climate suggest that the largest increases in near-surface air temperature will occur at high latitudes, possibly as a result of feedback mechanisms. We will therefore return in later chapters to the question of whether

1.1 The environment of the polar regions

such mechanisms are responsible for recent Arctic and Antarctic climate change, and the role that they may play in the future.

During the long polar winter, environmental conditions are quite different from the summer months. The lack of solar radiation, coupled with the relatively dry atmosphere (especially in the Antarctic) results in rapid cooling near the Earth's surface and the creation of an atmospheric temperature inversion, where the temperature increases with height for several hundreds of metres or more. The most pronounced temperature inversions are found on the Antarctic Plateau, where the temperature difference between the surface and an elevation of a few hundred metres can be in excess of 25 °C. Over the Arctic Ocean in winter there is still a sizeable flux of heat from the ocean into the lowest layers of the atmosphere, either through the sea ice or more often via the leads (the ice-free areas between the ice floes). This limits the strength of the temperature inversion in these maritime areas.

This section has been concerned with climatic and environmental factors that are common to both polar regions. However, in the next section we focus on some of the striking differences in the climates of the Arctic and Antarctic that result from the markedly different land/sea distributions and orographic conditions in the two polar regions.

1.1.1   The Arctic

The Arctic (see the front end paper) consists of the Arctic Ocean, a number of islands and archipelagos, and the northern parts of North America, Greenland, Russia and Scandinavia. The huge ice sheet of Greenland extends to over 3 km in height and contains about $3 \times 10^6 \, km^3$ of ice. It has a major impact on the atmospheric circulation of the North Atlantic and its climatological effects can be felt well into Europe. However, the rest of the orography of the Arctic is relatively low, and there are extensive ocean areas around the Arctic, such as the Greenland Sea, so that mid latitude weather systems can penetrate to high latitudes. As maritime air masses move northwards and are cooled, extensive areas of cloud form, influencing the radiation regime. The Arctic is therefore characterised by rather cloudy conditions, although as the air is fairly cold it is unable to hold a great deal of water vapour, so precipitation amounts tend to be relatively low. For example, Barrow, Alaska, only receives about 115 mm of precipitation per year, compared with London, which on average gets almost 600 mm a year.

The proximity of the Arctic to major industrial centres in Europe and Eurasia means that atmospheric pollutants, such as sulphates, can readily reach northern high latitudes. This phenomenon, which is known as Arctic haze, is seasonal in nature with a peak in the spring once the Sun returns after the long Arctic winter. The pollutants are apparent as a brownish haze that can reduce the visibility to less than 30 km and the aerosols that make up Arctic haze can have an impact on climate forcing, and also contaminate the fragile Arctic ecosystems and food chains.

At the heart of the Arctic is the Arctic Ocean, which covers an area of $14.09 \times 10^6 \, km^2$, making it the smallest of the world's oceans. It lies almost entirely within the Arctic Circle and is mostly surrounded by land, with its only outlets being the Bering Strait between Alaska and Russia, the Davis Strait between Greenland and Canada, and the Denmark Strait and the Norwegian Sea between Greenland and Europe. However, the only deep passage out of the Arctic Ocean is the Fram Strait between Greenland and Spitsbergen, which has a major influence on the exchange of water masses with the rest of the world's oceans. The ocean has an average depth of 3658 m off the continental shelf and the deepest point is the Fram Basin at a depth of 4665 m. The Arctic Ocean has the widest continental shelf of all the oceans. The Arctic Ocean is divided by the Lomonosov Ridge into two major basins, consisting of the Eurasian Basin and the North American Basin, both of which have a depth in excess of 4300 m.

The Arctic Ocean contains three primary water masses. In the near-surface layer, which is up to 200 m thick, there is Polar Surface Water, above warm Atlantic Water, with Arctic Ocean Deep Water below about 800 m. Polar Surface Water has a low salinity as a result of the very large input of fresh water from the major river systems that issue into the Arctic Ocean, such as the Lena, Yenisey and Mackenzie.

Sea ice is a major feature of the Arctic and it has a profound impact on the meteorology and oceanography of the entire Arctic region. Sea ice forms when the upper layer of the ocean freezes during the winter to form new or first-year ice. It is during late winter, around the March/April period, that Arctic sea ice reaches its maximum extent of about 15×10^6 km^2, covering most of the Arctic Ocean and extending into the neighbouring seas. During the summer months there is extensive melting of sea ice and by September it has retreated to cover about half the area of its late-winter peak. Multi-year sea ice can reach thicknesses of several metres. In recent years there has been a well-publicised decrease in the amount of Arctic sea ice, and climate projections for the next century suggest that the area covered by sea ice will continue to shrink. This subject will be discussed in detail in Chapters 6 and 7.

The near-surface ocean currents and the motion of the sea ice within the Arctic Ocean are essentially the same, and are characterised by the anticyclonic Beaufort Gyre poleward of North America and the Transpolar Drift that extends from the northern coast of Siberia to the Fram Strait and down the east coast of Greenland.

The land surrounding the Arctic Ocean is characterised by tundra, boreal forests, peat-lands and permafrost (permanently frozen ground). The boreal forest or taiga consists of coniferous trees and is located on the edges of the Arctic, with the northern limit roughly following the July mean 13 °C isotherm. The forest edge therefore extends from roughly 68° N in the Brooks Range of Alaska to 58° N on the west coast of Hudson Bay. In Siberia the tree limit is further south, some 500 km inland of the Siberian Sea.

North of the boreal forest is the treeless Arctic tundra, consisting of low shrubs, mosses and lichens, which can exist in an environment with a short growing season. Although the tundra has low species diversity, it supports large populations of wild and semi-domestic animals. Figure 1.1 shows a tundra landscape in the Denali National Park, Alaska.

The tundra is characterised by persistent winter snow cover and frozen ground. Snow depth is typically 30–40 cm, reflecting the limited capacity of the air to hold moisture and the remoteness of many parts of the Arctic from open water that can supply moisture to the air

Fig. 1.1   The tundra landscape in the Denali National Park, Alaska.

masses. During the short Arctic spring much of the snow cover melts, feeding the many rivers that discharge fresh water into the Arctic Ocean.

Permafrost, as well as seasonally frozen ground, is found across much of the Arctic. Permafrost consists of gravel and finer material that remains below the freezing point throughout two or more consecutive years. This layer of permanently frozen subsoil can be overlaid with unfrozen soil or organic material. When water saturates the upper surface, bogs and ponds may form, providing moisture for plants. The presence of permafrost means that there are no deep root systems in the vegetation of the Arctic tundra; however, a wide variety of plants are still able to resist the cold conditions. Ice-rich permafrost is found beneath about 80% of Alaska and it physically supports much of the state's infrastructure and natural ecosystems. Figure 1.2 illustrates an active ice wedge, which shows permafrost visible at the surface.

Permafrost is present across almost a quarter of the Earth's land area, with some 16.7×10^6 km^2 being present in northern Russia and Scandinavia, and 10.2×10^6 km^2 in North America. Most of the Arctic islands also contain permafrost. The thickness of permafrost varies from about 1500 m in some parts of Siberia and almost 750 m in northern Alaska, to only a few metres on the edge of the Arctic.

A major difference between the Arctic and Antarctic is that the Arctic is home to many indigenous peoples, while the Antarctic is only occupied by scientists and tourists who visit the continent for periods of days or weeks, up to several years. In the north, Inuit, Saami, Athapaskans, Nenets and other peoples have followed a traditional way of life for generations, often subsisting by hunting on land or through fishing. Many of the activities of these

Fig. 1.2 An active ice wedge exposed by hydraulic trenching at the northwest arm of Kaminak Lake, Kivalliq, Nunavut, Canada. (Photograph courtesy of Dr William W. Shilts.)

peoples are very dependent on the climate and, as discussed below, climate change in the future may have a radical impact on their way of life.

1.1.2 *The Antarctic*

Politically, the Antarctic is defined as the area south of 60° S and it includes the Antarctic continent, a number of the sub-Antarctic islands and a large part of the Southern Ocean (see back end paper).

The Antarctic continent is about 40% larger than the USA, covering an area of 13.6×10^6 km^2, which is about 10% of the land surface of the Earth. This figure includes the area of the ice sheet, the floating ice shelves and the areas of fast ice (sea ice frozen along the coast).

The continent is dominated by the Antarctic Ice Sheet, which contains around 30×10^6 km^3 of ice or about 70% of the world's fresh water. The Antarctic has the highest mean elevation of any continent on Earth, and reaches a maximum elevation of over 4000 m in East Antarctica.

The ice sheet is made up of three distinct morphological zones, consisting of East Antarctica (covering an area of 9.90×10^6 km^2), West Antarctica (1.96×10^6 km^2) and the Antarctic Peninsula (0.39×10^6 km^2). The orography rises very rapidly inland from the coast and the continent has a domed profile, with much of it being above 2000 m in elevation.

The high plateau of East Antarctica is where the lowest recorded temperature on Earth has been measured. At the Russian Vostok Station on 21 July 1983 the temperature dropped to $-89.2\,°C$, although as parts of the continent are at an even higher elevation it is not inconceivable that a near-surface temperature several degrees colder might one day be measured in the Antarctic. The extremely low temperature occurred at Vostok because of the very high elevation, the lack of cloud and water vapour in the atmosphere, and the isolation of the region from the relatively warm maritime air masses found over the Southern Ocean. The high plateau of East Antarctica is located slightly away from the South Pole and this has climatological implications for the atmospheric circulation over the Southern Ocean.

The very cold temperatures in the interior of the Antarctic, coupled with its isolation from warm, moist air masses mean that precipitation amounts here are very low, with only about 50 mm water equivalent falling per year. The Antarctic is therefore a desert and the driest continent on Earth. But the low temperatures mean that there is very little evaporation and sublimation, so that even this small amount of precipitation builds up year by year to form the ice sheet.

West Antarctica is lower than East Antarctica, with a mean elevation of 1119 m. However, some areas do reach more than 2000 m in height, with exposed mountain peaks rising above the ice sheet to more than 4000 m. East and West Antarctica are separated by the Trans-antarctic Mountains, which extend from Victoria Land to the Ronne Ice Shelf and rise to a maximum height of 4528 m.

The Antarctic Peninsula is the only part of the continent that extends a significant way northwards from the main ice sheet. It is a narrow mountainous region with an average width of 70 km and a mean height of 911 m. The northern tip of the peninsula is close to 63° S, so that this barrier has a major influence on the oceanic and atmospheric circulations of the high southern latitudes.

In the Antarctic coastal region temperatures are much less extreme than on the plateau, although at most of the coastal stations temperatures rarely rise above freezing point, even in summer. The highest temperatures on the continent are found on the western side of the

Fig 1.3   A photograph of the edge of part of the Larsen B Ice Shelf before its disintegration.

Antarctic Peninsula where there is a prevailing northwesterly wind, and here temperatures can rise several degrees above freezing during the summer months.

The Antarctic Ice Sheet has a maximum thickness of about 4700 m and this huge mass of ice gradually flows down towards the edge of the continent, moving fastest in a number of ice streams that travel at speeds of up to 500 m per year. Once the ice streams reach the edge of the continent they either calve into icebergs, which move northwards, or start to float on the ocean as ice shelves. The ice shelves constitute 11% of the total area of the Antarctic, with the two largest shelves being the Ronne-Filchner and the Ross Ice Shelves, which have areas of 0.53×10^6 km^2 and 0.50×10^6 km^2 respectively. The ice shelves are several hundreds of metres thick and the ocean areas under them are important for the formation of cold, dense Antarctic Bottom Water (AABW), which is discussed in Section 1.2. Figure 1.3 shows part of the Larsen B Ice Shelf before its disintegration.

The Antarctic continent is surrounded by the sea ice zone where, by late winter, the ice covers an average area of 19×10^6 km^2, which is more than the area of the continent itself. At this time of year the northern ice edge is close to 60° S around most of the continent, and near 55° S to the north of the Weddell Sea. Unlike the Arctic, most of the Antarctic sea ice melts during the summer so that by autumn it only covers an area of about 3×10^6 km^2. Most Antarctic sea ice is therefore first-year ice, with the largest area of multi-year ice being over the western Weddell Sea. Consequently most sea ice around the continent is relatively thin, with an average thickness of about 1–2 m.

Permafrost is much less extensive in the Antarctic compared with the Arctic, because of the very large area covered by the ice sheet. However, across parts of the Antarctic Peninsula and on the sub-Antarctic islands there are extensive areas that are free of ice and snow during the summer months. Here permafrost and seasonally frozen ground are significant features of the environment and they are important in the support of terrestrial ecosystems. Permafrost is also found in the McMurdo Dry Valleys and along the narrow coastal zone of East Antarctica. The Dry Valleys are a particularly interesting area where liquid water is rare, yet there are extensive ground ice glacial sediments that may be millions of years old. Permafrost is also found under the Antarctic Ice Sheet itself.

The major societal difference between the Arctic and Antarctic is the lack of indigenous peoples in the south. Here the only year-round residents are scientists who work on the research stations. Typically only a few hundred scientists overwinter on the stations, but this number swells to several thousand during the summer Antarctic season. In recent years there has been a very large increase in the number of tourists visiting the continent, either by cruise ship or flying to the Antarctic from South America. The International Association of Antarctic Tour Operators, who aim to promote safe and environmentally responsible private-sector travel to the Antarctic, reports that during the 2006–07 Antarctic summer 29 530 tourists visited the continent. This compares with only 6000 in the early 1990s. Such a huge increase in the number of tourists visiting the continent, and the possibility of these numbers increasing markedly in the coming years, will inevitably put pressure on the fragile environment of the continent.

1.2 The role of the polar regions in the global climate system

The global climate system is driven by solar radiation, most of which on an annual basis arrives at low latitudes. Over the year as a whole the Equator receives about five times as much radiation as the poles, so creating a large Equator to pole temperature difference. The atmospheric and oceanic circulations respond to this large horizontal temperature gradient by transporting heat polewards. In fact the climate system can be regarded as an engine, with the low latitude areas being the heat source and the polar regions the heat sink.

In the summer, when the Sun is above the horizon for long periods, the polar regions receive more solar radiation than the tropics, but the highly reflective ice- and snow-covered surfaces reflect much of this radiation back to space, aided by the relatively cloud-free atmosphere that also contains little water vapour. This is one of the important feedback mechanisms operating in the polar regions where a cooling can be enhanced much more than over the unfrozen ocean or bare ground.

Large parts of the Northern Hemisphere are affected by seasonal snow cover that can respond rapidly to changes in temperature, so they are very sensitive to climatic changes. By contrast, the Antarctic continent is covered by a large ice sheet that has an extent and albedo that is relatively unchanging on the scale of decades and centuries. It is therefore the ocean areas around the Antarctic and the coastal regions of the continent that are the most sensitive to climatic variability and change on shorter timescales.

Both the atmosphere and ocean play major roles in the poleward flux of heat, with the atmosphere being responsible for 60% of the heat flux, and the ocean the remaining 40%. In the atmosphere, heat is transported by both transient eddies (depressions) and the mean flow. The depressions carry warm air poleward on their eastern sides and cold air towards lower latitudes on their western flanks. The atmosphere is able to respond relatively quickly to changes in the high or low latitude heating rates, with storm tracks and the mean flow changing on scales from days to years.

The contrasting orography of the two polar regions is very important in prescribing the atmospheric and oceanic circulations of the Northern and Southern Hemispheres. The Antarctic continent is located close to the pole, with few other major orographic features being present in the Southern Hemisphere. The mean atmospheric flow and ocean currents are therefore very zonal (east–west) in nature. For example, the Antarctic Circumpolar Current (ACC), which is one of the major oceanographic features of the Southern Ocean, can flow unrestricted around the continent, isolating the high latitude areas from more temperate mid and low latitude surface waters. This has only been the case since about 30 million years ago when the Drake Passage, which connects the South Atlantic Ocean to the South Pacific Ocean, opened up. Prior to that time the ocean currents had more of a meridional component that allowed greater penetration poleward of more temperate water masses.

The orography of the Northern Hemisphere is dominated by the major ice sheet of Greenland and the mountain ranges of the Himalayas and the Rocky Mountains. This produces a larger meridional component to the atmospheric flow than is found in the Southern Hemisphere, giving a more pronounced exchange of air masses between the Arctic and lower latitudes than occurs around Antarctica. The effects of the large land masses on the near-surface ocean currents are even more pronounced. Ocean currents, such as the Gulf Stream–North Atlantic Current system, bring warm waters to a large part of the eastern subpolar North Atlantic, and into the Nordic Seas. This current transports more warm water to higher latitudes than in any other ocean and influences the sea ice distribution in the Arctic Ocean.

The large Antarctic Ice Sheet plays a central role in determining the atmospheric circulation of the high southern latitudes. The intense surface cooling that occurs on the high Antarctic Plateau drives a persistent downslope, katabatic flow that brings cold surface air down to the coastal region. This flow plays a very important part in shaping the high latitude circulation of the Southern Hemisphere, and its influence extends into mid latitudes.

The global thermohaline circulation (THC) is the ocean system that links the major oceans of the world (Fig. 1.4). It is driven by differences in the density of sea water, which in turn is controlled by temperature and salinity. The polar regions are closely coupled to the rest of the climate system via the THC, with the system providing a direct link between the Arctic and Antarctic. The seas around the Antarctic are particularly important because of the production of AABW, which is the densest water mass found in the oceans. AABW is formed by deep winter convection in the Antarctic coastal region, particularly in the Weddell and Ross Seas. The water mass is formed as cold air from the Antarctic rapidly cools the

1.2 The role of the polar regions

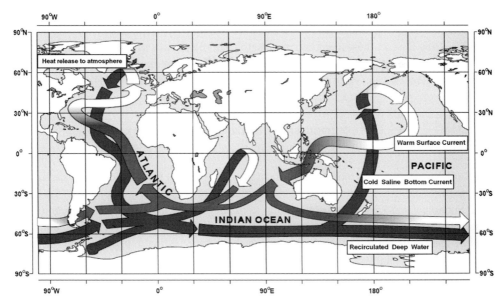

Fig. 1.4 A schematic illustrating the global thermohaline circulation. (After W. Broecker, modified by E. Maier-Reimer.)

surface waters, promoting downward convection. Brine rejection during the formation of sea ice on the ocean surface is also very important, as is melting under the ice shelves. AABW flows out into the world's oceans and is found below 4000 m in all the ocean basins. As can be seen in Fig. 1.4, AABW flows northwards in the Atlantic Ocean, reaching the Arctic, where it rises and releases heat into the atmosphere. From the above it can be appreciated that changes in Antarctic climate that might affect ice shelves, sea ice production and atmospheric circulation could have implications for the global ocean system.

Changes in ocean conditions in the Arctic can also have widespread implications since the North Atlantic Meridional Overturning Circulation (MOC) provides a means of linking high and low northern latitudes. For example, ocean salinity anomalies in the central Arctic have been shown to propagate into the Greenland Sea, where they can cause major modifications to ocean stratification, which in turn can affect conditions further south.

A further important part that the polar oceans play in the global climate system is via their role in the carbon cycle. Carbon can be transferred in both directions between the atmosphere and the ocean, but at high latitudes the low sea surface temperatures favour the uptake of carbon dioxide. The Southern Ocean is a major sink of CO_2 with the THC transporting the dense carbon-rich water to the deeper layers of the ocean.

While the atmosphere can respond very rapidly to changes in heating rates, the ocean responds more slowly. The circulation of the upper layers can change over months to years, but the deep ocean and the THC requires decades to centuries to respond.

The Arctic and, to a much lesser extent, the sub-Antarctic islands play a further important part in the global climate system through their role as sources and sinks of important greenhouse gases, which are held in the permafrost or seasonally frozen ground. The frozen ground of the Arctic tundra and taiga contains methane (CH_4), ozone (O_3) and carbon dioxide (CO_2). In fact these areas contain one third of the world's soil-bound carbon and the ground can be a source or sink of such gases. When permafrost melts, it releases carbon into the atmosphere, and can contribute to the global increase in greenhouse gases.

1.3 Possible implications of high latitude climate change

1.3.1 Introduction

If greenhouse gas concentrations continue to increase over the next century near-surface air temperatures are expected to rise more in the polar regions than in any other part of the Earth (see Section 7.4). This will have serious implications for the cryosphere, oceanic and atmospheric circulations, the marine and terrestrial environments and the indigenous peoples of the Arctic. In this section we consider how major climatic changes may have an impact on the high latitude environment and conditions at low latitudes.

The Antarctic and Greenland Ice Sheets respectively contain 90% and 9% of the world's glacier ice. If both these ice sheets melted completely they would respectively contribute ~60 m and ~6 m to sea level rise (figures from the US National Snow and Ice Data Center). The central parts of the Antarctic Plateau and Greenland Ice Sheet are always extremely cold, so increases of temperature of several degrees or more will not result in any melting at these locations. So major disintegrations of these ice sheets are unlikely to occur, even on the scale of centuries, although melting will contribute to sea level rise. However, around the peripheral parts of Greenland and the Antarctic temperatures can be close to freezing point, making them very sensitive to small increases in temperature. In addition, changes in water masses flowing under the ice shelves and tide water glaciers may result in increased basal melt. Although some melting may take place in these regions, a major concern is the impact on the flow velocity of the ice streams that drain from the interior of the ice sheets, which may result in much more significant discharge of ice into the ocean.

Sea level is currently rising at around 3 mm a year, much faster than the average rate of 1 mm per year that has taken place over the last 5000 years. Predictions of sea level rise for the next century are discussed in Section 7.10. These are computed from the expected thermal expansion of the ocean as ocean temperatures rise, changes in river discharge into the ocean, and melting of mid latitude glaciers. However, these predictions do not take into account dynamic discharge of ice from the large ice sheets of Greenland and Antarctica. If sea level continues to rise at its present rate, or at an even greater level, it will pose a very significant problem for low-lying areas across the world.

1.3.2 The Arctic

In recent decades there have been major changes in the Arctic environment, with rising near-surface air temperatures leading to large decreases in perennial sea ice extent, a reduction in

1.3 Implications of high latitude climate change 13

the amount of snow cover, melting of permafrost and decreases in river and lake ice extent. Melting of glaciers in many parts of the Arctic is contributing to rising sea levels worldwide, while in areas such as Alaska, loss of glacier ice can have pronounced regional effects through the contribution of their runoff to ocean currents and marine ecosystems in the Gulf of Alaska and Bering Sea. Such changes have been reported extensively in the scientific literature and the popular press, but have also resulted in pervasive reports by indigenous Arctic communities. These peoples of the high latitude region have reported warmer and increasingly variable weather, and also changes in terrestrial and marine ecosystems, which have had an impact on their way of life. If, as predicted, temperatures continue to rise in the Arctic, possibly at an accelerated rate, we can expect to see an amplification of these changes in the coming decades.

Many climate projections for the next century suggest that there will be a marked decrease in the extent of Arctic sea ice. While this will not affect sea level directly as the ice is already floating, there may be changes to the thermohaline circulation because of the freshening of the waters in the upper layer of the ocean, since there will be less brine rejection through the formation of sea ice. Sea ice retreat also allows larger storm surges to develop in areas of greater open water, increasing erosion through greater wave activity.

Although loss of Arctic sea ice could have serious implications for the ocean circulation, there could be some benefits in terms of navigation around the northern parts of Russia and Canada. With the longer ice-free summer season there could be substantial benefits to marine transport and offshore gas and oil operations, which could have major implications for international trade.

As discussed earlier, any melting of the Greenland Ice Sheet has the potential to raise sea level. But it has also been suggested that this flux of fresh water into the North Atlantic could have an impact on the thermohaline circulation, possibly weakening the Gulf Stream and North Atlantic Drift. This has the potential to seriously affect the climate of northern Europe, possibly lowering surface temperatures by several degrees. The possibility that European temperatures could drop in a world that is generally warming because of increasing greenhouse gases has received wide publicity, and is discussed in more detail in Chapter 7.

With rising temperatures there will be a northward shift of the climatic zones with a poleward extension of the boreal forests and the tree line. This will result in a transformation of Arctic landscapes, with the northern edge of the boreal forest advancing into the tundra. But in areas such as Alaska there are fears of a loss in the moisture needed for forest growth, along with insect-induced tree mortality, increased risk of large fires and changes to the reproduction of some trees, such as white spruce.

Loss of permafrost in the Arctic is already having a profound effect on the environment because of the severe damage that can be caused to buildings and infrastructure that rely on it to provide solid foundations. Oil pipelines running across areas of permafrost have fractured, resulting in significant pollution and environmental damage. The Trans-Alaska Pipeline, which runs for almost 1300 km across Alaska, was a hugely expensive enterprise, but breaks in the pipeline and other repair costs due to melting permafrost could become very significant in the future. The short-term risk of disruption to operations of the Trans-Alaska Pipeline is judged to be small, although costly increases in maintenance due to increased ground instability are likely. We can expect to see increases in such potentially damaging slumping of land in the coming decades if temperatures continue to rise.

Further thawing and melting of permafrost are likely to bring destruction of trees and loss of boreal forests, the release of water and the expansion of lakes, grasslands and wetlands. This will result in changes to habitats and ecosystems, such as loss of habitat for terrestrial birds and mammals such as caribou. But there will also be additional habitat for aquatic birds and mammals.

Loss of permafrost will also result in increased erosion and soil instability, especially on the coast and along the banks of rivers. This could result in the blocking of streams that are important for salmon spawning. There could also be increased occurrence of landslides and development of talik (a year-round thawed layer of what was formerly permafrost), and increased water table depth.

The melting of Arctic permafrost also has serious implications for the carbon cycle and the levels of greenhouse gases in the atmosphere. In the past, permafrost has been a sink of carbon, which is locked into the frozen ground. However, with melting and the resulting warmer soils, there could be an increase in the speed of decomposition and the release of CO_2 and CH_4 into the atmosphere.

1.3.3 The Antarctic

As discussed in Section 6.2 the pattern of climate change across the Antarctic in recent decades has been complex with major surface warmings observed around the Antarctic Peninsula and into West Antarctica, but little indication of statistically significant changes across the rest of the continent. However, as discussed in Section 7.4, over the next century the majority of climate models predict warming across most of the Antarctic if greenhouse gas concentrations continue to rise. The largest surface changes are expected to occur over the Southern Ocean as sea ice extent decreases. As discussed earlier, the Antarctic coastal region is one of the primary areas for the production of AABW, the densest water mass found in the world's oceans. With the production of less sea ice being likely over the next century there is concern there may be major implications for the ocean circulation.

Loss of sea ice around the Antarctic could also cause large-scale changes in marine ecosystems, including threats to populations of marine mammals, such as penguins and seals. Of the six species of Antarctic seal, four breed on the sea ice during spring and they may be affected by a large reduction in sea ice extent.

On the continent itself many animals and plant species require rather specific climatic conditions in which to flourish, with temperature and the amount of liquid precipitation being particularly important, although the level of UV-B (290–320 nm wavelength) radiation is also a significant factor. The amount of UV-B radiation received in the spring has increased markedly in recent years with the appearance of the Antarctic ozone hole, with consequent impact on the biota.

Climate change is expected to have a major impact on the terrestrial biota of the Antarctic. Studies have suggested that increasing temperatures and greater water availability could extend the active season, increase development rates and reduce the life cycle duration. In addition, there could be altered distribution of species and exotic colonisation.

The rapid warming on the Antarctic Peninsula over the last 50 years has already resulted in some of these impacts becoming apparent, such as an increase in the life cycle of the South Georgia diving beetle in response to higher lake temperatures, increases in population numbers and changes in range of native Antarctic flowering plants (*Deschampsia antarctica* and *Colobanthus quitensis*) and the decrease in number of Adélie penguins.

As discussed earlier, it is very unlikely that there will be a major disintegration of the Antarctic Ice Sheet during the next few centuries; however, some parts of the Antarctic have shown rapid glaciological changes over the last few years, indicating a sensitivity to climatic factors. In particular, there is extensive research taking place into the West Antarctic Ice Sheet (WAIS), much of which is grounded below sea level. A complete loss of the WAIS would result in a 5 m sea level rise so there is obvious concern over the disintegration of even a small part of the ice sheet. Satellite data have recently revealed a significant thinning of part of the WAIS in the region of the Amundsen Sea Embayment (75° S, 105° W). This area includes Pine Island Glacier, the largest glacier in West Antarctica, which is up to 2500 m thick and is grounded over 1500 m below sea level. In the 8 years from 1992 the glacier retreated inland by over 5 km with the loss of 31 km^3 of ice. The exact reasons for the loss of ice from the Pine Island Glacier at this time are not understood, although regional oceanic water mass changes under the ice are believed to have played an important role. However, if the present rate of thinning continues it is thought that the whole Pine Island Glacier could be lost to the ocean within a few hundred years. An important research target must be to understand why this glacier is currently shrinking, whether this is a result of anthropogenic activity and whether in a warming world other glaciers could start to accelerate and drain more ice from the interior of the WAIS with consequent impacts on sea level.

Over the next century it is predicted that there will be greater precipitation over the Antarctic, offsetting to a small extent sea level rise. However, this is expected to be a less important factor in net sea level change than greater ice discharge into the Southern Ocean from accelerating glaciers along the coast of West Antarctica.

The loss of ice shelves around the Antarctic Peninsula in recent years has often been reported and attributed to human activity. There is certainly evidence that the warming during the summer on the eastern side of the peninsula is at least in part a result of anthropogenically forced atmospheric circulation changes, in particular the loss of stratospheric ozone. If greenhouse gas levels continue to increase we may well see further loss of ice shelves in this region. Of course ice shelves are already floating on the ocean and their loss does not result in a change in sea level. However, there is evidence from the Antarctic Peninsula that glaciers flowing into the ice shelves accelerate once the shelves have disappeared, so contributing to sea level rise.

The water masses under the ice shelves are important for the production of AABW, which is a major element of the global thermohaline circulation. Thus, if there was a significant loss of ice shelves around the continent in a general warming environment, changes occurring in the Antarctic coastal region have the potential to influence conditions as far away as the Arctic.

2

Polar climate data and models

2.1 Introduction

In this chapter we describe the main types of data available for the study of climate change within the polar regions. In comparison with most other regions of the Earth the time-series of 'traditional', in-situ instrumental observations is relatively short, particularly in Antarctica where most stations have only been operating for about 50 years. With short records it is more difficult to determine whether recent trends are significant, particularly for regions where there is high natural climate variability, such as the Antarctic Peninsula. One way of extending climate records is to use 'proxy' climate data; for example, the commercial whaling expeditions in both the Arctic and Antarctic provide historical data about the position of the sea ice edge, where the greatest amount of hunting took place.

The inhospitable nature of the polar regions means that conducting science in such areas can be very expensive. Thus, there are relatively few surface meteorological stations compared with the mid latitude and tropical areas. This is illustrated in Fig. 2.1, which shows the coverage of surface, ship and aircraft observations assimilated into the European Centre for Medium-range Weather Forecasts (ECMWF) model at 00 GMT 12 July 2010. In recent years advances in technology have allowed the deployment of autonomous automatic weather stations (AWSs) and these are particularly useful in the polar regions as they can be sited in very remote locations. The majority of the synoptic reports in Antarctica situated away from the coast are from such AWSs.

Data availability from the polar regions increased markedly with the advent of space-borne remote sensing systems, which first became operational in the late 1960s and early 1970s. Early instruments provided relatively coarse resolution images, which were used for mapping and initial studies of weather systems over the Arctic and Southern Ocean. Today there are many types of instruments available that can provide data on a large number of geophysical parameters for use in meteorological and glaciological studies. The former are assimilated into numerical weather prediction (NWP) models and the availability of remotely sensed data in previously data-sparse regions has dramatically improved the accuracy of weather forecasts at high latitudes. Long time series of the NWP model data using stable assimilation and forecast schemes, called a reanalysis, now comprise one of the

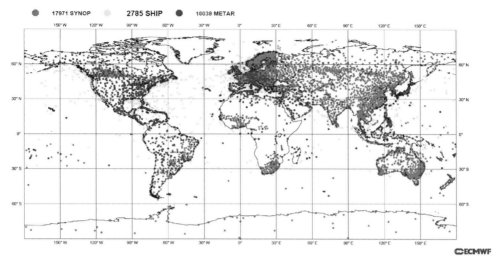

Fig. 2.1 Geographical coverage of meteorological observations available at 00 GMT 12 July 2010. Synoptic reports are shown as mid-grey dots, ship reports in light grey and aircraft reports in black.

key climatological research tools in the polar regions: they provide spatial fields of many meteorological parameters at multiple heights on a regular grid.

One unique data source that the polar regions have when considering long-term climate change is the proxy climate data contained within the ice sheets and icecaps. The European Project for Ice Coring in Antarctica (EPICA) ice core from Dome C, Antarctica, encompasses ~750 kyr and contains records of how the concentrations of atmospheric greenhouse gases CO_2 and CH_4 have varied with temperature through the last eight glacial cycles. In Greenland the North Greenland Ice Core Project (NGRIP) core contains ice deposited as snow as far back as 123 kyr BP. Oceanic sediment cores also provide important historical climate data, including proxies for the extent of sea ice.

2.2 Instrumental observations

2.2.1 Introduction

Meteorological measurements in the Arctic and Antarctic were first made during the many national expeditions to the two polar regions, which began during the nineteenth century (see Serreze and Barry, 2005 and King and Turner, 1997 for summaries of the early meteorological records in the Arctic and Antarctic, respectively). More systematic data gathering required international cooperation, which led to the first International Polar Year (IPY) in 1882–83, in which 12 stations were established in the Arctic (Fig. 2.2), while in the Southern Hemisphere a small number were located on sub-Antarctic islands. Today, data from both manned stations and an increasing number of AWSs, including a network

Fig. 2.2 Primary Arctic stations for the 1882–83 International Polar Year.

encompassing the Greenland Ice Sheet (Steffen and Box, 2001), are communicated to the world's major weather forecast centres via satellite, allowing their assimilation into NWP models. Other recent advances include the collective activities of the IPY and detailed field campaigns, such as the Surface Heat Budget of the Arctic Ocean (SHEBA) experimental campaign (Uttal et al., 2002).

The collection of climatological data from the Antarctic continent only began in earnest with the International Geophysical Year (IGY) of 1957–58. Data prior to the IGY came from expedition reports and a few stations from the northern Antarctic Peninsula region that began operating after the Second World War: these early data have been analysed by Jones (1990). During the IGY a maximum of 42 stations existed on the continent itself, primarily on the coast but also with several inland on the Antarctic Plateau, with a further 21 situated on Antarctic and sub-Antarctic islands. Most of the IGY stations are still operating today, although not all are occupied year-round. Figure 2.3 shows maps of the Arctic and Antarctic with the locations of stations with long climate records indicated. Today, several countries operate arrays of AWSs, which allow the collection of regular data from the relatively inaccessible Antarctic interior.

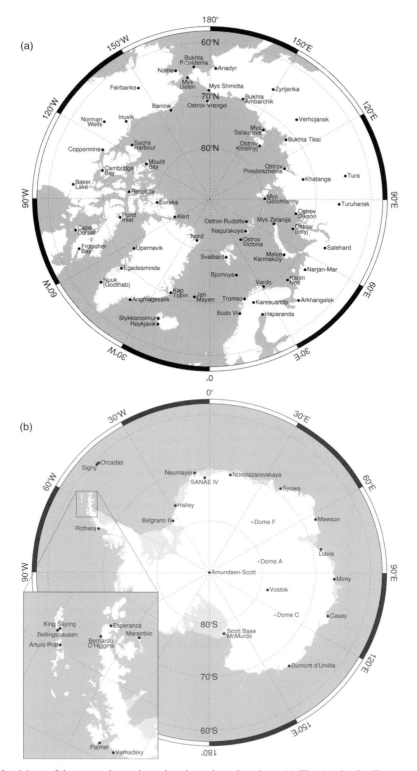

Fig. 2.3 Maps of the two polar regions showing selected stations. (a) The Arctic. (b) The Antarctic.

2.2.2 Surface data

Until relatively recently, most surface meteorological measurements in the polar regions have been obtained, as elsewhere, by using instruments housed in conventional Stephenson screens (louvre-sided boxes) at manned stations. These are situated at a nominal height of 1.5 m above the surface. Automation started in the 1980s but measurements are often still taken from Stephenson screens in order to check the accuracy of automatic instruments.

Pressure

Atmospheric pressure is recorded using barometers and is conventionally reduced to mean sea level. While this is appropriate for coastal stations, problems arise for stations located on the Antarctic Plateau. First, the exact procedure varies between nations and this can make a significant difference when surface pressure is less than 650 hPa, as is the case at Vostok. In addition, it is standard practice to assume a (theoretical) isothermal atmosphere below the station when reducing surface pressure to sea level: on the Antarctic Plateau there are often strong surface temperature inversions – surface temperatures are much colder than those a few hundred metres above the station – leading to anomalously strong high pressure values.

Temperature

Historically, near-surface temperature in the polar regions has been measured as elsewhere, using mercury or alcohol-in-glass thermometers (necessary for temperatures less than −30 °C) housed in a Stephenson screen. One problem is if the screen louvres become blocked with blowing snow. A second difficulty results from the inability of the Stephenson screen to effectively shield the thermometers from high levels of solar radiation in summer: this can be reflected off the snow surface though the gaps in the screen louvres and can lead to rapid, high-temperature spikes. From the 1980s, with the move to automated systems, there has been a switch to platinum resistance thermometers placed in artificially aspirated radiation shields. In the Arctic regular subsurface 'soil' temperatures at an array of different depths are analysed to study the spatial and vertical distribution of permafrost.

Wind

Wind speed and direction are generally measured using a cup anemometer and wind vane. Alternatives include the use of propeller and ultrasonic anemometers. The icing of mechanical wind vanes can lead to a loss of data: this may encompass several months for instruments that are not visited regularly. However, while ultrasonic anemometers are not affected by icing, since they have no moving parts, the accuracy of the data can be compromised by the effects of blowing snow.

Humidity

Humidity is generally very low in both the Arctic and Antarctic because of the low temperatures: the saturation vapour pressure of water is strongly influenced by temperature. This means that conventional psychrometric techniques are of little use. Modern humidity probes on automated systems include carbon film hygristors and polymer capacitive sensors and are capable of measuring humidity at temperatures down to ~–40 °C. Frost point hygrometers are more accurate but are generally considered too expensive for routine measurements.

Radiation

The automation of sensors has allowed the continuous recording of the surface radiation budget on the polar ice sheets (e.g. Steffen and Box, 2001; van den Broeke *et al.*, 2004a). Currently, there are four stations from each polar region operating in or planned to be in the Baseline Surface Radiation Network (BSRN) (Ohmura *et al.*, 1998). However, most measurements of the surface radiation budget components have been obtained during regional or local field campaigns. The Dutch Institute for Marine and Atmospheric Research, Utrecht (IMAU) instrumentation comprises a net radiometer, which houses two instruments for measuring the downward and upward shortwave radiation flux and two different sensors for similarly measuring the longwave radiation flux. The main practical problems encountered with these instruments have been with ice accretion and rime formation, which particularly impact the upward-looking sensor windows. However, by using theoretical assumptions and parameterised values any resultant errors can be detected and their effects minimised (van den Broeke *et al.*, 2004a).

Aerosols

Aerosols are now often measured routinely at sites within the Arctic and also Antarctica. As an example, there has recently been an intensive campaign (CHABLIS: Chemistry of the Antarctic Boundary Layer and its Interface with Snow) to study the atmospheric chemistry of the coastal Antarctic boundary layer, including detailed studies of seasonal oxidant chemistry, annual variation in the boundary layer NOy (reactive forms of nitrogen) budget, and elucidating air/snow transfer processes. The CASLab (Clean Air Sector Laboratory) is located at Halley Station on the Brunt Ice Shelf (75.5° S, 26.4° W) (Fig. 2.4). Instruments and techniques used at the laboratory include sampling firn air at various depths to 20 m, snow pit sampling, long path absorption spectroscopy, laser induced fluorescence, gas and liquid chromatography, filter, denuder and flask sampling and in-situ liquid flow methods.

Aerosols are particularly important in the Arctic because of the 'Arctic haze' that can occur when atmospheric pollutants, such as sulphates, are transported to high latitudes from the major industrial centres of Europe and Eurasia. Aerosol concentrations are therefore measured at a number of sites across the Arctic using filters that are subsequently analysed, and also via instruments such as micro pulse lidar. Measurements have also been made from research aircraft to derive profiles of aerosol concentrations through the troposphere.

Manual observations

While it has been possible to automate the majority of meteorological data collection, there are several important parameters that are still obtained primarily by human observers. One

Fig. 2.4 The CASLab clear air facility at Halley research station. (Photograph by C. Gilbert, British Antarctic Survey.)

of these is an assessment of cloud: its coverage, height and type. In the polar regions the situation is complicated by 'unique' types of cloud. Curry et al. (1996) describe a number of 'unusual' cloud types that may form above the Arctic Ocean. One, which is difficult to observe correctly, comprises tenuous low-level cirrus (ice crystal) clouds, which give 'clear-sky' or 'diamond-dust' precipitation, and which are also common over the Antarctic Plateau. Radok and Lile (1977) found that in one year's worth of accumulation at Plateau Station (79.2° S, 40.5° E) up to 87% of the total accumulation (25.4 mm water equivalent) comprised such ice crystals.

Other manual observations include snow depth (although automatic instruments are becoming more widespread), visibility, present and past weather types. The latter two may be subjective decisions – for example the precipitation rate – and long-term time series often demonstrate marked steps in the frequency of a particular weather type coincident with an observer change. Similarly, on King George Island at the tip of the Antarctic Peninsula, where there are several bases in close proximity operated by different nations, the present weather reports at a given moment can vary quite widely, such as confusing low cloud and fog or blowing snow and actual precipitation. A further quantity that is measured at a number of stations is total column ozone. For this a Dobson spectrophotometer can be used, which measures the relative intensities of UV radiation at pairs of selected wavelengths.

Automatic weather stations

The harsh polar environment, particularly on top of the high ice sheets, means that it is often not possible to operate manned stations. However, the development of low power computer components and the availability of the Argos data collection system on the National Oceanic and Atmospheric Administration (NOAA) series of polar-orbiting satellites have made possible the development of AWS units capable of providing near real-time data from these remote regions.

The Antarctic Meteorological Research Center (AMRC) at the University of Wisconsin-Madison has deployed the largest number of AWSs in Antarctica and also has a network of stations near the summit of the Greenland Ice Sheet. Its basic AWS unit measures air temperature, wind speed and wind direction at a nominal height of three metres above the surface, and air pressure at the electronics enclosure at about 1.75 m above the surface. The heights are nominal because of snow accumulation that may occur: some sites must be visited every year to raise the mast and ensure the instruments stay above the snow surface. If the NOAA satellite is within line of sight of the AWS unit, the transmission will be received and stored by the Argos Data Collection System. Data transmission from the AWSs is staggered so that data collection is maximised. At high latitudes there are sufficient satellite overpasses to ensure that data losses are minimal. The AWS unit is powered by six to twelve 12-volt gel-cell batteries charged by one or two solar panels. At the South Pole, 12 batteries and two solar panels are sufficient to operate the AWS unit through the year, while, further north, six batteries and one solar panel are adequate on the Ross Ice Shelf. The IMAU at Utrecht University employs lithium batteries in their AWSs (Fig. 2.5) and wind energy to heat the Argos transmitter, which was found to stop working at $-50\,^{\circ}\mathrm{C}$.

In the 1990s automatic geophysical observatories (AGOs) began to be deployed across the Antarctic continent. These are primarily for use by the geospace community to examine the electrodynamics of the polar cap region of the Earth's upper atmosphere. However, AGOs also incorporate standard meteorological instruments and thus expand the AWS network further.

While many of the AWS units are deployed to support individual experiments or glaciological activities (e.g. van den Broeke *et al.*, 2004b), some of those in the Antarctic have now been operating for a sufficiently long time, since the mid 1980s, to provide useful climatological data. Although there may be disruptions to the time series, due to power and instrument failures, gaps in pressure and temperature data can be filled using artificial neural network techniques (Reusch and Alley, 2002). The biggest problem remains the wind vane: very strong katabatic winds and the build-up of hoar frost on the instrument during winter often lead to a lack of wind data in this season. Nevertheless, by 1994 more AWS observations from Antarctica were on the Global Telecommunication System (GTS) than from manned stations (Turner *et al.*, 1996) and the network has expanded considerably since then.

On Greenland there are several networks of AWSs, the largest being the ice-sheet-wide GC-Net of 18 stations (Steffen and Box, 2001). The instruments are primarily there to support glaciological analysis of the ice sheet, particularly the modelling of snow and ice melt, but will eventually provide a surface climatology against which climate models can be assessed.

Precipitation

As the dominant term in the mass balance equation of the polar ice sheets (cf. Section 6.9), precipitation is a very important parameter to measure. However, it is very difficult to

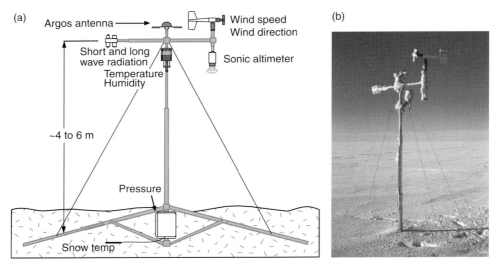

Fig. 2.5 (a) Schematic of the IMAU automatic weather station (AWS) used in the polar regions. (Courtesy C. Tijm-Reijmer, University of Utrecht, the Netherlands.) (b) An AWS on Berkner Island, Antarctica, illustrating the problem of riming.

determine using conventional methods. Conventional gauges give a significant undercatch of solid precipitation, with catch efficiency primarily dependent on the wind speed. Although data may be adjusted (e.g. Yang *et al.*, 2001), alterations tend to be made on a climatological basis and vary according to gauge type and reporting practice. Serreze and Barry (2005) suggest that this has created artificial discontinuities across national borders in the Arctic. In the Antarctic, there has been an increasing use of acoustic depth gauge instruments attached to AWSs that measure the distance to the snow surface (see Fig. 2.5). These have the advantage of being able to discriminate both positive and negative changes – the latter includes melting or the removal of uncompacted snow by the wind – in surface elevation and, combined with other meteorological measurements, can distinguish between precipitation actually falling at the site and blowing snow.

Cloud observations

Accurate cloud observations in the polar regions are made particularly difficult during the polar night, although cloud searchlights can be used to estimate their height, and in blizzard or 'white-out' conditions caused by blowing snow. For example, Intrieri and Shupe (2004) report that diamond-dust conditions were over-reported by 45% during the polar night.

Observing cloud height and type at the high latitude synoptic reporting stations presents a number of problems. During the winter months in conditions of complete darkness, the fractional cloud cover can be estimated from the extinction of the stars but it is difficult to estimate cloud type and height. When the sun is above the horizon, synoptic observations of cloud are easier to make and the data can be given greater credence.

Over the high interior of the Antarctic there is frequently a thin veil of suspended ice crystals, the effects of which can often be seen by the presence of various optical phenomena, such as halos. When such thin cloud is present an observer is often faced with the problem of whether to report zero or eight oktas (eighths of the sky covered) of cloud. In addition, surface-based instruments, such as lidars, which use changes in the backscatter of atmospheric particles to determine the height of clouds, have also been used at several surface stations for a number of years.

The in-situ observations of cloud from high latitude stations can be useful for obtaining an overall picture of the type of clouds found at a location and the approximate fractional cloud cover (Warren *et al.*, 1986). But polar synoptic reporting stations are largely limited to the coastal rim of the Antarctic and the land areas surrounding the Arctic Ocean. Moreover, different interpretations between observers of how cloud should be reported can mean that analysis of time series of cloud observations may be compromised as a result of jumps in particular elements. A more systematic and consistent means of observing polar clouds is from the polar-orbiting satellites (cf. Section 2.4.2).

2.2.3 Upper air climate data

Many high latitude research and meteorological observing stations also have daily or twice daily radiosonde launches in addition to maintaining a surface meteorological observing programme. The radiosondes make measurements of temperature, pressure and humidity, which are sent by radio back to the ground station. Wind speed and direction during the ascent are computed by tracking the radiosonde with Global Positioning System (GPS) sensors. The high latitude stations use standard radiosonde packages, although as discussed below, some special procedures must be carried out to ensure that the balloons do not burst in the very cold polar stratosphere.

Occasionally, ozonesondes are launched at some research stations as part of investigations into high latitude, stratospheric ozone depletion. These instrument packages produce a profile of ozone concentration through the troposphere and stratosphere, which is a supplement to the surface-based Dobson spectrophotometer measurements of total ozone amount.

Arctic radiosonde data have been collated into the Historical Arctic Rawinsonde Archive (HARA) (Kahl *et al.*, 1992). Most of the stations began operating in the 1950s although some started in the previous decade. The HARA includes data for 95 individual stations north of 65° N, with two thirds of these having at least 10 years of data. There is a reasonable distribution of stations across the Arctic: even above the Arctic Ocean there are 16 850 temperature profiles measured from Russian drifting ice stations between 1954 and 1990 (Kahl *et al.*, 1993). There are far fewer radiosonde stations in the Antarctic: since the IGY long-term measurements have been undertaken at 17 Antarctic stations, although by 2008 regular radiosonde flights were only being carried out at 12 stations. Large spatial data gaps exist in the West Antarctic coastal region and the entire Antarctic Plateau, where Amundsen–Scott and Dome C are the only operational upper-air stations. However, in

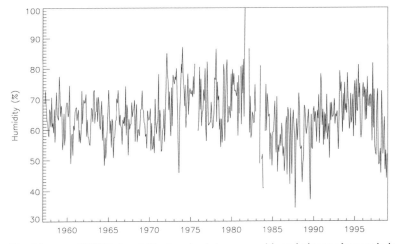

Fig. 2.6 Monthly mean 850 hPa humidity showing inhomogeneities relating to changes in instrument type at Halley station.

coastal East Antarctica there are sufficient radiosonde stations to estimate atmospheric water-vapour transport into the region using the 'aerological method' (Bromwich, 1988; Connolley and King, 1993) as has also been done across 70° N into the Arctic (Serreze et al., 1995). The East Antarctic radiosonde stations have demonstrated the largest regional mid-tropospheric warming in the observational period (Turner et al., 2006a).

There are unique practical problems with regard to operating radiosonde programmes in the polar regions. Very strong near-surface winds can make balloon launching impossible but it is important that non-flights are minimised so the resultant data are not biased towards lower wind speed conditions. Similarly, it must be ensured that balloon bursting at very cold temperatures in the stratosphere – above 30 hPa – does not lead to a warm bias in the data at this altitude. This can be achieved by dipping the balloons in oil before release (e.g. Roscoe et al., 2003). Another issue, common to all long-term radiosonde time series is the potential effect of temporal inhomogeneities resulting from changes in radiosonde type as technology has improved (Fig. 2.6). The principal issues result from changes in sensors and changes in solar radiation corrections to the data. In the polar regions these can have particular effects on humidity, which is very difficult to measure at low temperatures. There have been many studies examining these issues and attempting to homogenise the long-term time series in order to produce reliable trends (e.g. Lanzante et al., 2003; Thorne et al., 2005). Unfortunately, the appropriate metadata, detailing changes in instrument type and observing practice, are not always available.

2.2.4 Ocean data

Ocean data from in-situ measurements comprise both meteorological data from oceanic regions and also data on the ocean itself. Surface observations from the Arctic Ocean began in 1937 with the first manned Soviet drifting station on sea ice, while the first submarine transect under Arctic sea ice occurred in 1958. Major deployment of surface buoys in both polar regions first took place in 1978–79 in support of the First GARP (Global Atmospheric Research Program) Global Experiment (FGGE) (Garrett, 1980), with over 300 systems operational in the Southern Ocean. As technology has advanced most oceanic measurements from the polar oceans, both on the surface and at depth, have been automated, although data from ships' cruises are still important, particularly repeat profiles such as the World Ocean Circulation Experiment (WOCE) SR1b transect across the Drake Passage. A combination of different data types has been utilised to produce various oceanographic atlases of both the Arctic Ocean (e.g. Environmental Working Group Joint US–Russian Atlas of the Arctic Ocean: http://nsidc.org/data/g01961.html) and Southern Ocean (e.g. WOCE Southern Ocean atlas: http://woceatlas.tamu.edu/).

Drifting buoys (Fig. 2.7) are still utilised to provide meteorological data from oceanic regions where other in-situ measurements are absent, although as satellite-borne sensors become more sophisticated their use may decline. The principal parameters they measure are similar to standard AWSs and can be of particular value in studying the atmospheric boundary layer above sea ice. If a buoy is located in a region of sea ice it can provide information on sea ice motion while an array of buoys can provide additional data on ice deformation. The divergent flow of sea ice around Antarctica means that most buoys drift northward and emerge into open water within several months. The position of the buoys and other drifting ocean sensors, described below, is obtained through GPS. Some oceanic data are placed on the GTS and can be used by operational forecasting models – such as the UK Meteorological Office Forecasting Ocean Assimilation Model (FOAM).

Ocean moorings are bottom tethered and can extend from the sea floor to the surface. A wide variety of instruments can be placed on these moorings to routinely obtain key physical (temperature, salinity, conductivity and currents), geochemical (sediment traps) and bio-optical (fluorometers, downwelling irradiance and upwelling radiance) measurements at one or more depths. Recent advances mean that high-resolution time series of atmospheric boundary layer and surface ocean CO_2 partial pressure (pCO_2) data can be used to evaluate the temporal variability in air–sea CO_2 fluxes. Ocean drifters carry similar types of instrument packages to the moorings, with a typical lifetime of only a few months in the Southern Ocean.

A relatively new type of ocean instrument is the Argo float, which is designed to float deep in the ocean, at about 2000 m, following the current there. Every 10 days, they come up to the surface, measuring temperature and salinity as they go, and telemeter their location and the profile data back via satellite. At the 'parking depth' the floats are neutrally buoyant by having a density equal to the ambient pressure and a compressibility that is less than that of sea water. When the float is due to rise, fluid is pumped into an external bladder and it rises to the surface over about 6 hours. After data telemetry the bladder deflates and the float returns to its original density and sinks to drift until the cycle is repeated. Argo floats are designed to make about 150 such cycles. This system allows, for the first time, continuous near real-time monitoring of the temperature, salinity and velocity of the upper ocean (Fig. 2.8).

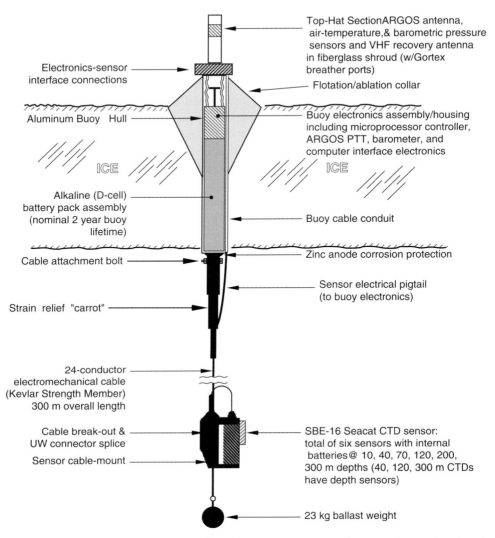

Fig. 2.7 A schematic of a polar ocean profile drifting buoy. (Courtesy of Dr J. Morison, University of Washington, USA.)

2.3 Meteorological analysis fields

2.3.1 Introduction

The production of surface and upper air meteorological analysis charts for the high latitude areas has been carried out for many years. When the first charts were prepared for the European region at the end of the nineteenth century some of the Arctic areas were analysed

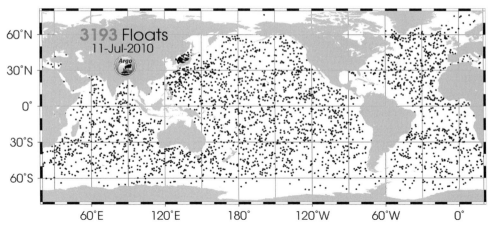

Fig. 2.8   The location of Argo floats on 11 July 2010. (Courtesy of the International Argo Project.)

in the northern part of these charts, although there were insufficient data to analyse areas such as the Arctic Ocean. The situation gradually improved during the twentieth century as there were a relatively large number of ship observations available, some of which were in the high latitude areas.

In the Antarctic routine surface and upper air chart production took place during the IGY of 1957–58, although the reliability of the analyses in the data-sparse areas of the Southern Ocean is questionable and such fields are not used in most climate change studies.

The major advance in high latitude analysis came with the introduction of atmospheric sounding instruments on the polar-orbiting satellites in the 1970s. In this section we consider the use of such data in conjunction with conventional in-situ observations in the production of routine high latitude analyses and the value of the fields for studies of climate change.

2.3.2   *Numerical weather prediction models*

Weather forecasting systems assimilate observations in order to forecast the future atmospheric state, usually across the entire globe. The accuracy of the forecast is dependent on using available observations to nudge the initial (first-guess) atmospheric state of the numerical weather prediction (NWP) system towards reality. There are, of course, a whole range of data types available – surface observations from ground stations, radiosonde data and satellite-derived data, both at the surface and throughout the atmospheric column – at different temporal and spatial resolutions. Surface observations assimilated in the polar regions generally comprise only pressure and humidity, while temperature, wind and humidity data are taken from radiosondes. Satellite data that are assimilated into the NWP schemes include cloud track winds, vertical temperature and moisture profiles and total column ozone. To assimilate all these data in an optimal way is a key component of the NWP process.

The spatial resolution of global NWP models has improved markedly as computer power has increased: the 2010 ECMWF operational model had ~16 km horizontal resolution. In the vertical axis, the atmosphere is divided into 91 vertical layers up to 0.01 hPa (about 80 km) just over the mesopause. The vertical resolution (measured in terms of geometrical height) is finest in the planetary boundary layer and coarsest in the stratosphere and mesosphere. In the lowermost troposphere, where the Earth's orography affects the dynamics, the levels follow the surface, whereas above they are identical to constant pressure surfaces.

Although global NWP has improved over recent decades, particularly in the Southern Hemisphere, there is a need for regional or mesoscale forecasts in the polar regions. Monaghan *et al.* (2003) showed that although global NWP models have good skill in predicting the free atmosphere, they are less good at forecasting surface weather conditions. An analysis of the output from the US Antarctic Mesoscale Prediction System (AMPS) indicated the importance of a high spatial resolution (~3 km) in improving the forecasts in regions of complex topography with relatively few observations available for assimilation (Bromwich *et al.*, 2005). In the Arctic a similar mesoscale model is being used to help predict the regional hydrological cycle, including river discharge into the Arctic Ocean.

2.3.3 Reanalyses

A reanalysis project using an NWP model first involves the recovery, compilation and quality control of an archive of various historical climate data sets. The reanalysis utilises all available observations, including those that were not available in real-time in the original forecast. These data are then assimilated into a near state-of-the-art NWP model with a constant assimilation and forecast system. The idea is to remove spurious climate jumps due to changes in the model formulation – as noted previously, operational NWP models are being constantly improved – and hence provide a time series appropriate for examining long-term climate change. However, changes in the availability of data types still cause inhomogeneities. In the polar regions this is particularly the case with the advent of satellite sounding data in the late 1970s. Before this time there were relatively few conventional (surface and radiosonde) observations to constrain the NWP model and the reanalysis products can be quite poor during this time. In the remainder of this section general descriptions of past, current and future reanalyses are given: detailed studies of their quality over the Arctic and Antarctic are presented in subsequent sections and reviewed by Bromwich *et al.* (2007).

The first major reanalysis was undertaken by the National Centers for Environmental Prediction (NCEP) and National Center for Atmospheric Research (NCAR) and work began in 1991 (Kalnay *et al.*, 1996; Kistler *et al.*, 2001). The NCEP/NCAR reanalysis begins in 1948 and has the advantage over some other reanalyses of being continually updated. The model used has a spectral resolution of T62 (equivalent to a spatial resolution of ~210 km) and 28 vertical levels, with the highest level at 10 hPa. There were three major errors made in the assimilation of data into the NCEP/NCAR reanalysis that had an impact on the polar regions. First, a snow cover corresponding to 1973 was also used for 1974–94. Clearly, this will have its largest impact on those Northern Hemisphere regions where the snowline varies about the 1973 limit (Kistler *et al.*, 2001). Moreover, annual snow cover data were unavailable prior to 1967 so a climatology was used. Note that this is also necessarily the case for sea ice extent in reanalyses prior to 1979. Second, Australian pseudo-observations of sea level pressure (PAOBS) – the product of human analysts who estimated pressure based on satellite data, conventional data and time continuity – over the Southern Ocean were assimilated with a 180° longitude error between 1979 and 1992. Due to the sparse nature of observations at high southern latitudes the impact of the error is largest in this region: fortunately, it seems that on monthly and longer timescales the resultant errors are small. Third, a formulation of horizontal moisture diffusion caused 'spectral snow' at high latitudes, with spurious wave patterns of high and low values of precipitation (e.g. Cullather *et al.*, 1996; Cullather and Bromwich, 2000). These errors and others were corrected in a second reanalysis (NCEP2) running from 1979 to the present (Kanamitsu *et al.*, 2002).

The other three principal reanalyses have been produced by ECMWF: first a 15-year reanalysis (ERA-15), which encompassed the period from 1979 to 1993 (Gibson *et al.*, 1996), and more recently a longer reanalysis from 1957 (the IGY) to 2002 (known as ERA-40) (Uppala *et al.*, 2005). ERA-15 has a spectral resolution of T106 (~110 km) and 31 vertical levels, again up to 10 hPa. ERA-40 is generally considered to be the best reanalysis for the polar regions at present. However, it still has some known issues in the polar regions. A lower tropospheric cold bias exists over sea ice from 1989 onwards in both polar regions, resulting from the assimilation of High Resolution Infra-red Sounder (HIRS) radiances. Fortunately, this was identified while production was ongoing and the problem resolved for data from 1997 onwards (Uppala *et al.*, 2005). At the time of writing ECMWF is undertaking an 'interim' reanalysis that covers the period from 1989 onwards, overlapping the earlier ECMWF 40-year reanalysis.

There have been other reanalyses undertaken too, although these have been less widely used by the scientific community than those described previously. Examples include the Goddard Earth Observing System (GEOS) reanalyses (Schubert *et al.*, 1993) and the 25-year Japanese Reanalysis Project (JRA-25), which uniquely includes Chinese snow cover data (Bromwich *et al.*, 2007).

A new reanalysis produced by NASA Goddard named MERRA (Modern Era Reanalysis for Research and Applications) began in 2006 and concentrates on the hydrological cycle. Some ocean reanalyses are also being produced now, although it remains to be seen how useful they will be in the data-sparse polar oceans.

Another reanalysis that has been carried out covers the period from the late nineteenth century to the present. This is based on surface pressure observations alone (Compo *et al.*, 2006), and, while it may be of little use in the polar regions for the early period when there are no observations to constrain the model, may provide useful information for the Arctic in the middle of the twentieth century. One of the authors of this book was involved with the digitisation of meteorological registers and barographs from whaling stations on South

32 *Polar climate data and models*

Georgia for this project: these date back to the turn of the twentieth century so this reanalysis might be able to provide useful early data from the Antarctic Peninsula region.

There is also an Arctic reanalysis being undertaken which will improve on global reanalyses products. Special attention is being paid to specific high latitude concerns such as the structure of the low-level temperature inversion, polar clouds and associated radiative fluxes, the vapour flux convergence and optimal use of satellite data.

The Arctic

In general, reanalyses are better at representing climate in the Arctic than the Antarctic, primarily due to the significantly greater observation density in the north, which is better able to constrain the background NWP model. There have been several studies examining moisture and energy fluxes into the Arctic, typically across $70°$ N. For example, Cullather and Bromwich (2000) showed comparable values of meridional moisture flux convergence calculated from radiosondes and both the NCEP/NCAR and ERA-15 reanalyses. More recently, Serreze *et al.* (2007a) compared the various components of the energy budget across $70°$ N in the NCEP/NCAR and ERA-40 reanalyses. The two reanalyses were broadly similar, differing primarily in the geopotential energy flux, which the authors suggested may relate to differences in the vertical resolution of the two NWP models or an incorrect calculation of the term at ECMWF.

However, at a regional scale there can be marked differences between different reanalyses and observations. Serreze *et al.* (2005) compared precipitation in the NCEP/NCAR, ERA-15 and ERA-40 reanalyses and the Global Precipitation Climatology Project version-2 (GPCP) fields (Adler *et al.*, 2003) against gridded gauge data, adjusted for undercatch of solid precipitation (see Section 2.2.2). The study revealed a significant positive bias in the NCEP/NCAR reanalysis data over land during summer, related to excessive convective precipitation and evaporation rates. Biases in ERA-40, ERA-15 and GPCP are generally smaller than in the NCEP/NCAR reanalysis. Interestingly, and somewhat disappointingly, the authors noted that ERA-40 was actually poorer than the older ERA-15 during summer: they speculated that this may have been associated with incorrect retrieval of satellite radiances, as mentioned in the previous section. Another important climate variable is the temporal variability of precipitation in the catchment areas of the large rivers that drain into the Arctic Ocean (cf. Section 3.9.1) and ERA-40 was found to show good agreement with observations (Serreze *et al.*, 2005; Su *et al.*, 2006). Su *et al.* (2006) found that the principal errors in ERA-40 were too early snow melt and a double peak in the runoff, as shown in the data for the catchment of the Lena (Fig. 2.9).

The radiative differences between reanalyses and observations in the Arctic were examined by a detailed comparison of clouds and their radiative impacts with observations at Barrow, Alaska (Bromwich *et al.*, 2007). ERA-40 was judged the best of the reanalyses examined, reasonably accurately predicting the variability in cloud fraction and resultant downwelling longwave radiation. However, it overpredicted the downwelling shortwave radiation, suggesting an incorrect parameterisation of the transmission of shortwave radiation through clouds because the latter are optically too thin.

2.3 Meteorological analysis fields

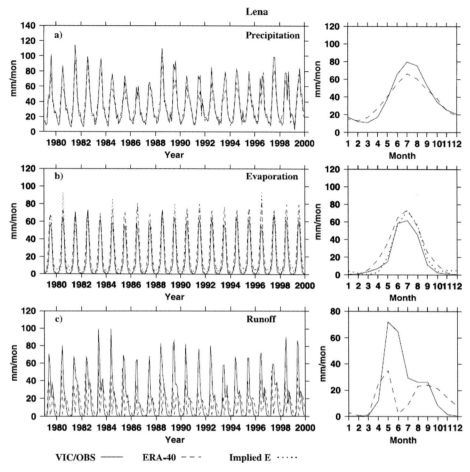

Fig. 2.9 Monthly time series (left) and seasonal means (right) of water budget components from a model/observations, ERA-40 and implied evapotranspiration. Units are in mm per month. (From Su et al., 2006.)

The Antarctic

The main issue with reanalysis data in high southern latitudes is the poor quality of the data prior to the assimilation of satellite sounder data. Figure 2.10 shows a comparison of the mean sea level pressure (MSLP) biases in the NCEP/NCAR and ERA-40 reanalyses at 40° S and 65° S compared with observations. Before 1979 the NCEP/NCAR reanalysis had pressures that were much too high, with a maximum bias at 65° S, particularly in austral winter (Hines et al., 2000; Marshall, 2003; cf. Fig. 2.10). Although there is less of a clear positive bias or annual cycle in ERA-40 pressures, values are clearly less accurate before the assimilation of satellite data: a marked improvement can be seen in 1975 with the

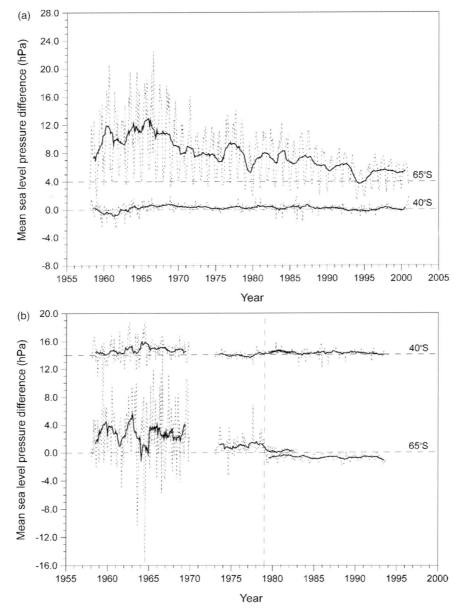

Fig. 2.10 Bias in zonal MSLP at 40° S and 65° S as derived from (a) NCEP/NCAR reanalysis and (b) ERA-40 and ERA-15 reanalyses. Note offsets used for clarity. (From Marshall, 2003.)

assimilation of Vertical Temperature Profile Radiometer (VTPR) data and then another subsequent improvement in 1979 with TIROS Operational Vertical Sounder (TOVS) data assimilation. Bromwich and Fogt (2004) concluded that only austral summer reanalysis data can be considered reliable before 1979. Bromwich *et al.* (2007) suggested that the poorer winter reanalysis may be attributed to the use of a climatology of sea ice extent (as there were no satellite-derived data) and fewer ship reports over the Southern Ocean. The improvements and resultant spurious decrease in pressure at 65° S, which is 8 hPa in the NCEP/NCAR reanalysis (Hines *et al.*, 2000) affects the derived Southern Annular Mode (SAM) index. Marshall (2003) demonstrated that trends in the SAM (cf. Section 6.3.3) derived from this reanalysis were three times as large as those similarly obtained from observations. Marked improvements in the accuracy of other parameters, such as near-surface temperature, and pressures and temperatures throughout the troposphere and lower stratosphere, are apparent in both the NCEP/NCAR and ERA-40 reanalyses.

In the earlier part of the reanalyses it appears that the few available observations were sufficiently different from the background climatology that they were rejected by the assimilation scheme. Moreover, Marshall and Harangozo (2000) revealed that there were periods of several years when data from some Antarctic stations were missing from the reanalysis archive in the 1950s and 1960s. An attempt was made to rectify this issue in the ERA-40 reanalysis but unfortunately not all data were entered into the assimilation system. Other problems with reanalyses products over Antarctica result from incorrect station heights being used: in a region with such sparse conventional data these errors can propagate outwards and significantly affect the regional climate. Bromwich *et al.* (2000) revealed that errors in West Antarctic precipitation in ERA-15 were due to an incorrect assimilation height for Vostok Station being used, which was propagated into the geopotential height fields through assimilation of the radiosonde data. Similarly, Marshall (2002) found that errors in the height of East Antarctic Australian AWSs assimilated into the NCEP/NCAR reanalysis caused a spurious decrease in regional geopotential height in the 1990s.

One key potential use of the reanalyses is to provide gridded accumulation data across the Antarctic Ice Sheet in order to examine changes in its mass balance (see Section 6.9.2), particularly as conventional ice-core observations are sporadic, difficult to obtain and encompass different time periods. However, although ERA-40 is generally considered to have the most accurate Antarctic precipitation of current reanalyses (Bromwich *et al.*, 2007), van de Berg *et al.* (2005) showed dramatic increases in ERA-40 accumulation (calculated as precipitation minus evaporation, abbreviated to P–E) over Antarctica occurred between 1979 and 1980, with a 50% rise in the continental interior. Thus, the circulation changes across the high southern latitudes caused by the assimilation of satellite data means that in most cases they can only be used for time-series analysis back to 1980. If future reanalyses are to prove useful further back it will be necessary to (i) ensure that all possible observations are assimilated and (ii) an assimilation scheme is used that is appropriate for regions of sparse data rather than one tuned for data-rich regions where the rejection of individual observations is not critical.

36 *Polar climate data and models*

2.4 Remotely sensed data

2.4.1 Introduction

The polar regions are remote, vast and inhospitable and it is therefore very expensive to collect in-situ measurements at regular spatial and temporal intervals from these areas. Thus, research in the Arctic and Antarctic has been especially helped by the advent of airborne and satellite remote sensing techniques. The first remote sensing data acquired in the polar regions were the various airborne surveys that were used to map the regions in detail for the very first time. Such data are now useful in examining long-term changes in glacier fronts as indicators of climate change (e.g. Cook *et al.*, 2005). Satellite imagery of the polar regions began to be collected in the 1960s. One problem specific to the polar regions is that some instruments flown on satellites in sun-synchronous polar orbits are unable to provide data at latitudes greater than ~80°. However, instruments flown in such orbits will provide a higher revisit capability than at lower latitudes because ground coverage from these orbits overlaps at high latitudes. Other potential problems associated with the satellite imagery from the polar regions are sensor saturation from highly reflective snow and ice surfaces, difficulty in distinguishing cloud over such surfaces and the absence of sunlight during the polar night. For the interested reader, recent volumes by Lubin and Massom (2006) and Massom and Lubin (2006) provide a detailed analysis of remote sensing in the polar regions.

2.4.2 Satellite imagery

Imagery

Many of the early high latitude cloud studies were concerned with simply identifying the presence of cloud and perhaps determining the phase. This was because the studies primarily used Advanced Very High Resolution Radiometer (AVHRR) imagery, which only had five channels in the visible and infrared. In recent years a number of new satellite instruments have been flown that have many more spectral channels. For example, the MODIS (Moderate Resolution Imaging Spectroradiometer) instrument has 36 channels that allow significantly more information to be determined about polar clouds (Key *et al.*, 2003). The disadvantage is that the time series of imagery is much shorter than that from AVHRR. Specialised satellite missions, such as Cloudsat, provide additional information on the properties of polar clouds. Cloudsat flies the Cloud Profiling Radar (CPR) instrument, which is a 94-GHz nadir-looking radar that measures the power backscattered by clouds as a function of distance from the radar. The CPR provides many products, including information on cloud liquid water content, optical depth and cloud type.

A major focus for cloud studies from satellite data has been the International Satellite Cloud Climatology Project (ISCCP). This has processed imagery from the geostationary and polar orbiting satellites and produced a series of data sets that will be discussed in Section 3.7.1.

2.4 Remotely sensed data 37

The detection of cloud from satellite imagery still presents a number of problems at high latitudes. In the extra-polar regions clouds are usually colder than the underlying surface and are often more reflective than the surface. Techniques have been developed to automatically detect cloud in sequences of satellite images and to produce cloud climatologies. Applying such techniques to the polar regions has proved more of a challenge, but in recent years there has been some success in developing robust methods of identifying cloud over the polar regions (Turner *et al.*, 2001).

High-resolution (~50 m) imagery from the Landsat series of satellites, using channels operating in the visible and near-infrared wavelengths, was first used as a tool to simply map the inhospitable polar regions. With repeat imagery, changes in the margins of ice masses could be examined, including the collapse of ice shelves on the Antarctic Peninsula (Doake and Vaughan, 1991). Feature tracking on glaciers, using crevasse patterns for example, allowed the estimation of mean ice velocities between two images (e.g. Lucchitta *et al.*, 1993). Examples of the use of AVHRR infrared imagery (~1 km) for polar meteorological studies includes the tracking of katabatic winds across the Ross Ice Shelf (Bromwich *et al.*, 1992) and the tracking of weather systems in the Antarctic Peninsula region (Turner *et al.*, 1998).

Synthetic aperture radar

The synthetic aperture radar (SAR) is an active microwave instrument, which generates a signal and then analyses information from the return of that signal from the Earth's surface. In some ways, SAR instruments are the ideal sensor for monitoring ice sheets and sea ice because they combine a high spatial resolution (~20 m) with the ability to 'see through' cloud, allowing the acquisition of data whenever the satellite orbit allows. However, there are limitations: the swath width is very narrow, data rates are very high and SAR 'imagery' is not simple to process and interpret. The principal SAR instruments flown in the past have been on the European ERS satellites and Canadian Radarsat-1, while at the time of writing there are several satellites operating with different instrument frequencies: examples include Envisat, Radarsat-2 and Terrasar-X. Radarsat-1 imagery was used for the Antarctic Mapping Mission in 1997, the first complete high-resolution mapping of the continent (Fig. 2.11). The satellite was rotated 180° in yaw to allow the radar to image to the left of the satellite track, allowing coverage of the South Pole.

The high resolution of SAR imagery also means that it can be used to monitor sea ice: accurate measures of sea ice concentration can be derived and, because individual flows can be tracked, sea ice kinematics and deformation can be monitored. Moreover, the different scattering properties of various types of sea ice, such as first-year and multi-year ice (cf. Section 3.8.1) means that ice classification algorithms can be developed.

At a reduced spatial resolution (~25 km), scatterometer instruments have also been effectively utilised in the polar regions. These instruments have been designed principally to capture the near-surface wind field over the oceans: for example, Marshall and Turner (1997) used the data to examine the strength of Antarctic mesocyclones. However, many uses have also been found for scatterometer data over ice sheets (e.g. Drinkwater *et al.*, 2001) and for monitoring sea ice (e.g. Nghiem *et al.*, 2006).

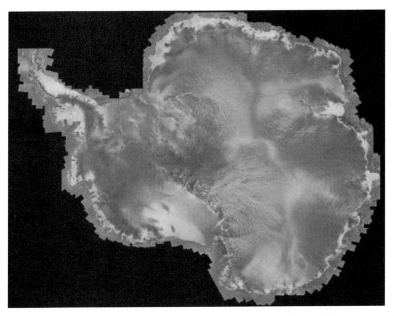

Fig. 2.11 Low-resolution version of the Radarsat mosaic of Antarctica, which was compiled from synthetic aperture radar images acquired in September and October 1997. (Photograph courtesy of the Canadian Space Agency.)

Radar interferometry is a technique that utilises phase information contained within the complex radar return signal to measure extremely small changes in surface elevation by means of the interference between two signals acquired on different dates or slightly different locations. Either ground elevation through parallax can be obtained, by assuming no surface movement between the two images, or the surface displacement – hence ice velocity of glaciers or ice streams – may be resolved if the effects of topography are corrected for. This latter technique is called differential interferometry. The two radar images are co-registered and processed to produce an interferogram, a representation of the coherence between the two images that is characterised by sequences of 'fringes', each indicative of a displacement of half a wavelength. By subtracting an artificial interferogram that accounts for the topography the surface displacement can be computed very accurately to within a few millimetres per day. Interferometry has been used widely in Antarctica, Greenland and for smaller glaciers elsewhere in the Arctic.

Altimeters

Another type of radar instrument that has proved important for polar studies is the radar altimeter. This accurately measures the distance between the satellite and the ground surface and hence repeated measurements can be used to monitor changes in the thickness and, with appropriate assumptions regarding changes in density, the mass balance of ice sheets.

2.4 Remotely sensed data 39

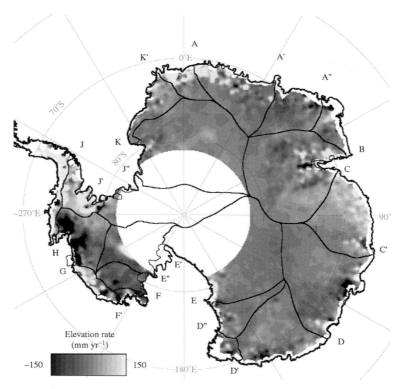

Fig. 2.12 Rate of elevation change of the Antarctic ice sheet, 1992–2003 from ERS altimeter data. (After Wingham et al., 2006. Used with permission of the Royal Society.)

Measurements between 1992 and 2003 obtained from the ERS-1/2 satellites have revealed a marked drawdown (thinning) of the Pine Island Glacier region in West Antarctica (Shepherd et al., 2001; Wingham et al., 2006) (Fig. 2.12) and a thickening in parts of East Antarctica (Davis et al., 2005). In Greenland altimeter data have revealed that the ice sheet has been thinning at the margins and growing inland (Zwally et al., 2005).

Gravity measurements

A newer type of instrument is the NASA–German Aerospace Center Gravity Recovery and Climate Experiment (GRACE), which provides measurements of the Earth's gravitational field at monthly intervals. After atmospheric and oceanic contributions are removed, changes in the gravitational field reflect changes in snow and ice mass over the polar regions (e.g. Chen et al., 2006). As much of the loss from Antarctica and Greenland during the period of observation is due to accelerated flow from glaciers and ice streams (e.g. Shepherd and Wingham, 2007) a synthesis of measurements from altimetry/gravimetry and interferometry is required to best monitor changes in mass.

Passive microwave instruments

The principal use of passive microwave instruments, which detect radiation emitted from the Earth, in the polar regions has been to map sea ice. The large contrast between the emissivity of open water (0.4 at 10 GHz) and sea ice (0.7–0.9 at 10 GHz) means that both the extent and concentration of sea ice can be monitored. In addition, changes in sea ice properties with age have allowed the discrimination of different types of ice, particularly by examining different polarisations of the emitted energy. The main satellite instruments used for observing sea ice are the Scanning Multichannel Microwave Radiometer (1978–87), the Special Sensor Microwave Imager (SSM/I) from 1987 onwards and the Advanced Microwave Scanning Radiometer (AMSR-E) from 2002 onwards (Comiso and Nishio, 2008). Thus, the sea ice data set is now sufficiently long to examine climatological changes (e.g. Zwally *et al.*, 2002a; Serreze *et al.*, 2007b). Additional information on sea ice thickness has been obtained from upward-looking sonar data from submarines (e.g. Rothrock *et al.*, 1999).

The MODIS instrument has been used to obtain wind speed and direction in the polar regions by using triplets of data, from overlapping swaths near the poles (Key *et al.*, 2003). Cloud features are tracked in the infrared band (11 μm) and water vapour features in the 6.7 μm band (Fig. 2.13). When these polar winds were included in the ECMWF 4D-var data assimilation scheme they had a positive impact over the polar regions, where previously the winds were largely determined by the forecast model (Bormann and Thépaut, 2004).

Fig. 2.13 Cloud-drift winds around the island of Svalbard from MODIS on 5 March 2001. The wind vectors are overlain on an 11-μm image. (Image courtesy of NOAA and the University of Wisconsin/CIMSS.)

2.4.3 Satellite sounding data

Satellite sounding instruments have made a huge difference in improving our knowledge of the climate of the polar regions and hence the forecasts of NWP models. Today, the Southern Ocean, previously a very large data void, is encompassed by sounder data with a spatial resolution of tens of kilometres. The wavelengths of these types of instrument are selected according to the absorption and emission properties of carbon dioxide, nitrous oxide, oxygen, ozone and water vapour. As the distribution of these gases is relatively constant and their radiative properties understood, variations in the radiance received by the satellite (the brightness temperature) can be related via mathematical inversion methods to the temperature of that part of the atmosphere from which most of the signal emanates. By using a number of different wavelengths that have different absorption characteristics a complete profile of radiances through the atmospheric column can be derived (Fig. 2.14).

The principal satellite sounding instruments are the TOVS sensor system, which has been flown on polar-orbiting satellites since 1978, and its successor ATOVS, which began operating in 1998. TOVS comprises three radiometer arrays: the HIRS/2, the Microwave Sounding Unit (MSU), and the Stratospheric Sounding Unit (SSU). In ATOVS the Advanced Microwave Sounding Unit-A (AMSU-A) and the Advanced Microwave Sounding Unit-B (AMSU-B) replace the MSU and the SSU, while HIRS/3 replaces the HIRS/2.

Unfortunately, comparison of TOVS-derived temperatures against radiosondes launched from Arctic sea ice demonstrated marked deficiencies in the original TOVS retrieval algorithms over sea ice during the winter (Francis, 1994). Near-surface temperature

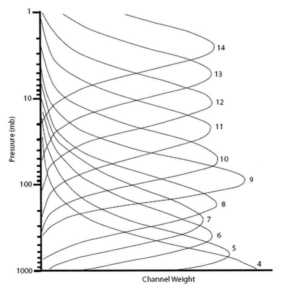

Fig.2.14 Weighting functions for channels 4–14 of the AMSU-A instrument.

inversions were missed, primarily because of surface temperature retrievals over sea ice being 5–15 °C too high. This was traced to a tendency for the algorithm to overestimate total column water vapour. However, a comparison of integrated water vapour from AMSU-B and limb-sounding observations by a GPS receiver on board Low Earth Orbiter (LEO) satellites (obtained using radio occultation techniques: relating changes in a radio signal as it passes through the atmosphere to variations in the physical properties of the atmosphere) showed good agreement between the two independent methods, indicating that modern satellite instruments can obtain accurate data on moisture over the Antarctic Plateau, which could be usefully assimilated into NWP forecasts (Johnsen *et al.*, 2004).

2.5   Proxy climate data

2.5.1   Introduction

The in-situ observational climate records from the polar regions are shorter than those from the rest of the world, especially in the Antarctic where the records only start around the middle of the twentieth century. Proxy data are therefore of particular value in extending the records back before the instrumental period.

In this section we examine the climate data that can be obtained from the ice core, ocean sediment core and dendrochronological records from the two polar regions.

2.5.2   Ice cores

The solid precipitation that accumulates on the Greenland and Antarctic Ice Sheets, together with numerous smaller Arctic icecaps, allows us to reconstruct past atmospheric chemistry and climate variability of the Earth. Ice cores provide the most highly resolved records of Pleistocene climate available: in Greenland the deepest core contains ice formed 123 kyr BP (Andersen *et al.*, 2004), while ice believed to be ~800 kyr old has been obtained from Antarctica (EPICA Community Members, 2004). We can thus examine the composition of the atmosphere in the so-called 'pre-industrial' period, before significant human impact began in the early nineteenth century. Analysis of the natural variability and range of the climate system allows us to set the significant climate changes of the twentieth and twenty-first centuries in a longer context. However, for such historical reconstructions to be meaningful the transfer functions linking the chemical composition of the snow and ice to that in the atmosphere need to be reasonably well constrained.

When snow is deposited on an ice sheet it becomes compressed by subsequent snowfall. It then becomes firn – snow that has survived one summer without being transformed to ice (density $< 830 \, \mathrm{kg \, m^{-3}}$) – and finally ice itself, when the interconnecting air passages between firn grains are sealed off, a process termed bubble close-off (density $> 830 \, \mathrm{kg \, m^{-3}}$). It is the air in these bubbles that contains a record of past atmospheric composition. Note that when trying to accurately date rapid climate changes the difference between ice age and gas age (at bubble close-off), which increases with lower accumulation and temperatures, must be accounted for: for example, in central Greenland the difference is ~800 years at the time of the Younger Dryas termination (~11.6 kyr BP) (Severinghaus *et al.*, 1998). The air bubbles record trace gas concentrations, chemical impurities of marine and terrestrial origin, volcanic (and recently anthropogenic) aerosols, cosmogenic isotopes (for example, from nuclear detonations), and extraterrestrial material.

The quantity most commonly measured in ice cores is the oxygen-isotope ratio ($\delta^{18}O$). This is the relative difference between oxygen-18 (heavy) and oxygen-16 (light) atoms and standard mean ocean water, expressed in per mil (parts per thousand, written as ‰). A physical linear relationship between $\delta^{18}O$ and the condensation temperature arises because water molecules containing heavy (light) oxygen atoms condense (evaporate) more easily. As precipitation-bearing air masses rise in the atmosphere or move poleward they cool and the heavy oxygen is removed preferentially so any remaining precipitation becomes progressively depleted of oxygen-18. When the climate is cooler this process takes place at lower altitudes and/or latitudes, and so precipitation over the polar regions has lower $\delta^{18}O$ values. Deuterium excess (*d*) is also used as an indicator of temperature but additionally reflects changes in sea surface temperatures (SSTs) at the source region of the core accumulation (e.g. Stenni *et al.*, 2001).

One problem with the use of $\delta^{18}O$ as an isotopic palaeothermometer is that factors other than the local environment can affect the temperature/isotope spatial slope. These include changes in the origin, pathways and seasonality of the precipitation and changes in the surface-inversion strength. Temporal variation in these and other factors mean that the present-day linear relation between $\delta^{18}O$ and temperature in Greenland may be inappropriate at other times (Cuffey *et al.*, 1995; Jouzel *et al.*, 1997): on the Antarctic Plateau the modern relationship has been shown to be invalid for deriving past temperatures during warmer interglacial periods (Sime *et al.*, 2009). Some of the issues affecting Greenland cores can be overcome by using nitrogen and argon isotopes, $\delta^{15}N$ and $\delta^{40}Ar$, which allows direct measurement of the magnitude of rapid temperature change (e.g. Severinghaus and Brook, 1999; Landais *et al.*, 2004). Borehole studies, which use temperatures from the ice core hole in conjunction with a combined ice flow / heat transport model, can be used to infer past temperatures directly (Dahl-Jensen *et al.*, 1998).

The greenhouse gases carbon dioxide (CO_2) and methane (CH_4) play a vital role in the global climate system with data derived from ice cores clearly indicating how these gases have both fluctuated with temperature through previous glacial cycles. The concentration of CO_2 in the atmosphere over the last 420 kyr, as derived from the Vostok core, has alternated between about 180 ppm in glacial maxima and 280 ppm in interglacials. Such changes would act as a major amplifier in the climate system and it is clearly important to determine the causes of the CO_2 increases at the end of glacial periods in order to accurately predict future climate. As the greenhouse gas concentrations represent a global signal, their variability can be utilised to constrain the relative timing of climate change in the two polar regions from independent $\delta^{18}O$ records (e.g. Blunier and Brook, 2001). Changes in the concentrations of the gases can provide evidence for changes in source regions (e.g. Chappellaz *et al.*, 1990) while analysis of the lags/leads between the concentrations of these gases and temperature proxies in ice cores provides insight into the physical processes through which variability in the global climate system is transferred to the polar regions (e.g. Delmotte *et al.*, 2004).

In addition to trace gases, the chemical impurities in the snow and ice can be examined: Legrand and Mayewski (1997) provide a review of this area of research, called glaciochemistry. Most of the analyses are undertaken under strict clean conditions in the laboratory. The outsides of the cores are often discarded as they are assumed to have been contaminated during the collection process. Analytical procedures have advanced such that it is now possible to rapidly undertake multiple glaciochemical analyses by using a combination of melting and continuous flow techniques (e.g. Röthlisberger *et al.*, 2000).

One of the principal uses of glaciochemistry is to provide an annual dating methodology so that individual years may be counted. In coastal Antarctica, methanesulphonic acid (MSA) – the atmospheric source of which is dimethylsulphide (DMS), which is emitted by marine biota and is thus related to regional oceanic primary productivity – shows a clear maximum (minimum) in summer (winter). However, in deeper layers the MSA can relocate to winter layers (Mulvaney *et al.*, 1992). There is also a strong seasonal cycle in sea salt aerosol that peaks in winter: this is now thought to be because the dominant source of sea salt is the highly saline brine and 'frost flowers' that cover newly formed sea ice (Wolff *et al.*, 2003). Many studies have used a combination of different chemical species to determine the annual cycle of accumulation: in this way if there is an abnormal signal in one year in one species then comparison with the others analysed will allow a more robust chronology. In addition, multiple chronologies allow the use of empirical orthogonal function (EOF) analysis to relate the glaciochemistry to different air mass types (e.g. Reusch *et al.*, 1999).

Electrical conductivity measurement (ECM) allows rapid, continuous, high-resolution measurements along the full length of the core and can be undertaken in the field using portable equipment. It is a proxy indicator of the acid content of the accumulation and has been used to look at a range of temporal variability, from millennial-scale changes in deep cores (e.g. Taylor *et al.*, 1993) to looking at individual precipitation events in shallow cores (e.g. Marshall *et al.*, 1998). Note that current spreading between ECM electrodes can reduce the resolution below annual thickness and therefore this technique may not be appropriate for annual layer counting in deep cores.

Visual stratigraphy has also been used to date ice cores, particularly in Greenland, where visible layers are strongly related to chemical impurities and dust. Modern line-scanning equipment in combination with sophisticated automatic image-processing techniques means that a chronology can be obtained relatively easily (e.g. Svensson *et al.*, 2005; Fig. 2.15). A limitation is that measurements need to be performed in the field shortly after the ice is recovered, otherwise the internal structure of the ice relaxes. In the NGRIP core there are also problems with diffuse/inclined layering and large crystals at the bottom part of the core.

Another component to producing a reliable chronology is the use of reference horizons, of which there are several types. First, there are 'cosmic events': for example, the relatively high concentration of cosmogenic beryllium (^{10}Be), formed from the impact of cosmic rays on oxygen and nitrogen, observed at ~41 kyr BP is associated with a reversal of the Earth's geomagnetic field (e.g. Schwander *et al.*, 2001). More recently, the nuclear bomb tests in the

2.5 Proxy climate data

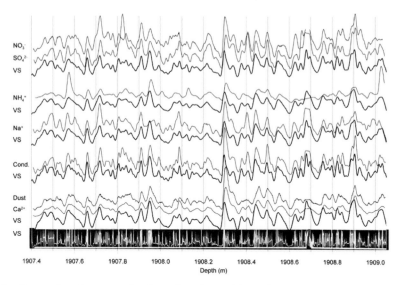

Fig. 2.15 Comparison between visual stratigraphy and impurity concentrations in the NGRIP ice core. (From Svensson *et al.*, 2005.)

1950s and 60s and the subsequent spread of radioactive debris (tritium) have produced other easily detectable layers. Similarly, large, explosive volcanic eruptions, which emit large amounts of ash and gas into the stratosphere, are associated with significant depositions of sulphuric acid that show up as distinct spikes in non-sea-salt sulphate (nss SO_4^{2-}) and ECM profiles. In documented historical times volcanic eruptions can be utilised to provide exact dating, whereas earlier records can only be dated as accurately as the overall ice core stratigraphy at that level but do allow the direct comparison of different ice cores (e.g. Stenni *et al.*, 2002).

A further constituent frequently measured in ice cores is dust, often represented by calcium (Ca^{2+}) and/or iron (Fe). Provenance and meteorological studies have revealed that the principal sources for dust in Greenland and Antarctica are the desert regions of China and Patagonia, respectively (Fischer *et al.*, 2007). Dust records are often used to infer past changes in atmospheric circulation and climatic conditions at the source region. For example, Röthlisberger *et al.* (2002) showed that the record from Dome C, Antarctica, is closely linked to palaeoclimatic records from southern Patagonia during the transition from the last glacial to the Holocene (~18–12 kyr BP). Wolff *et al.* (2006) suggested that the dust record principally reflects changes in atmospheric circulation influencing Patagonian wind strength and aridity. At a longer timescale there is significant variability in dust flux through the ice ages, with maximum values occurring towards the end of a glacial period.

One potential problem with ice cores was highlighted by the Greenland Ice-core Project (GRIP) and Greenland Ice Sheet Project 2 (GISP2) cores from central Greenland. These cores, drilled 28 km apart, correlate well from the surface down to ~2700 m. However, below this, in the bottom 10% above the bedrock, there are major differences in $\delta^{18}O$ and electrical conductivity measurements (Grootes *et al.*, 1993). The discrepancy was subsequently found to be due to stratigraphic distortion in the ice near the pressure melting point at the bottom of the ice sheet (Chappellaz *et al.*, 1997), a characteristic not evident in the more recent NGRIP core.

In addition to the very deep cores drilled in both Greenland and Antarctica, there have also been many shallower ice cores obtained. These are usually annually resolved, by using the seasonal variation of chemical species within the accumulation to discriminate the annual cycle. The shortest are often 10 m deep and are taken primarily to enable the temperature at the bottom of the hole to be recorded: the 10 m temperature provides a good proxy for mean annual surface temperature. A spatial array of shallow cores allows detailed examination of regional changes in accumulation and the quantification of the effects of factors like topography and distance from the coast and natural climate variability such as the El Niño–Southern Oscillation (ENSO) (e.g. Kaspari *et al.*, 2004). They can provide ground truth for satellite observations or climate-model-based precipitation estimates. If long enough, they also enable the instrumental record to be placed in an historical context of the last few centuries. In Antarctica, the International Trans-Antarctic Scientific Expedition (ITASE) (e.g. Kaspari *et al.*, 2004) has involved many nations undertaking traverses across the continent along which many hundreds of shallow ice cores were obtained. Similarly, in Greenland many shallow cores were obtained as part of the Program for Arctic Regional Climate Assessment (PARCA) (McConnell *et al.*, 2001; Fig. 2.16). A smaller network of cores is also being developed across the Canadian Arctic icecaps.

2.5.3 *Sediment cores*

Sediment cores taken from the ocean floor are typically only of the order of metres to hundreds of metres in length but can encompass millions of years of data as sedimentation rates are so much lower than snow accumulation on the ice sheets. However, this means that the temporal resolution of such cores is also much lower. Certain rates are necessary ($\sim$20 mm kyr^{-1}) to enable sampling at sufficiently high resolution for accurate comparison with other cores.

Away from the continental margins marine sediments principally comprise the microscopic fossilised remains of plankton. The most valuable fossils for palaeoclimate studies are from tiny animals and plants with calcium carbonate shells, namely foraminifera and coccoliths, respectively, and those with silicon dioxide (opal) shells, namely radiolarians (animals) and diatoms (plants). These fossils are important because they contain oxygen: the $\delta^{18}O$ of the fossils has been shown to be dependent on both the temperature and the oceanic $\delta^{18}O$ in which they form, with the latter a function of global ice volume and salinity. The $\delta^{18}O$ is higher, as compared with an international standard, when ocean waters are colder and ice sheet volume greater and vice versa.

2.5 Proxy climate data

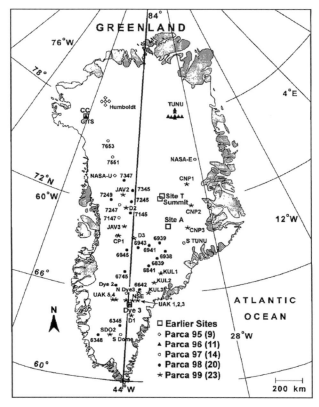

Fig. 2.16 Locations of PARCA cores together with other selected Greenland core sites. (From Mosley-Thompson *et al.*, 2001.)

During the cooler ice ages, the water vapour containing oxygen-18 atoms rained out of the atmosphere at lower latitudes leaving that containing oxygen-16 atoms to condense and precipitate onto the polar ice sheets where it was stored; hence, the $\delta^{18}O$ of water remaining in the ocean became increasingly higher and the exact oxygen ratios relate to the volume of water stored in the ice sheets. Conversely, as temperatures rose during interglacials the melting ice sheets added fresh water into the ocean, returning light oxygen to and reducing the salinity of the global oceans.

Unfortunately, the measurement of $\delta^{18}O$ in the fossil shells is far more complicated because the biological and chemical processes that form the remains skew the oxygen ratio in different ways depending on temperature. Different species also require different offsets to the measured $\delta^{18}O$ values. However, by measuring other chemicals in the shells these uncertainties can be minimised. Benthic records of $\delta^{18}O$ are likely to produce a better global signal than planktonic records as local variability in the deep ocean is markedly less than near the surface. Producing a 'stack' or average $\delta^{18}O$ signal improves the signal to noise ratio further (e.g. Lisiecki and Raymo, 2005).

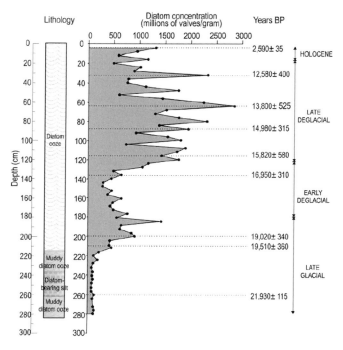

Fig. 2.17 Sediment record and diatom valve concentrations from a marine core obtained in the southwest Atlantic Ocean with dates marked. (From Allen *et al.*, 2005.)

Dating of the longer cores can be achieved with a combination of fixed points, such as magnetic reversals that are recorded in the sediments, and then tuned to fit an appropriate age model. In more recent deposits radiocarbon dating of bulk organic matter can be undertaken, while comparison with other cores can be achieved using biostratigraphic markers (e.g. Allen *et al.*, 2005; Fig. 2.17).

Changes in the concentration of different organisms in the core can provide important information about oceanic conditions at the time of deposition. For example, based on data obtained from sediment traps in the Southern Ocean, Gersonde and Zielinski (2000) derived an empirical relationship in the relative abundance of different diatom species to determine the presence or not of Antarctic sea ice. When organic matter is missing then geochemical and geophysical parameters can be utilised to interpret past conditions. Domack *et al.* (2005) obtained marine cores from the northwestern Weddell Sea, above which the Larsen B Ice Shelf was situated before its retreat and final collapse in 2002. They found three distinct stratigraphic units: at the bottom was a homogeneous, poorly sorted sediment (diamicton), indicative of glacial deposition; above this a stratified deposit, suggesting deglaciation; and, finally, a laminated sandy mud, representing sub-ice-shelf conditions. Using cores from the Arctic Ocean, Moran *et al.* (2006) suggested that larger grain sizes and concomitant increases in bulk density are attributable to more extensive sea ice and intensified iceberg

2.5 Proxy climate data

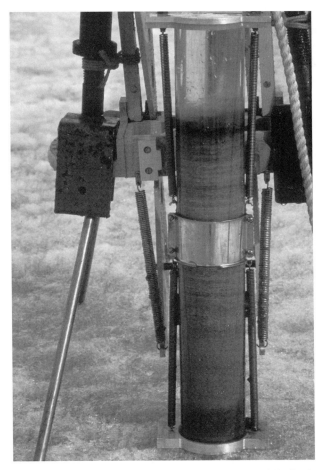

Fig. 2.18 Surface corer and recently obtained sediment core from a lake in the Larsemann Hills region of East Antarctica. (Photograph by C. J. Ellis-Evans, British Antarctic Survey).

activity during glacial periods. Note that these cores were obtained from beneath the zone of permanent sea ice by using two ice breakers to keep the drill site free of ice. In addition, near-shore cores can include pollen and spores from terrestrial species of flora (e.g. de Vernal and Hillaire-Marcel, 2008; see Section 4.2.2).

Sediment cores have also been acquired from terrestrial lakes in both polar regions (Fig. 2.18): such features are abundant in the high Arctic where alternative long-term climate data are sparse. Douglas *et al.* (1994) demonstrated a marked change in diatom assemblages at the beginning of the nineteenth century, having been stable over the previous few millennia: these authors proposed that a longer growing season due to warmer temperatures was the cause. Modern calibration data allowed accurate Holocene temperatures in Arctic Siberia to be reconstructed using records of pollen and chironomids (midges)

50 *Polar climate data and models*

contained within lake sediments (Andreev *et al.*, 2004). In East Antarctica, Hodgson *et al.* (2005) used a multi-proxy data set of lithology (macroscopic composition), geochronology (based on radiocarbon measurements), diatoms, pigments and carbonate stable isotopes to provide information on lake levels, salinity, productivity and the prevalence of open water back to 40 kyr BP.

2.5.4 Dendrochronology

Dendrochronology is the method of scientific dating based on the analysis of tree-ring growth patterns. Variations in the width of annual growth rings may provide qualitative indices of temperature and precipitation during the growing season. In the polar regions this methodology is limited to regions of the Arctic near the tree line in Scandinavia, Siberia and North America. In such regions trees are at the limits of their distribution and are subject to considerable stress from the climate. As a consequence, they have slow growth rates, have higher longevity and their growth rate is particularly strongly related to the local climate, all attributes that make them well suited to dendrochronological analysis (Briffa, 2000). Arctic tree-ring chronologies can be several thousand years in length, with the longest, a 7400 year time series of proxy summer temperatures from northern Swedish Lapland, encompassing much of the Holocene (Grudd *et al.*, 2002).

2.5.5 Dating of driftwood

Driftwood, which can be dated using radiocarbon analysis, has been collected from many regions of the Arctic. Temporal variations in the regional patterns of driftwood size and source regions have been used to propose changes in the path of the Transpolar Drift, one of the major currents of the Arctic Ocean (cf. Section 5.6.2), during the Holocene (Dyke *et al.*, 1997). Driftwood dates have also been utilised in attempting to define the history of Lake Agassiz, the vast meltwater lake that formed to the south of the Laurentide Ice Sheet in North America at the end of the last glacial period (e.g. Broecker, 2006).

2.5.6 Historical records

In some parts of the polar regions there are historical records that provide qualitative proxy climate data prior to the instrumental period. One example is the set of descriptions of weather and sea ice extent from Iceland (Ogilvie, 1992). The various sources begin in the twelfth century but become extensive after ~1600. Two of the primary sources are the 'annals' kept by individuals and official records made by district sheriffs: both describe the main events of the year including the weather and sightings of coastal sea ice. They enable a sea ice index to be produced for the seventeenth century onwards as shown in Fig. 2.19 (Ogilvie and Jónsdóttir, 2000). The ACSYS (Arctic Climate System) study has produced an historical ice chart archive for the North Atlantic part of the Arctic Ocean back

2.6 Models

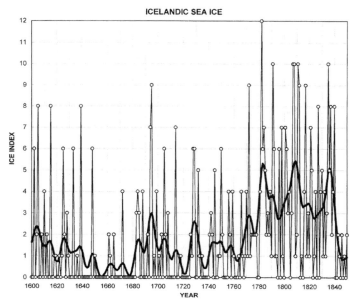

Fig. 2.19 A sea ice index for Iceland showing variations in the incidence of ice off the coasts during AD 1600–1850. Data have been smoothed to highlight lower frequency variations using a 15-year low pass filter. (From Ogilvie and Jónsdóttir, 2000.)

to 1533. This is principally based on reports from vessels entering the Arctic to explore or hunt whales and seals, which became sufficiently dense after ~1850 to create an almost continuous time series of maps for the April–August summer period (Divine and Dick, 2006). Whaling records are fewer and more recent in the Antarctic, beginning at the turn of the twentieth century, but have still been utilised to estimate sea ice conditions prior to the satellite era. De la Mare (1997) used catch data to suggest that the summer ice edge was ~3° further north in the mid 1950s compared with the early 1970s (cf. Section 6.5.2).

2.6 Models

2.6.1 Introduction

Computer modelling is an extremely powerful means of investigating the Earth's climate system, why it changed in the past and how it might change in the future. Indeed models are the only means available for predicting how the Earth will evolve over coming decades, centuries and millennia. But the climate system is extremely complex and to simulate it realistically powerful computers are needed, so high-resolution modelling of the whole Earth system has a relatively short history.

The models now employed to investigate environmental issues embrace a very wide range of tools from simple programs simulating the broadscale workings of the climate system, through models of intermediate complexity, atmosphere-only models, models of the ocean and ice sheet models to highly sophisticated global coupled, atmosphere–ocean general circulation models (AOGCMs). With the rapid increase in the power of personal computers (PCs) it is now possible to run virtually all models on a single PC or cluster of PCs, although state-of-the-art AOGCMs need to be executed on supercomputers if they are to be run for extended periods and produce output on a realistic timescale.

All the models are based on the laws of physics, and provide an objective means of investigating the environment under various forcing scenarios. One of their great strengths is that they provide a means of investigating the *mechanisms* responsible for current and past conditions and allow us to understand how particular elements of the climate system may evolve in the future.

In order to have confidence in the performance of any model it is essential to verify its performance under known conditions. The simplest way to do this is to attempt to simulate the present climate (or often pre-industrial climate, i.e. at a time when greenhouse gases were at a fairly low concentration). Many models are therefore run for extended periods with the present or pre-industrial greenhouse gas concentrations and aerosol loadings prescribed. The resulting mean conditions can then be compared with available observational data. It is also possible to ensure that the models have the correct seasonal cycle of such quantities as near-surface air temperature, cloud cover, sea ice extent, etc. Additionally, it is important to ensure that any sophisticated model of the Earth's climate correctly simulates the major modes of climate variability, such as the ENSO, the North Atlantic Oscillation (NAO) and the SAM.

In recent years many models have been run to simulate conditions in the past when the Earth's orbit around the Sun (see Section 3.4.1) and orbital inclination were different from the current values. Simulations have been carried out for conditions at various points during the Holocene, when we have a relatively good understanding of climatic conditions because of the large amount of proxy data that is available. The Last Glacial Maximum (LGM) some twenty thousand years ago is also an important period that has been simulated extensively.

In this section we consider the various types of environmental, climate and ice sheet model that are available. We describe a number of the models that have been used to study high latitude climatic conditions and the major ice sheets and consider how they represent Arctic and Antarctic conditions.

2.6.2   Climate models

Climate models are a relatively new tool for studying the polar regions and it is only in the last couple of decades that they have had the horizontal resolution and a sufficiently accurate representation of physical processes to be of value in the study of past, present and future climate. In the following we examine the four main types of climate model – box, energy balance, intermediate complexity and general circulation models. We also consider high-resolution ocean models that have proved of value in investigations of Southern Ocean variability.

2.6 Models

Box models

These are some of the simplest climate models that split the different components of the Earth system into a relatively small number of boxes, consisting, for example, of the ocean, atmosphere, land and sea ice. Such models can be of zero, one or two dimensions depending on the problem being investigated. They can be run on desktop computers and are particularly valuable for investigating processes and interactions between different elements of the Earth system. They are especially useful for looking at long-term variability in the climate system, which in terms of computer resources would be much more expensive to investigate with more complex models.

Figure 2.20 illustrates the components of a simple box model that has been used to study the global carbon cycle. In this model the boxes represent different elements of the climate

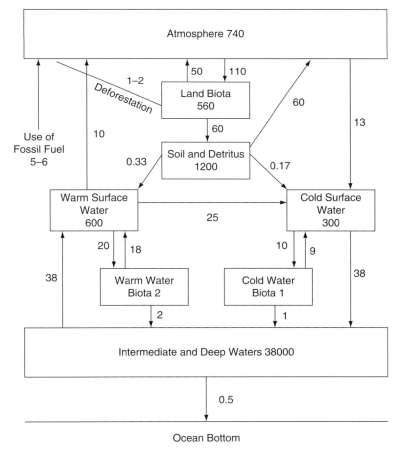

Fig. 2.20 The components of a simple global carbon box model. Units are gigatons in reservoirs (numbers in boxes) and gigatons of carbon per year in fluxes (numbers besides arrows). (Adapted from www.cgrer.uiowa.edu/research/research_highlights/sgcm/Sgcm.html.)

system that can store carbon, such as the atmosphere, land biota and various water masses. The arrows represent fluxes of carbon between different reservoirs. The model can be run to examine various change scenarios, such as carbon dioxide emission and ocean carbon uptake.

Box models have been used in a number of investigations into the climate of the polar regions. For example, Stephens and Keeling (2000) used a seven-box model of the world's oceans to investigate the impact of a decrease in Antarctic sea ice on atmospheric CO_2 levels. Chemical box models have also been used to investigate the reasons for the loss of stratospheric ozone since the 1980s – the so-called 'ozone hole' and to investigate decadal timescale variability in the Arctic climate system (Dukhovskoy *et al.*, 2004).

Energy balance models

These are slightly more complex than box models and are based on energy being conserved within the Earth system. They consider incoming solar radiation, emitted terrestrial radiation and the albedo of the various types of surface, and respond thermodynamically, but not dynamically, to changes in radiative forcing. The models can be used to investigate feedbacks, such as the impact of a reduction in sea ice extent on the high latitude climate. The most sophisticated energy balance models have been extended to two or more dimensions and coupled to ocean and ice sheet models (e.g. Gallée *et al.*, 1991).

Energy balance models do have a number of limitations. They tend to only have temperature as a predicted variable, although some models now include a representation of the hydrological cycle and have been linked to ocean models (Weaver *et al.*, 2001). The models also have rather limited internal variability and are heavily dependent on the external forcing applied. At present they also don't include an explicit representation of the circulation of the atmosphere.

Intermediate complexity models

These models fall between the rather simple energy balance models and the complex, computationally expensive AOGCMs. They include the physics of energy balance models, but also usually include a representation of the circulation of the atmosphere, and sometimes the ocean. In addition, such models can also have other components added that are of specific relevance to high latitude studies, such as sea ice and the major ice sheets. One of the greatest strengths of such models is that they can be run for many thousands or hundreds of thousands of years, a task which can be prohibitively expensive with AOGCMs. However, they have rather coarse horizontal resolution, with model grid boxes often being tens of degrees in latitude and longitude. In addition, the rapid advances in computer power over the last few years have meant that full AOGCMs can now start to be run for millennia on a realistic timescale.

There are currently several models of intermediate complexity in use. Gallée *et al.* (1991) developed a two-dimensional, zonally averaged model that was used to simulate the last

2.6 Models

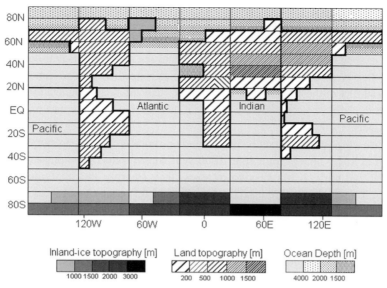

Fig. 2.21 The geography of the Potsdam CLIMBER model of intermediate complexity.

glacial cycle. Another model with a simple energy balance and coupled atmosphere and ocean has also been developed for palaeoclimate investigations by Stocker *et al.* (1992).

One of the best known intermediate complexity models is the CLIMBER-2 model developed at Potsdam (Claussen *et al.*, 2002). This model has been of great value in studies of the stability of the thermohaline circulation (Ganopolski *et al.*, 1998) and in investigating the potential for abrupt changes in the Earth system (Stocker and Schmittner, 1997; Rahmstorf and Ganapolski, 1999). Figure 2.21 illustrates the relatively coarse horizontal resolution of this model via the geography field, which shows the land/sea mask, orographic heights and ocean depths. With a model grid of only 7 by 18 boxes there clearly cannot be explicit representation of atmospheric weather systems. However, the limited horizontal resolution does allow integrations on the millennial scale to be carried out.

Although many studies with intermediate complexity models have been concerned with simulation of past climatic events, they have also been used to investigate recent Arctic climate variability (for example, see Mysak *et al.*, 2005).

General circulation models

General circulation models (GCMs) solve the full set of dynamical/physical equations applying to the atmosphere and ocean to simulate climatic conditions. These include Newton's laws of motion, the second law of thermodynamics, the gas law and conservation of mass and moisture. They also include parameterisations for processes operating on horizontal scales smaller than the grid length of the model, such as clouds, precipitation, blowing snow and sea ice. Typically such models have a horizontal resolution of 2.5–5° (approximately 200–500 km) with upwards of 20 levels in both the atmosphere and ocean.

The early models were relatively simple atmosphere-only models, and in these it was necessary to specify the sea surface temperature and sea ice extent. But as computer power increased they were coupled to ocean models and were then able to better simulate the natural variability of the climate system. In recent years interactive vegetation, aerosols and the carbon cycle have been added to some GCMs, so that they are now more like Earth system models.

GCMs use a timestep of typically an hour and on supercomputers can carry out a one year integration in a few hours. Some long 1000 year control runs of the present climate have been carried out with GCMs in order to assess their stability, but these are major undertakings and it is not practical to investigate millennial timescale climate variability via long runs over such periods. For such investigations a number of runs are carried out at selected points across the period of interest using a 'time slice' approach with appropriate forcing being applied at each point in time. A frequently studied episode is the Mid Holocene of about 6 kyr BP (see Section 5.4.1), since temperatures were higher than today yet solar forcing was approximately the same as current levels. However, the accuracy of such integrations is clearly dependent on our knowledge of the external forcing to apply, such as the orbital data, levels of volcanic aerosols and solar output. In an ideal situation only the above parameters would have to be applied in order to simulate the climate at a particular point in time, but in reality it is also necessary to specify the extents of the ice sheets, sea ice, levels of gases such as CO_2 and CH_4, and vegetation.

All models are now global in extent (although regional climate models can be embedded within them) in order to avoid problems at the lateral boundaries, so they allow interactions between the climates of the polar regions and lower latitude areas. However, GCMs still have a number of difficulties at high latitudes. This partly arises because for many years model development was concentrated on improving the performance of the models over the tropics and mid latitude areas. Sea ice presents several problems for climate models as it has high spatial variability and there are complex interactions with the atmosphere, ocean and between individual ice floes at a scale well below the horizontal resolution of the models.

Many early models represented only the atmosphere and had sea ice extent and concentration specified with climatological values. Nevertheless, these models still allowed valuable experiments to be carried out to investigate the atmospheric impact of environmental changes, such as a reduction in sea ice extent. But far more realistic simulations could be carried out when the atmosphere, ocean and sea ice were coupled. Today AOGCMs carry information on the sea ice concentration, ice thickness and depth of the snow on top of the ice. It is also possible to simulate features such as sea ice ridging and rafting. However, large regional errors in the sea ice extent can occur if the large-scale near-surface wind forcing in the model is incorrect (Turner *et al.*, 2006b).

The major ice sheets are in many ways easier to represent in AOGCMs than sea ice as their elevation does not change over the typical integration period of a century. A more important limitation is the fact that the rapid changes in elevation found around the edges of the Antarctic and Greenland icecaps cannot be accurately included because of the need to keep numerical stability in the models. Areas such as the Antarctic Peninsula therefore tend to be rather smooth, with consequent effects on the atmospheric circulation and the distribution of weather systems. In areas such as this, high horizontal resolution, limited-area models are of great value.

The centres responsible for the development of the major climate models tend to put most effort into refining the performance of their models over the major population centres of the mid latitudes and the tropics. So polar atmospheric and oceanic processes are sometimes less well handled in the models than their lower latitude counterparts. For example, clear sky precipitation (also known as diamond dust), which consists of ice crystals falling out of what appears to be a cloudless sky, is an important factor in the hydrological cycle of the polar regions. This process is not represented explicitly in the current generation of climate models and may therefore have an impact on the ability of the models to simulate correctly the present or future hydrological conditions at high latitudes. Similarly, polar clouds may not be represented correctly in the models, yet it has been shown that the use of more realistic polar cloud schemes can have a beneficial effect on the model simulation (Lubin *et al.*, 1998).

An important factor in the climate system of the Arctic is the layer of permafrost that is under much of the land area of the high northern latitudes. In model simulations of a world with increased greenhouse gas it is essential to have a realistic representation of permafrost and seasonally frozen ground so that the release of important greenhouse gases can be represented correctly.

Regional climate models

Local climate change is influenced greatly by small-scale features, which are not well represented in the coarse horizontal resolution global models. Such models with higher resolution cannot practically be used for long simulations because of time constraints. To overcome this, regional climate models (RCMs) with a higher resolution (typically 50 km or less) are constructed for limited areas and run for shorter periods (20 years or so). For the Antarctic the area used can cover the whole continent or be a subregion, such as the Antarctic Peninsula. The areas used by the Antarctic Mesoscale Prediction System, a regional weather forecasting system, are shown in Fig. 2.22. RCMs take their input at their lateral boundaries and sea surface conditions from the global AOGCMs.

One of the best known RCMs is the mesoscale model developed by Pennsylvania State University/NCAR and known as MM5. This is a limited-area, high-resolution model designed to simulate or predict mesoscale atmospheric circulation. A polar version of this model (Polar MM5) has been developed at the Byrd Polar Research Center, Ohio State University, and used successfully in both polar regions (Bromwich *et al.*, 2001; Guo *et al.*, 2003). Another RCM that is frequently used in polar climate studies is the Polar Weather Research and Forecasting model (Hines and Bromwich, 2008).

A well-documented suite of limited-area models used in polar studies is the Regional Atmospheric Climate Model (RACMO and RACMO 2), which has been developed at the

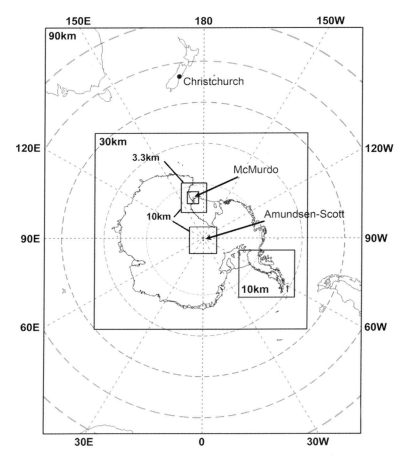

Fig 2.22 The spatial domains used by the Antarctic Mesoscale Prediction System, which is based on the Polar Mesocale Model 5 model. (From http://polarmet.mps.ohio-state.edu/PolarMet/pmm5.html.)

University of Utrecht (van Lipzig *et al.*, 1999). This model has been used to create a new, improved near-surface wind field for the Antarctic (van Lipzig *et al.*, 2004).

The Arctic Regional Climate Model Intercomparison Project (ARCMIP) has recently become a focus for the assessment and improvement of RCMs run in high northern latitudes. Output from these models has been compared and evaluated using observations from satellites, in-situ measurements and field experiments (e.g. Tjernstrom *et al.*, 2005).

Ocean general circulation models

Eddies in the ocean are an order of magnitude smaller than their counterparts in the atmosphere and are therefore very difficult to represent in the current generation of GCMs. For this reason a number of ocean general circulation models (OGCMs), which are three-dimensional representations of the ocean and sea ice, have been developed that are in many ways the ocean counterpart of an atmosphere-only general circulation model. OGCMs are useful by themselves for studying ocean circulation, interior processes and variability, but they depend on being supplied with data about surface air temperature and other atmospheric properties, in particular surface wind speed and direction.

The first high-resolution model to be developed that covered the Southern Ocean was the Fine Resolution Antarctic Model (FRAM) (Webb *et al.*, 1991). This had a quarter degree horizontal resolution in the north–south direction and half a degree in the east–west direction. It had 32 levels in the vertical ranging in thickness from 20 m near the surface to 250 m at a depth of 5500 m.

The OCCAM (Ocean Circulation and Climate Advanced Modelling) project has developed several high-resolution models that have been applied to the Arctic Ocean and the Southern Ocean (Screen *et al.*, 2009).

Many other models are now available for use globally or for specific regions. MICOM (Miami Isopycnic Coordinate Ocean Model) is a three-dimensional general circulation model that has been applied to the Antarctic coastal region and used to investigate the ocean circulation beneath the ice shelves (Holland *et al.*, 2003).

The current generation of ocean models have a high degree of success in simulating the surface and subsurface flow in the Arctic Ocean and across the Southern Ocean. However, as with modelling of the atmosphere, it is essential to carry out model intercomparison projects so that the strengths and weaknesses of the models can be assessed and comparisons made with the best observational data. One such initiative has been the Pilot Ocean Model Intercomparison Project of the World Climate Research Programme's CLIVAR project.

2.6.3 Ice sheet models

A high priority within polar science is to produce reliable predictions of how the Antarctic and Greenland Ice Sheets will evolve over the next few centuries since significant melting will have serious consequences for global sea level. However, this is a very difficult task since we currently have a poor understanding of the factors that control the flow of the ice streams and the stability of the ice shelves.

One powerful tool that we have to investigate changes in the ice sheets are models that can be run on single computers or clusters of machines. These are large-scale, three-dimensional thermomechanically coupled models that can include a range of processes, including basal sliding, glacial isostasy and the treatment of marine calving. These models can be run to simulate past glaciations or to determine how the ice sheets might evolve in the future.

The models are forced with temperature and precipitation fields, which are usually obtained from global climate model runs for past era or future conditions where greenhouse gas levels have been specified.

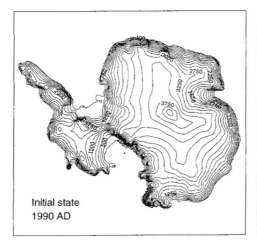 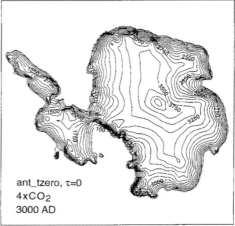

Fig. 2.23 The current elevation of the Antarctic Ice Sheet (left) and one prediction of the elevation in AD 3000 in a $4 \times CO_2$ modelling experiment. (From Huybrechts and de Wolde, 1999.)

Ice sheet models have been used to reconstruct the Quaternary ice sheets during the last glacial cycle (Zweck and Huybrechts, 2005); however, at the moment, there is no ice sheet model that can both predict the retreat of the Antarctic Ice Sheet since the end of the last ice age and correctly represent current variability.

The predictions of ice sheet evolution are important for the Intergovernmental Panel on Climate Change (IPCC) and a number of estimates of change have been made. Figure 2.23 shows the current elevation of the Antarctic Ice Sheet (left) and one prediction of the elevation in 3000 AD from a modelling experiment that assumed that atmospheric CO_2 concentration would increase by a factor of four by that time (Huybrechts and de Wolde, 1999). This shows very little change across the high, cold interior of the continent, with the ice retreat limited to the relatively warm fringes. We will discuss the predictions of such models in Chapter 7.

Although the complex ice sheet models are the primary tool in investigating past change and possible future evolution, simpler models can still be of great value in ice sheet research. Energy balance models such as that of Braithwaite and Olesen (1990) can be very valuable in studying ice sheet processes and do not require the large amounts of computer power needed by the more complex models.

3

The high latitude climates and mechanisms of change

3.1 Introduction

The climates of the polar regions are characterised by long periods of continuous sunlight in summer and perpetual darkness in winter that lead to large annual cycles in many aspects of the environment. The temperatures are very low in winter and only moderate in the summer because of the low elevation of the Sun and the highly reflective nature of the snow and ice surfaces. In fact the cryosphere is a major factor in defining the climates of the high latitude areas and the interactions of the ice and snow with the ocean and atmosphere will be discussed extensively in the following sections.

Many factors are responsible for high latitude climate variability and change, which can occur on a range of timescales. On long timescales, major changes in global climate are driven by orbital and solar variability. These affect the seasonal and latitudinal distribution of energy received from the Sun. Oxygen isotope data from ocean floor sediments indicate periods when the polar ice sheets were significantly more expansive than at present, particularly in the Northern Hemisphere (glacials) and when they were of similar size to the present (interglacials). Changes in the output of the Sun can result in high latitude climatic fluctuations with periods of reduced irradiance, such as the Maunder Minimum of the seventeenth and eighteenth centuries, being detectable in some aspects of the polar climates.

Variability in ocean circulation can affect the climate system on decadal to millennial timescales. The global ocean circulation is complex: non-linear interactions with the atmosphere and cryosphere can result in local, regional or global variability and change. The thermohaline circulation (THC) is the primary route by which water masses move around the world's oceans and it provides a means for linking the low and high latitude regions. Ocean signals can move in both directions and provide one means for exporting high latitude climatic signals to lower latitudes. A number of studies have noted an out-of-phase climatic relationship between the two polar regions, which may be a result of oceanic variability, although such a mechanism is still not fully understood (see Section 4.4).

On the subdecadal timescale the largest climatic cycle on Earth is the El Niño–Southern Oscillation (ENSO). Although ENSO has its origins in the tropical Pacific Ocean its impact can be felt in many parts of the world, including at high latitudes. However, as discussed later in this chapter, many of its influences are non-linear and there are interactions between the high latitude expressions of ENSO and other climate cycles.

Since the start of the Industrial Revolution in the eighteenth century there has been an increasing anthropogenic influence on the climate system via higher concentrations of a number of greenhouse gases, aerosols and halogens which are responsible for stratospheric ozone depletion. Separating natural climate variability from the various anthropogenic influences has proved a challenge, but the influence of humankind is now generally accepted (IPCC, 2007), although understanding all the variability seen in the climate systems of the two polar regions still remains a challenge.

In this chapter we describe the mean states of various climatic parameters of the polar regions and the mechanisms that cause the climates to change. We consider the mechanisms that maintain the large-scale atmospheric circulations of the polar regions, then examine the climatological features, providing a brief overview of the present climates, and examine the mean atmospheric and oceanographic conditions of the Arctic and Antarctic. We also examine the variability of the high latitude climate systems in the relatively data-rich twentieth century (change over this period is examined in Chapter 6). This has been a period of rapid change in some parts of the polar regions when evidence for humankind's influence on the environment has been increasing (Gillett *et al.*, 2009). While the ideal would be to have a high-quality data set of climate parameters in the pre-industrial era, this is clearly not possible: our benchmark for examining past climate variability, as determined from the various proxy records, and future high latitude change, as determined from models, has to be the recent climate record.

We present mean fields of some of the principal climatic variables, such as near-surface air temperature, pressure at mean sea level, sea ice extent and concentration, precipitation and ocean currents. Whereas, originally, many fields of atmospheric and oceanic parameters were created by manually analysing the often sparse observational data available, in recent years use is made of fields from sophisticated reanalysis schemes and numerical models. As discussed in Section 2.3, series of global, atmospheric analyses have been created from the historical record of observations by using current, state-of-the-art data assimilation and analysis schemes. In this chapter we will use time series of reanalysis fields to create mean fields that demonstrate the main features of the polar atmospheric circulation. We will also illustrate the major modes of high latitude atmospheric variability as represented in the reanalysis fields.

Of course the atmospheric analyses are only as good as the data that are available for their creation. In-situ meteorological observations are available for about a century for much of the Northern Hemisphere and a number of stations in the Arctic have records extending back this far. However, in the Antarctic most of the time series of meteorological parameters start during the International Geophysical Year of 1957–58. In addition, there are very few in-situ observations over the Southern Ocean, especially during the winter, so that the atmospheric reanalysis fields for the high southern latitudes are not reliable prior to the late 1970s when satellite sounder data became available.

The situation is somewhat different for the oceans where computer analyses of the available observations are only just starting to be carried out. To show the mean ocean circulation we have used traditional manual analyses and the output of ocean models that were run for extended periods. Because long, reliable series of ocean analyses are not available our understanding of ocean variability is much poorer than that for the atmosphere.

For quantities such as the broadscale sea ice extent, it is only possible to create mean fields for the satellite era, which for this parameter started in the late 1970s. Since the record is so short it is difficult to obtain a clear picture of decadal variability.

The climatological fields shown in this section will provide a reference for when we come to consider in depth how the polar climates have varied on historical timescales when anthropogenic influences were not a factor on the climate system. The data will also be of value for considering how the climate might change in the future under various scenarios of increasing greenhouse gas emissions.

3.2 Factors influencing the broadscale climates of the polar regions

3.2.1 Introduction

In Chapter 1 we outlined the broadscale characteristics of the polar regions in terms of the solar radiation received and the impact of the surface conditions on the surface radiation balance. In this section we examine in more detail the radiative conditions at high latitudes, then consider the mechanisms that maintain the large-scale atmospheric circulations of the two polar regions.

3.2.2 The radiation regime

The most important factor in dictating the polar climates is the large annual cycle of solar radiation, which results in long periods of winter darkness and perpetual sunlight in summer. The actual amount of solar radiation that arrives at the top of the atmosphere has been investigated by the Earth Radiation Budget Experiment (ERBE) instrument that has been flown on polar-orbiting satellites. Figure 3.1 shows the top of the atmosphere radiation received as a function of latitude and time of year. This figure shows clearly the large differences in the annual cycles of received radiation between high and low latitudes. Close to the Equator the Sun is nearly overhead throughout the year and the top of the atmosphere radiation only varies from about 400 to 450 W m^{-2}. However, in the polar regions the long periods of complete darkness and perpetual sunlight mean that the received radiation varies from zero in winter at the poles to over 500 W m^{-2} in the summer. At the poles themselves, there are six months of darkness and six months of continuous solar input, with only one sunrise and one sunset. Despite the fact that the Sun is never more than about 23° above the horizon in summer, the length of the days mean that the summer high latitude areas receive more radiation at this time of year than anywhere else on Earth. But over the year as a whole, the Equator receives over five times the solar radiation of the two poles.

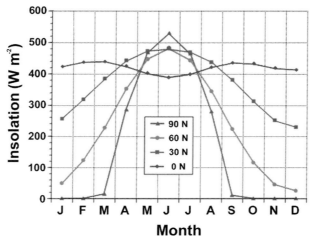

Fig. 3.1 The annual cycle of top of the atmosphere radiation (W m^{-2}) at four latitudes.

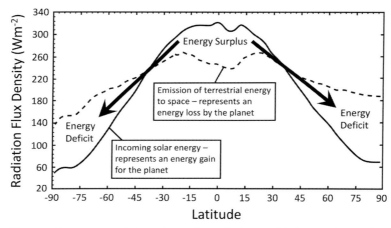

Fig. 3.2 The zonal mean annual incoming solar radiation (solid line) and outgoing longwave emission (broken line) as a function of latitude. Both quantities are given in W m^{-2}.

At high latitudes the very reflective nature of the surface means that the radiation available to heat the surface and lowest layers of the atmosphere (the net radiative flux) is relatively limited. Although the surface in the polar regions is cold, it still emits large amounts of longwave radiation to space, so that this heat loss, coupled with the large percentage of solar radiation that is reflected back to space in summer means that the surface can cool. Figure 3.2 shows the zonal mean annual incoming solar radiation and outgoing longwave emission as a function of latitude. Poleward of about 40° in both hemispheres there is a net loss of energy, with low latitude areas having a surplus. The tropics do not heat up in the long term and the mid latitude regions are not observed to cool, so there is a transport of heat poleward in both hemispheres.

The isolation of the Antarctic interior from the relatively warm air masses associated with depressions, coupled with the highly reflective surface and the much higher elevation, means that the temperatures above the continent are 11–12 °C colder than at similar latitudes in the Arctic. Temperatures at most latitudes in the Southern Hemisphere are therefore lower than at comparable latitudes in the north, and climatic zones are slightly further north. In both hemispheres there is an Equator to pole decrease of tropospheric temperature that is accompanied by a comparable drop in the height of pressure surfaces, which induces a pressure gradient. The latitudinal gradient of solar heating also results in density gradients in the ocean, which give a net poleward transport of sensible heat via the ocean.

3.2.3 The poleward heat flux

The polar regions are the two sinks in the atmospheric heat engine and there is a year-round flux of heat polewards via the ocean and atmosphere. Figure 3.3 shows the mean annual poleward transport of heat contributed by the atmosphere and the ocean. Figures are shown for the NCEP/NCAR and ECMWF reanalyses. The ocean transport was calculated from the surface heat flux inferred from the atmospheric energy budget. In the Southern Hemisphere the atmospheric heat transport is virtually everywhere greater than the contribution from the ocean, while north of the Equator the ocean transport dominates from the Equator to 17° N in the NCEP reanalysis. The peak of transport is found in both hemispheres close to 35° from the Equator. At those latitudes the atmospheric component is about 78% of the total flux in the Northern Hemisphere and 92% in the Southern Hemisphere.

The different distributions of land masses in the Northern and Southern Hemispheres have a major impact on the atmospheric and oceanic circulations in the two hemispheres,

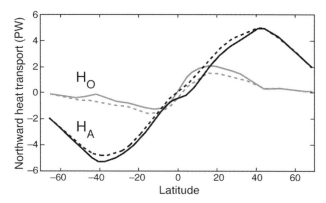

Fig. 3.3 The northward heat flux via the atmosphere (H_A) and the ocean (H_O) as derived from NCEP reanalysis (solid line) and ECMWF reanalysis (broken line) data. Units are PW = 10^{15} W. (From Czaja and Marshall, 2006.)

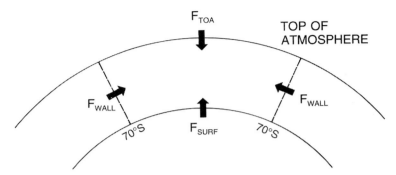

Fig. 3.4 A schematic view of the South Pole atmospheric cap and the fluxes of energy into it. (From King and Turner, 1997.)

with the mean ocean and atmospheric flows in the south being much more zonal than in the north. The Equator to pole temperature difference in the Southern Hemisphere is almost 40% greater than in the north, producing stronger mid latitude westerlies.

The poleward atmospheric transport of heat is largest during winter, when there is a strong net radiation deficit at high latitudes, and when there is the largest temperature difference between the equatorial and tropical latitudes. In the summer the Equator to pole temperature difference is much reduced and there is consequently a smaller poleward heat flux. The Arctic and Antarctic are polewards of the major storm tracks of mid latitudes, but many weaker depressions are found at high latitudes, especially around the coast of the Antarctic and over the Arctic Ocean.

The geographical distribution of the poleward transport of heat is quite different for the Arctic and Antarctic. In the Arctic the greatest transport occurs close to the Greenwich Meridian and the date line, as these areas are close to the main storm tracks. The large surface heat fluxes from the areas of open water in these regions are also an important factor in maintaining the poleward heat flux. In the Antarctic the poleward heat flux is much more evenly distributed around the continent because of the more uniform density of depressions across the Southern Ocean.

3.2.4 *The atmospheric heat budgets of the polar regions*

To examine the high latitude atmospheric heat budgets a number of studies have considered a simple model of the polar cap, bounded by the Earth's surface, the top of the atmosphere and a boundary along a latitude circle, such as 70° (Figure 3.4). The heat budget within this volume can then be determined by considering the energy fluxes through the surfaces at the boundaries. This approach has been adopted by Nakamura and Oort (1988) and Masuda (1990) for both polar regions, Genthon and Krinner (1998) and Serreze and Barry (2005) for the Arctic. The Nakamura and Oort study used the available radiosonde data to calculate the poleward heat transport, but more recently reanalysis data sets have been produced by some

3.2 Influences on the polar climates

Table 3.1 *Components of the energy budget of the Arctic and Antarctic. Seasonal means for an area poleward of 70° north and south. All values are in W m^{-2}*

Season	F_{TOA} Arctic[a]	F_{TOA} Antarctic[b]	F_{WALL} Arctic[a]	F_{WALL} Antarctic[c]	F_{SURF} Arctic[a]	F_{SURF} Antarctic[b]
Summer	−25	−32	83	31	−60	−29
Autumn	−163	−130	109	79	28	12
Winter	−164	−131	120	117	41	7
Spring	−100	−67	100	102	16	0
Year	−113	−90	103	81	6	−3

[a] From Serreze and Barry (2005);
[b] after Nakamura and Oort (1988);
[c] from Genthon and Krinner (1998).

of the major analysis and climate centres (see Section 2.3.3). These provide reasonably consistent fields of the major meteorological variables extending back several decades and have been used to compute heat fluxes in studies such as that of Genthon and Krinner (1998).

The budget can be considered as:

$$\Delta E/\Delta t = F_{TOA} + F_{WALL} + F_{SURF} \qquad (3.1)$$

where E is the moist static energy within the polar cap and $\Delta E / \Delta t$ represents the change in storage. F_{TOA} is the net radiation at the top of the atmosphere, F_{WALL} the net poleward energy flux across an imaginary wall at 70° and extending through the full depth of the atmosphere and F_{SURF} the net heat flux at the Earth's surface. The sum of the terms on the right-hand side of Equation (3.1) determines whether the polar cap as a whole is gaining or losing energy. We now consider each of these terms separately. Estimated seasonal values of the various terms are presented in Table 3.1 for the two polar regions from Nakamura and Oort (1988), Genthon and Krinner (1998) and Serreze and Barry (2005). It should be noted that since the estimates in Table 3.1 come from a variety of sources the sums of the three terms on the right-hand side of Equation (3.1) do not sum to zero. In particular, F_{SURF} is especially difficult to estimate and the values in Table 3.1 should be used with caution.

The net radiation balance at the top of the atmosphere (F_{TOA})

This is the difference between the incoming shortwave radiation (F_{SW}) arriving at the top of the atmosphere and the emitted terrestrial longwave radiation (F_{LW}), plus reflected solar radiation. It can be expressed as:

$$F_{TOA} = (1 - A)F_{SW} - F_{LW} \qquad (3.2)$$

68 *The high latitude climates and change mechanisms*

where A is the average albedo (reflectivity) for the high latitude area enclosed by 70°. This includes the albedo of the surface, clouds and radiation scattered by the atmosphere. Serreze and Barry (2005) show figures of the Arctic planetary albedo for the months of March to September, with values varying from 45% to 71%. The albedo is higher during the spring and autumn when there is more fresh snow on the ground, and slightly lower in the summer when more old snow is present. The albedo values for the two polar regions are much larger than for other parts of the Earth (the average planetary albedo for the Earth is 30%) as the albedo of the ice/snow surface is high. Only the ice-free ocean areas and the snow-free land have lower albedos.

The F_{TOA} flux can be measured by satellite instruments, such as the ERBE instrument discussed above, and seasonal values for the two polar regions are included in Table 3.1. During winter incoming solar radiation is negligible and there is a strong top of the atmosphere net radiation deficit as a result of the emission of longwave radiation to space. As the incoming solar radiation for the two polar regions is almost identical, the differences in F_{TOA} for the Arctic and Antarctic come from the different surface conditions in the two polar regions. Since surface temperatures are higher in the Arctic than the Antarctic the north has greater net emission to space than the Antarctic. In summer the shortwave and longwave fluxes are much more in balance, and in the Arctic in July the top of the atmosphere net radiation balance is only 2 W m^{-2}.

The net poleward transport of sensible and latent heat through the imaginary wall along a line of latitude at 70° (F_{WALL})

This is the net horizontal transport of energy into the polar cap, which can be considered as:

$$F_{WALL} = SH + LH + GP \tag{3.3}$$

where SH is the sensible heat, LH the latent heat and GP the potential energy. Each of these terms has contributions from the transient eddies, the stationary eddies and the mean meridional circulation.

Over the year as a whole F_{WALL} is about 27% larger for the Arctic than the Antarctic, which goes some way to compensate for the greater emission to space of radiation from the Arctic. However, there are considerable variations in the ratio of F_{WALL} from the two polar regions over the year. In winter and spring the values are comparable for the two polar regions, but the values for the Arctic are larger during the summer and autumn.

Genthon and Krinner (1998) calculated the annual mean, zonally averaged and vertically integrated transport of these quantities across 70° S, and these values are presented in Table 3.2. Over the year as a whole the largest contribution to F_{WALL} comes from the transport of geopotential energy, followed by the sensible and then the latent heat. Virtually all the contribution to the geopotential term comes from the mean meridional circulation, while the transient eddies are most important in transporting sensible and latent heat polewards.

3.2 Influences on the polar climates 69

Table 3.2 *Annual mean, zonally averaged, and vertically integrated meridional energy (SH, sensible heat; LH, latent heat; GP, geopotential; MSE, moist static energy) transport across 70° S by the total atmospheric circulation (TE + SE + MMC), the transient eddies (TE), the stationary eddies (SE), and the mean meridional circulation (MMC). Units: $W\,m^{-2}$ of surface area south of 70° S. Transport is negative northward, so a positive flux heats the polar cap.*

	TE+SE+MMC	TE	SE	MMC
SH	19.6	32.1	15.8	−28.3
LH	14.7	13.2	3.8	−2.2
GP	46.5	0.7	−0.4	46.2
MSE	80.8	45.9	19.2	15.7

Figures from Genthon and Krinner (1998).

Table 3.3 *The annual and seasonal mean zonally averaged and vertically integrated meridional energy (SH, sensible heat; LH, latent heat; GP, geopotential) transport across 70° S by the total atmospheric circulation (TE + SE + MMC). Units: $W\,m^{-2}$ of surface area south of 70° S. Transport is negative northward, so a positive flux heats the polar cap.*

	Spring	Summer	Autumn	Winter	Year
SH	44.8	−22.2	14.3	46.9	19.6
LH	14.4	10.7	17.1	16.8	14.7
GP	42.8	42.7	47.8	53.2	46.5
Total	102.0	31.1	79.2	117.0	80.8

Figures from Genthon and Krinner (1998).

The seasonal data on the main contributions to the transport are shown in Table 3.3. The largest sensible heat transport occurs during the winter, followed closely by the spring. But in summer the transport is actually *Equatorwards*, as there is a minimum in depression activity around the Antarctic during that season (Jones and Simmonds, 1993). Similarly, the latent heat transport is at a minimum during the summer, although the variability across the year is much less than with sensible heat. The geopotential transport is fairly uniform over the year, although with a maximum during the winter.

The net flux at the Earth's surface (F_{SURF})

This is the difference between the net radiative flux from the surface, and the atmospheric fluxes of sensible and latent heat to the surface. This term is small over the year as a whole, and Serreze and Barry (2005) give an annual mean for the Arctic of only 6.25 W m^{-2}. However, their monthly data varies from 40 W m^{-2} in November to −85 W m^{-2} in July.

During the winter there is a large input of heat to the cold atmosphere from the relatively warm ocean. Much of this is through leads (ice-free areas) between the sea ice floes, which can account for 90% of the total flux.

F_{SURF} is markedly different in the two polar regions, as most of the Antarctic south of 70° S consists of the ice-covered continent, while 72% of the equivalent area of the Arctic is ocean. In the Arctic this term is large as sea ice extent is limited in the summer and autumn, and there is also a large flux of heat from the ocean. In the Antarctic the term is largest in the autumn when the sea ice is at a minimum, but decreases during the winter as the ice becomes more extensive.

The change in storage of energy in the polar atmosphere.

While the long-term storage of heat in the polar atmosphere is close to zero, over the annual cycle there is a change in heat held at high latitudes. With no incoming solar radiation during the winter this is the time of maximum heat loss to space. But with the return of the Sun in spring the atmosphere starts to gain heat, while there is a corresponding loss in autumn. The annual cycle of heating across the Antarctic and Arctic is reflected in the changes throughout the year of temperature, humidity and the heights of atmospheric pressure surfaces above the continent. For example, the Antarctic 'core-less winter', which is characterised by a rapid drop in temperature through the autumn and equally marked warming in the spring, is a result of the rapid changes in insolation at these times of year.

3.2.5   The water vapour budget

The polar regions are characterised by low temperatures, especially during winter, and the air cannot hold a great deal of water vapour. However, water vapour is still extremely important as it is the source of all precipitation, it is a major greenhouse gas and is critical in consideration of the mass balance of the major ice sheets.

The water vapour budget of the polar regions can be considered in a similar way to the heat discussed above, except that there is no exchange of water vapour at the top of the atmosphere, limiting the fluxes to the surface and the lateral boundary around a polar cap. On an annual timescale the storage of moisture in the atmosphere is negligible and the transport of moisture into the polar cap is balanced by precipitation less the evaporation.

Precipitation is very difficult to measure in the polar regions because of the effects of blowing snow – snow falls into the gauges, but is easily blown out in even moderately strong winds. So accumulation is often measured via snow stakes or pits. The average accumulation for the Antarctic has been estimated in a number of studies and is around 180–190 mm yr^{-1} (see Turner et al., 1999). Precipitation cannot easily be measured directly over the ocean and reanalysis data sets have to be used to estimate this quantity.

In the Antarctic the many depressions that are found over the Southern Ocean are thought to make the greatest contribution to the precipitation on the continent. These systems often have strong northerly winds on their eastern flank that can bring relatively warm, moisture-laden air towards the continent. As these air masses are forced up the steep coastal slope the air is cooled, resulting in extensive cloud and precipitation. As discussed in Section 3.7, most Antarctic precipitation falls in the coastal region. Depressions are also responsible for much of the precipitation over the Arctic, since active weather systems can penetrate into the Arctic Ocean from the North Atlantic and via the Bering Strait.

Evaporation and condensation do not make major contributions to the surface energy balance at high latitudes, but they can be of importance in determining the surface mass balance. At locations where the surface wind is strong, for example at the bottom of glacial valleys, evaporation during the summer has been reported to remove about a third of the annual precipitation. Data from Antarctic automatic weather stations have also suggested that 45% of the precipitation could be removed on the Ross Ice Shelf.

Blowing snow is extremely important in studies of ice sheet mass balance and understanding the detailed information locked into ice cores. Blowing snow has been included in some atmospheric models, but their horizontal resolution is still rather coarse and they are not able to replicate the fine-scale structure of the blowing snowfield. Various estimates have been made of the flux of snow off the Antarctic continent, but it is not thought to be more than 10% of the measured accumulation.

Snow is also extremely important when it falls onto sea ice. The high albedo of fresh snow can shield the sea ice from the heating effects of solar radiation when sunlight returns in the spring. But the weight of snow can also push sea ice into the ocean creating a form of ice called 'snow ice'.

3.3 Processes of the high latitude climates

3.3.1 *High latitude feedbacks and amplification*

The polar regions are very sensitive to feedbacks, which are processes where an initial atmospheric, oceanic or cryospheric perturbation to the climate system can result in an amplified or damped response to the same or a different climate element. Many features of the high latitude environment can be involved in feedbacks, including aerosols, ozone, water vapour, clouds, sea ice, snow cover, vegetation and the atmospheric and oceanic circulations.

Positive feedback processes, where a small change results in a larger response, receive a great deal of attention, and can be important in long-term climate change. However, the damping effects of negative feedbacks can also play an important role. In the following sections we discuss some of the most important feedback processes and provide examples from the Arctic and Antarctic.

Ice-albedo feedbacks

This is the most important positive feedback process at high latitudes and one that is thought to have been responsible for some of the large changes observed across the Arctic in recent decades. It assumes that a small increase in near-surface air temperature will result in a

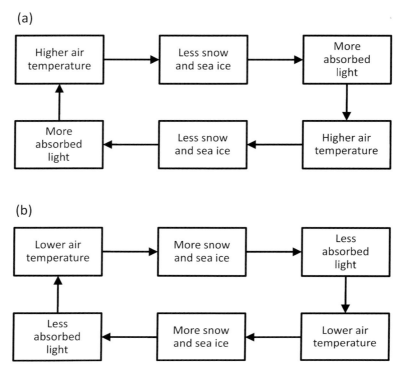

Fig. 3.5 The ice-albedo feedback mechanism, giving (a) enhanced warming, (b) enhanced cooling.

reduction in the area of high albedo snow or sea ice, exposing the lower albedo ground or unfrozen ocean. Greater amounts of solar radiation can then be absorbed by the ground or ocean, so resulting in further melting of snow or sea ice. The mechanism is illustrated schematically in Fig. 3.5(a).

The ice-albedo feedback mechanism can also work in reverse and amplify a regional cooling. This occurs when a drop in air temperature results in an expansion of the area covered by snow or sea ice, so increasing the albedo across a region and resulting in the absorption of less solar radiation (Fig. 3.5b).

Curry *et al.* (1996) discussed the nature of the ice-albedo feedbacks within the snow and sea ice, as well as those associated with the areal coverage. For example, higher temperatures over sea ice can lead to the formation of melt ponds, which have a lower albedo than the sea ice itself, and so will lead to absorption of more radiation into the ice pack. Higher temperatures will also cause greater melting of snow cover on the sea ice, exposing the lower albedo sea ice below. Clearly there are complex processes and interactions taking place within the ice and snow at high latitudes, which are essential to include in climate models. However, many of these take place on very small horizontal scales that are extremely difficult to reproduce in the current generation of GCMs.

The ice-albedo feedback mechanism only works during the period when there is solar radiation, but it can be extremely important during the spring when the Sun returns and there is still extensive ice and snow. Key factors in ice-albedo feedbacks in the sea ice zone are the timing of seasonal transitions, the evolution of summer melt ponds and the duration of summer melt. In a warming world it is usually assumed that there will be a longer ice melt season, with an earlier start to the summer melt and a later start to the winter freeze-up. In addition, it is expected that there will be more ponding on sea ice. All these factors will result in stronger feedbacks. However, there might be more subtle changes in the nature of the sea ice that could have an impact on the feedbacks. For example, there could be a higher percentage of first-year ice compared with multi-year ice, and there may be an increase in snowfall at high latitudes (see Section 7.4), so changing the surface albedo. However, while a deeper snow layer on the sea ice would reduce surface melt early in summer, it could also result in greater pond coverage and potentially greater surface ablation later in the season.

Changes in the ice-albedo feedback could also have an impact on the interactions between the marine and terrestrial environments. For example, earlier melting of the snow cover over land would result in significant increases in the total heat input to the ground. In coastal areas, this would result in more rapid melting of fast ice hard up against the shore. This could extend the ice-free season and expose the coast to more storms and erosion. Changing environmental factors could therefore produce complex changes in the air–sea-ice conditions and the possible feedbacks may therefore be highly non-linear.

As discussed in Chapter 7, many climate models predict that over the next century the greatest temperature increases will be at high latitudes, and particularly in the Arctic. This is felt to be due, at least in part, to ice-albedo feedbacks. We will return in Chapter 7 to consider the reasons for the large predicted warmings at high latitudes.

Cloud-radiation feedbacks

In the polar regions the observed net cloud radiative forcing at the surface is usually positive, with an increase in clouds resulting in warming. However, there is a complex feedback between the surface temperature, clouds and the net radiative balance of the surface. It is based on the assumption that an increase in greenhouse gases or aerosol amount will result in a perturbation to the surface radiation balance, which will change the characteristics of the snow or ice surface. This could be via modifications to the albedo, snow and ice extent, thickness or other characteristics of the surface. Such surface changes would then modify the radiation emitted from the surface, as well as the fluxes of sensible and latent heat into the lowest layers of the atmosphere. This in turn will modify the cloud properties, such as the optical depth and the cloud fraction, changing the radiative fluxes. This feedback mechanism is illustrated schematically in Fig. 3.6.

This is a complex series of non-linear interactions concerning not only cloud amount, but also cloud properties, such as optical depth, phase and cloud thickness. It therefore has implications for global change studies. While many modelling studies have suggested that increases in greenhouse gases will result in changes in cloud amount and surface

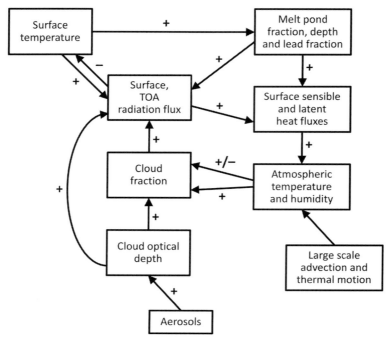

Fig. 3.6 The cloud-radiation feedback mechanism. (Adapted from Curry et al., 1996.)

temperature, the sign of the feedback is not clear. However, the balance of opinion at present is that the feedback is positive.

Dynamical considerations

Other recent theoretical research has given insight into polar amplification and feedback mechanisms. Cai (2005) used a climate model to examine the dynamical amplification of polar warming. He found that the additional atmospheric poleward heat transport caused by an anthropogenically induced radiative forcing would lead to a polar warming amplification by redistributing part of the extra energy intercepted by the low latitude regions to high latitudes. This in turn strengthened the water vapour feedback at high latitudes – this is a positive feedback whereby higher temperatures allow the atmosphere to hold a greater amount of water vapour that reduces the loss of longwave radiation to space and so increases tropospheric temperatures. He found that the dynamical amplifier contributed about a quarter of the total high latitude surface warming in winter, which compared with only a tenth for the globe as a whole.

Modelling experiments without sea ice-albedo feedbacks (Alexeev, 2003; Alexeev et al., 2005) have provided further insight into the means by which high latitude climate change may be amplified in an environment of increasing greenhouse gas. This comes about due to a larger meridional heat transport and an increase in the moisture of the high latitude atmosphere, resulting in a longwave forcing on the surface heat budget of the polar regions.

Stratospheric ozone depletion

The identification of the rapid reduction in amounts of springtime stratospheric ozone above the Antarctic in the early 1980s (the Antarctic 'ozone hole') was one of the most significant environmental discoveries of the late twentieth century. Subsequent research showed that the destruction of the ozone was a result of increasing levels of chlorofluorocarbons (CFCs), which were involved in a complex series of chemical reactions that destroyed ozone above the Antarctic. The 'hole' is at its maximum in the spring when sunlight returns and ultraviolet light breaks the bond holding the chlorine atoms to the CFC molecules. The free chlorine atoms participate in a series of chemical reactions that destroy the ozone and also return the free chlorine atoms to the atmosphere unchanged, where it can destroy further molecules. The ozone destruction is therefore a feedback process where a relatively small number of CFC molecules can destroy a huge number of ozone molecules. For more information on the processes associated with stratospheric ozone depletion see Section 3.4.4.

3.3.2 Air–sea-ice interactions

Interactions between the atmosphere, unfrozen ocean and sea ice are extremely important in the polar regions and closely link to feedback mechanisms. In the polar regions during the winter the ocean under the sea ice has a temperature between zero and $-2\,°C$, yet the air above the ice can have a temperature as low as -20 or $-30\,°C$. This presents a great potential for large fluxes of heat and moisture between the ocean and atmosphere.

Some of the most dramatic examples of air–sea interactions occur during periods of strong offshore or off-ice flow in winter. This is illustrated in Fig. 3.7 through a satellite image showing the cloud distribution associated with strong northerly flow in winter over the area between Svalbard and northern Norway. In this area relatively warm ocean currents can bring waters with a sea surface temperature several degrees or more above freezing up close to the ice front. Air–sea temperature differences can therefore be $20\,°C$ or more, and with strong winds of gale force or above the sensible heat fluxes can be in excess of $300\,W\,m^{-2}$ with latent heat fluxes of close to $200\,W\,m^{-2}$ (Turner *et al.*, 1993). Under these conditions as soon as the air crosses from the ice to the ocean the fluxes of heat and moisture can result in the development of convective clouds, often in the form of streets that can extend for hundreds of kilometres. This can be seen in Fig. 3.7 where in the top-left corner the air passes across the ice edge and within a very short distance picks up enough moisture and the convection becomes developed enough for the cloud streets to form. Where cloud streets from different ice edges approach, a convergence line may develop, characterised by a more organised line of cloud, as is the case in the middle of Fig. 3.7. A similar band of

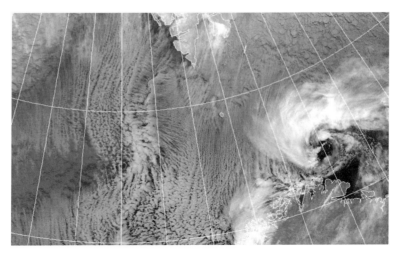

Fig. 3.7 An infrared satellite image of the area between Svalbard and northern Norway taken at 02:37 GMT 27 January 1982. (Image courtesy of the NERC Satellite Receiving Station, University of Dundee.)

cloud may form at the leading edge of a cold, polar air outbreak when it is often referred to as an Arctic front.

In the Southern Hemisphere the locations of the major land masses are such that warm surface water masses do not penetrate to such high latitudes as in the north, and during the winter the Antarctic continent is surrounded by a ring of sea ice. Nevertheless, strong southerly winds can advect large amounts of sea ice away from the coast exposing the ice-free ocean and creating a coastal lead or polynya. Under these conditions the air–sea temperature differences are very large and there is rapid freezing of the surface of the ocean and the formation of large amounts of new sea ice. In fact when offshore flow persists this can result in the continuous production of sea ice and the advection of this ice towards the north. Due to the form of the coastline and the large embayments that exist around the Weddell and Ross Seas, coupled with the climatological cyclonic atmospheric circulations that exist in these areas, the western sides of these regions are particularly favoured for sea ice generation. Renfrew *et al.* (2002) estimated that the coastal polynya at the edge of the Ronne Ice Shelf was the source of around 1.11×10^{11} m^3 of ice each year, or about 6% of the total ice production of the whole Weddell Sea. Figure 3.8 shows a thermal infrared satellite image of the semi-permanent polynya that exists during the winter off the Ronne Ice Shelf over the southern Weddell Sea. This image essentially shows the skin temperature of the ice and ocean surface, with the white colours indicating the cold temperatures of the sea ice and the darker shades showing the much higher temperatures across the water and thin sea ice of the polynya itself. Renfrew *et al.* (2002) investigated the surface energy budget of the coastal polynyas in this area using a combination of meteorological observations, satellite measurements and simple physical models. The study showed that for most of the seven-year period

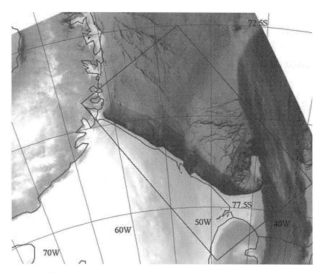

Fig. 3.8 An infrared satellite image of the southern Weddell Sea and the Ronne Ice Shelf at 19:30 GMT 21 July 1998. (From Renfrew et al., 2002.)

investigated the summer season oceanic warming (as a result of absorption of shortwave radiation) was in approximate balance with the wintertime oceanic cooling. The winter seasons were characterised by periods of high heat fluxes interspersed by quiescent episodes that were controlled by the dynamics of the polynya and the broadscale atmospheric circulation. Sensible heat fluxes were found to be responsible for 63% of the energy exchange, with latent heat fluxes and radiative fluxes being responsible for only 22% and 15% of the transfer respectively.

The continuously changing atmospheric circulation above sea ice results in the frequent opening and closing of the leads between the ice floes. The leads and other areas of open water are important regions for air–sea interactions and studies have suggested that if the ice concentration is less than four tenths the heat flux into the atmosphere is almost equal to that from the open ocean (Worby and Allison, 1991).

Certain coastal areas are prone to the frequent development of polynyas because of the climatological offshore wind flow. However, occasionally a major polynya can develop within the main body of the pack ice. This was the case with the Weddell Polynya (Fig. 3.9), which existed during the winter periods between 1974 and 1976 (Carsey, 1980; Moore et al., 2002) and covered an area of approximately 350 000 km^2. This became a region of significant air–sea interaction involving the densification of the surface waters, convective overturning of the water column and the formation of large amounts of Antarctic Bottom Water (AABW). The atmospheric consequences included the air temperature above the polynya being around 20 °C warmer than usual, a 50% increase in cloud in the region and periods when the air–sea fluxes were 5–10 times larger than normal (Moore et al., 2002). Various

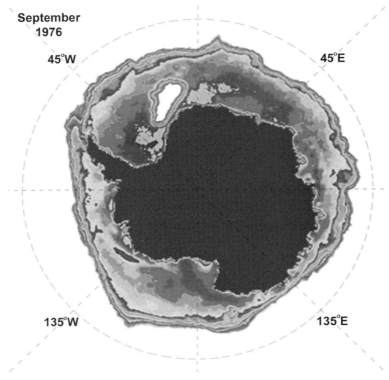

Fig. 3.9 The Weddell Polynya (shown in white) as seen in passive microwave satellite imagery from September 1976.

mechanisms have been proposed for the formation of the Weddell Polynya. Motoi et al. (1987) suggested that it developed as a result of the presence of a high-salinity mixed layer in summer. The high salinity resulted in deep convection driven by surface cooling alone, with the upward transfer of heat and salt prohibiting sea ice formation during the winter. Since 1976 the Weddell Polynya has not reappeared, although much smaller polynyas have been sighted on a recurring basis in this area.

Areas of intense air–sea interactions can also result in the development of weather systems. In particular, the regions where cold air flows over relatively warm oceans are prone to the formation of polar lows, which are intense, short-lived mesoscale depressions (Rasmussen and Turner, 2003). These areas include the Norwegian and Barents Seas of the North Atlantic, the Gulf of Alaska and the Sea of Japan. A developing polar low can be seen on the right-hand side of Fig. 3.7. Polar lows are very important to weather forecasting at high latitudes, but their greatest climatological impact may be in terms of their effect on the ocean. The large fluxes of heat and moisture from the ocean to the atmosphere result in significant cooling of the upper layers of the ocean and its destabilisation. If a region is affected by several active polar lows over an extended period then oceanic deep convection

3.3 Processes of the high latitude climates 79

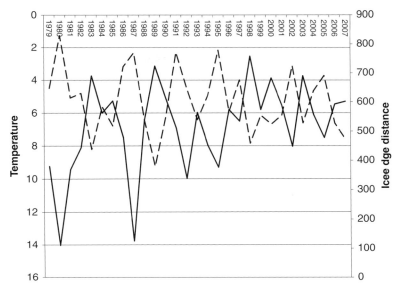

Fig. 3.10 Winter season (June–August) mean temperatures for Faraday/Vernadsky Station on the western side of the Antarctic Peninsula (solid line) and the mean ice edge location north of Alexander Island (km) along 70° W (broken line). The figure illustrates the large interannual variability in temperature as a result of cryospheric feedbacks and the high anticorrelation between temperature and sea ice extent over the Bellingshausen Sea.

may be triggered, possibly affecting the deeper layers of the ocean. However, more observational and modelling studies are needed to determine if this is the case.

When the atmospheric circulation is strongly meridional cold, polar air masses may reach mid latitudes, increasing the air–sea temperature differences. This can result in the formation of cloud streets and deep convective cloud extending into these regions from polar latitudes. In extreme conditions this may lead to the formation of polar lows in mid latitudes, as observed in areas such as New Zealand, and even over the UK where polar lows can bring heavy snowfall (Suttie, 1970).

The strong air–sea interactions at high latitudes involving the cryosphere lead to greater interannual variability of the polar climate compared with lower latitudes. One area where this is very apparent is on the western side of the Antarctic Peninsula. Faraday Station (now the Ukrainian Vernadsky Station) is located about halfway down the peninsula on the maritime western side, close to the location of the sea ice edge during the winter. The Antarctic Peninsula extends north–south so that the annual cycle of sea ice growth/ retreat takes the ice past the station. The sea ice acts as a very effective cap on the ocean during the winter months so that the exposure or covering of the ocean by sea ice has a huge impact on the near-surface temperatures at Faraday/Vernadsky. As can be seen from Fig. 3.10, the winter season (June–August) mean temperatures at the station varied from almost −2 to −14 °C over the period 1979 to 2007. The warm winters reflect years when

80 *The high latitude climates and change mechanisms*

there was little sea ice over the Bellingshausen Sea and large fluxes of heat from the ocean to the atmosphere. By contrast, the cold winters are associated with years of extensive ice and a greater degree of isolation of the atmosphere from the huge available source of heat in the ocean.

In summary, the climates of the polar regions are highly influenced by the complex interactions between the atmosphere, ocean and ice, and as will be discussed in depth in later chapters, climate variability in the past has been extensively affected by links with the cryosphere. We will also address the question of how air–sea interactions and feedbacks may influence the high latitude climates over the next century.

3.4   The mechanisms of high latitude climate change

3.4.1   Orbital and solar changes

Introduction

Louis Agassiz introduced the idea of 'ice ages', based on his observations of striated rocks in the Alps, which he summarised in his book *Etudes sur les glaciers* (Agassiz, 1840). An astronomical theory – that climate responded to changes in insolation forcing due to orbital variations – to explain these ice ages was first proposed by Adhémar (1842) and subsequently enhanced by Croll (1875), who included the effects of orbital eccentricity and obliquity in addition to precession of the equinoxes. However, it was the pioneering calculations of how incoming solar radiation varied with both latitude and season over hundreds of thousands of years by Milankovitch (1930), which really advanced the astronomical theory and led to it commonly being referred to as the Milankovitch Theory. As the glacial–interglacial (cold–warm) cycles are closely associated with the build-up and melting of the Laurentide and Fennoscandian Ice Sheets at high northern latitudes, summer insolation at these latitudes (usually 65° N) has often been cited as the critical controlling factor.

The theory only really became accepted following the paper by Hays *et al.* (1976). These authors utilised a frequency analysis of palaeoclimate records contained within radiometrically dated marine sediment cores and concluded that, 'Changes in the Earth's orbital geometry are the fundamental cause of the succession of Quaternary ice ages.' However, there remain a number of uncertainties with the Milankovitch Theory, discussed below, one of which has led to alternative theories in which variation in the inclination of the Earth's orbital plane is important.

Changes in orbital eccentricity

The Earth's orbit around the Sun varies between a mildly elliptical orbit (eccentricity (e) = 0.058) and an almost circular one (e = 0.005), with a mean eccentricity of 0.028 and a current value of 0.016 (Fig. 3.11). The major variation in eccentricity ($\pm$ 0.012) has a period of 413 kyr, while several smaller variations, notably a doublet with periods of 95 and 125 kyr, combine to produce a secondary periodicity at ~100 kyr. At the moment, there is a 7% change in the insolation incident at the top of the atmosphere between the

3.4 The mechanisms of high latitude climate change

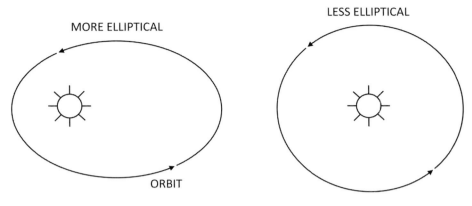

Fig. 3.11 Changes in the eccentricity of the Earth's orbit about the Sun.

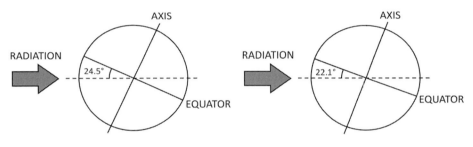

Fig. 3.12 Changes in the axial tilt of the Earth's orbit about the Sun.

aphelion – Earth is furthest from the Sun (152 million km) on 4 July – and the perihelion – Earth is nearest the Sun (147 million km) on 3 January. This difference would increase to 23% when the orbit is at its most eccentric. Note also that a change in orbital eccentricity does impact the total annual incoming solar radiation integrated across all latitudes over a year, but by only a very small amount: the amplitude is only 0.1% of the current value. Eccentricity also exerts a control on the amplitude of the precession cycle.

Changes in obliquity

The tilt of the Earth's axis of rotation, the angle between the Earth's equatorial and orbital planes, varies between 22.1° and 24.4°, with the present value being 23.4°. The frequency of this cycle is the difference between the precessional frequencies of these two planes, and it has a periodicity of 41 kyr (Figure. 3.12). Insolation changes associated with this 'tilt' cycle increase poleward in both hemispheres: for example, decreases in the axial tilt result in

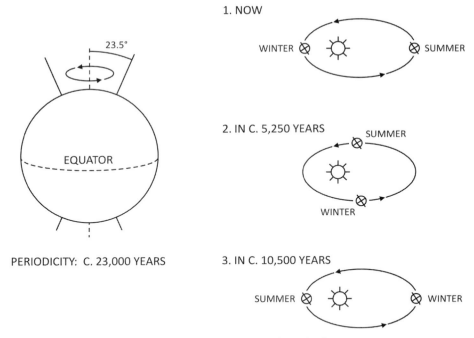

Fig. 3.13 Changes in the precession of the Earth's orbit about the Sun.

cooler summers and warmer winters in the polar regions but do not change significantly the amount of solar radiation at low latitudes.

Precession of the equinoxes

The precession of the equinoxes is the directional change of the Earth's axis of rotation relative to the Sun due to the gravitational pull of the Sun and Moon on its equatorial bulge. Precession causes the four cardinal points (the equinoxes and solstices) on the Earth's elliptical orbit to move around the orbital path (in a clockwise motion if looking down on the North Pole). Simultaneously the elliptical orbit is slowly rotating in the opposite direction. Thus, while the period of the axial precession is 25.8 kyr the period of the precession of the equinoxes is 21.7 kyr.

Its effect is to change the season relative to the location of the Earth on its elliptical orbital path and hence the Earth–Sun distance (Fig. 3.13). Put another way, it alters the timing of the perihelion and aphelion. The precessional cycle dominates insolation variability at middle and low latitudes and is out of phase between the Northern and Southern hemispheres. Currently, the perihelion occurs in the northern winter, so that this hemisphere has milder seasons with relatively warm winters and cooler summers. In the Southern Hemisphere there will be greater contrast between the seasons because the perihelion occurs in summer (therefore warmer) and the aphelion is in winter (therefore colder). Note that if the orbit were circular then precession would not affect climate as each season would always occur at the same distance from the Sun.

Changes in orbital inclination

The inclination of the Earth's orbital plane (relative to the equatorial plane of the Sun) varies with a period of ~70 kyr and relative to the invariant plane (representing the angular momentum of the solar system) with a period of ~100 kyr. This latter cycle is thought to affect the accretion of interplanetary dust particles in the Earth's atmosphere (e.g. Farley and Patterson, 1995), which in turn has been postulated to affect the climate system (Muller and MacDonald, 1997).

The Milankovitch Theory: problems and explanations

There are several difficulties in reconciling the Milankovitch Theory with palaeo-observations (ocean floor sediments and ice cores; cf. Chapter 2). Principally, the strongest climate signal over the past few hundred thousand years has been the 100-kyr cycle, yet the insolation changes resulting from eccentricity variability are far smaller than those of obliquity or precession (only 0.15% of the amplitude of the latter). Beginning with the paper by Hays *et al.* (1976) there have been many theories and models developed to explain how these 100-kyr cycles exist based on non-linear feedbacks to astronomical forcing (e.g. Imbrie *et al.*, 1993; Paillard, 1998; Shackleton, 2000). Note that other theories suggest that the 100-kyr cycle arises primarily from random internal climate variability (Wunsch, 2003). A relatively recent and comprehensive theory linking orbital forcing with observations was proposed by Ruddiman (2003), based on new information about the relative timing of changes in CO_2. He hypothesised that the 100-kyr cycle resulted from external orbital pacing of climate, rather than an internal oscillation (Fig. 3.14). A statistical analysis by Huybers and Wunsch (2005) also demonstrated a strong relationship between the obliquity cycle and glacial terminations, with the Earth tending to a glacial state near some but not all obliquity maxima. Tziperman *et al.* (2006) used the concept of non-linear phase locking as the pacemaker mechanism, which works via the dependence of the period of a non-linear oscillator such as Milankovitch forcing on its amplitude. This mechanism is able to predict the quantisation of the obliquity period into the dominant 100-kyr period and reveals why this period is nearly independent of the climate mechanism responsible for the cycle. However, while this concept may explain the timing of the ice ages the exact physical mechanisms responsible remain uncertain.

An alternative explanation for the 100-kyr cycle was proposed by Muller and MacDonald (1997). These authors suggested that it is caused by the periodicity of inclination of Earth's orbital plane relative to the invariant plane, which in turn controls the accretion of extra-terrestrial dust. Vaporisation of larger dust particles in the upper atmosphere and their consequent effect on the radiation balance is the speculated mechanism for a control on climate. Berger *et al.* (1994) suggested that the poor generalised fit of the inclination of the Earth's orbit relative to the amplitude of $\delta^{18}O$ in marine sediments relative to a Milankovitch forcing is sufficient to reject the former hypothesis. Furthermore, Huybers and Wunsch

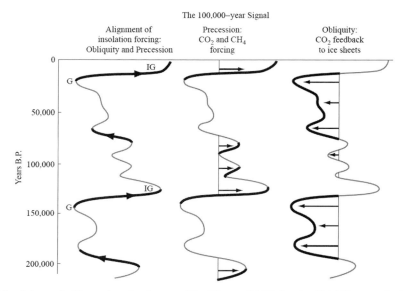

Fig. 3.14 Schematic illustrating the theory of Ruddiman (2003) for the 100 000 year glacial cycle. G, glacial; IG, interglacial.

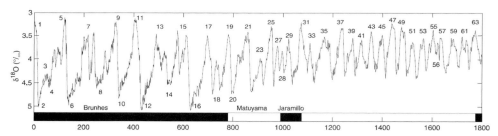

Fig. 3.15 Benthic $\delta^{18}O$ stack constructed from 57 globally distributed records (Lisiecki and Raymo, 2005). The numbers refer to marine isotope stages (MIS).

(2005) demonstrated that the statistical evidence that Muller and MacDonald (1997) used to dismiss the Milankovitch Theory is flawed. Finally, Winckler and Fischer (2006) demonstrated that comic dust particle concentration in an Antarctic ice core remains relatively constant between a glacial and interglacial period.

Another observation not immediately explained by Milankovitch Theory is the change from a dominant frequency of 41 kyr to one of 100 kyr at ~900 kyr BP, an event termed the Mid Pleistocene transition (MPT). Ruddiman (2003) believed that long-term cooling was a plausible explanation for the MPT, citing an increase in the $\delta^{18}O$ of benthic forminifera indicative of increased AABW and hence cooler temperatures at high southern latitudes in addition to the increase in global ice volume, with the appearance of large Northern Hemisphere ice sheets (Fig. 3.15). Raymo (1997) suggested that prior to ~900 kyr BP temperatures at high northern latitudes were cold enough to let ice sheets grow at many 41-kyr (obliquity) and 23-kyr (precession) insolation minima, but sufficiently warm to melt this ice during subsequent insolation maxima. However, after this time temperatures were sufficiently cooler for some ice to persist through weak insolation maxima, hence producing 100-kyr cycles because ice sheets would persist through clusters of low-amplitude insolation maxima at precession period and then melt during the next group of high-amplitude precession maxima. These maxima occurred 100 kyr apart due to eccentricity modulation of precession. Recently, Raymo *et al.* (2006) suggested that at the MPT Southern Hemisphere temperatures had cooled sufficiently that the previous terrestrial margin of the East Antarctic Ice Sheet (EAIS) expanded northwards such that the margins became marine based. Thus, sea level changes driven by ice sheet fluctuations in the Northern Hemisphere, controlled primarily by summer insolation forcing (precession), became the primary control on Antarctic ice volume through iceberg calving at the EAIS margins.

The Stage 5 or causality problem centres on the timing of Termination II, the penultimate interglacial at ~130 kyr BP. According to orbital theory, the transition into a termination should occur with maximum Northern Hemisphere summer insolation – in order to melt the Northern Hemisphere ice sheets – as it does with Termination I at ~12 kyr BP. Yet, a number of independent dated global palaeorecords have indicated that the ice sheets collapsed (and hence sea level increased) at approximately 135 kyr BP, some 8 kyr prior to the maximum in Northern Hemisphere summer insolation. However, recent, highly accurate ^{230}Th (thorium) dated oxygen-isotope records from Chinese stalagmites have demonstrated that the timing of the last four terminations is entirely consistent with Northern Hemisphere summer insolation being the principal forcing mechanism of the initial retreat of the northern ice sheets (Cheng *et al.*, 2009).

The Stage 11 problem or paradox results from the ~400 kyr eccentricity variation being undetected in climate proxies: moreover, the maximum glacial–interglacial contrast (from MIS 12 to MIS 11 at Termination V) occurred when Northern Hemisphere summer insolation was at a minimum. It appears that a fundamental climate change occurred at this time, known as the mid-Brunhes event (MBE). Subsequently, the late Pleistocene had stronger but shorter interglacial periods (EPICA Community Members, 2004). Berger and Wefer (2003) postulated that at the MBE internal oscillations in the climate system became co-equal with Milankovitch forcing in governing ice-age dynamics. The overall cooling trend (see above) combined with various feedback mechanisms, including erosion in regions of ice-sheet expansion that provided a threshold for ice-sheet sustainability, meant that the climate system could exhibit maximum contrast during the MBE, when insolation forcing was small. A 'synthetic' oxygen-isotope curve based only on Milankovitch forcing (July insolation at 65° N) demonstrates a similar pattern to proxy records: therefore, it is not a change in forcing but a change in the response (Berger and Wefer, 2003).

Other 'problems' include the unsplit peak problem, the fact that modern palaeorecords may be of sufficient quality to discriminate the 95-kyr and 125-kyr frequencies of eccentricity variation. Also, some data records contain significant variability at non-orbital frequencies, although they are usually considered as either harmonics or combination tones of the orbital periods (Elkibbi and Rial, 2001). Thus, at the present time, the Milankovitch or astronomical theory is unable to predict accurately every significant climate change during the Pleistocene. It remains to be seen whether, as dating techniques and our understanding of the key physical mechanisms involved in modulating climate continue to improve, these issues can be resolved in the future.

3.4.2 Heinrich events

There is only space for a summary of these features of the palaeoclimate record here: a recent review of current knowledge is given by Hemming (2004) while the individual events are discussed in more detail in Section 4.2. Heinrich (1988) was the first to document significant ice-rafted debris (IRD) sediments in the North Atlantic and subsequently the periods during which they were formed have been termed Heinrich events. They are thought to be related to massive ice rafting that occurred at irregular intervals during the last glacial. There were six principal Heinrich events between 60 and 15 kyr BP, typically about 7 kyr apart. In all but one, geochemical studies point to the Laurentide Ice Sheet as the principal contributor of these sediments, although IRD from Europe is found in the same sediment pulses (Bond et al., 1997). In addition to the abundance of IRD these layers reveal a decrease in foraminiferal shells, indicative of a cold climate. The events themselves are short lived, lasting for approximately 500 ± 250 years (Hemming, 2004) (Fig. 3.16). Model results (e.g. Ganopolski and Rahmstorf, 2001) and sedimentary evidence indicate that the North Atlantic 'conveyor' component of the THC shuts down in response to a massive freshwater perturbation that stops the overturning circulation. In the model this THC mode is unstable and the 'conveyor' is able to start spontaneously with formation of North Atlantic Deep Water (NADW) occurring south of Iceland. The impact of Heinrich events can be observed globally in terms of sea level variability. Coral data indicates sea level rises of 10–15 m associated with some events (Lambeck and Chappell, 2001).

The cause of Heinrich events remains unsolved, even whether they are driven by internal dynamics of the Laurentide Ice Sheet or by some external climate forcing. Three principal mechanisms have been proposed as causes for Heinrich events. First, the 'binge-purge' model proposed by MacAyeal (1993). In this the Laurentide Ice Sheet expands at a variable rate until some critical threshold is reached and the ice sheet becomes unstable leading to the purging of the ice sheet through the Hudson Strait. The cause of the flow instability may be due to ice stream surging, ice shelf break-up and tidewater instability (Clark et al., 1999). Second, Johnson and Lauritzen (1995) suggested that a series of jökulhlaups – massive floods that occur when a glacier-dammed lake is drained – from a lake in the Hudson Bay region, possibly subglacial, are responsible for the Heinrich events. However, Bond et al. (1992) demonstrated that the events were preceded by cooling and thus may be responding to, rather than be responsible for, a significant climatic shift. This is also suggested by the mix of provenance in the IRD deposits, suggesting a widespread change in climate.

3.4 The mechanisms of high latitude climate change

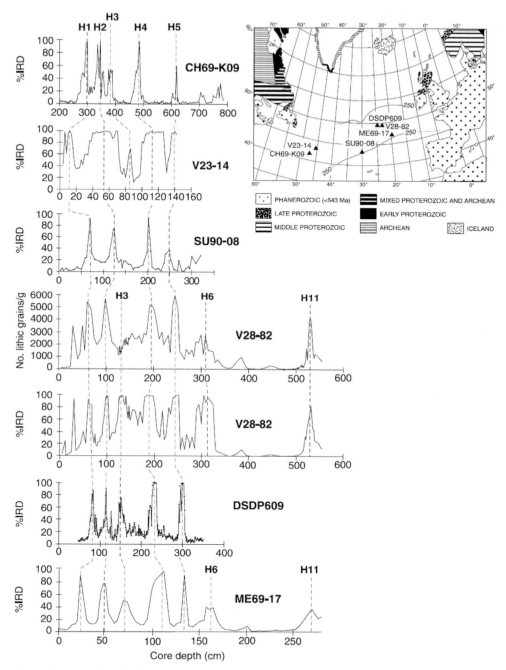

Fig. 3.16 Ice-rafted debris (IRD) data for North Atlantic sediment cores with Heinrich layers. The map shows the location of the cores. (From Hemming, 2004.)

The third hypothesis allows for such an external forcing. Hulbe *et al.* (2004) proposed that an ice shelf fringed the Laurentide ice sheet margin during extreme cold conditions. This would be vulnerable to sudden climate-driven disintegration during any externally forced climate amelioration, as has occurred recently in the Antarctic Peninsula. The ice shelf itself might be the source of the icebergs that carried the IRD into the Atlantic: alternatively, with the buttress effect of the ice shelf removed, the ice streams of the Laurentide Ice Sheet might have surged, leading to increased iceberg production. However, it still remains difficult to tie this theory in with the Greenland temperature record (Bond *et al.*, 1992).

3.4.3 Dansgaard–Oeschger events

Johnsen *et al.* (1992) and Grootes *et al.* (1993) demonstrated that $\delta^{18}O$ records from Greenland ice cores revealed large, abrupt temperature fluctuations throughout most of the last glacial period: a total of 25 such Dansgaard–Oeschger (DO) events, comprising a warm interstadial (usually numbered) and a cold stadial, occurred between 14 and 115 kyr BP (Fig. 3.17). Temperature changes between the interstadial and the intervening stadials is ~16 °C (Landais *et al.*, 2004). Since the publication of these papers similar patterns of temporal variability have been observed in other palaeorecords of temperature throughout the Northern Hemisphere; for example, from stalagmite data in France (Genty *et al.*, 2003) and North Atlantic sea-floor sediments (Shackleton, 2000). In addition, changes in CH_4 trapped in the Greenland ice show a similar variability (Chappellaz *et al.*, 1993) with interstadials (stadials) having increased (decreased) concentrations. Near-simultaneous global fluctuations in this greenhouse gas allow an accurate comparison of records from both polar regions. Blunier and Brook (2001) found an approximate antiphase between temperatures in the Arctic and Antarctic – during Greenland interstadials (stadials) Antarctic temperatures tended to be cooler (warmer). Using higher resolution Antarctic data EPICA Community Members (2006) corroborated this finding and demonstrated that the amplitude of Antarctic warm events is linearly dependent on the duration of the concurrent Greenland

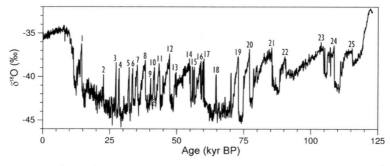

Fig. 3.17 Dansgaard–Oeschger events as observed in the NGRIP Greenland ice core $\delta^{18}O$ record.

stadial. Note that an analysis by Steig and Alley (2002) suggests that the actual relationship is more complex than a simple 'bipolar seesaw'.

In simple form, each DO event comprises a rapid warming over a few decades and a subsequent cooling encompassing several centuries. The rapid warming phase is indicative of a change in atmospheric or near-surface ocean dynamics rather than at depth. Landais *et al.* (2004) showed that the relationship between $\delta^{18}O$ and temperature varied over the duration of a single DO event, probably due to changes in precipitation origin and seasonality. According to these authors a DO stadial was due to a decrease in northward oceanic heat transport, associated with a weakened THC, resulting from freshening due to the discharge of icebergs into the North Atlantic from the Northern Hemisphere ice sheets. Subsequent retreat of these ice masses reduced the freshwater input allowing the THC to restart, and causing a contraction of cold air within the polar vortex. This allowed a rapid meridional heat input into the Arctic, as revealed by the Greenland $\delta^{18}O$ records. The modelling study of Ganopolski and Rahmsdorf (2001) showed only an 8 °C change between DO events and stadials: Landais *et al.* (2004) believed that this was because of the simplified atmosphere in the model and that the remainder of the 16 °C temperature rise would have been due to atmospheric feedbacks such as changes in the storm tracks.

Heinrich events only occurred during stadials, and frequently followed IRD events from sources other than the Laurentide Ice Sheet. Shaffer *et al.* (2004) suggested that subsurface oceanic warming during the stadials would lead to ice-shelf melting and break-up leading to surging of the ice sheets and hence an increase in iceberg production. However, the recent collapse of the Larsen B Ice Shelf in the Antarctic Peninsula has been attributed primarily to changes in atmospheric dynamics rather than oceanographic heat (e.g. Marshall *et al.*, 2006): therefore it is not clear whether the ocean or atmosphere or a combination of the two may have been the principal driver in this scenario.

3.4.4 Atmospheric gases and aerosols

Greenhouse gases

Greenhouse gases are transparent to incoming shortwave solar radiation but opaque to the outgoing longwave infrared radiation emitted by the Earth's surface. Water vapour is actually the main greenhouse gas but is not well mixed, with its distribution influenced by climatic conditions, especially temperature. The principal well-mixed greenhouse gases include carbon dioxide (CO_2), methane (CH_4) and nitrous oxide (N_2O), which have sufficiently long lifetimes to be relatively homogeneously mixed throughout the troposphere. In contrast, another greenhouse gas, ozone, has a much shorter lifetime and has a particular role to play in the polar regions. Thus it is considered separately. Most greenhouse gases were present in the atmosphere prior to anthropogenic activity but since the Industrial Revolution in the eighteenth century anthropogenic emissions of these gases have increased dramatically: for example, since 1750 atmospheric CO_2 has increased by 30% (see Fig. 3.18) and CH_4 by 150%. Palaeo ice core records from Antarctica have

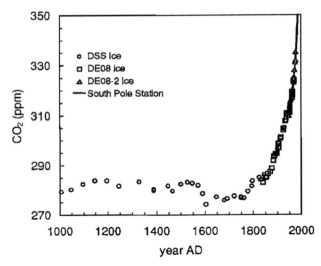

Fig. 3.18 Carbon dioxide mixing ratios (ppm) from three ice cores collected at the South Pole and the modern atmospheric record from the same location. (From Etheridge *et al.*, 1996.)

revealed that current levels of CO_2 are unprecedented over the last 800 kyr (Lüthi *et al.*, 2008).

Palaeo ice core records from the polar regions have also described a very close relationship between global greenhouse gas concentrations and regional temperatures over the last 800 kyr (see Chapter 2 for how proxy temperatures are 'stored' in the ice). Unfortunately, the N_2O record is poor due to bacterial alteration and post-depositional chemical reactions (e.g. Sowers, 2001). However, while there are issues with Greenland ice cores with in situ CO_2 enrichment (Smith *et al.*, 1997), records from Antarctic sites are, with exceptions (Ahn *et al.*, 2004), considered accurate. These demonstrate that CO_2 concentrations are generally higher during interglacials (up to 280–300 ppmv (parts per million by volume)) than glacials (as low as 180–200 ppmv) (e.g. Petit *et al.*, 1999, see Fig. 3.19), the difference between them equivalent to 170–190 Gt of atmospheric carbon.

Initial forcing due to the direct radiative effects of CO_2, CH_4 and N_2O is estimated to produce global warming of 0.95 °C (Petit *et al.* 1999). This initial forcing is amplified by rapid feedbacks due to associated water vapour and albedo modifications (sea ice, clouds, etc.). Modelling results suggest that the total response is 2–3 °C or about half that of globally averaged glacial–interglacial temperature.

Much uncertainty still exists regarding the roles of greenhouse gases during the transition into and out of ice ages, that is whether they act as a forcing or feedback. Ruddiman (2003) suggested that the exact role of CO_2 and CH_4 in the climate system depends upon which of the orbital variations described in Section 3.4.1 is acting to alter insolation changes. During obliquity cycles the ice sheets control changes in atmospheric CO_2 because they vary in-phase. However, during precession cycles CO_2 leads ice volume and is similarly phased

3.4 The mechanisms of high latitude climate change

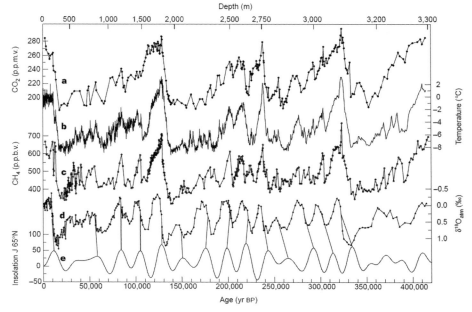

Fig. 3.19 Vostok time series and insolation. Series with respect to time (GT4 timescale for ice on the lower axis, with indication of corresponding depths on the top axis) of: a, CO_2; b, isotopic temperature of the atmosphere; c, CH_4; d, $\delta^{18}O_{atm}$; and e, mid-June insolation at 65° N (in Wm^{-2}). (From Petit et al., 1999, which gives more details on this figure.)

to Southern Ocean SSTs, so that greenhouse gases are playing a forcing rather than feedback role. Sigman and Boyle (2000) proposed that the most viable hypotheses for the cause of glacial–interglacial CO_2 change involve the extraction of carbon from the surface ocean by biological production. However, a more recent study by Kohfeld et al. (2005) indicated that sequestration of carbon by iron fertilisation of marine biota from increased atmospheric dust deposition could only have contributed less than half of the measured glacial–interglacial variation in atmospheric CO_2, and that physical mechanisms must have been the dominant processes. The reduced atmospheric CO_2 subsequently acts as a positive feedback on ice sheet size.

Generally, during a termination, temperature slightly leads both key greenhouse gases by a few millennia but essentially they move together. Carbon dioxide tends to have an approximately linear increase while CH_4 has a two-step rise, with a slow increase followed by a more rapid jump. However, it has already been noted that at Termination V (~420 kyr BP) CO_2 increases 5 kyr prior to CH_4 in contrast to the younger terminations (EPICA Community Members, 2004).

Greenhouse gases within Greenland ice cores also show marked variability at millennial and century scales, during DO events (Section 3.4.3). Severinghaus and Brook (1999) showed that CH_4 is near-synchronous with Greenland temperatures but lags the latter by a few decades and thus is a response rather than driver of these abrupt Northern Hemisphere climate changes. Work by Sowers (2006) demonstrated that the rapid increases in CH_4 resulted from increasing emissions in response to abrupt changes in the hydrologic cycle that are teleconnected to Arctic temperatures but that marine clathrates – a form of water ice that contains a large amount of CH_4 and is widespread in ocean floor sediments – were stable during abrupt warmings. Thus, changes in wetland area were primarily responsible for the rise in CH_4. Modelling studies by Brook *et al.* (2000) and Dällenbach *et al.* (2000) indicated that primary source regions for CH_4 changes varied temporally with both tropical and northern extratropical wetlands being important at different times. Relatively rapid changes in CH_4 are also apparent in Antarctic ice cores (see Fig. 3.19).

Stratospheric ozone

Ozone (O_3) is an important greenhouse gas that occurs in both the stratosphere (90%) – the ozone layer – and troposphere (10%). Stratospheric ozone is formed naturally when ultraviolet (UV) radiation breaks an oxygen molecule (O_2) into separate oxygen atoms (O), which then combine with further oxygen molecules to form ozone (O_3). It is unique in that it absorbs incoming solar UV radiation across a limited range of wavelengths (240–320 nm). As well as protecting life from this potentially harmful UV-B radiation, this absorbed radiation is a source of heat in the stratosphere that helps to maintain the positive, stable temperature–altitude relationship observed there and hence plays a role in the Earth's radiation budget. Ultraviolet absorption and ozone levels peak at ~20 km in polar regions. The residence time of ozone in the atmosphere is relatively short, varying from weeks to months. Some stratospheric ozone is transported down into the troposphere and can influence ozone amounts at the Earth's surface. In the polar regions it can then be depleted by natural processes involving high concentrations of bromine monoxide derived from frost flowers (ice crystals), which grow on frozen leads in regions of sea ice that surround Antarctica in winter and spring (e.g. Kaleschke *et al.*, 2004).

The amount of ozone above a point on the Earth's surface is measured in Dobson Units (DU). In the tropics values are typically ~260 DU throughout the year but in the polar regions there is a strong natural annual cycle in ozone, with maximum values in late winter and early spring (Fig. 3.20). Amounts of ozone are higher in the polar stratosphere than in the tropics primarily because of the Brewer–Dobson circulation. This involves the vertical propagation of atmospheric planetary waves, generated by large-scale temperature contrasts and orographic barriers, into the stratosphere. These act to slow down the polar night jet, strong circumpolar winds that isolate the polar vortex region, thus deforming the vortex and leading to sudden warming of the stratosphere. This creates a radiative and resultant meridional mass imbalance to which the Brewer–Dobson circulation is a response.

In this circulation, ozone-rich air rises very slowly in the tropics and then moves poleward before descending in the winter hemisphere, causing ozone to accumulate there and hence giving the maximum values in both the Arctic and, prior to the formation of the ozone hole, in the Antarctic at this time of year. However, planetary wave generation is much greater in the Northern Hemisphere and hence the Brewer–Dobson circulation is stronger there.

3.4 The mechanisms of high latitude climate change

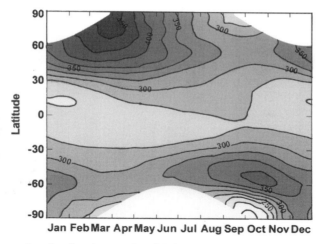

Fig. 3.20 The annual cycle of total ozone for all latitudes based on TOMS 1979–92 data.

Thus, pre-depletion spring values in the Arctic tended to be much higher than in the Antarctic, 450 DU and 350 DU respectively, although both polar regions had significantly greater amounts of ozone than in the tropics.

In recent decades our understanding of the role that ozone plays in the climate system has increased rapidly because of the need to understand the physical and chemical mechanisms through which anthropogenic activity is responsible for the seasonal depletion of ozone over Antarctica and, to a lesser extent, over the Arctic. The Antarctic 'ozone hole' has been described as the first unequivocal demonstration of anthropogenic climate change. The influence of ozone depletion on polar climate is described in Chapter 6, while here we describe briefly the principal mechanisms through which it occurs.

The causes of ozone depletion have been traced to the atmospheric emission of halogen source gases that transport chlorine and bromine into the stratosphere. Chlorine is contained in chlorofluorocarbons (CFCs), once used in most refrigeration and air conditioning units, carbon tetrachloride and methyl chloroform. Bromine is contained in halons, used in fire extinguishers and methyl bromide, used in agriculture. These unreactive halogen source gases accumulate in the troposphere and are subsequently transported into the stratosphere and move poleward via the Brewer–Dobson circulation.

Ozone depletion has been most pronounced above Antarctica, and the ozone hole develops within the polar vortex (Fig. 3.21). The strong horizontal wind shear associated with the vortex edge isolates this region from the middle latitudes. In the upper stratosphere and above, air rich in inorganic chlorine species derived from CFCs spirals slowly poleward and downward into the polar vortex to altitudes at which polar stratospheric clouds (PSCs) form. Reactions on these cloud particles convert the unreactive inorganic chlorine species into reactive forms which then catalyse ozone loss in sunlight.

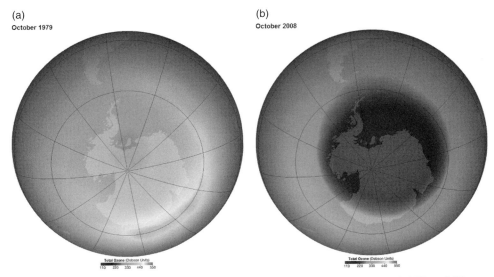

Fig. 3.21 Monthly mean October total ozone concentration (DU) for (a) the pre (1979) and (b) post (2008) ozone hole period. (From http://ozonewatch.gsfc.nasa.gov.)

PSCs form when stratospheric temperatures drop below −78 °C, which, in general, occurs predominantly during winter in the polar regions. At these temperatures nitric acid (HNO_3) and sulphur-containing gases condense with water vapour to form both the solid and liquid particles of which Type 1 PSCs are composed. Type 2 PSCs are less common and are formed of water-ice crystals at even lower temperatures. It is the heterogeneous chemical reactions on the surfaces of these PSC particles that lead to ozone depletion. The normally inert chlorine compounds ($ClONO_2$ and HCl), described as reservoir gases because they do not react directly with ozone, and their bromine counterparts are converted into reactive forms such as chlorine monoxide (ClO), which rapidly and efficiently destroys ozone. Ultraviolet sunlight is necessary for these chemical reactions and hence maximum loss occurs during spring in high latitudes, although depletion begins at the sunlit edge of the polar night (e.g. Lee et al., 2000) and subsequently propagates poleward. Once formed, PSC particles move downwards because of gravity, thereby removing nitric acid from the ozone layer, a process termed denitrification. This slows down the rate of removal of the reactive chlorine monoxide and hence further increases ozone destruction. In late spring, temperatures increase, thereby ending PSC formation and thus the most effective chemical cycles that destroy ozone. Also, the polar vortex breaks down, allowing ozone-rich air once again to be transported into the polar regions.

The Arctic winter stratosphere is generally warmer than its Antarctic equivalent so in some winters temperatures remain sufficiently high to prevent the formation of PSCs. Furthermore, in winters when they do form, the Arctic polar vortex is less well defined than its southern counterpart, meaning that temperatures are more variable and PSC formation is reduced. However, significant ozone reductions can occur. Sinnhuber *et al.* (2000) reported that in March 2000 the mean Arctic total column ozone value was 365 DU, in contrast to pre-1990 March data of 450 DU. Model data indicated 70% ozone loss inside the polar vortex and suggested that widespread denitrification was responsible for continued ozone depletion into late spring.

The ozone layer is also influenced by natural variability in solar output and volcanic eruptions. However, global total ozone levels only vary by 1–2% over the eleven-year solar cycle, which is less than the interannual variability of ~5%. The effect of volcanoes on ozone is discussed in the next section. Most models suggest that following the Montreal Protocol, which limits the production and consumption of halogen-containing gases, ozone levels should return to 'natural' levels by ~2070. But, the exact time for recovery is dependent on the impact of global warming: this will reduce stratospheric temperatures and hence mean PSCs can form more easily (e.g. Shindell, 2003). This is discussed further in Section 7.9.2.

Aerosols

Aerosols are liquid or solid particles suspended in the air and they have both a direct and indirect radiative forcing on climate. The direct forcing is through their scattering and absorption of solar and infrared radiation in the atmosphere. The indirect forcing is by their increasing droplet and ice particle concentrations, which can alter warm, ice and mixed-phase cloud formation and processes and precipitation frequency. The magnitude of the impact of the aerosol forcing is highly dependent on their particle size distribution, which in turn is influenced by their chemical composition. Thus, measurements of both the physical and chemical properties of aerosols are necessary to accurately predict their climate forcing. Such observations are limited and hence, while the overall impact of aerosols on radiative forcing is estimated to be strongly negative (IPCC, 2007), this prediction has significant uncertainty attached to it: indeed it is currently one of the largest uncertainties in predicting future climate. The two principal methods of determining aerosol forcing are (i) forward calculations, based on a knowledge of pertinent aerosol physics and chemistry, and (ii) inverse calculations, based on the premise that the observed warming is caused by total positive anthropogenic forcing. Anderson *et al.* (2003) showed that significant inconsistency exists between the two methods, with the forward calculations showing the biggest negative effect. Another consideration is that, unlike greenhouse gases, aerosols have short atmospheric lifetimes and are not well mixed: thus they may have their biggest impact on radiative forcing on regional rather than global scales.

In the Arctic aerosols transported from mid latitude industrial source regions contribute to the significant air pollution: in recent decades this has given rise to so-called 'Arctic haze', which is most noticeable in late winter and spring. For example, Herber *et al.* (2002), who used the standard definition of a haze event as an increase in total aerosol optical depth of two standard deviations, demonstrated that at a site in Svalbard haze occurs with a 40% frequency in spring but less than 4% frequency in summer. A climatology of atmospheric transport into the Arctic (Stohl, 2006) revealed three distinct pathways: low-level transport followed by ascent in the Arctic; low-level transport alone; and uplift at mid latitudes and descent over the Arctic. Pollution from northern Eurasia is greatest because the air masses have similarly low potential temperatures (θ) to Arctic haze – and hence can penetrate the domes of cold air with constant θ that overlies the Arctic in winter and spring – because their preferred pathway into the Arctic is over snow-covered surfaces that cool the polluted air non-adiabatically. The marked seasonality is caused by the more northward migration of the polar front to 70° N in summer. This separates mild mid latitude air from colder polar air to the north and effectively isolates the Arctic by preventing low-level transport from the south. Garrett and Zhao (2006) estimated that the increase in cloud longwave emissivity resulting from haze at Barrow, Alaska, results in a surface warming of 1.0–1.6 °C. Even in the less polluted Antarctic, Barbante *et al.* (1998) demonstrated that the increase and subsequent decrease of atmospheric lead following its removal from petrol can be observed in Antarctic snow. Around Antarctica and in summer in the Arctic aerosols are primarily derived from natural sources. Bigg and Leck (2001) described two direct sources for cloud condensation nuclei (CCN) from open water and leads in the pack ice. The first is dimethyl sulphide (DMS), which is produced by planktonic algae, particularly in the biologically active marginal ice zone, and promotes particle growth after oxidation into a sulphate aerosol. The second source is bubble bursting from the surface microlayer of the ocean, which entrains small insoluble organic particles into the air; some of these may be large enough to act as CCN. Until recently the former source was considered much more important (e.g. Charlson *et al.*, 1987) but Leck and Bigg (2005) argued, based on their Arctic studies, that the latter is actually the dominant source.

Observations in the Southern Ocean indicate a marked seasonal cycle in DMS production (e.g. Berresheim, 1987) with a clear maximum in summer. This coincides with the annual cycle in CCN observed at Amundsen–Scott Station at the South Pole (Bodhaine *et al.*, 1986), which has a maximum greater than $100\,\mathrm{cm}^{-3}$ in summer, an order of magnitude greater than that measured during winter. However, at Dumont d'Urville, in coastal East Antarctica, there are still significant amounts of DMS in the air during winter: Jourdain and Legrand (2001) suggest that residual emissions are from the marginal ice zone and that DMS has a longer atmospheric lifetime in winter because photochemical oxidation is reduced during the polar night. Over much longer time periods the relationship between methanesulphonate (MSA), an atmospheric oxidation product of DMS that is preserved in ice cores, and glacial cycles is difficult to interpret. In Antarctica, cores from the high plateau have demonstrated high (low) concentrations of MSA during glacials (interglacials) (e.g. Legrand *et al.*, 1991) while at Siple Dome, in West Antarctica (and also in Greenland), the opposite is true (Saltzman *et al.*, 2006), with a positive correlation between MSA concentrations and temperature. Wolff *et al.* (2006) point to post-depositional changes in MSA to explain the former relationship: these authors used non-sea-salt sulphate (nss SO_4^{2-}) instead as a proxy for marine biogenic emissions in the Dome C core and found no significant changes through glacial cycles.

In contrast to CCN, the observed annual cycle in the aerosol scattering coefficient at the South Pole has a maximum in late winter (Bodhaine *et al.*, 1986), which coincides with that of sea salt. The winter record at the pole is characterised by a series of individual events.

The authors found that these events resulted from the movement of cyclones within the circumpolar trough, which every two weeks or so established rapid transport paths from the coast to the pole. The advection of warmer air weakened the surface temperature inversion and thus allowed the detection of sea salt aerosol at the surface. Although weather systems around Antarctica tend to be stronger in winter than summer, and hence bubble bursting and subsequent transport of the sea salt are likely to be greater, modelling studies suggested that this could not explain the magnitude of the winter peak in the annual cycle. The greater sea ice extent in winter means that the aerosol would have to be carried much further. Moreover, the two- to fivefold increase in sea salt concentrations relative to the current climate that are observed in Antarctic ice cores during the Last Glacial Maximum (LGM) cannot be reproduced by models (e.g. Mahowald *et al.*, 2006). However, Wolff *et al.* (2003) proposed that the primary source of sea salt aerosol to Antarctica is newly formed sea ice, which is covered in highly saline brine and, in some cases, equally salty frost flower crystals that can be readily dislodged by strong onshore winds. These authors believe that chemical analysis of some ice cores from the Antarctic interior suggests that sea ice is the main source of sea salt aerosol there, in addition to coastal regions.

Dust is another important aerosol, with the major source regions being deserts and semi-arid regions. On century-scale and longer timescales dust concentrations in polar ice cores can be used to infer past changes in atmospheric circulation in both the deposition and source regions. For example, Röthlisberger *et al.* (2002) demonstrated that dust concentration variability within a core obtained from Dome C, Antarctica, is closely linked to palaeoclimatic records from southern Patagonia. During the transition from the last glacial period into the Holocene (11.7 kyr BP), the amount of dust, as represented by non-sea-salt calcium, decreased by a factor of 24 while there was relatively little contemporaneous change in sea salt, thus reflecting a change in climate at the source region rather than coastal Antarctica. Dust flux variability in the Greenland GISP2 core over the same period was interpreted by Mayewski *et al.* (1993) as changes in circulation intensity and/or aridity over the continental source areas. The increased dust concentrations in the Younger Dryas (YD) (Fig. 3.22) are indicative of the outwash plain south of the Laurentide Ice Sheet becoming a source region although the majority of dust is of Asian origin. The authors suggest that such long-range dust transport would require a southward migration of the polar front. Thus, dust fluxes can be used as a proxy for polar atmospheric circulation patterns: Fig. 3.22 reveals that the Arctic atmosphere was significantly more dynamic in the Younger Dryas than the Holocene.

Hu *et al.* (2005) undertook a detailed modelling study of the impact of different aerosols in the Arctic, examining the climatic impact of haze sulphate, black carbon (soot), sea salt, organics and dust. The model data were compared directly with measurements acquired during the SHEBA (Surface Heat Budget of the Arctic Ocean) campaign. The model results were shown to be strongly dependent on aerosol composition: Table 3.4 compares temperature and cloudiness observations with the output of three model runs: (i) without any aerosol forcing; (ii) with sulphate forcing only; and (iii) with forcing from the five aerosol types listed above. It is clear that the latter forcing produces the most realistic model simulation. The modelling work of Hu *et al.* (2005) indicates that while most aerosol

Table 3.4 *Intercomparison of monthly averaged values of aerosol forcing and cloudiness observations for May 1998*

Variable	Control	Sulphate	Five aerosol species	Observations
Surface temperature (K)	259.46	255.98	259.62	258.79
Cloudiness	0.979	0.888	0.890	0.895

Data from Hu *et al.* (2005).

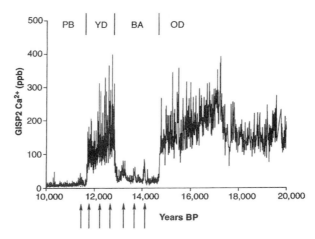

Fig. 3.22 Calcium concentration (ppb) covering the period ~10 000 to 20 000 years ago as determined from the GISP2 ice core from Greenland. Arrows mark periods of increased calcium concentration referred to in the text of Mayewski *et al.* (1993), from which the figure is taken.

types are associated with radiative cooling – through the enhanced scattering of solar radiation and increasing cloud reflectivity and increasing cloud lifetimes – the low reflectivity of soot actually has a heating effect through the absorption of sunlight: the magnitude of this forcing was estimated to have been at a maximum of 3 W m^{-2} in 1906–10 (McConnell *et al.*, 2007). Lubin and Voglemann (2006) utilised multisensor radiometric data to reveal that enhanced aerosol concentrations over the North Slope of Alaska altered the microphysical properties of clouds in such a way – the indirect effect – as to cause an average increase of 3.4 Wm^{-2} in surface longwave fluxes under clouds and hence a net warming. In addition, the net effect of aerosols in the Arctic is significantly affected by natural variability and complex dynamic regional feedbacks.

The two aerosols emitted by volcanic emissions that have the largest impact on climate are dust and gaseous sulphur (mostly SO_2). Although globally much more anthropogenic SO_2 is produced, the radiative effect of volcanoes is similar because the sulphur is input higher into the atmosphere and has a longer residence time. Major eruptions, such as Pinatubo in June 1991, put aerosols into the stratosphere and cause a significant cooling for a couple of years after. The biggest seasonal impact in the polar regions is observed in winter (e.g. Herber *et al.*, 1996). Moreover wintertime diabatic cooling of the stratosphere causes subsidence over the polar regions and the downward motion of the aerosol into the troposphere.

Stenchikov *et al.* (2002) modelled the effect of the Pinatubo eruption (Fig. 3.23) on the Northern Annular Mode (NAM), the principal mode of variability in the Northern Hemisphere extratropics. They showed that if the full radiative effects of the aerosol cloud were included it produced a pronounced and statistically significant positive phase of the NAM for one or two Northern Hemisphere winters following the eruption. The authors also examined separately the stratospheric and tropospheric components of the aerosol cloud. For the stratosphere, the post-eruption aerosol heating in the tropics produced a stronger meridional temperature gradient, which in turn led to stronger zonal winds (the polar night jet) that inhibited the propagation of planetary waves. The polar vortex (of which the NAM is essentially a measure) was thus strengthened, which, combined with associated ozone loss (see below), cooled the polar stratosphere. Interestingly, the tropospheric component of the aerosol also led to a strengthening of the NAM and was indeed able to produce a positive NAM in isolation.

Volcanic aerosol also contributes to significant ozone losses, which in turn influences polar climate. Angell (1997) found a 5–6% decrease in north polar ozone following Pinatubo. There are three principal mechanisms involved. First, the absorption of solar and terrestrial radiation by volcanic aerosol radiatively warms the atmosphere, which in turn affects the meridional transport of trace gases including ozone. Second, backscattering of solar radiation by the aerosol changes the photolysis area of ozone. Third, and most importantly in the polar regions, heterogeneous chemical reactions in the lower stratosphere that destroy ozone are enhanced on the surface of volcanic aerosol particles. A modelling study by Yang and Schlesinger (2002) produced ~10% ozone depletion in high latitudes in late winter/early spring for three years following the Pinatubo eruption.

A signal of past volcanic eruptions is stored in the ice sheets of the polar regions and such signals can be utilised to provide absolute dating of an ice core if the date of the eruption is known from historical records or to aid in the relative dating of different cores (e.g. Stenni *et al.*, 2002). An eruption is recognised as a spike in the record of nss SO_4^{2-} content or electrical conductivity. Robock and Free (1995) attempted to obtain hemispheric aerosol loading patterns using an array of ice cores. They found that by using a large number of cores they were able to reduce the noise of the local non-volcanic signal but still couldn't get quantitative data. In the Northern Hemisphere there was a greater divergence between the volcanic signal in different cores than in the Southern Hemisphere. However, the average signal in the north coincided well with temperature records, which the mean Southern Hemisphere signal did not.

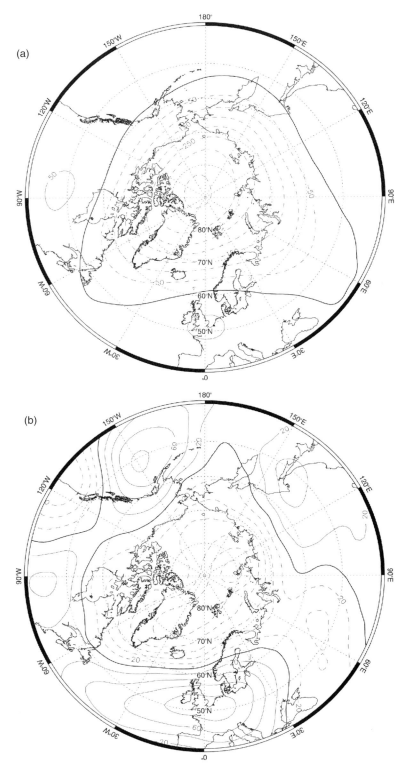

Fig. 3.23 Geopotential height anomalies at 50 hPa (a) and 500 hPa (b) in winter (DJF) of 1992/93 following the 1991 Pinatubo eruption. Data are derived from the ECMWF ERA-40 reanalysis, based on the mean values from 1971/72 to 2000/01. Units are in metres.

3.4.5 *The effects of extra-polar climate variability*

The annular modes

A significant proportion of low-frequency (> 10 days) extratropical atmospheric variability is accounted for by a relatively small number of teleconnection patterns (e.g. Quadrelli and Wallace, 2004): a teleconnection is a high simultaneous correlation of some meteorological parameter (positive or negative) between widely separated regions. Scientific knowledge of such features began with the pioneering work of Walker and Bliss (1932), who first defined the North Atlantic Oscillation (NAO). More recently Thompson and Wallace (1998, 2000) have introduced the 'annular' modes based on the leading empirical orthogonal function (EOF) of sea level pressure or geopotential height at some tropospheric pressure level. The leading or first EOF describes the spatial pattern that represents the principal mode of variability. The annular modes in the Northern and Southern Hemispheres are known as the Arctic Oscillation (AO) and Antarctic Oscillation (AAO) (Thompson and Wallace, 1998; Gong and Wang, 1999), respectively, but more recently the terms Northern Annular Mode (NAM) and Southern Annular Mode (SAM) (Limpasuvan and Hartmann, 1999) have come into favour and will be used in this book.

The NAM and SAM, as defined by Thompson and Wallace (2000), are essentially zonally symmetric or annular in structure, comprising synchronous pressure anomalies of opposite sign in mid and high latitudes. Thus, they can be considered as an index of the strength of the mid latitude westerlies: when the index is high (low) there is low (high) pressure over the polar regions giving strong (weak) westerlies. However, Fyfe and Lorenz (2005) have proposed that annular modes are best characterised as a north–south shift in the mid latitude jet, the result of both latitudinal shifts in the jet, similar to the ideas of Namias (1950), and independent fluctuations in jet strength.

Some workers have questioned whether the annular modes really exist as separate dynamical entities. Cash *et al.* (2002) revealed that in an aquaplanet GCM the model annular mode represents a zonally homogeneous distribution of zonally localised events with a similar meridional structure rather than a zonally symmetric mode of variability per se. That is, the annular modes are simply a statistical feature resulting from the EOF analysis. Deser (2000) suggested that the annular structure of the NAM is a reflection of the dominance of its Arctic centre of action rather than any coordinated behaviour of the Atlantic and Pacific centres of action, which have weak correlations: thus if using MSLP fields, the NAM is virtually indistinguishable from the principal mode of variability in the Atlantic sector (the NAO). Moreover, Ambaum *et al.* (2001) showed similar findings and by using additional parameters other than pressure, such as streamfunction, demonstrated that the NAO paradigm is both more physically relevant and robust. They make the important argument that EOF values at a point depend on the whole data set rather than simply local factors: thus pressure at two points that are of similar sign in the NAM – the northern Atlantic and Pacific – need not necessarily correlate positively. In defence of the NAM, Wallace (2000) argued that the annular modes are distinctive because of their polar

The high latitude climates and change mechanisms

symmetry and large areal extent of their primary high latitude centres of action rather than the strength of associated mid latitude teleconnections.

The Northern Annular Mode and the North Atlantic Oscillation

Wallace (2000) argued that the two paradigms of the NAM and NAO cannot be equally valid and a consensus is required as to which to use. However, at the time of writing no such consensus had been reached and, with the caveats described previously, the NAM and NAO will be considered here as essentially the same pattern of atmospheric variability.

The NAO teleconnection has long been known: van Loon and Rogers (1978) quote the diary of a missionary from the 1770s who wrote that, 'In Greenland all winters are severe, yet they are not alike. The Danes have noticed that when the winter in Denmark was severe, as we perceive it, the winter in Greenland in its manner was mild, and conversely.' Essentially the NAO comprises a latitudinal redistribution of atmospheric mass between the Arctic and mid latitude Atlantic, the variability of which has a significant and widespread influence on the weather and related human activity across much of Europe (Fig. 3.24). Here there is only sufficient space for a brief description of the NAO/NAM and its impacts on Arctic climate: the interested reader is referred to the comprehensive monograph by Hurrell *et al.* (2003).

Basically, the NAO describes the difference between sea level pressure in the subpolar Atlantic (the Icelandic low) and subtropical Atlantic (the Azores high): if the difference is relatively large (small) the NAO is in its positive (negative) phase. There is no single definition of the NAO: an index based on the standardised pressure difference between two stations near the two climatological centres of action – usually Akuryri and Ponta Delgades (Rogers, 1984) or Lisbon and Stykkisholmur (Hurrell, 1995) – has the advantage of providing an NAO time series as long as the available station records. However, the centres of action have an annual cycle to their geographical locations, which the fixed stations are unable to capture. In particular, the subtropical centre migrates northwestward from winter to summer. Hence, Portis *et al.* (2001) produced an NAO index based on the standardised pressure difference at the locations of maximum negative correlation. This definition reveals that the NAO is present throughout the year, whereas the station-based indices show a much stronger teleconnection in winter than the other seasons. The principal disadvantage of this definition, also true for EOF-based definitions of the NAO, including the AO, is that the gridded data sets from which they are derived are only currently available for ~50 years, although a coarse data set is available back to the late nineteenth century.

The various indices suggest that the NAO does not vary on any preferred timescale, consistent with it arising primarily from internal climate dynamics. This is confirmed by spectral analysis, which reveals the NAO has an autocorrelation period of about 10 days (Feldstein, 2002). However, it is thought that external forcing does account for some of the NAO variability. One such mechanism is the interaction between the troposphere and lower stratosphere. Model studies indicate that changes in ozone and greenhouse gas concentration affect the strength of the winter polar vortex above the Arctic (e.g. Gillett *et al.*, 2003), which in turn appears to influence the troposphere via refraction of planetary longwaves at the tropopause. The ocean is also likely to have some influence on the NAO. At shorter timescales changes in the rate and location of tropical heating through altered SST distribution will impact the atmospheric circulation (e.g. Kushnir et al., 2002), while at longer timescales broadscale ocean variability – such as changes in the thermohaline circulation (Bellucci and Richards, 2006) – will modify oceanic temperature gradients, which feed back onto the NAO itself. Note, however, there are also periods when the seasonal NAO is primarily of one sign over a decade or more, such as the negative period in the 1960s and subsequent positive period in the 1990s.

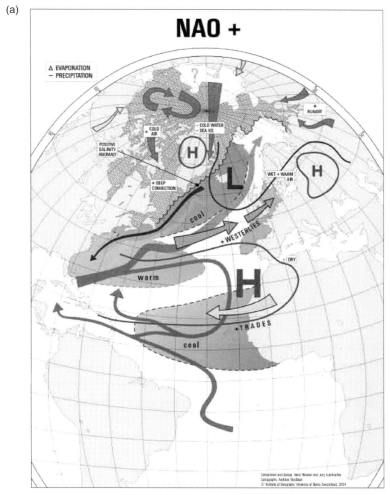

Fig. 3.24 Schematics showing the NAO in its (a) positive and (b) negative phases. (From www.giub.unibe.ch/klimet/wanner/nao.html.)

(b)

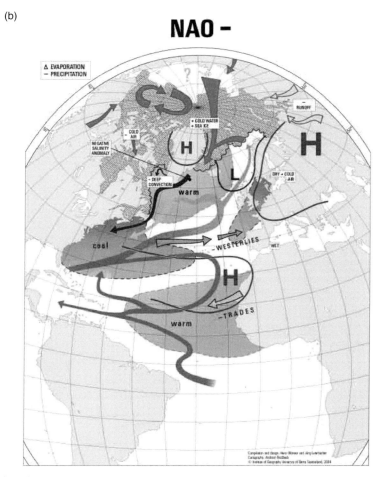

Fig. 3.24 (cont.)

When the NAO is positive an enhanced westerly flow exists across the North Atlantic, with a northeast shift of the winter Atlantic storm track. Enhanced cyclone activity in the Greenland/Iceland/Norwegian (GIN) and Barents Seas advects warm and moist maritime air masses over Scandinavia and the northwest Russian Arctic. Stronger northerly winds over Greenland and northeast Canada carry cold air southward and decrease SSTs over the northwest Atlantic. Hurrell (1996) revealed that the NAO accounts for 31% of hemispheric wintertime interannual surface temperature variance from 1935 to 1994. However, this relationship is not temporally stable: Jones *et al.* (2003) demonstrated that correlations between surface air temperature and the NAO across the Arctic were more coherent and stronger in the latter half of the twentieth century compared with the previous 50 years. Between strongly negative and positive NAO, the total net moisture flux through 70° N, as estimated from radiosonde data, increases from 4.4 to 7.6 $\times$ 10^7 kg s^{-1}. The largest longitudinal change occurs through the Nordic Seas–Scandinavia (10° W–50° E), where the fraction of the total increases from 0 to 58%. Winter precipitation also increases dramatically in this sector, by up to 0.15 m (Dickson *et al.*, 2000). Relatively little change occurs across the rest of the Arctic.

Atmospheric changes associated with NAO variability impact the oceanographic component of the northern high latitude climate. Anomalies of SSTs in the North Atlantic are driven by changes in the surface wind and air–sea heat exchanges associated with NAO variability (e.g. Visbeck *et al.*, 2003). This relationship is strongest when the latter leads the former by several weeks: large-scale extratropical SSTs respond to atmospheric forcing on monthly to seasonal timescales. The Arctic warming associated with a positive NAO is the combined result of a warmer and stronger inflow of Atlantic Water along both the main inflow branches (Barents Sea and Fram Strait), these are 1–2 °C warmer and 0.13 Sv stronger (1 sverdrup = 10^6 m^3 s^{-1}) for a one sigma change in the NAO (Dickson *et al.*, 2000). Variations in the winter NAO index explain ~60% of the variance in the monthly depth-integrated temperature of the Fram Strait since 1965.

On longer timescales other non-local ocean dynamics become more important. Beneath the surface the impact of high-frequency atmospheric variability diminishes rapidly, although limited subsurface observations indicate fluctuations coherent with low frequency winter NAO variability down to 400 m. However, interdecadal variability in the formation of intermediate and deep waters through convective renewal at depth in the North Atlantic region, which contributes significantly to the production and export of NADW and thus helps drive the global THC, is related to NAO variations. Winter convection in the Labrador (GIN) Seas, which have got colder and fresher (warmer and saltier) deep waters, respectively, is positively (negatively) correlated to the NAO (Dickson *et al.*, 1996).

Salinities along the principal inflow branches have declined from the 1960s to 1990s, consistent with increasing freshwater accession, sea ice reduction and an increasing volume flux of ice from the Arctic (Dickson *et al.*, 2000). Observations indicate that during a positive NAO there is more (less) sea ice extent in the Labrador Sea (Greenland Sea). For example, during 1958–97 winter sea ice concentration in the latter region was significantly anticorrelated with the NAO, with the strongest relationship ($r = -0.64$) in the 'Odden' region, where a large tongue of sea ice builds northeastward into the Greenland Sea during winter months, occurring between 8° W–5° E and 73–77° N (Deser *et al.*, 2000): see also Section 3.8.3. Furthermore, winter NAO variability explains ~60% of the variance in the annual volume flux through Fram Strait since 1976, with a 1 sigma change in NAO equivalent to 200 km^3 change in ice flux. These changes result through the atmospheric forcing of sea ice, dynamically via the surface winds and thermodynamically through changing surface temperatures.

The Southern Annular Mode

The different land–sea distribution in the Southern Hemisphere, with the unbroken circumpolar Southern Ocean, means that there is less doubt concerning the physical reality of the

SAM, as compared with its northern analogue. When pressures are below (above) average over Antarctica the SAM is said to be in its high (low) index or positive (negative) phase. The SAM is equivalent barotropic and is revealed as the leading EOF in many atmospheric fields (see Thompson and Wallace, 2000 and references therein). Model experiments demonstrate that the structure and variability of the SAM result from the internal dynamics of the atmosphere (e.g. Hartmann and Lo, 1998; Limpasuvan and Hartmann, 2000). Poleward eddy momentum fluxes – synoptic-scale weather systems – interact with the zonal mean flow to sustain latitudinal displacements of the mid latitude westerlies. The SAM contributes a significant proportion of Southern Hemisphere climate variability (typically ~35%) from high-frequency (Baldwin, 2001) to very low frequency timescales (Kidson, 1999), with this variability displaying a red (lower frequency) noise distribution.

Similar to the NAM, gridded reanalysis data sets have been utilised to derive time series of the SAM (e.g. Thompson et al., 2000; Renwick, 2004). However, the poor quality of reanalyses at high southern latitudes prior to the assimilation of TOVS satellite sounder data in 1979 (Hines et al., 2000) means that long-term SAM time series cannot be derived this way. Based on a definition by Gong and Wang (1999), Marshall (2003) produced a SAM index based on 12 appropriately located station observation time series in the extratropics and coastal Antarctica. Unfortunately, this index itself is limited to the period following the International Geophysical Year (IGY) of 1957/58. However, attempts have been made to reconstruct century-scale records based on proxies of the SAM. Goodwin et al. (2004) use sodium concentrations from the Law Dome ice core while Jones and Widmann (2003) employ tree-ring width chronologies: both these studies stress the decadal variability in their derived SAM time series.

The SAM has shown significant positive trends during autumn and summer over the past few decades (Thompson et al., 2000 see Chapter 6 for details; Marshall, 2003), resulting in a strengthening of the circumpolar westerlies. These trends have contributed to the spatial variability in Antarctic temperature change (e.g. Kwok and Comiso, 2002a; Thompson and Solomon, 2002; Schneider et al., 2004; Marshall et al., 2006), specifically a warming in the northern peninsula region and a cooling over much of the rest of the continent. The SAM also impacts the spatial patterns of precipitation variability in Antarctica (Genthon et al., 2003).

The imprint of SAM variability on the Southern Ocean system is observed as a coherent sea level response around Antarctica (Aoki, 2002; Hughes et al., 2003) and by its regulation of Antarctic Circumpolar Current flow through the Drake Passage (Meredith et al., 2004). Modelling studies indicate that a positive phase of the SAM is associated with northward (southward) Ekman drift in the Southern Ocean (at 30° S) leading to upwelling (downwelling) near the Antarctic continent (~45° S) (Hall and Visbeck, 2002; Lefebvre et al., 2004). These changes in oceanic circulation impact directly on the thermohaline circulation (Oke and England, 2004) and may explain the recent patterns of observed temperature change at Southern Hemisphere high latitudes described by Gille (2002). Although the SAM is essentially zonal, a wave number 3 pattern is superimposed (Fig. 3.25), with a marked

3.4 The mechanisms of high latitude climate change

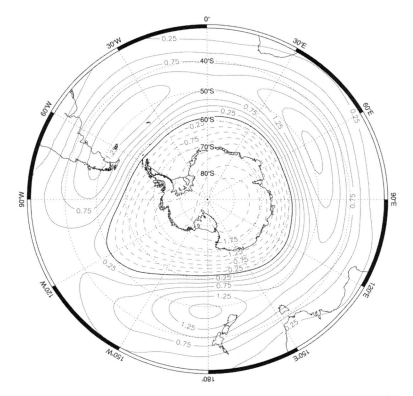

Fig. 3.25 The Southern Annular Mode, defined as the rotated EOF 1 from ERA-40 500 hPa geopotential height monthly anomaly data for 1979–2001.

low pressure anomaly west of the peninsula when the SAM is positive leading to increased northerly flow and reduced sea ice in the region (Liu *et al.*, 2004). Raphael (2003) reported that diminished summer sea ice may in turn feed back into a more positive SAM. However, modelling work by Marshall and Connolley (2006) indicated that increased SSTs at high southern latitudes will warm the atmosphere and, through thermodynamics, cause the atmospheric centre-of-mass to rise and geopotential height to increase thus producing a more negative SAM.

3.4.6 ENSO and the Pacific teleconnections

The ENSO phenomenon involves large SST anomalies in the eastern tropical Pacific, which lead to substantial anomalies in evaporation and the reorganisation of cumulus convection throughout the Pacific, providing a deep anomalous tropical heat source. This in turn produces a Rossby wave response, which propagates into the extratropics.

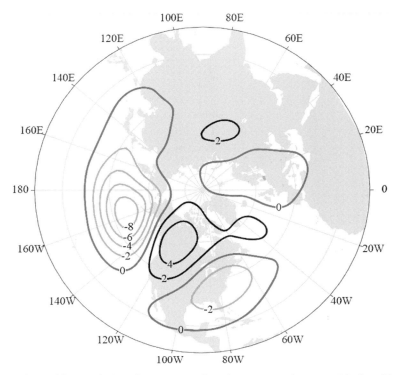

Fig. 3.26 The Pacific–North American pattern plotted as a regression map. Black, mid-grey and light-grey contours on the maps indicate positive, zero and negative isopleths, respectively. Contour interval is 2 decametres.

The Pacific–North American (PNA) pattern

The PNA teleconnection pattern, first defined by Wallace and Gutzler (1981) is a quadripole of height anomalies and comprises the four centres of action shown in Fig. 3.26. It represents 'variations in the waviness of the atmospheric flow in the western half-hemisphere and thus the changes in the north–south migration of the large-scale Pacific and North American air masses and their associated weather' (Kushnir *et al.*, 2002). Pressure over the North Pacific Ocean, near the Aleutian Islands, varies out of phase with that further south, west of Hawaii. When the Aleutian Low is deeper than average the PNA is said to be in its positive phase and vice versa. The PNA pattern is in part set up by the Tibetan Plateau and Rocky Mountains jutting into the atmosphere's westerly flow. Many studies have noted that SST anomaly patterns associated with a warm ENSO and a positive PNA are comparable, in that they both involve a deepening of the Aleutian Low. Feldstein (2000) states that the PNA attains its large interannual variance through external forcing associated with tropical Pacific SST anomalies. However, in contrast, Straus and Shukla (2002) demonstrated that the PNA is not forced by ENSO and is indeed a robust internal pattern of Northern Hemisphere extratropical climate variability: SST anomalies caused by ENSO can vary significantly between different events whereas those associated with the PNA pattern are spatially stable.

Similar to the NAO/NAM, the PNA is strongest in winter, when it explains more than half of the interannual pressure variability in the regions near the four centres of action. On moving from intraseasonal to interdecadal timescales the NAM and PNA patterns together account for an increasing proportion of total winter climate variance, from over one third to virtually all the variance. In combination, these two leading MSLP principal components also account for substantial fractions of the variance of winter monthly mean surface air temperature and precipitation throughout most of the Northern Hemisphere. Thus they probably play an important role in driving climate variability at century timescales and longer (Quadrelli and Wallace, 2004).

Positive PNA events are associated with an intensified Aleutian Low and therefore greater than average storminess in the Gulf of Alaska, which, together with ridging over western Canada and Alaska, drives warm winds northward over these latter regions, the so-called Arctic Express. Although this warm air does not penetrate far out into the Bering Sea, Azumaya and Ohtani (1995) indicated that the PNA had some impact on cold bottom water temperatures on the Bering Sea shelf. During a negative-phase PNA there is significant atmospheric blocking in the northern Pacific and broadly opposite regional temperature and precipitation anomalies occur.

The Pacific–South American (PSA) pattern

The PSA pattern (Mo and Ghil, 1987) is similar to a reflection of the PNA in the Equator but its variability is considered to be primarily related to ENSO. Nonetheless, it is prominent at many timescales, including that of the Madden–Julien Oscillation (Renwick and Revell, 1999), even in the absence of a strong ENSO signal. Thus, like the PNA, the PSA may also be an internal mode of climate variability. It is generally described as a wave-like anomaly pattern emanating from the subtropical western Pacific, characteristic of Rossby wave propagation associated with the vorticity generated by adiabatic heating from tropical temperature anomalies (e.g. Hoskins and Karoly, 1981). The PSA pattern comprises low pressure east of New Zealand, high pressure in the south-east Pacific and low pressure over South America/South Atlantic and its evolution shows a coherent eastward propagation (Fig. 3.27). The latter two 'centres of action' influence Antarctic climate and are discussed in more detail later in this section. Note that Mo and Higgins (1998) state that the PSA actually comprises two separate modes in quadrature with each other, each associated with enhanced convection in different parts of the Pacific and suppressed convection elsewhere.

An alternative mechanism for the PSA teleconnection was proposed by Liu *et al.* (2002) and previously seen in the work of Chen *et al.* (1996). They suggest that the increased convection associated with El Niño events alters the mean meridional atmospheric circulation through longitudinal changes to the Hadley circulation and subsequent alterations of the subtropical jet position and strength. In the extratropics, the Ferrel cell is influenced directly by storm-track changes that alter the meridional eddy heat flux divergence and

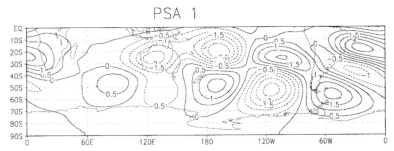

Fig. 3.27 EOF 1 for the 200-hPa eddy streamfunction. The EOF was normalized to 1 and time 100. Contour interval is 0.5 non-dimensional unit. (From Mo and Higgins, 1998.)

convergence and a shift in the latent heat release zone. Yuan (2004) suggests that the two mechanisms operate in phase and are comparable in magnitude.

While there are general patterns of high latitude climate anomalies that co-vary with ENSO, teleconnection correlations tend to be small. A modelling study by Lachlan-Cope and Connolley (2006) indicated that this is because (i) Rossby wave dynamics are not well correlated to 'standard' definitions of ENSO because the relationship between upper level divergence and SSTs via deep convection is complex and (ii) natural variation in the zonal flow of Southern Hemisphere high latitudes can swamp the ENSO signal. An example of the latter is shown in the relationship between ENSO and the Southern Hemisphere jets. Chen et al. (1996) examined the El Niño/La Niña pair from 1986–89 and concluded that the subtropical jet (STJ) was stronger and the polar front jet (PFJ) weaker during El Niño events and vice versa during La Niña events. However, in a study encapsulating a much longer period, Bals-Elsholz et al. (2001) found that split-flow conditions in the South Pacific are not strongly modulated by ENSO. Moreover, Fogt and Bromwich (2006) demonstrated that a weaker high latitude teleconnection in spring during the 1980s compared with the following decade was due to the out-of-phase relationship between the PSA and SAM at this time: subsequently, the tropical and extratropical modes of climate variability had an in-phase relationship.

Although apparent throughout the year, the PSA demonstrates the strongest teleconnections to the Antarctic region in austral spring and summer. During these seasons an El Niño (a La Niña) event is associated with significantly more (less) blocking events in the southeast Pacific (Renwick and Revell, 1999). This pressure anomaly in the Amundsen–Bellingshausen Sea (ABS) is primarily responsible for the Antarctic Dipole (ADP) in sea ice (Yuan and Martinson, 2001: see Section 3.4.7). In addition the ADP sea ice anomalies are reinforced by ENSO-related storm-track variability, which influences the Ferrel cell by changing meridional heat flux divergence/convergence and shifting the latent heat release zone, as mentioned previously, which in turn modulates the mean meridional heat flux that impacts sea ice extent. Correlations with ENSO indices imply that up to 34% of the variance in sea ice edge is linearly related to ENSO (Yuan and Martinson, 2000) while Gloersen (1995) showed that different regions of sea ice respond to ENSO at different periodicities. However, the strongest sea ice correlations associated with ENSO occur at 120–132° W (ABS) lagging the tropical temperature anomaly by 6 months; so the austral spring/summer ENSO signal is observed in the subsequent ice growth period in autumn/winter (Yuan and Martinson, 2000). Note that the temporal quasi-periodic nature of both ENSO and the ice anomalies prevents the identification of direction of causality. Kwok and Comiso (2002b) state that recent trends in sea ice extent in both the Bellingshausen Sea and Ross Sea are related directly to ENSO variability.

There have been a number of studies claiming that a signal relating to ENSO variability can be observed in Antarctic accumulation at certain locations. Both Legrand and Feinet-Saigne (1991) and Meyerson *et al.* (2002) suggested that MSA concentrations in ice cores from the South Pole co-varied with ENSO. However, the description of the physical mechanisms by which the ENSO signal arrives in the Antarctic interior remains somewhat qualitative (Turner, 2004). MSA is principally derived from DMS production in the marginal ice zone at southern high latitudes so a potential mechanism linking ENSO and MSA concentrations in Antarctic accumulation is the ADP. Back trajectory analysis by Harris (1992) suggested that the majority of air flow reaching the South Pole reflected storm activity in the Ross Sea. In an El Niño year sea ice tends to be reduced in this region (Yuan, 2004), whereas Meyerson *et al.* (2002) claim that increased sea ice extent is associated with a higher frequency of El Niño events. It seems likely that the observed relationship may be due to the sea ice edge, the MSA source region, being closer inshore in an El Niño year and hence the acid may more easily be transported to the South Pole (Abram *et al.*, 2007). Isotopes in both Antarctic precipitation and in shallow ice cores have also been found to be correlated to ENSO (Bromwich *et al.*, 2000; Ichiyanagi *et al.*, 2002).

In addition, the relationship between ENSO and recent Antarctic accumulation has been studied using the gridded reanalysis data sets. The primary cause of Antarctic precipitation variability is due to changes in storm tracks over the Southern Ocean (Cullather *et al.*, 1998). These are impacted by the shifts in strength and position of the Southern Hemisphere jets, which, as described previously, are influenced by the phase of ENSO (e.g. Marshall and King, 1998; Yuan, 2004) so an ENSO signal in Antarctic accumulation might be expected. At a continent-wide scale, Genthon *et al.* (2003) revealed that using detrended data, the Southern Oscillation Index (SOI) and first EOF of Antarctic precipitation had a correlation coefficient of 0.66 for 1980–98, similar to the SAM. The authors note, however, that the magnitude of correlation varies considerably through time. Similarly, Cullather *et al.* (1996) demonstrated a strong positive correlation between moisture convergence into West Antarctica (120–180° W) and the SOI – that is, more (less) accumulation is associated with La Niña (El Niño) years – in the 1980s (Fig. 3.28). However, in the 1990s a sudden reversal to a correlation in the opposite sense occurred. The authors relate the change to a strong East Antarctic ridging pattern in the more recent period. In a recent paper, Fogt and Bromwich (2006) indicate that the change reflects a shift from an out-of-phase relationship between the PSA and SAM during austral spring in the 1980s to a positive one in the subsequent decade. Note that Genthon *et al.* (2005) showed that any correlation between

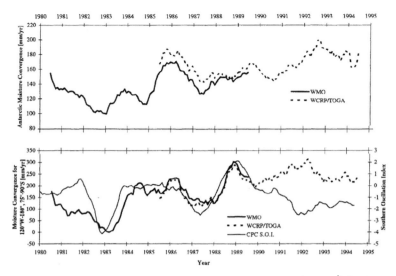

Fig. 3.28 Twelve-month centred running-mean moisture convergence (mm yr^{-1}) for Antarctica (top) and West Antarctic sector (bottom) in comparison with the Southern Oscillation Index (in hPa). (From Cullather et al., 1996.)

accumulation and ENSO in the ERA-40 reanalysis declines rapidly on moving eastward from 120° W. They also suggested that local depositional noise in ice cores may be sufficient to almost fully obscure the larger regional signals related to atmospheric modes of variability.

3.4.7 The Antarctic Circumpolar Wave (ACW) and Antarctic dipole (ADP)

The Antarctic Circumpolar Wave (ACW), as first defined by White and Peterson (1996), is an apparent easterly progression of phase-locked anomalies in Southern Ocean surface pressure, winds, SSTs and sea ice extent (Fig. 3.29). As such, it thus represents a coupled mode of the ocean–atmosphere system. The ACW has a zonal wave number of 2 (a wavelength of 180°) and the anomalies propagate at a speed (6–8 cm s^{-1}) such that they take 8–10 years to circle Antarctica giving the ACW a period of 4–5 years. A similar feature has also been identified in sea surface height using satellite altimeter data (Jacobs and Mitchell, 1996). Given the much shorter response times of the atmosphere, these authors proposed that the ocean plays an important part in creating and maintaining the ACW. As there is little Antarctic multi-year ice, Gloersen and White (2001) suggested that the memory of the ACW in the sea ice pack is carried from one austral winter to the next by the neighbouring SSTs. White et al. (2004) showed a complex tropospheric response to sea ice anomalies that, for example, explained anomalous poleward surface winds and deep convection observed with negative sea ice edge anomalies.

3.4 *The mechanisms of high latitude climate change* 113

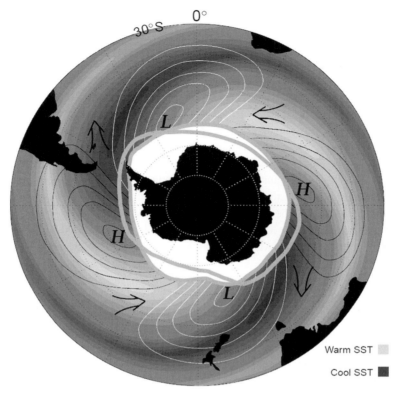

Fig. 3.29 Simplified schematic summary of interannual variations in sea surface temperature (light grey, warm; dark grey, cold), atmospheric sea level pressure (bold H and L), meridional wind stress (denoted by the arrows), and sea ice extent (grey lines). Sea ice extent is based on an overall 13-year average (White and Peterson, 1996). (From www.clivar.org/publications/other_pubs/clivar_transp/ pdf_files/av_d5_991.pdf.)

While some authors have suggested that an ACW signal can be observed in an Antarctic ice core over the last 2000 years (Fischer *et al.*, 2004), since the initial discovery of the ACW others have questioned its persistence: a number of observational and modelling studies have indicated that the ACW is not apparent in recent data before 1985 and after 1994 (e.g. Connolley, 2002), somewhat fortuitously the period that White and Peterson (1996) chose for their original analysis. In addition, there has been some discussion on one of its key characteristics, whether it really has a wave number 2. Many studies (e.g. Christoph *et al.*, 1998; Cai *et al.*, 1999; Weisse *et al.*, 1999) have indicated that an ACW-like feature is apparent in GCM control runs but that it has a preferred wave number 3 pattern. Venegas (2003), using frequency domain decomposition, suggested that the ACW comprises two significant interannual signals that combine constructively/destructively to give the observed irregular fluctuations of the ACW on interannual timescales, as also seen in GCM studies (Christoph *et al.*, 1998). The two signals comprise (i) a 3.3 year period of

114 *The high latitude climates and change mechanisms*

zonal wave number 3 and (ii) a 5-year period of zonal wave number 2, which was particularly pronounced during the period studied by White and Peterson.

However, most of the ACW debate has centred on its forcing mechanisms and, as a consequence, the very nature of its existence or not. For example, White *et al.* (2002) state explicitly that the ACW exists independently of the tropical standing mode of ENSO and that its eastward propagation depends upon atmosphere–ocean coupling rather than advection by the Antarctic Circumpolar Current (ACC): both these points have been refuted in the literature. Yuan and Martinson (2001) described an Antarctic dipole (ADP) in the interannual variance structure in the sea ice edge and SST fields of the Southern Ocean, which is characterised by an out-of-phase relationship between anomalies in the central/eastern Pacific (Amundsen–Bellingshausen Sea) and the Atlantic (Weddell Sea) sectors. Thus the ADP has the same wavelength as the ACW, but key differences are that the associated variability is twice that of the ACW and that the dipole consists principally of a strong standing mode together with a much weaker propagating motion. Moreover, Yuan and Martinson (2001) state that the ADP is clearly associated with ENSO events. Other authors have found that ENSO variability is important in driving a geographically fixed standing wave in Southern Ocean interannual variability, centred in the South Pacific and linked to the PSA pattern (e.g. Christoph *et al.*, 1998).

Furthermore, several papers mention the ACC as the means of anomaly propagation. In an ocean model forced by realistic stochastic (random) anomalies Weisse *et al.* (1999) demonstrated that the ocean acts as an integrator of short-term atmospheric fluctuations (white noise) and turns them into a red response signal (lower frequency signal), and subsequently the average zonal velocity of the ACC determines the timescale of the oceanic variability and thus the propagation speed and period of the ACW (Cai *et al.*, 1999; Venegas, 2003). However, the results of the study of Park *et al.* (2004) indicated that any such propagating anomalies comprised only ~25% of interannual Southern Ocean SST variability and are often rapidly dissipated in the Indian Ocean, are intermittent in phase and frequently do not complete a circumpolar journey. Hence, they questioned the very existence of the ACW, as originally described by White and Peterson (1996).

3.5 Atmospheric circulation

3.5.1 *Arctic*

Surface pressure and upper height fields

The Arctic winter circulation is dominated by three subpolar centres of action: (i) the Siberian High over east-central Asia; (ii) the Icelandic Low in the North Atlantic, to the southeast of Greenland; and (iii) the Aleutian Low in the North Pacific (Fig. 3.30). The Siberian High is a cold, shallow anticyclone, driven largely by radiative cooling and generally constrained by topography with a maximum area in February (Serreze and Barry, 2005). It is centred at 45–50° N, 90–110° E but when it periodically intensifies outbreaks of cold air move eastward. In winter both the Icelandic Low and Aleutian Low form through a

3.5 Atmospheric circulation

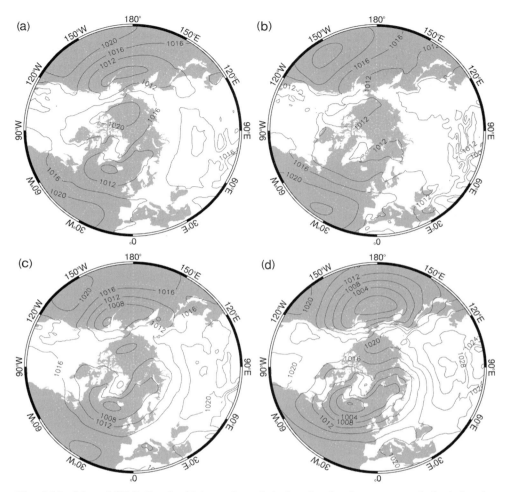

Fig. 3.30 Mean MSLP for the high northern latitudes for the four seasons derived for the period 1979–2009 from the ECMWF reanalysis and operational analyses. (a) Spring (March–May), (b) Summer (June–August), (c) Autumn (September–November), (d) Winter (December–February).

number of different but related processes. Primarily, they are located in regions of strong baroclinicity (horizontal temperature-gradient), which are associated with cyclonicity. In addition, both lows are located downstream of the major mid-tropospheric stationary troughs where cyclogenesis is favoured through upper-level divergence (Serreze and Barry, 2005). The Icelandic Low and Aleutian Low form part of the primary North Atlantic and North Pacific cyclone tracks, respectively (see below). In addition, the orographic effects of the steep southern Greenland Ice Sheet may lead to lee cyclogenesis (e.g. Tsukernik *et al.*, 2007) in the region of the Icelandic Low, which may extend into the Barents and Kara seas.

During spring and summer the general Arctic circulation, including the three centres of action, weakens. The Aleutian Low shifts to the northeast in spring and is no longer present as a climatological mean feature in summer, while to the north high pressure cells develop over the Beaufort Sea and the pole region. A springtime weakening of the Siberian High occurs in response to increased solar heating and snow melt (Serreze and Barry, 2005). The Icelandic Low is also much weaker and broader with low pressures extending west over the Canadian Arctic. The circulation in autumn represents a transition from summer to winter conditions.

Cullather and Lynch (2003) examined the annual cycle of pressure fields across the Arctic. Their study indicated that the Canada Basin–Laptev Sea side is dominated by the first harmonic, with an amplitude of ~5 hPa and maximum (minimum) pressure occurring in March (September). In contrast, along the periphery of northern Greenland and extending to the pole, a weak semi-annual cycle is found with maxima in May and November. Dynamically, this progression of the annual cycle can be attributed to the transfer of atmospheric mass from Eurasia and into the Canadian Archipelago in spring and the reverse flux in autumn.

The mid-tropospheric circulation (Fig. 3.31) is dominated by a well-developed cyclonic vortex during much of the year, which is strongly asymmetrical in winter. Colder and drier conditions within the vortex are isolated from external sources of advection and communicated to the surface through radiative processes. At 60° N there is a strong wave number 3 component – with troughs over eastern North America, and eastern and western Asia. These are a response to dynamical forcing by major orographic features like the Rocky Mountains and Himalayas, thermal forcing associated with the large-scale land–ocean distribution and radiative forcing (Serreze and Barry, 2005). In summer, the vortex is much weaker and more symmetrical due to the more even latitudinal distribution of radiation and temperature.

A vortex is also the dominant feature in the stratosphere: it is more symmetrical than in the mid-troposphere because only the long planetary-scale Rossby waves can propagate up to high levels. Nevertheless, Waugh and Randel (1999) demonstrated that the Arctic vortex can be displaced far off the pole and also elongated. Distortion increases throughout its life cycle from autumn to early winter. There are two preferred centres in early winter (60°–120° E and 60°–90° W) and the vortex can switch from one to the other in a few days. This large variability is due in part to the occurrence of 'stratospheric warmings', events in which the vortex becomes very distorted. While Waugh and Randel (1999) postulate that such observed longitudinal shifts may be related to changes in tropospheric circulation, Baldwin and Dunkerton (2001) showed that variations in the strength of the stratospheric circulation, appearing first above ~50 km, descend to the lowermost stratosphere and subsequently into the troposphere. Here, the anomalous regional circulation maps onto the NAM pattern (cf. Section 3.4.5) during the 60 days after the onset of these events (Fig. 3.32). However, these authors suggest that in winter the stratosphere is originally modified by waves originating in the troposphere, which alter the conditions for planetary-wave propagation in such a way as to draw mean-flow anomalies poleward and downward.

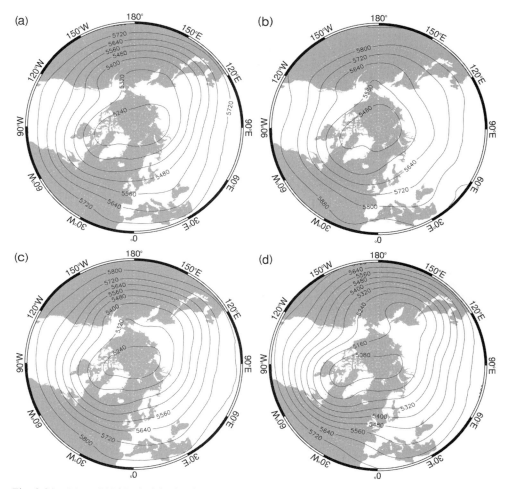

Fig. 3.31 Mean 500 hPa height for the high northern latitudes for the four seasons derived for the period 1979–2009 from the ECMWF reanalysis and operational analyses. (a) Spring (March–May), (b) Summer (June–August), (c) Autumn (September–November), (d) Winter (December–February).

Depression tracks

The path of cyclones, known as depression or storm tracks, can be obtained using three broad methodologies. First, individual weather systems can be identified by their cloud signatures in satellite imagery or meteorological charts and then their subsequent path determined from further multi-temporal imagery. The polar regions are well suited to this because polar orbiting satellites provide many overlapping passes at high latitudes. Although attempts have been made to automate this process, most studies to date have been done manually, and the labour-intensive nature of this process means that the time periods covered are relatively short. For example, Turner *et al.* (1998) examined 12 months

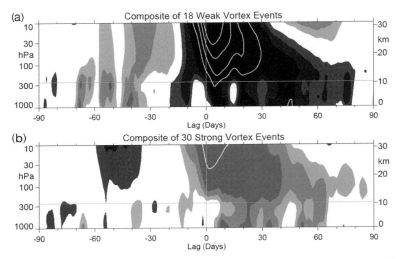

Fig. 3.32 Composites of time–height development of the Northern Annular Mode for (a) 18 weak vortex events and (b) 30 strong vortex events. The events are determined by the dates on which the 10-hPa annular mode values cross −3.0 and +11.5, respectively. The indices are non-dimensional; the contour interval for the grey shading is 0.25, and 0.5 for the white contours. Light shading represents positive values and dark shading represents negative values. Values between −20.25 and 0.25 are unshaded. The thin horizontal lines indicate the approximate boundary between the troposphere and the stratosphere. (From Baldwin and Dunkerton, 2001.)

of AVHRR data in the Antarctic Peninsula sector during which 504 synoptic-scale lows were identified.

The availability of gridded data from NWP models and reanalyses allows entirely automated methods to be used that can provide climatological information on depression tracks, including trends. System-centred tracking is achieved by searching for local minima in certain fields, such as MSLP (e.g. Jones and Simmonds, 1993), or maxima in vorticity (e.g. Hoskins and Hodges, 2005).

Cyclones in the Arctic region tend to be more (less) intense, shorter (longer) lived, and fewer (greater) in number during winter (summer). In winter the greatest number are found over the North Pacific and particularly the North Atlantic, while they are more spatially variable around the subpolar regions in summer (Zhang *et al.* (2004). Hoskins and Hodges (2002) found that few synoptic systems can be tracked along the length of the Pacific storm track. It is the systems that are generated in the central-east Pacific that occlude on the northwest coast of North America and, similarly, weather systems generated off the east coast (of the USA) move into the northwestern Atlantic. Winter cyclone activity is a maximum in the region of the Icelandic Low and Baffin Bay (Serreze and Barry, 2005). Cyclones can also propagate over the central Arctic Ocean. There are also more cyclones originating in the mid latitudes propagating into the Arctic Ocean in winter than in summer: these are often deeper than those generated locally, particularly those coming from the North Atlantic. These move along three preferred storm tracks: from the North Atlantic Ocean to

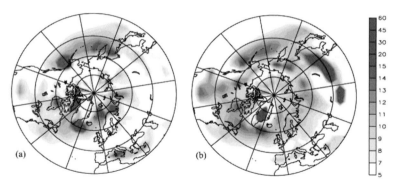

Fig. 3.33 The long-term mean (1948–2002) cyclone centre count in (a) winter and (b) summer. Units are counts per 10^5 km^2. (After Zhang et al., 2004.)

the GIN and Barents Sea, from the northwestern Pacific Ocean to the Bering Strait/Chukchi Sea, and from northern Europe to the Eurasian shelf seas (Zhang et al., 2004). In summer the cyclone tracks are diverse and no dominant northward pathway is evident. There is also a major increase in cyclone activity over land areas, but also a cyclone maximum over the central Arctic Ocean, which arises primarily from the migration of systems generated over Eurasia in conjunction with the weakened North Atlantic track.

Cyclogenesis and cyclolysis

An additional type of information derived from depression tracking studies is the locations of cyclogenesis and cyclolysis, or when a weather system is first and last identified, respectively. By collating data from all cyclones examined, regions of preferred cyclogenesis and cyclolysis may be determined.

The winter Icelandic Low and the region extending east into the Barents Sea is an area of frequent cyclogenesis and subsequent system intensification during winter. Development is influenced by the bifurcation of lows moving in from the south and southwest, by the high orographic barrier of Greenland, and the distortion of temperature and wind fields and lee-side vorticity production off the southeast coast of Greenland (Serreze and Barry, 2005). In addition, because of strong baroclinicity across the sea ice margin, the region is also one of rapid deepening of lows and redevelopment of extratropical systems migrating from the south (Serreze et al., 1997). Other regions of cyclogenesis include one to the north of Greenland associated with katabatic flow (Hoskins and Hodges, 2002), the lee of the Canadian Rockies and locally over the Arctic Ocean (Zhang et al., 2004).

The North Atlantic storm track is weaker in spring in terms of cyclone frequency and cyclogenesis, although it penetrates further into Eurasia (Serreze and Barry, 2005) (Fig. 3.33). In contrast, cyclogenesis increases over east-central Eurasia and northwestern Canada. In summer the North Atlantic track continues to weaken but in this season there is a seasonal maximum cyclogenesis over land areas; by autumn the North Atlantic track has become dominant once again.

Cyclolysis occurs on the northern side of the Pacific storm track and then on the west coast of North America and the Gulf of Alaska (Hoskins and Hodges, 2002). Other regions of frequent cyclolysis include an area from the Great Lakes to Hudson Bay and extending eastward to Iceland – especially the southern tip of Greenland – and northern Siberia.

3.5.2   The Antarctic

Surface pressure and upper height fields

The Antarctic continent is surrounded by the circumpolar trough (CPT), a belt of low pressure over the Southern Ocean (Fig. 3.34). To the north there is higher pressure in the subtropics while to the south reducing Antarctic surface pressures to sea level results in an apparent 'anticyclone'. Thus, climatologically Antarctica is ringed by strong meridional pressure gradients and the associated strong westerlies. However, the CPT is a region of high cyclonic activity with, typically, four to six synoptic-scale weather systems existent at any one time. Thus, as van Loon (1972) stated, the CPT is 'only a statistical mean . . . and isobars aligned west–east are the exception on a synoptic map'.

Within this broad description of southern high latitude pressure, climatological variability takes place at several recognised spatial and temporal frequencies. In terms of the zonal harmonic waves, which are quasi-stationary, wave number 1 (ZW1) and wave number 3 (ZW3) that are associated with the distribution of Southern Hemisphere land masses are the most important, contributing ~90% and 8% of the total variance, respectively (van Loon and Jenne, 1972). Variability of ZW1 is associated primarily with pressure anomalies in the South Pacific, a region described as a 'pole of variability' (Connolley, 1997). The influence of ZW3 is apparent through the climatological low-pressure centres in the three major ocean basins: in the Amundsen Sea (~140° W: the Amundsen Sea low), the eastern Weddell Sea (~20° E) and off Wilkes Land in East Antarctica (~110° E). ZW3 plays a major role in driving sea ice variability (Raphael, 2007). The favoured longitudinal position of blocking events in the CPT – persistent positive pressure anomalies associated with a split in the mean circumpolar westerly flow – have also been related to both ZW1 and ZW3 (Renwick, 2005). At interannual and decadal timescales ENSO plays a role in affecting the pressure pattern around Antarctica. This is particularly true in the South Pacific, through which the PSA pattern of alternate positive and negative pressure anomalies lies. Thus, ENSO variability has been found to influence both ZW1 and ZW3 (e.g. Raphael, 2003, 2004).

Immediately above Antarctica (500 hPa) there is a weak cyclonic vortex (ZW1), which, on average, is displaced towards the Ross Ice Shelf (Fig. 3.35). The vortex becomes stronger with altitude. However, ZW3 actually has its maximum at the tropopause in summer while in winter it lies above, in the lower stratosphere (van Loon and Jenne, 1972). The winter stratosphere, in particular, is dominated by the polar vortex, which has relatively small spatial variability compared with the Arctic: Randel and Wu (1999) described an off-pole maximum in October linked with the intensification of ZW1 and a 30° westward shift in the centre in late winter followed by an eastward return during spring.

3.5 Atmospheric circulation

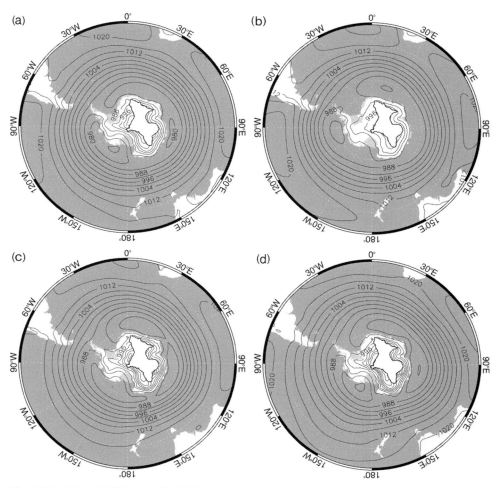

Fig. 3.34 Mean MSLP for the high southern latitudes for the four seasons derived for the period 1979–2009 from the ECMWF reanalysis and operational analyses. (a) Spring (September–November), (b) Summer (December–February), (c) Autumn (March–May), (d) Winter (June–August).

Although the first harmonic in pressure, the annual cycle, shows large interannual variations, these tend to cancel out in the long-term means (van Loon and Rogers, 1984), so that the second harmonic, the semi-annual oscillation (SAO) dominates the total intra-annual variance, explaining up to 80% of the total. The SAO, which forms in response to differences in heat storage between Antarctica and the surrounding oceans, causes a half-yearly wave in both the position and strength of the CPT. During the equinoctial seasons the CPT is deepest and also further south, the latter revealed by coastal Antarctic stations having pressure maxima (minima) in the solsticial (equinoctial) seasons (e.g. van den Broeke, 1998) with an amplitude of the SAO of 2–4 hPa. In response to the variation in the

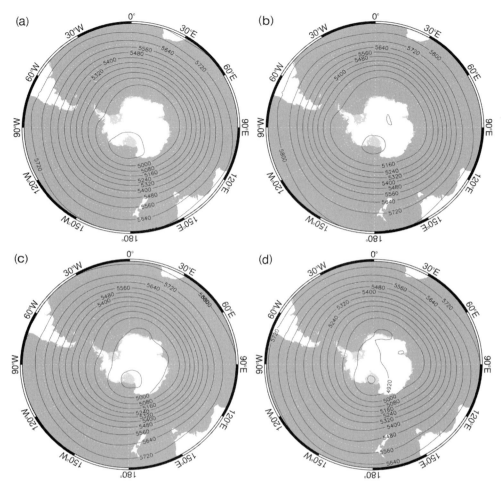

Fig. 3.35 Mean 500 hPa height for the high southern latitudes for the four seasons derived for the period 1979–2009 from the ECMWF reanalysis and operational analyses. (a) Spring (September–November), (b) Summer (December–February), (c) Autumn (March–May), (d) Winter (June–August).

meridional pressure gradient, the zonal westerlies show equinoctial maxima that are 20–30% stronger than those in summer and winter. To the north of ~60° S the phase of the SAO is reversed. Hence, in conjunction with pressure rises over the three mid latitude continents, the expansion of the CPT in autumn causes an amplification of the wave-3 structure of the circulation around Antarctica (Fig. 3.36).

In contrast, the annual cycle in surface pressure is the dominant harmonic on the Antarctic Plateau, with a clear winter minimum. Van den Broeke *et al.* (1997) suggested that this may be due to variations in the katabatic wind drainage. Katabatic winds result from the continuous cooling of air overlying the domed Antarctic continent. Negative buoyancy gradient accelerates the air downslope towards the coast with the topography and Coriolis

3.5 Atmospheric circulation

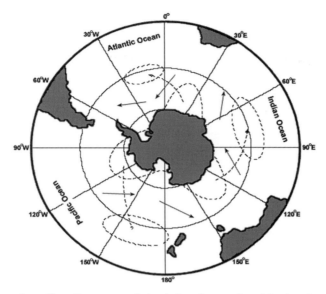

Fig. 3.36 Schematic outline of pressure and circulation changes from March to June. Dashed arrows connect locations of largest pressure changes and solid arrows indicate mean circulation changes. (After van den Broeke, 1998.)

force defining the exact path travelled. This results in a non-homogeneous drainage pattern that exhibits several regions of confluence near the coast (e.g. Parish and Bromwich, 1987; van Lipzig et al., 2004). As the katabatic winds are forced by radiational cooling at the surface, which is greatest during the darkness of the polar night, they are strongest and have the largest downslope component in midwinter. They thereby effectively remove air from the Antarctic Plateau in the winter months, leading to the minimum pressure in this season. The annual cycle in surface pressure over Antarctica implies a net transport of air into the continent in early winter and in spring, with a corresponding export of atmospheric mass, primarily via katabatic drainage, to lower latitudes in late winter and summer. At the coast the katabatic winds, which have very high directional consistency, are responsible for the maintenance of coastal polynyas (e.g. Bromwich and Kurtz, 1984) and also interact with weather systems over the Southern Ocean. However, Parish and Cassano (2003) use the lack of seasonal variation in the directional constancy, even in summer when the katabatic component is small, to postulate that the role of katabatic winds in the Antarctic boundary layer may be overemphasised. Their numerical simulations revealed that the adjustment process between the continental ice surface and the ambient pressure field may be the primary cause of the surface wind field over the Antarctic continent.

Depression tracks

An example of depression tracks at high southern latitudes derived from an automated scheme is shown in Fig. 3.37a. An alternative methodology is to utilise Eulerian storm-track

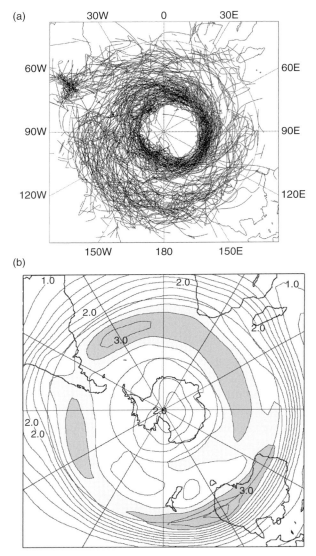

Fig. 3.37 (a) Depression tracks for winter 1985–89. (From Jones and Simmonds, 1993.) (b) Bandpass-filtered (2–6 day) variance converted to standard deviation for ξ_{250} for winter 1958–2002. (From Hoskins and Hodges, 2005.)

diagnostics by identifying the variance of vorticity at synoptic timescales (2–6 days): Fig. 3.37b is an equivalent figure to 3.37a using this method. Note the significant difference in results in the CPT with the system-centred approach finding a maximum while the Eulerian method indicates relatively little cyclone activity here. The satellite-imagery study of Turner et al. (1998) reveals that there is indeed a maximum in cyclone activity here.

There is a marked difference between the main CPT storm track in austral summer and winter. In the summer it is nearly circular and confined to high latitudes south of 50° S. In winter (Fig. 3.37a) the storm track is more asymmetric with a spiral from the Atlantic and Indian Oceans in towards Antarctica and a pronounced STJ-related storm track at ~35° S over the Pacific, spiralling towards southern South America (e.g. Hoskins and Hodges, 2005). A study by Simmonds and Keay (2000a) found that the mean track length of winter systems (2315 km) was slightly longer than in summer (1946 km). The equinoctial seasons have depression-track patterns intermediate between summer and winter. The CPT has a maximum in storm activity in the Atlantic and Indian Ocean regions at all times of year. While the previous description relates to the mean climatology, individual weather systems can behave differently; for example, many studies have shown cases where cyclones have a strong northward track, particularly when associated with cold-air outbreaks from the Antarctic continent. The automated procedures can also be used for locating and tracking anticyclones. In such a study, Sinclair (1996) found that south of 50° S anticyclones were generally rare but their tracks were associated with the main regions of blocking in the South Pacific as described previously.

Cyclogenesis and cyclolysis

Using data from the NCEP-NCAR reanalysis, Simmonds and Keay (2000a) showed there to be a net creation of cyclones (i.e. cyclogenesis > cyclolysis) north of 50° S and net destruction south of this. However, most Southern Hemisphere cyclogenesis actually occurs at very high latitudes (~60° S) in the CPT but rates of cyclolysis here are even higher. Turner *et al.* (1998) found that approximately half the systems in the Antarctic Peninsula region formed within the CPT and half spiralled in from further north, and work by Simmonds *et al.* (2003) also revealed the northern part of the peninsula to be a region of high cyclogenesis. Some of these systems formed through lee cyclogenesis – a dynamical process associated with the passage of air over a barrier – to the east of the Antarctic Peninsula. Another important region of cyclogenesis within the CPT is the Indian Ocean sector, with a maximum at 65° S, 150° E (Hoskins and Hodges, 2005) at the edge of the sea ice. Cyclolysis is generally confined to the CPT, with maxima in the Indian Ocean and also the Bellingshausen Sea, where the steep, high Antarctic Peninsula often prevents weather systems passing further to the east: such areas are known as 'cyclone graveyards'.

The detailed study of Hoskins and Hodges (2005) revealed that the main regional depression tracks around Antarctica can be envisaged as a series of overlapping plates, each composed of cyclone life cycles. Cyclolysis in one plate appeared to be important for cyclogenesis in the next plate further east through downstream development in the upper-troposphere spiral storm track (Fig 3.38). So for example, the cyclogenesis upstream of the Ross Sea feeds the track towards the Antarctic Peninsula, which, in combination with two more northerly cyclogenesis regions east of South America, feeds the storm track across the Atlantic sector.

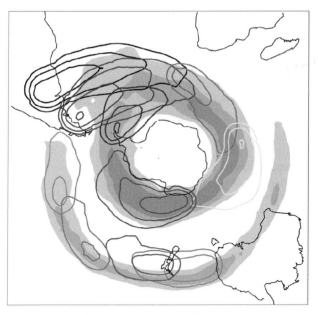

Fig. 3.38 Track density contours for the main cyclogenesis regions: contours are 0.5, 1.0, 2.0 and 4.0 tracks per 5° radius per month. (After Hoskins and Hodges, 2005.)

3.6 Temperature

3.6.1 Arctic

Surface

Although a significant part of the Arctic region is ocean, there are sufficient data from Russian North Pole drift stations and buoys, via the International Arctic Buoy Programme, for an accurate description of surface air temperature (SAT) across the entire region. Rigor *et al.* (2000) produced seasonal maps of Arctic temperature based on a blended data set of station and ocean data (Fig 3.39). These reveal the annual range in temperature is much greater at inland continental sites than those in maritime and coastal regions.

The stratospheric Arctic vortex influences the tropospheric circulation field, which controls the Arctic winter SAT pattern through changes in atmospheric temperature, humidity and cloudiness. These impact the regional downward longwave radiation and thus the thermal energy budget (Overland *et al.*, 2000). For example, the coldest winter temperatures are recorded in northeastern Siberia with mean values less than −40 °C in January, in association with a strong anticyclone. Over northern North America the polar vortex is modified by a strong trough–ridge system – related to dynamical forcing of the planetary waves by the orography of the Rocky Mountains – this means that Ellesmere Island experiences colder temperatures than the Chukchi Sea to the west of the Greenland Sea to

3.6 Temperature

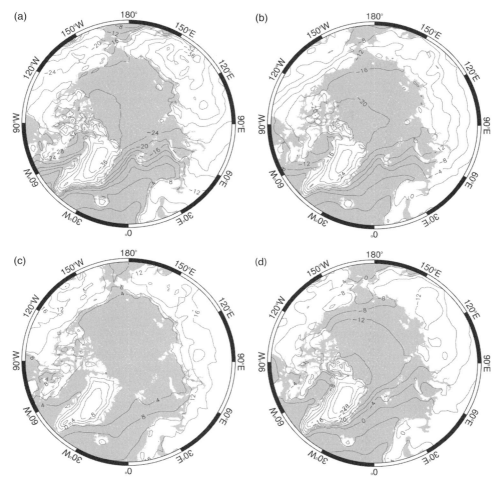

Fig. 3.39 Mid-season mean fields of surface air temperature (°C) derived from ECMWF interim reanalysis fields for 1990–2009. (a) January, (b) April, (c) July and (d) October.

the east (Overland *et al.*, 2000). In addition, the formation of areas of open water locally over the Arctic Ocean allows air–sea fluxes of sensible and latent heat into the boundary layer keeping SATs much higher. The open water results from strong near-surface winds associated with frequent storms, which are guided by the planetary waves (Serreze and Barry, 2005). In winter, over the land the radiation deficit at the surface is balanced by turbulent heat flux from the atmosphere whereas at the top of the boundary layer this flux divergence is zero (by definition). Hence the boundary layer cools relative to the surface producing a stably stratified temperature profile and surface inversion. Such strong low-level

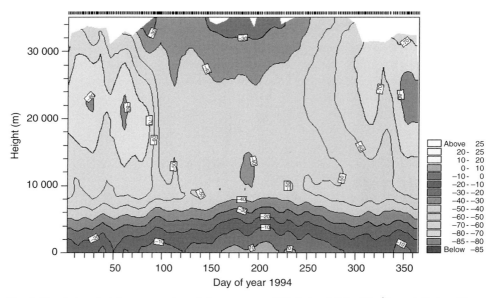

Fig. 3.40 A time–height cross-section of air temperature (°C) during 1994 at Ny Ålesund Station. (From www.awi.de/en/infrastructure/stations/awipev_arctic_research_base/observatories_and_facilities/atmospheric_observatory/meteorological_measurements/upper_air_soundings_at_koldewey/)

temperature inversions can extend to about 1200 m during January–March, with an inversion strength of ~11–12 °C (Serreze et al., 1992). Especially strong inversions are found inland in the deep intermontane basins and valleys of eastern Siberia and northwest Canada–Alaska: in these regions of high pressure surface winds are unable to break the inversions.

In spring the Arctic vortex weakens, allowing meridional temperature advection into the Arctic from lower latitudes. During spring and summer the lowest SATs are found over the Arctic Ocean and the Canadian Arctic Archipelago. In coastal regions temperatures increase to the melting point over the coasts by the end of May and most of the central Arctic by mid-June. Station data reveal an isothermal melt period, when the SAT reaches the ice melt point until all the snow and ice in a region have melted; for example at Indian Mountain in the Alaskan interior, the melt period begins in March but SATs do not become greater than 0 °C until the end of April (Rigor et al., 2000). July temperatures at the coast average about 5 °C with a rapid increase inland (Serreze and Barry, 2005). The freeze isotherm advances more slowly than the melt isotherm: the formation of sea ice begins at the pole in mid-August and it returns to the marginal seas a month later (Rigor et al., 2000). Thus, the melt season over the central pack ice is approximately 60 days long, compared with about 100 days over the marginal seas.

Upper air

The annual cycle of temperature in the middle troposphere shows a well-developed kernlose (or coreless) pattern (see data from Ny Ålesund Station: Fig. 3.40), similar to what is observed in Antarctica. Chase *et al.* (2002) proposed that winter Arctic mid-tropospheric temperatures are regulated by moist adiabatic ascent from air masses in regular contact with high latitude open ocean regions south of the pack ice: Tsukernik *et al.* (2004) identify the Norwegian Sea as a key region for such convective warming. Despite continued net radiative loss throughout the winter, temperatures across the Arctic seldom fall below −45 °C at 500 hPa, indicating that the system is in quasi-radiative equilibrium (Chase *et al.*, 2002).

In the Arctic winter stratosphere there are low temperatures in the vortex core of the lower stratosphere that increase outward from the axis of rotation. In contrast, in the upper stratosphere (pressure < 10 hPa) there are higher temperatures in the vortex core that decrease outward. This vertical change in temperature structure is attributed to warming through large-scale subsidence in the mesosphere during the polar night (Serreze and Barry, 2005).

3.6.2 *Antarctic*

Surface

Several factors combine to cause Antarctica to be the coldest continent. Antarctica is located at high latitudes, which means that it receives less solar radiation per year than lower latitudes and thus acts as a heat sink for the Southern Hemisphere. The high albedo of the snow and ice, which covers more than 99% of the continent, reflects 80–90% of incident solar radiation, hence enhancing the cold temperatures. The extreme dryness of the air and lack of cloud in the interior causes any heat that is radiated back into the atmosphere to be lost to space instead of being absorbed by water vapour and cloud droplets within the atmosphere. Furthermore, Antarctica is also the highest continent, with a median elevation of 2150 m a.s.l. At its highest the interior Antarctic Plateau exceeds 4000 m a.s.l. Its domed shape, with relatively steep gradients near the coast, prevents the frequent intrusion of warmer maritime air from coastal cyclones. The coldest temperatures on Earth are found during the polar night on top of the plateau, where no direct sunlight is received during winter months and where slopes are sufficiently shallow to prevent mixing through significant katabatic wind flow. The lowest measured temperature of −89.2 °C was recorded at Vostok station (78.5° S, 106.9° E, 3488 m a.s.l.) on 21 July 1983 (Turner *et al.*, 2009a) but it is estimated that temperatures on parts of the plateau may occasionally dip below −100 °C.

Radiosonde soundings reveal a strong temperature inversion above the Antarctic surface that is maintained by radiative cooling of the surface and lower atmosphere as described previously. Thus the inversion is strongest on the plateau during the polar night (winter), when there is no incoming solar radiation and surface cooling is at a maximum. However, significant inversions can occur at any time of year, even in coastal regions. When strong katabatic winds occur they mix the cold air vertically through a deep layer destroying the inversion: therefore near-surface temperatures co-vary strongly with wind speed.

Antarctica may be separated into three distinct zones based on the annual temperature cycle: examples from stations located in each of these are shown in Fig. 3.41. In the northern

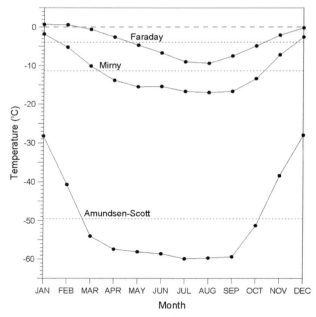

Fig. 3.41 The annual cycle of temperature at three Antarctic stations: Faraday on the western Antarctic Peninsula (65.2° S, 64.3° W); Mirny in coastal East Antarctica (66.6° S, 93.0° E) and Amundsen–Scott at the South Pole. Dashed lines are plotted at the annual mean. (From *Encyclopedia of the Antarctic*. Reproduced courtesy of Routledge.)

part of the Antarctic Peninsula (Faraday), monthly mean temperatures exceed zero in summer, and here the annual cycle of temperature has a form found across many regions of the Earth, with a broad maximum in summer and a slightly shorter winter minimum. However, on the Antarctic Plateau (Amundsen–Scott Station) the form of the cycle changes completely. As well as temperatures being much colder, summer is now very short (December/January) with rapid changes in temperature defining the ends of a long or coreless winter period that encompasses several months with broadly similar mean temperatures. This rapid cooling and warming is due to the abrupt removal and return of solar radiation south of the Antarctic Circle at the beginning and end of the polar night. The annual temperature cycle at Mirny, representative of East Antarctic coastal stations, has a profile in between the other two regions mentioned: while monthly temperatures are much closer to those on the peninsula, the winter period is much longer, being more similar to the polar plateau.

The significant modification of the seasonal cooling that produces the coreless winter is thus observed across much of Antarctica. It is a consequence of the semi-annual oscillation (SAO), a twice-yearly expansion and contraction of the circumpolar low-pressure belt around Antarctica (Section 3.5.2). The magnitude of temperature modification by the SAO appears largely determined by the strength of the winter surface temperature inversion.

3.6 Temperature

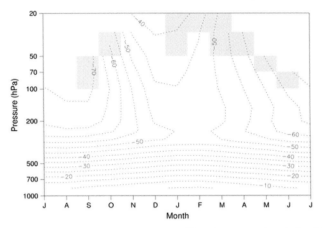

Fig. 3.42 Annual cycle of upper-air temperatures at Bellingshausen station derived from radiosondes (1969–99). Note missing data at high altitudes during winter; grey areas are based on less than 70% of years.

Hence the impact is largest on the plateau, followed by coastal East Antarctica and then peninsula stations: this can be observed in Fig. 3.41 with regard to the duration of the coreless winter at the three stations. Heat stored within the winter snowpack also dampens any temperature fluctuations. In coastal East Antarctica the coupling between the SAO and near-surface temperature can be explained by seasonal changes in the meridional circulation brought about by the former. Its influence on winter temperatures on the plateau is less well understood although stronger katabatic wind drainage during this season may be important.

Antarctic temperatures are also influenced by climate variability in other parts of the Earth as evidenced through so-called teleconnections (see Section 3.4.5). The principal mode of variability in the Southern Hemisphere is the SAM, which comprises zonally symmetric synchronous pressure anomalies of opposite sign in Antarctica and the mid latitudes. Statistical studies suggest SAM variability impacts temperatures across much of Antarctica, although not necessarily in the same direction (see later). Also important is the PSA pattern: this reflects the phase of tropical ENSO variability and primarily affects Antarctic Peninsula winter temperatures.

Upper air

Above the inversion, temperatures in the Antarctic troposphere follow a broadly similar seasonal pattern to that observed at the surface at East Antarctic coastal sites. The range between maximum and minimum monthly mean temperatures is ~10 °C at 500 hPa (~5 km a.s.l.) across the whole continent, as illustrated by data obtained from Bellingshausen radiosonde flights (Fig. 3.42). The circumpolar vortex isolates most of the continent from the transport of warm air from lower latitudes until late November, when the marked meridional temperature gradient is reduced by increasing solar radiation.

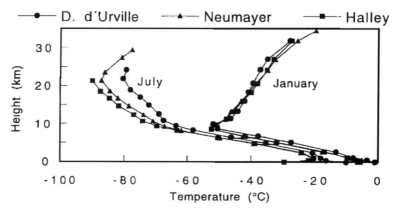

Fig. 3.43 Mean monthly profiles of January and July temperature at three coastal East Antarctic stations. (From König-Langlo et al., 1998.)

The tropopause marks the boundary between the troposphere and the stratosphere, defined by a minimum in the lapse rate. During summer, the tropopause is well pronounced above Antarctica (see Fig. 3.43), coinciding with the temperature minimum at an altitude of ~9 km. During the polar night (winter) the negative radiation balance and isolating effect of the polar vortex results in the marked seasonal cooling of the stratosphere. The tropopause becomes somewhat vague and can disappear completely with no distinct layer of stable thermal stratification between the troposphere and stratosphere, which allows a stratosphere–troposphere exchange of air masses (König-Langlo et al., 1998).

Stratospheric temperatures below −80 °C are typical for conditions inside the vortex while temperatures outside are some 10 °C higher. In the Antarctic, interannual variability in stratospheric temperatures is generally much weaker than in the Arctic and is strongest during spring due to variations in the timing of the Antarctic vortex break-up. For example, an anomalously warm spring season was observed in 2002, in conjunction with the split vortex (Fig. 3.44).

3.7 Cloud and precipitation

A knowledge of cloud and precipitation distribution across the polar regions and the mechanisms responsible for changes in these quantities is essential for our understanding of ice sheet mass balance. Snowfall onto the surface is the means by which ice sheets grow and many climate models predict that there will be greater snowfall across the Antarctic and Greenland Ice Sheets over the next century, so mitigating to some extent the expected rise in sea level. But many factors influence the distribution and variability of cloud and precipitation, such as changes in storm tracks, sea ice extent, availability of cloud condensation nuclei and the seasonal temperatures.

3.7 Cloud and precipitation

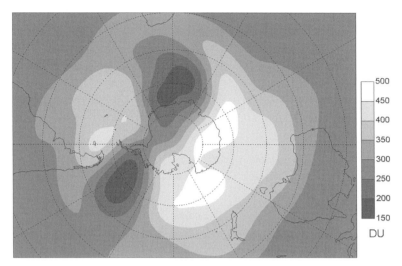

Fig. 3.44 Satellite-derived total ozone values (Dobson units, DU) on 25 September 2002 showing the split vortex.

In this section we consider our understanding of the mechanisms responsible for the distribution and variability of polar cloud and precipitation and present mean fields for these quantities.

3.7.1 Cloud

Clouds are important in the polar regions through their role in the radiation balance. With their high albedo they reflect a high percentage of incoming solar radiation back to space and reduce the radiation absorbed by the surface. Their impact is particularly apparent over the ice-free ocean areas and the snow-free tundra, which have a low albedo. Here surface temperatures are lower when cloud is present (see Section 3.3.1).

Although polar cloud always has a fairly high albedo, the exact value changes with the age of the cloud and its microphysical properties, but typical values are in the range 80–90%. Satellite imagery of the polar regions demonstrates that cloud can have an albedo that is higher or lower than that of a snow or ice covered surface, which can make identification of the cloud difficult in satellite imagery.

As well as reducing the shortwave radiation received at the surface, clouds also increase the downward component of longwave radiation since they are effective radiators. Over high albedo surfaces, such as sea ice and snow surfaces, the gain of downward longwave radiation outweighs the loss of incoming solar radiation, so that the presence of clouds has the net effect of warming the surface. However, cloud has less of an impact on the radiation balance at the top of the atmosphere over snow and ice surfaces since the cloud and surface have similar reflectivities.

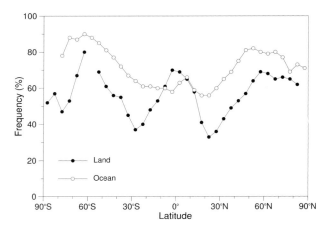

Fig. 3.45 The zonal mean annual cloud percentage from land and ship observations over 1971–96. (Data from *Climatic Atlas of Clouds Over Land and Ocean* by S.G.L Warren and C.J. Hahn. See 'Introduction' at www.atmos.washington.edu/CloudMap/)

Distribution

Figure 3.45 shows the zonal mean annual cloud percentage from land and ship observations. The broadscale distribution of cloud is characterised by a peak close to the Equator, minima at the latitudes of the subtropical high-pressure belts and then increases in cloud amount poleward into mid latitudes. The most cloudy latitude belt is close to 60° S, where there is extensive depression activity and a plentiful supply of moisture from the Southern Ocean. Poleward of this latitude the cloud amounts decrease very rapidly, with there being a rapid decrease from the coast of the Antarctic up onto the plateau of the continent. At the South Pole there is just over 50% cloud cover over the year as a whole. In the Northern Hemisphere the mean cloud fraction decreases poleward of the main storm track at 40–50° N, to a minimum at the North Pole of 70%. There is more cloud in the Arctic than the Antarctic because of the extensive open water during the summer when there is little sea ice.

The zonal cross sections of mean cloud fraction for the summer and winter seasons are indicated in Fig. 3.46. In the Southern Hemisphere there is little difference in cloud amount between the two seasons close to the main storm track over the Southern Ocean. However, the observations from the Antarctic staffed stations indicate that there is most cloud during the summer (60–70%), with a drop to less than 50% during the winter. Temperatures are very low during the austral winter and the air can hold very little moisture; however, as discussed in Chapter 2, it is very difficult for an observer to see clouds in the darkness and care should be exercised in taking these figures too literally.

In the high latitude areas of the Northern Hemisphere there is a much greater difference between the amounts of cloud in the summer and winter. During the winter the extensive sea ice severely limits the flux of moisture from the ocean, although depressions that penetrate

3.7 Cloud and precipitation

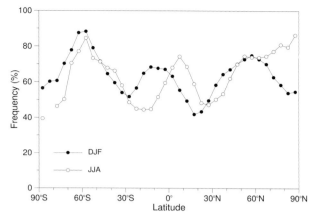

Fig. 3.46 The zonal mean summer and winter cloud percentages from land and ship observations over 1971–96. (From *Climatic Atlas of Clouds Over Land and Ocean* by S.G. Warren and C.J. Hahn. See http://www.atmos.washington.edu/CloudMap)

into the Arctic Basin can bring significant cloud. The mean cloud fraction in the winter is therefore a little over 50%. However, in the summer this rises to almost 90% as the numerous leads allow a greater flux of moisture into the atmosphere, while passage of the air over the remaining sea ice can result in rapid cooling of the air and the formation of cloud.

Cloud formation, type and properties

In the Northern Hemisphere the land–sea distribution is such that most depressions enter the high latitude areas via the Norwegian Sea and the Bering Strait and bring extensive cloud into these areas. The depressions can bring some middle and high level cloud into the Arctic Basin, but they also have extensive low cloud associated with them. Figure 3.47 illustrates the zonal mean annual percentage cover of high, middle and low cloud as derived from land and ship observations. It shows the preponderance of low and middle cloud in the Arctic and the lack of high cloud. Of course care needs to be taken when interpreting such figures, since extensive low cloud will obscure high cloud, which will therefore be reported less frequently by observers at the surface.

The mid latitude maxima of cloud amount shown in Fig. 3.45 are associated with the main mid latitude storm tracks. In the Southern Hemisphere many depressions follow a track towards the southeast spiralling in towards the coast of the continent and then become slow-moving or track towards the east at the latitude of the circumpolar trough (Section 3.5.2). In addition, many synoptic and mesoscale depressions actually develop within the trough and much of the cloud is associated with depressions. As relatively warm mid latitude air masses move south they cross the increasingly cold waters of the Southern Ocean, and in many areas, the sea ice. This cools the air masses and is responsible for much of the cloud that is found around the Antarctic coastal region.

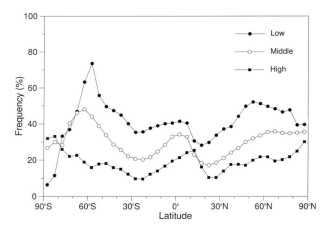

Fig. 3.47 The zonal mean annual percentage cover of high, middle and low cloud from land and ship observations over 1971–96. (From *Climatic Atlas of Clouds Over Land and Ocean* by S.G. Warren and C.J. Hahn. See http://www.atmos.washington.edu/CloudMap)

One of the most striking features in Fig. 3.47 is the very rapid decrease in the amount of low cloud over 65–70° S. This corresponds to the transition from the coastal region of the Antarctic up onto the high plateau of the continent. Much of the interior of the Antarctic is at elevations of 2–4 km and a lot of the cloud there is still fairly near the surface. However, because of its height above mean sea level it is reported as medium level.

The polar regions are generally characterised by a lack of deep convection. This is especially the case in the Antarctic where cumulonimbus has not been reported at the research stations. However, satellite imagery does reveal extensive shallow cumulus cloud over the polar oceans in both hemispheres, especially when there are outbreaks of cold air over the open ocean. One scenario in which cumulonimbus cloud is found in the Arctic is in association with polar lows. These active mesoscale depressions are found in certain high latitude areas of the Arctic where very cold air masses flow over relatively warm waters that have been advected northwards. The deep instability results in cumulus and cumulonimbus clouds. Such systems are generally limited to the winter season and they occur in the Norwegian and Barents Seas, the Labrador Sea and occasionally off the northern coasts of Canada, Alaska, Russia and Japan.

Research is starting to be carried out on the microphysical properties of polar cloud, but our knowledge of cloud at high latitudes is still far behind comparable studies in mid latitudes (Lachlan-Cope, 2010). This is largely because of the difficulties of using research aircraft in the polar regions, especially during the winter months. However, some information on polar clouds has been gained from multi-spectral satellite data and surface-based instruments, such as Lidars. The 3.7 μm channel on the AVHRR satellite imager has revealed the extent to which supercooled water droplets are present in Antarctic clouds and the prevalence of mixed phase clouds. Such imagery during the periods when solar radiation is present is particularly valuable for detecting the phase of clouds, along with details of the droplet size. The lidar data collected in the Antarctic have provided information on the height of the cloud base and cloud top, along with the ice crystal form and size distribution. However, such records are relatively short and can only provide a snapshot of conditions.

One type of cloud that has proved to be of particular importance is polar stratospheric cloud (PSC). As their name implies, these clouds form at levels of about 15 000–25 000 m in the stratosphere, which is usually very dry and devoid of clouds. However, during the polar winter temperatures can drop low enough to allow clouds to form at this level. The clouds form when temperatures fall below −78 °C and the clouds are usually observed just after sunset, when they are characterised by their colourful appearance and hence their alternative name of Mother of Pearl clouds. PSCs have been found to play a critical role in the springtime loss of stratospheric ozone because of the chemical reactions that take place on the surface of the ice crystals (see Section 3.3.1).

3.7.2 Precipitation

Precipitation is defined as the fall-out of water onto the Earth in the form of rain, snow, hail or sleet. In the tropics and mid latitude areas most precipitation comes from fairly deep layers of cloud associated with depressions or convective clouds. In the polar regions depressions are certainly responsible for a large percentage of the precipitation, but there is also a gradual fall-out of ice crystals from an apparently cloud-free sky, which is known as clear sky precipitation or diamond dust. This is discussed in detail below.

The forms of high latitude precipitation

There are many frontal depressions in the Antarctic coastal zone at the latitude of the circumpolar trough. These bring much of the precipitation to the coastal region and the immediate inland area. However, because of the steep orography at the edge of the continent, few depressions penetrate into the interior.

More active depressions reach high latitudes in the Arctic than the Antarctic as a result of the storm tracks that pass through the Norwegian Sea and to a lesser extent the Bering Strait. Frontal precipitation can be a significant contributor to the annual accumulation in these sectors of the Arctic.

The polar regions are generally characterised by precipitation from layer cloud; however, some convective precipitation does occur. As discussed earlier, deep convective clouds can form in certain sectors of the Arctic where cold air masses flow over relatively warm ocean areas. Here polar lows can develop during winter and bring heavy snow showers to areas such as the Norwegian Sea, the Bering Strait and along the north coast of Norway. Such situations probably give the heaviest precipitation that occurs in the polar regions.

On the high plateau of Antarctica and in many parts of the Arctic a feature of the climate is diamond dust. This is a gradual fall-out of ice crystals from thin cloud or an apparently clear sky. On the Antarctic plateau this is an almost daily occurrence (Rusin, 1961) and is a feature of all seasons. There has not been a great deal of research into the formation of clear sky precipitation. However, Bromwich (1988) suggested that at levels above 3000 m in the Antarctic it was a result of radiative cooling that maintained saturation of the air and which resulted in the formation of ice crystals. The accumulation rate for clear sky precipitation is very low and has been estimated by Sato *et al.* (1981) for the summer season at South Pole as 0.1 mm per hour.

The polar regions are often assumed to be characterised by extensive snowfall. However, the low temperatures limit the amount of moisture that the air can hold so total precipitation amounts tend to be much less than at lower latitudes. As discussed above, the largest amounts of precipitation are associated with active synoptic-scale weather systems and polar lows.

While snow is certainly the most common form of precipitation at high latitudes, rain is a feature of many parts of the Arctic and even some areas of the Antarctic. The western side of the Antarctic Peninsula is particularly prone to rainfall, especially in summer, as temperatures here frequently rise above freezing. However, during periods of strong northerly flow rain can even occur in the middle of winter as far south as Rothera Station (67.6° S). The more meridional flow in the Northern Hemisphere allows warm air masses to penetrate well into the Arctic Basin throughout the year. Areas such as northern Norway have highly variable temperatures in winter with a mix of snow and rain.

Precipitation distribution

In light of the difficulties in measuring precipitation at high latitudes, as discussed in Chapter 2, here we present a number of estimates of this quantity based on a range of tools and data sets. First we show the mean annual total precipitation for the globe as determined from the ECMWF 40-year reanalysis project (Fig. 3.48). This reveals that the highest precipitation totals are found close to the Equator and in the South Pacific Convergence Zone that extends from Indonesia in a southeasterly direction into the South Pacific. In the North Pacific and North Atlantic there are precipitation maxima along the main storm tracks that extend towards the northeast across the ocean basins.

The lowest precipitation amounts at high latitudes are on the plateau of Antarctica and in the central area of Greenland. Here the reanalysis has average precipitation rates of less than 0.2 mm per day (or 73 mm yr^{-1}). Much of the Arctic Ocean and the coastal region of the Antarctic has annual precipitation in the range 73–365 mm yr^{-1} with greater amounts in the Norwegian/Barents Sea and along the southern coast of the Bellingshausen Sea. Here annual accumulations are over 1 m yr^{-1}.

A more detailed view of precipitation across the Arctic is shown in Fig. 3.49, which provides estimates of annual precipitation from a new climatology produced by Serreze and Hurst (2000). This field was produced by taking the gridded data of Legates and Willmott (1990) and adding the North Pole precipitation data and gauge-corrected observations from

3.7 *Cloud and precipitation* 139

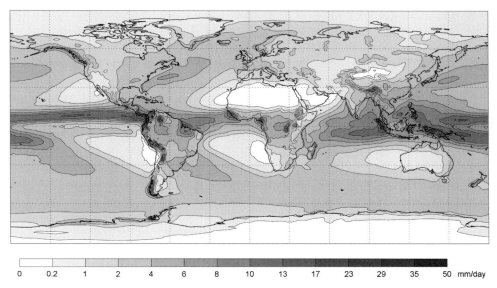

Fig. 3.48   Annual mean precipitation (mm per day). (From the ECMWF reanalysis.)

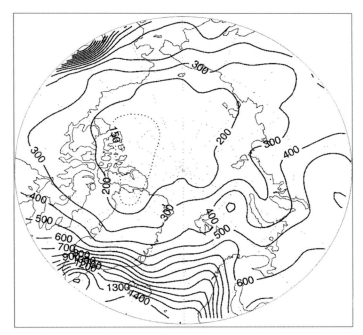

Fig. 3.49   Annual mean precipitation (mm) from an analysis produced by Serreze and Hurst (2000). The contour interval is 100 mm except for the dotted contour, which encloses areas with precipitation totals less than 150 mm.

land stations. The new analysis is in broad agreement with the ERA-40 field in having less than 300 mm yr^{-1} of precipitation across much of the Arctic ocean. However, the data set provides greater detail of the spatial distribution. The data suggest that the lowest Arctic precipitation (less than 150 mm yr^{-1}) is found along the coast of the Canadian Arctic. Much of the central Arctic receives 150–200 mm yr^{-1}, while the largest totals are in the northern North Atlantic. There is a strong north–south gradient of annual precipitation between Iceland and Svalband with annual totals decreasing from around 1400 to less than 400 mm yr^{-1} across this distance. A comparison of Figs. 3.48 and 3.49 suggests that ERA-40 has slightly larger annual precipitation totals in this area than the Serreze and Hurst (2000) analysis.

Figure 3.50 shows the January and July patterns of precipitation from the same analysis. In January precipitation varies from a maximum of 180 mm southeast of Greenland to around 20 mm over the Canadian Arctic archipelago, the central Arctic Ocean and east-central Eurasia. In July the maxima over the Atlantic and Alaska are not present. At this time of year the precipitation is more uniformly distributed across the Arctic, but there is a noticeable increase in precipitation over the land compared with January. Serreze and Hurst (2000) suggest that this is a result of the greater level of cyclogenesis over land on the Arctic front, which is north of the polar front.

In Antarctica most of the research stations are distributed around the edge of the continent so it is not possible to produce a map of precipitation distribution of the type that has been prepared for the Arctic. Instead the analyses tend to be of net surface mass balance since that is the quantity determined from analyses of ice cores and snow pit data. A study by van de Berg *et al.* (2006) has compared the output of the RACMO2 regional model against the available in-situ estimates of mass balance, and the distribution of mass balance across the Antarctic from their model data is shown in Fig. 3.51. The largest annual accumulation amounts are found on the western side of the Antarctic Peninsula (Fig. 3.52) where annual totals in excess of 3000 mm yr^{-1} water equivalent are thought to occur: the burial of AWSs on masts 8 m high under a year's worth of snow suggests the model estimates are reasonable. With the forced ascent of air masses up the steep orography of the peninsula much of the moisture is lost as snowfall and on the eastern side of the barrier there is a snow shadow, with the low-lying Ronne Ice Shelf receiving less than 500 mm accumulation a year.

Beyond the Antarctic Peninsula the largest amounts of accumulation are found just inland of the coast of West Antarctica. Here annual totals are in excess of 1000 mm. The RACMO2 model indicates that a number of small areas along the coast of East Antarctica also have more than 1000 mm of accumulation a year.

The most striking feature of the spatial distribution of Antarctic mass balance is the rapid drop of accumulation totals into the interior of the continent. Inland of the coast the totals soon drop to around 200 mm yr^{-1} and in the high interior values of less than 50 mm yr^{-1} are found across an extensive area.

3.7 Cloud and precipitation

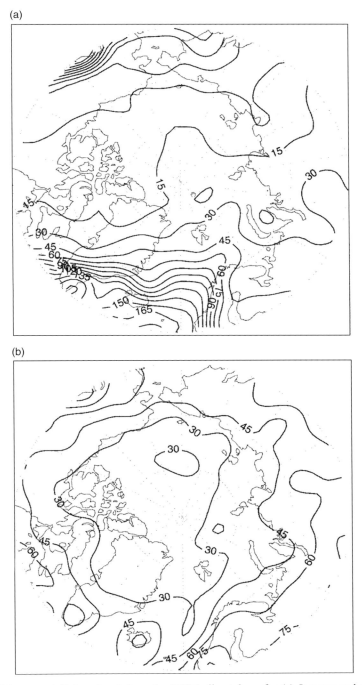

Fig. 3.50 Mean precipitation (mm) from the LWM climatology for (a) January and (b) July. The contour interval is 15 mm.

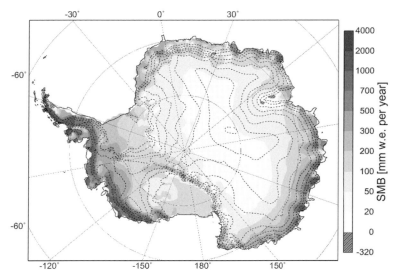

Fig. 3.51 Antarctic surface mass balance from the RACMO model. (From van de Berg *et al.*, 2006.)

3.8 Sea ice

As discussed in earlier sections, sea ice is extremely important in the climates of the polar regions through the part it plays in insulating the atmosphere from the huge source of heat in the ocean, its role in the formation of bottom water and the part it plays in feedback and amplification processes. In this section we examine the form, climatological distribution and motion of sea ice in the Arctic and Antarctic. This is largely based on the period since the early 1970s when the first reliable maps of sea ice distribution and concentration from satellite measurements could be prepared. In later chapters we consider how sea ice distribution has changed in the past and how it is expected to evolve over the next century.

3.8.1 The nature of sea ice

Sea ice forms from sea water when it freezes at subzero temperatures. The exact freezing point depends on the salinity of the water, but is typically around −1.9 °C. There are numerous categories of sea ice, but the broad stages of evolution are:

New ice. Consisting of recently formed ice made up of ice crystals that are only weakly frozen together.

Nilas. A thin crust of ice on the surface of the ocean that is up to 10 cm thick. It bends easily due to the action of waves and swell and under pressure grows in a pattern of interlocking fingers.

Young ice. Ice in transition from nilas to first year ice. It is typically 10–30 cm in thickness.

3.8 Sea ice

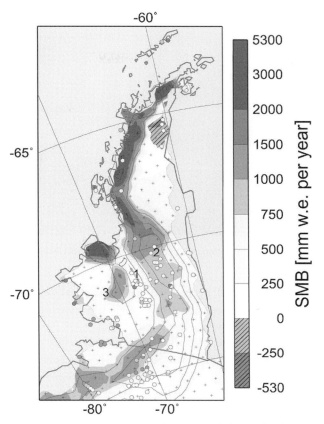

Fig. 3.52 Model-simulated and observed surface mass balance in the Antarctic Peninsula. Observations are indicated by circles. A light grey is used for model sea grid points. The crosses mark land grid points. Model elevation is drawn in thin black lines with 500 m interval. The thick grey line indicates the actual edges of land and ice shelves and also the southern border of the peninsula. The numbers mark locations in the peninsula: 1, King George VI Sound; 2, Dyer Plateau; 3, Alexander Island. (From van de Berg *et al.*, 2006.)

First-year ice. Ice developing from young ice that is not more than one winter's growth. It has a thickness of 30 cm or more.

Multi-year ice. Ice that has survived at least one summer's melt. It is generally smoother on its upper surface than first-year ice.

Observations of sea ice show that it is usually composed of a complex mixture of different ice types, consisting of ice floes (Fig. 3.53) varying in horizontal extent from a few metres, up to several kilometres or more. The varying mechanical and thermal forcing on the ice means that the pack is continually varying, even during winter.

The ice floes are separated by ice-free areas called leads, which often have a linear form. The passage of weather systems and the changing divergent and convergent nature of the

Fig. 3.53 Ice floes observed from a ship. (From Chris Gilbert, British Antarctic Survey.)

atmospheric and oceanic flow result in the opening and closing of the leads. Sea ice attached to the coast is called fast ice, and under conditions when the wind is offshore, the ice can be moved away from the coast creating a coastal lead.

The presence of ice-free leads within the pack means that the ice concentration can vary considerably. Even during the middle of winter the ice concentration is rarely 100% over a large area. Zwally *et al.* (1983a) found that for the Antarctic in September only half the pack ice had a concentration over 85%, with the remainder having concentrations between 15% and 85%.

Convergent forcing on the ice may also result in the creation of pressure ridges, which are raised areas between floes. Mechanical forcing may also lead to floes being forced over one another, a process known as rafting. Larger areas of open water, known as polynyas, may also be created within the pack. Some areas where polynyas develop frequently are discussed in the following section.

An extremely important quantity is the total thickness of sea ice. This consists of the part that is below the sea surface (the draft) and the portion above the water (the free board). The combined thickness can be measured accurately by drilling cores through the ice during ocean cruises. However, such work is very time consuming and there are limits on the number of cores that can be collected. Some ice thickness data can be collected using upward looking sonar (ULS), either from a submarine or with the ULS fixed to the ocean floor. Satellite instruments are starting to be used to estimate sea ice thickness from space, but there are a number of difficulties in achieving this. Space-based microwave and laser radar altimeters are being used to measure the free board of sea ice, but this can often be only a few tens of centimetres in height, which is close to the limit of accuracy of the instruments. Nevertheless, instruments with improved accuracy have great potential for sea ice thickness monitoring in the future, especially in the Arctic where ice thicknesses are generally larger.

3.8 Sea ice 145

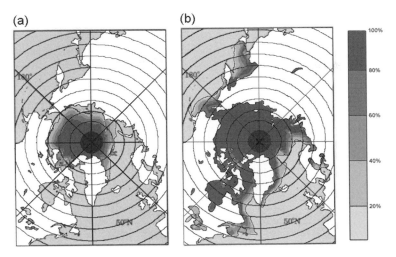

Fig. 3.54   Mean Arctic sea ice extent and concentration (1979–96) in (a) September and (b) March. (Adapted from Parkinson *et al.*, 1999.)

3.8.2   Sea ice motion

Satellite data have revealed a great deal about the complexity of sea ice motion and some general comments can be made about the motion field. Sea ice is advected by the surface wind and the ocean currents, although the wind forcing tends to be dominant, with the ice moving at about 2% of the wind speed.

Sea ice responds to the wind forcing as individual weather systems cross the sea ice zone. In sequences of satellite images the sea ice motion can often be seen as a poleward–equatorward oscillation as the storms move eastwards. Depressions are associated with low-level convergence that closes leads, while a corresponding divergence and reduction in sea ice concentration is associated with anticyclones.

3.8.3   Climatological occurrence

The Arctic

The distribution of Arctic sea ice is inextricably linked to the geography of the high latitude areas of the Northern Hemisphere, as much of the Arctic Ocean is bounded by the continents of North America, Europe and Asia. The constraints imposed by the surrounding land limit the drift of ice and also influence the distribution of ice types and thicknesses.

The Arctic Ocean is characterised by extensive multi-year or perennial sea ice, which survives the summer melt season. This can be seen in Fig. 3.54a, which shows the mean sea ice extent and concentration in September, close to the time of minimum ice extent. In this figure an area is considered to be ice covered if the concentration is at least 15%. At this time of year the sea ice zone covers about 7 million km^2 or about 60% of the area that it had at the

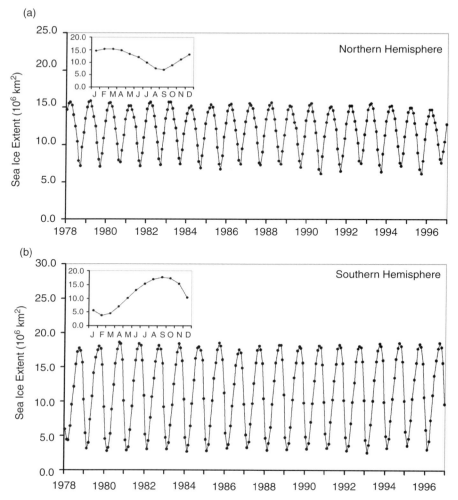

Fig. 3.55 The annual cycles of Arctic (a) and Antarctic (b) sea ice extent for 1978–98. Units are 10^6 km^2. (From Zwally et al., 2002a.)

winter maximum in March (Fig. 3.54b). At this time of year there is still extensive ice across the central part and western side of the Arctic Ocean, but with ice-free conditions north of Norway and western Russia. These result from the northward flowing Norwegian Current carrying relatively warm water up the west coast of Norway into the Barents Sea. In a similar way, there is little ice in the northern part of the Davis Strait and over the eastern Sea of Okhotsk because of the northward flowing warm currents of the West Greenland Current and the West Kamchatka Current respectively.

The annual cycle of Northern Hemisphere sea ice extent is shown as the inset in Fig. 3.55a. From the sea ice minimum in September the ice extent increases rapidly through

3.8 Sea ice 147

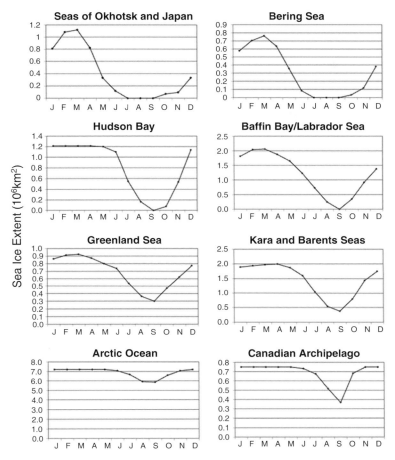

Fig. 3.56 The annual cycle of ice in various areas of the Arctic for 1979–96. (From Parkinson, 1999.)

the autumn and winter period until December, after which the growth slows. The ice reaches a maximum in March when it covers an area of 15.4 million km^2. The March to September period is characterised by ice melt, with the maximum decay occurring from June to August.

However, the annual cycle varies considerably across the Arctic (Fig. 3.56). For example, the Seas of Okhotsk and Japan and the Bering Sea have no sea ice cover during the summer months, but extensive ice during the winter. Moreover, in an individual year, sea ice extent can often have an out-of-phase relationship in these two regions, which is at least in part due to changes in the location and intensity of the Aleutian Low.

Hudson Bay is almost fully surrounded by land and its location means that the continental climate results in it being ice covered for several months of the year. In most areas of the bay the ice lasts from December until May, and in some years until June. Ice in Baffin Bay and

148   *The high latitude climates and change mechanisms*

the Labrador Sea has a similar occurrence to Hudson Bay, but starts to melt earlier in the summer. Several sectors of the Arctic retain ice throughout the year, but the Arctic Ocean and the Canadian Archipelago have the smallest annual cycle of sea ice extent.

The maxima and minima of sea ice extent in both polar regions lag the minima and maxima of insolation respectively by about 3 months. This is because of the large thermal inertia of the ocean.

There is a large interannual variability of Northern Hemisphere sea ice, as can be seen from the monthly extent data presented in Fig. 3.55a, where the summer minimum and winter maximum can differ markedly from the average. Parkinson *et al.* (1999) found that for the Northern Hemisphere ice extent, the summer minima was more variable than the winter maximum. The greatest percentage interannual variability is found in the Kara and Barents Seas. In contrast, Hudson Bay has a small interannual variability as it is always fully ice covered in winter, although there is greater variability in summer and autumn. Baffin Bay and the Labrador Sea are never fully ice covered and the region also has an opening to the North Atlantic, so the area has a greater interannual variability of winter ice cover. The ice extent in this area exhibits a cyclical nature (Parkinson, 1995) and the large amounts of ice in 1983 and 1984 have been linked to the major El Niño event in that year (Mysak *et al.*, 1996). Sea ice in the Greenland Sea is characterised by high interannual variability, both in the maximum and minimum. This is a result of the high variability of storm activity in the area, particularly in the region of the Odden, a large tongue of sea ice that is occasionally observed during the winter in the east Greenland Sea (Fig. 3.57). It has a length of about 1300 km and can cover an area as large as 330 000 km^2. It is a highly variable feature (Shuchman *et al.*, 1998) with large changes in size, shape and length being observed in the 20 years of data examined by Comiso *et al.* (2001).

In recent years the Arctic Ocean has been fully ice covered in at least one month of each year, and often for at least 3 months, which is usually February to April. There is usually near full ice coverage from December to April, with ice growth in the autumn being more rapid than the spring decay.

In areas such as the Arctic Ocean and the Canadian Archipelago, which are mostly ice covered throughout the year, the ice-free area within the pack peaks in summer, when the ice concentrations are lowest, and has its lowest values in winter, when the ice concentrations are high. But in areas such as the Bering Sea and the Seas of Japan and Okhotsk, where most of the ice melts in summer, the seasonal cycle is similar to the cycles of ice extent and area in terms of peaking in winter and decreasing to zero in summer.

Figure 3.58 shows the mean sea ice drift and surface water circulation of the Arctic. The most prominent feature is the Beaufort Gyre, which is an anticyclonic circulation of the sea ice in the western part of the Arctic Ocean. The second major feature is the Transpolar Drift, which crosses the Arctic Ocean from north of Siberia, exiting through the Fram Strait. In fact the ice transport out of the Fram Strait accounts for 90% of the ice leaving the Arctic Ocean. The ice drift then continues south into the East Greenland Current, which transports ice and cold, Arctic water down the east coast of Greenland.

3.8 Sea ice

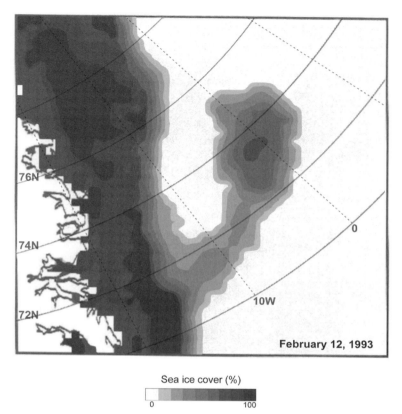

Fig. 3.57 Sea ice concentrations within the Odden ice tongue on 12 February 1993.

Although we have a good record of sea ice extent and concentration dating back to the 1970s, there is much less information on sea ice thickness, since this quantity cannot yet be routinely derived from satellite data, although it is hoped that this can be achieved with future missions. Our knowledge of sea ice thickness is therefore based on in-situ measurements derived from cores or taken via ice profiling sonar (ULS) on submarines or fixed to the floor of the ocean.

The Scientific Ice Expeditions (SCICEX) programme of the 1990s provided an opportunity to use US submarines for Arctic research and the ULS data from three cruises gave a perennial ice cover with mean drafts of between 1 m and 3 m. In the Arctic the region with the thickest sea ice is found north of Canada, where the ice drift towards the land results in extensive ridging of the pack. In the peripheral seas there is greater divergence and consequently more open water, resulting in more thin ice.

There are too few observational data to allow us to produce a reliable annual cycle of ice thickness from direct measurements, but Rothrock *et al.* (1999) computed this as part of

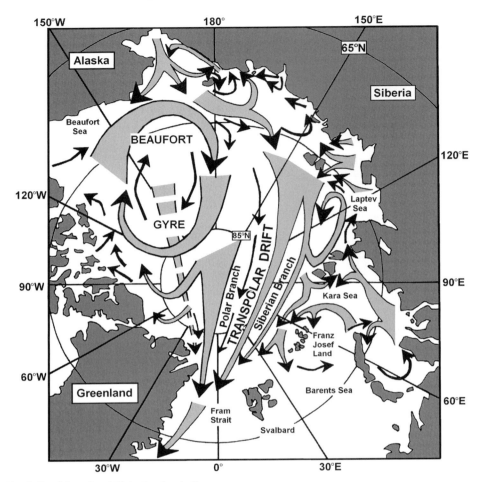

Fig. 3.58 Mean ice drift in the Arctic Ocean.

their study into the application of ULS data (Fig. 3.59). Not surprisingly, the annual cycle has the same form as that of the sea ice extent, with minima in September and maxima in April. They assumed that the draft values were the thickness divided by 1.12 and their work suggested that the thickness (draft) maximum was about 4.4 m (4.0 m) and the minimum 3.0 m (2.6 m).

On short timescales individual weather systems influence the movement of the sea ice. However, on longer timescales the major modes of atmospheric variability can have a large effect on ice extent. For example, sea ice extent in the North Atlantic and Labrador Seas sectors of the Arctic is strongly influenced by the NAO (see Section 3.4.5). In its positive (negative) phase there are strong (weak) winds across the North Atlantic and a deep (weak) Icelandic Low. Contrasting positive ice anomalies in the Labrador Sea sector

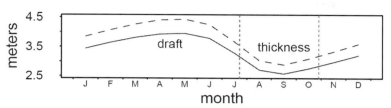

Fig. 3.59 The modelled seasonal cycle of Arctic ice thickness and draft. (From Rothrock et al., 1999.)

and negative anomalies in the Greenland/Barents Seas can be associated with changes in the NAO.

The Antarctic

The more zonally symmetric orography and atmospheric and oceanic circulations of the Southern Hemisphere compared with the north means that the annual cycle of sea ice growth and melt is much simpler. At the minimum of ice extent in February (Fig. 3.60) most of the ice around the coast of East Antarctica (which extends further north than much of West Antarctica) has melted, with the remaining ice being off the coast of West Antarctica and especially in the western Weddell Sea. At this time of year the area covered by ice is only 13% of that found in September. The annual cycle of ice extent (see the insert in Fig. 3.55b) shows a near linear growth from March to July and then a slower advance to the ice maximum in September. Over the year the extent varies from 3.5×10^6 km^2 in February to 19.0×10^6 km^2 in September.

There are regional variations in the annual cycle of sea ice extent, as can be appreciated from Fig. 3.61, although these are small compared with those found in the Arctic, where regional topographic factors are a very large influence on the cycle of ice. All five regions examined have the minimum sea ice extent in February, but then the rate of advance varies by region. The Ross Sea has a very rapid increase in extent during the autumn, while the advance is slower for the Indian Ocean sector. In contrast, this latter area has the most rapid decrease of ice extent between the peak in September and the end of the year.

Since the Antarctic continent is isolated with no land areas in the vicinity, the spread of ice during the winter is divergent and away from the coast. The lack of constraint on the ice is reflected in the annual range of ice coverage, which at about 16×10^6 km^2 is much larger than occurs in the Arctic. The season of ice growth is longer than the retreat period, so the maximum extent is found about a month after the end of the winter season. The most rapid retreat of ice is over October to December.

The variability of sea ice extent over 1978–97 can be appreciated from Fig. 3.55b. Many factors influence the extent of ice in a particular year, including surface air temperature, the wind field and ocean currents. But around the Antarctic the large-scale atmospheric flow and in particular the position of the circumpolar trough has a major influence on the ice extent.

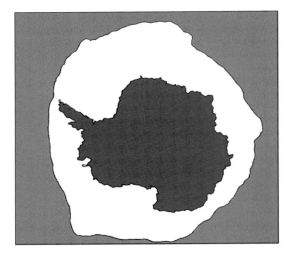

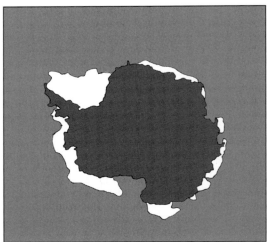

Fig 3.60   The mean sea ice extent for 1978–87. (Top) September, (bottom) February.

A number of variations in ice extent over the record are apparent in Fig. 3.55. Zwally *et al.* (1983b) noted that some anomalies persist for periods of 3–5 years, although decadal-scale changes have been smaller. On a regional basis the 3–5 year variability in extent is large, but the variability in different regions tends to be out of phase, so that the variability in the extent of total sea ice cover is much smaller. The interannual variability of the annual mean ice extent is only 1.6%, compared with 6–9% for the five areas shown in Fig. 3.61.

Sea ice motion over the Southern Ocean is dictated by a number of factors. The ACC, which is driven by the prevailing westerly winds, is a key feature of the Southern Ocean. It plays a major part in the thermohaline circulation, which distributes heat and fresh water around the world's oceans. Because the ACC is just north of the Antarctic sea ice zone it

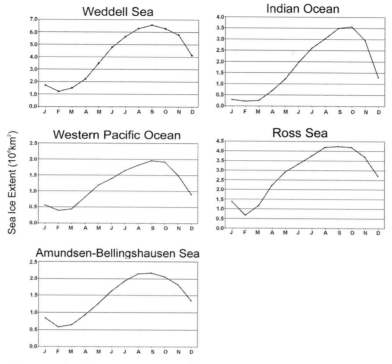

Fig. 3.61 The annual cycles of sea ice extent for various sectors of the Antarctic. (From Zwally et al., 2002a.)

moves the outer part of the sea ice cover towards the east, as demonstrated in Fig. 3.62, which shows the mean ice motion in September as derived from passive microwave satellite imagery. Around much of the continent the ice at the edge of the pack moves towards the east; however, there are a number of locations where the flow is markedly different from this. One location is near the Greenwich Meridian where the flow is from the south, reflecting the mean atmospheric flow in this area. Close to the coast of the Antarctic the movement of sea ice is mainly towards the west, driven by the coastal easterly winds on the southern side of the CPT.

Antarctic sea ice is primarily first-year ice since most of the pack melts during the summer and autumn months. Sea ice thicknesses are therefore mostly less than 1 m. The main exception is over the western Weddell Sea where the greatest amount of multi-year sea ice is found. Here helicopter-borne electromagnetic measurements of total (ice plus snow) thickness were made in 2004 (Haas et al., 2008). These data revealed that along the Antarctic Peninsula at 67° S extensive heavily deformed ice was present with a modal thickness between 3.0 and 5.0 m. East of 55° W, thick second-year ice was encountered with modal thicknesses around 3.0 m. This consisted of 1.6 to 2.0 m of ice and 0.6 to 1.0 m of

154   *The high latitude climates and change mechanisms*

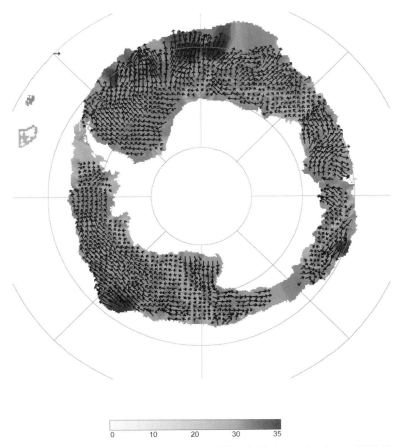

Fig. 3.62 Mean sea ice motion during September. Gridded fields interpolated out of SSM/I point data with drift arrows from SSM/I and buoys where available. The background shades of grey give the parameters magnitude, the length of the scale arrow (top left) indicates $10\,\text{cm}\,\text{s}^{-1}$ drift velocity or $10\,\text{m}\,\text{s}^{-1}$ wind velocity. note: Interpolated values are only representative in the regions with drift arrows. (From C. Schmitt, Ch. Kottmeier, S. Wassermann, M. Drinkwater, *Atlas of Antarctic Sea Ice Drift*, 2004.)

snow. East of 57° W, a band of first-year ice had modal thicknesses of 2.0 m, including 0.3 m of snow.

3.9 The ocean circulation

3.9.1 *The Arctic*

The Arctic Ocean is strongly and very stably stratified, principally due to vertical changes in salinity, the 'cold Arctic halocline' that inhibits vertical mixing. The surface layer extends to a depth of about 45 m and has a salinity of 28–31 psu (practical salinity unit: the conductivity ratio of a sea water sample relative to a standard KCl sample). In contrast, from ~400 m downwards the salinity is uniform at 34.5–35.0 psu. Near-surface temperatures are around $-1.8\,°C$, the salinity-adjusted freezing point, which allows sea ice to form readily. Below this temperatures increase rapidly reaching a maximum greater than $0\,°C$ at 300–500 m depth, the so-called Atlantic Layer.

Across the western Arctic Ocean surface waters circulate in a large clockwise rotational pattern – the Beaufort Gyre (see Fig. 3.58) – that is driven by the atmospheric circulation. This current, which also turns the Arctic sea ice, takes about four years for a complete rotation. The other predominant current in the Arctic Ocean is the Transpolar Drift Stream (TDS) that carries water and ice from the Laptev Sea north of Siberia, across the polar region and subsequently down the east coast of Greenland through Fram Strait, where it forms the East Greenland Current. The TDS flows in response to freshwater input from Siberian rivers (see below), and a predominant westerly wind that both act to push the Arctic surface water into the North Atlantic. Its exact track across the Arctic Ocean is influenced by the phase of the NAO (Mysak, 2001). Within the Barents Sea, a smaller, counterclockwise gyre flows in response to the local quasi-stationary climatological low-pressure system.

The North Atlantic Current, the northern extension of which comprises the several branches of the West Spitsbergen Current, provides about 60% of the inflow to the Arctic Ocean bringing warmer water from the Atlantic Ocean. Its average volume transport has been estimated as 10 Sv ($1\,\mathrm{Sv} = 10^6\,\mathrm{m}^3\,\mathrm{s}^{-1}$), with marked variability in the latter (Schauer *et al.*, 2004). Some water also moves into the Arctic Ocean from the Bering Sea and the Pacific Ocean, by way of the Bering Strait. In addition, there is significant input of fresh water into the Arctic Ocean from rivers: the Arctic Ocean contains 1% of global sea water but 11% of global river flow (Serreze and Barry, 2005). Two thirds of this discharge, ~2400 km^3 yr^{-1} (McClelland *et al.*, 2006), is provided by just four rivers; the Ob, Yenisey, Lena in Eurasia and the Mackenzie in Canada. Peak river flow occurs in June, following the onset of snowpack melt with 60% of river discharge occurring from April to July (Lammers *et al.*, 2001). As a consequence, salinity in summer is much less than during winter, especially in the regions local to river discharge, along the Siberian continental shelves and Mackenzie Delta. In addition, relatively low salinity waters from the Pacific flows into the Arctic Ocean through the Bering Strait, giving low salinity in the Beaufort and Chukchi Seas. Precipitation and brine rejection during sea ice formation also contribute to the fresh surface waters of the Arctic Ocean.

Water flows from the Arctic Ocean into the Pacific and Atlantic Oceans. The greatest volume of water leaves the Arctic Ocean through Fram Strait between Greenland and Spitsbergen as the cold East Greenland Current. Its average volume has been estimated as 12 Sv (Schauer *et al.*, 2004) so there is a net transport to the south in the region. A weaker cold current, the Labrador Current, flows south through the Canadian Archipelago. Finally, a considerable amount of water also flows from the Arctic Ocean into the northern Barents Sea forming the East Spitsbergen and Bear Island Currents. Note that only surface waters are exchanged via ocean currents because submarine ridges prevent the exchange of very deep waters. The lack of movement in the deep waters has caused a stagnant pool of very cold water to accumulate at the bottom of the Arctic Basin.

The flux of sea ice and relatively low salinity water through Fram Strait in the East Greenland Current has implications for deep-water formation (DWF) in the northern North Atlantic, which in turn influences the Atlantic Meridional Overturning Circulation (AMOC).

North Atlantic Deep Water (NADW) forms in the North Atlantic as warm, saline waters from the Gulf Stream move northward, cooling, becoming denser and thus sinking. Its mean characteristics are a temperature of 2.5 °C and a salinity of 35.03 psu. A best estimate gives a production rate of 15 ± 2 Sv (Ganachaud and Wunsch, 2000). However, NADW is actually made up of three different water masses. Labrador Sea Water (LSW), the upper one, comprises about 20% of NADW and is formed primarily due to deep winter convection in the upper 200 m of the Labrador Sea. Here it tends to accumulate and recirculate within the basin. The other 80% of NADW forms as two water masses in the Greenland Sea. The second, lower water mass is Denmark Strait Overflow Water, which overflows the 600 m sill depth moving south in the East Greenland Current and entrains water from its surroundings. The third water mass is known as Iceland–Scotland Overflow Water (or Northeast Atlantic Deep Water) and leaves the Greenland Sea Basin further east, as the name indicates between Iceland and Scotland: here it becomes much warmer and saline after having incorporated surrounding Atlantic waters. These descending water masses eventually form part of the Deep Western Boundary Current that flows at depth along the western North Atlantic.

There are two principal processes that lead to DWF in the North Atlantic. The first is open ocean deep convection and other turbulent vertical mixing processes and the associated heat loss to the atmosphere. For deep convection to occur the ocean needs to be preconditioned: cyclonic near-surface winds push surface waters towards the coasts, causing surface divergence that leads to the upwelling of denser deep waters and weakens the local vertical density gradient. Subsequently, individual weather systems can cause strong buoyancy loss to the atmosphere and remove the remaining oceanic stratification. The resultant strong vertical mixing has a spatial and temporal scale of 50 km and a few days, respectively (Kuhlbrodt et al., 2007). There is a strong positive relationship between the NAO and convection in the Labrador and Irminger Seas (e.g. Dickson et al., 1996). A modelling study by Marshall et al. (2001) indicated that meridional shifts in the zonal winds produce anomalous ocean gyre strength in these regions.

The second process for DWF involves salt which is input when sea ice freezes because strongly saline water (brine) is released during this process, increasing the density of surface waters: the Barents Sea is particularly saline because it receives relatively little river input compared with other Arctic shelf regions. Brine rejection in itself can be sufficient to initiate convection in shallower continental shelf regions. In the Nordic Seas NADW formation down to the depths of the sills in the Greenland–Iceland–Scotland ridge over which it has to flow to reach the North Atlantic – 630 m for the Denmark Strait and 840 m for the Faroe Bank Channel – is sufficient to create the overflow waters involved in the AMOC.

3.9 The ocean circulation

3.9.2 Antarctic

Characteristics of the Southern Ocean

The Southern Ocean encompasses an area of 20.3×10^6 km^2 and extends northwards from the Antarctic coastline to ~60° S, although physically its northern boundary is often taken as the Antarctic Polar Front, described later. The Southern Ocean comprises part of the Scotia Sea, the Weddell Sea, part of the South Indian Ocean, the Ross Sea, part of the South Pacific Ocean, the Amundsen Sea, and the Bellingshausen Sea (often linked together as the Amundsen–Bellingshausen Sea). The relatively narrow Drake Passage between Tierra del Fuego and the northern Antarctic Peninsula links the first and last of these regions, allowing a circumpolar ocean circulation around Antarctica that links all three ocean basins via the ACC. Generally, isopleths (lines of constant water property; e.g. temperature, salinity and density) lie approximately parallel to lines of latitude although bottom topography plays a significant steering role and can dominate locally, such as in the regions of the Scotia Arc system and Southwest Indian Ridge (15–35° E). However, the gradient of these isopleths is not uniform and several frontal regions exist that separate different water masses and are therefore marked by strong gradients in water properties – such as temperature, salinity and biological productivity – and pronounced isopycnal (surfaces of constant potential density) tilting throughout the deep water column that drives strong geostrophic flows. The marked temperature and sea surface height gradients allow the position of the fronts to be located using satellite data. Note that the frontal structure of the Southern Ocean can be extremely complex, with individual fronts splitting into separate branches or merging together.

The Subtropical Front (STF) or Subtropical Convergence (Fig. 3.63) demarcates the northern limit of Sub-Antarctic Mode Water (SAMW). It is marked by a northward temperature increase of 10 °C to 12 °C and salinity rise from 34.6 psu to 35.0 psu at 100 m depth (Orsi *et al.*, 1995). Its latitude varies from a minimum of ~47° S south of Australia to ~25° S west of Chile. The STF is associated with an eastward geostrophic current that attains near 30 Sv near South Africa, but, because the STF does not pass through the Drake Passage, it cannot be considered as part of the ACC.

The Subantarctic Front (SAF) defines the northern limit of the ACC and the southern boundary of SAMW formation (see below). To the south, the Antarctic Polar Front (PF) or Antarctic Convergence is a zone where cold, dense northward moving Antarctic Surface Water (AASW) sinks below the relatively warmer sub-Antarctic water before continuing northwards as Antarctic Intermediate Water (AAIW). The convergence is driven by a southward increase in the wind stress causing convergent Ekman flux in the surface layer of the ocean and forcing water to sink. South of the wind stress maximum the surface flow is divergent and thus drives upwelling of deep water. Analysis of satellite data indicates that seasonal variability in the PF location is not spatially coherent around the Southern Ocean. Meanders and eddies on the front can break away as cold core vortices containing AASW.

In addition, the Southern Antarctic Circumpolar Current Front (SACCF), which is a relatively narrow, fast-moving jet, lies south of the PF. In contrast to the other fronts, the

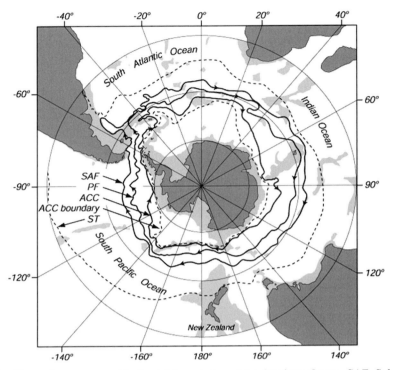

Fig. 3.63 The main oceanographic frontal boundaries of the Southern Ocean. SAF, Subantarctic Front; PF, Polar Front; ACC, Antarctic Circumpolar Current; ST, Subtropical Front. (From http://oceanworld.tamu.edu/resources/ocng_textbook/chapter13/chapter13_04.htm)

SACCF does not separate distinct surface water masses but it does mark the southern extent of Upper Circumpolar Deep Water at around 400 m leading to strong subsurface temperature, salinity and oxygen gradients (Orsi et al., 1995).

Water masses

Antarctic Surface Water (AASW) is found in the upper 200 m south of the PF. It is cold and fresh (salinity <34.5 psu) and relatively high in oxygen and nutrients compared with sub-Antarctic surface waters. Its properties vary with the annual cycle of sea ice advance and retreat: when sea ice forms brine rejection increases the salinity and density with the reverse happening during retreat. During the summer a distinct temperature minimum at the freezing point (−1.8 °C) occurs in the AASW at ~200 m depth. This marks the base of the winter mixed layer and the water in this part of the AASW is known as Winter Water.

Antarctic Circumpolar Deep Water (CDW) forms the core of the ACC and is both warmer and more saline than the AASW that it underlies. It flows south at depth, as much as several thousand metres, and rises to the surface south of the PF. Below the surface the temperature rises, to a maximum of around 2 °C at about 500 m and decreases to close to 0 °C near the ocean floor. Mean salinity of CDW is 34.70–34.76 psu and reaches a maximum between 700–1300 m depth. CDW can be split into Upper Circumpolar Deep Water (UCDW) and Lower Circumpolar Deep Water (LCDW). This split occurs in the southwest Atlantic, where North Atlantic Deep Water (NADW) encounters the ACC. UCDW is characterised by low oxygen and high nutrient concentration, due to the breakdown of sinking organic matter from the surface of the ocean, while LCDW has higher salinity (>34.70 psu) and low nutrient concentration. The denser LCDW penetrates south under the AASW and can spread as far as the continental shelves. Here it can mix with shelf waters to form Antarctic Bottom Water (AABW) (see below). The distinct (low oxygen) signal of the less dense UCDW 'disappears' at the southern edge of the ACC where it meets the Weddell Gyre where entrainment with the rising LCDW and the mixed layer takes place.

Antarctic Intermediate Water (AAIW) is a freshwater mass that encompasses the majority of Southern Hemisphere oceans at 800–1000 m depth. It is formed when Sub-Antarctic Mode Water (SAMW) is modified by intermediate depth convection west of Chile. Sources for SAMW and AAIW are water that flows southward in the Southern Hemisphere's subtropical gyre western boundary currents, water from south of the ACC that is pushed northward across it and upwelling from the underlying ACDW (Talley, 2002). AAIW (together with SAMW) is believed to be an important sink for anthropogenic CO_2. AAIW is characterised by a temperature of ~2.2 °C and a salinity of ~33.8 psu where it is formed, which both increase as it travels northward across the STF.

Antarctic Bottom Water (AABW) is found at the lowest depths of the ocean, below about 3000 m. It is defined by Orsi *et al.* (1999) as the total volume of southern bottom waters that are denser than the ACDW: thus defined it is a very dense water mass (neutral density >28.27 kg m^{-3}) with temperatures ranging from 0 to -0.8 °C and salinity from 34.6 to 34.7 psu. It is not circumpolar in extent: the two primary zones for AABW formation are the Weddell Sea (80%) and Ross Sea with their clockwise gyres. In addition, Rintoul (1998) presented evidence for a third source along the Atlantic coast south of Australia. Despite all having high densities, AABW formed in these different regions has very distinct characteristics, allowing it to be traced northwards. Production is closely linked to brine rejection during sea ice formation. The two major source waters involved in the formation of AABW at the continental shelf break are warm remnants of CDW (temperature >0 °C), imported from the ACC; and extremely cold shelf High Salinity Shelf Water (HSSW) (temperature <-1.7 °C), primarily formed during winter by strong interactions in polynyas between air and sea ice (e.g. Renfrew *et al.*, 2002). The formation of AABW, denser than either component, occurs because of the non-linearity of the equation of state for sea water. In addition, HSSW can sink southward down the continental shelf under the floating Filchner Ronne and Ross Ice Shelves, where it causes melting, forming a fresher, colder water mass termed Ice Shelf Water (ISW), defined as water with potential temperature below its freezing point temperature (~1.95 °C). ISW is observed to flow out from beneath the Filchner Ice Shelf and descend the continental slope northwards as plumes of relatively cold water; here they too mix with CDW and eventually lead to the formation of AABW (Foldvik *et al.*, 2004). Note that a regional terminology is

160   *The high latitude climates and change mechanisms*

sometimes used for water masses: for example Weddell Deep Water (WDW) and Weddell Sea Bottom Water (WSBW), the precursor to AABW. The transport of AABW into the ACC is believed to be between 10–30 Sv (Talley, 2002). By exporting AABW the Southern Ocean ventilates the vast majority of deep waters in the rest of the world ocean and hence this process is a key component of the THC.

The Antarctic Circumpolar Current

The Antarctic Circumpolar Current (ACC) (Fig. 3.63) is an eastwards-flowing current around Antarctica with a length of ~24 000 km. Its northern limit varies between 56° S through the Drake Passage to approximately 40° S east of South America but is generally around 45° S. Its southern limit is sometimes the shelf break of Antarctica such as along the western Antarctic Peninsula but in most areas it is separated from the continent by gyres. It serves as a conduit for heat and salt between the three major ocean basins. This strong zonal current also acts to limit meridional exchange so reducing heat flux to the ocean and Antarctica to the south but there is a prominent meridional circulation associated with the ACC, as described previously, and thus it is an important component of the THC. The ACC is strongly constrained by the distribution of both land and bathymetric features. For example, after constricted flow through the Drake Passage, it is split by the Scotia Arc into the northward Falklands Current and a second deeper branch that passes further to the east. Another split occurs in response to the Kerguelen Plateau in the Indian Ocean while marked deflections take place south of New Zealand around the edge of the Campbell Plateau and also at the mid-ocean ridge in the southeast Pacific.

The ACC is driven primarily through wind stress from the circumpolar westerly winds. This force, which accounts for 80% of the global ocean wind stress, together with the Coriolis effect, drives a net northward surface flow known as Ekman flux. This tilts the isopycnals and leads to eastward geostrophic flow in the ocean. The tilting of the isopycnals is balanced by baroclinic instability which transfers the energy into eddies. These cause a net southward flow and transfer energy to the abyssal ocean where bottom form stress can also drive a southward flow. These together balance the northward Ekman flux and drive meridional overturning flow.

In general the flow of the ACC is concentrated in the fronts described above, with speeds in the fronts reaching approximately $1 \, \text{m s}^{-1}$ in regions where the fronts are narrow (Fig. 3.64). Between the fronts are regions of slower eastward flow or even westward flow. Eddies also spin off the fronts due to instabilities of the flow and these have strong, but largely closed and transient, flows between the fronts.

In many regions there is a semi-annual oscillation in the latitude of the ACC that can be related to the atmospheric SAO in pressure. However, there is no clear pattern in interannual variability. Whitworth and Peterson (1985) calculated transport through the Drake Passage using several years of data from an array of current meters together with measurements of bottom pressure measured by gauges on either side of the passage. They found that the average transport through the Drake Passage was 125 ± 11 Sv, and that the transport varied

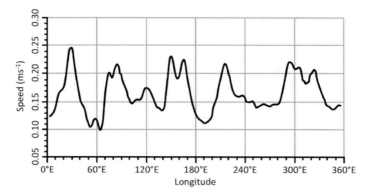

Fig. 3.64 Spatial variation of the Antarctic Circumpolar Current mean speed along the current axis.

from 95 Sv to 158 Sv with a standard deviation of ~10 Sv. Subsequent studies have found a similar flow through the passage though with narrower variability ranges.

The coastal current and gyres

A coastal westward flowing current around Antarctica is driven by the prevailing atmospheric easterlies. It is limited to south of 65° S and has a transport of ~10 Sv. In the Weddell Sea this current is diverted northwards by the Antarctic Peninsula and then joins the ACC east of the Drake Passage. The resultant cyclonic circulation is called the Weddell Gyre, which has a volume transport of more than 50 Sv and is important in determining regional sea ice extent. A similar gyre pattern also exists in the Ross Sea. Martinson and Iannuzzi (2003) suggested a strong link between Weddell Gyre variability and ENSO, with 'spin-up' occurring during the El Niño phase: changes in the STJ and PFJ positions resulted in changes in the cyclonic forcing of the gyre.

3.10 Concluding remarks

This chapter has discussed some of the many factors that influence and modulate the climates of the polar regions. The high latitude areas are particularly sensitive to forcing from the extra-polar areas and also exhibit large climate variability as a result of interactions between the atmosphere, ocean and ice. Here we have presented a number of mean fields of atmospheric and oceanic parameters derived from recent data. In the following chapters we consider how the climates of the polar regions have changed in the past, before examining how they may change over the next century if greenhouse gas concentrations continue to increase.

4

The last million years

4.1 Introduction

The period examined in this chapter extends back one million years, a time span that was chosen as encompassing the mid and late Quaternary and being slightly longer than the oldest Antarctic ice obtained at the time of writing: this was drilled at Dome C and extends back to ~800 kyr BP (years before AD 1950) (Parrenin *et al.*, 2007). It also comprises the modern half of the Pleistocene ('most recent') epoch, a term originally coined by English geologist Charles Lyell in 1839. Throughout this period there has been the cyclic growth and decay of major Northern Hemisphere ice sheets through the exchange of mass between these ice sheets and the oceans.

4.1.1 On the notation used

Examining climate on this long-term timescale, using oxygen-isotope (δ^{18}O) stratigraphy analysis of marine sediment cores and ice cores, has led to the development of a particular notation that allows direct comparison between records with differing sedimentation/accumulation rates. Emiliani (1955) introduced marine isotope stages (MIS) based on δ^{18}O records derived from deep sea sediment cores. These MIS are time periods with boundaries at the mid-point between isotopic temperature maxima and minima of successive stages (Fig. 4.1). Beginning with MIS 1, which characterises the Holocene, odd and even stages represent interglacial and glacial periods further back in time, respectively: the exception is MIS 3, which was incorrectly identified as an interglacial when first defined and actually forms part of the last glacial with MIS 2 and MIS 4. Shackleton (1969) subdivided the different stages using letters whereby letters a, c, etc. represent warmer substages, and b, d, etc. cooler substages, although this notation is now generally limited to MIS 5, including MIS 5e for the last interglacial. Subsequently, the concept of climatic events was introduced (e.g. Pisias *et al.*, 1984), which are points in time rather than periods. They are denoted by a decimal value: the integer portion refers to the MIS, the first decimal place to a substage – like the MIS an odd (even) number refers to colder (warmer) periods – and a second decimal place is used similarly if the resolution allows.

4.1 Introduction

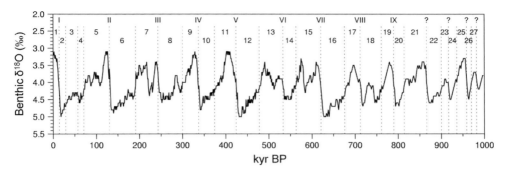

Fig. 4.1 The dates of the terminations (roman numerals) and time periods of the marine isotope stages over the last one million years. Note that Termination IX is the earliest defined explicitly in the literature, so earlier terminations are identified simply as question marks.

In addition, both the Arctic and Antarctic have their own specific notation associated with features observed in the ice and/or sediment cores from these regions. In the Arctic, each Heinrich event (H1–H6) (see Section 3.4.2) and Dansgaard–Oeschger (DO) event (DO 1–DO 25) (see Section 3.4.3) are given their own number, increasing with years BP. In the longer Antarctic ice cores, which encompass more than one glacial cycle, the Terminations (end) of each glaciation (originally observed as a rapid decrease in the $\delta^{18}O$ values in marine sediments at the end of each glacial period) are labelled with roman numerals (Fig. 4.1). Furthermore, in the relatively recent period (100 kyr), when the temporal resolution from ice cores is higher than further back in time due to the ice being less compressed than at depth, much higher frequency events have been identified: for example, Antarctic warm events (A1–A7; Blunier and Brook, 2001) and Antarctic Isotope Maxima (AIM 1–AIM 12; EPICA Community Members, 2006). These have been shown to co-vary with the DO events of the north and support the concept of the 'bipolar seesaw' (cf. Section 4.4).

4.1.2 The frequency of ice ages

Approximately 900–1000 kyr BP there was a change in the dominant orbitally driven time period of climate change from 41 kyr (obliquity) to 100 kyr (eccentricity), known as the Mid Pleistocene transition (MPT) (Fig. 4.2). This was not due to a change in insolation and hence represents a fundamental change in the climate system (e.g. Clark *et al.*, 2006). The mechanisms behind the MPT are not well understood and several hypotheses have been proposed. For example, Raymo *et al.* (2006) postulated that prior to the MPT the obliquity signal dominates records of ice volume of the northern and southern terrestrial ice masses because the precession signal (23 kyr) is out of phase between the two hemispheres whereas the obliquity signal is in phase in the polar regions. However, a long-term cooling (~5 °C) of the Earth – possibly forced by a reduction in CO_2 concentrations (Berger *et al.*, 1999) – allowed Northern Hemisphere ice sheets to become sufficiently large to survive summer insolation minima and also caused a transition from a primarily terrestrial to a marine-based

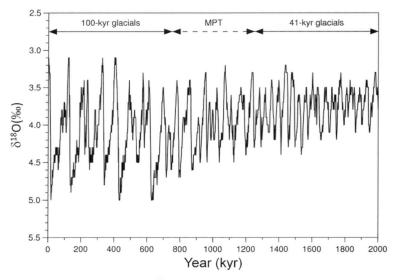

Fig. 4.2 Globally distributed benthic $\delta^{18}O$ stack from 0–2 million years (data from Lisiecki and Raymo, 2005). Note the switch from a 41-kyr glacial period to 100-kyr through the Mid Pleistocene Transition.

East Antarctic Ice Sheet (EAIS) at the MPT. As a consequence, it is generally considered that sea level changes, driven by Northern Hemisphere ice sheet fluctuations, became the controlling force on Antarctic ice volume and caused the ice sheet volume of both hemispheres to co-vary. A modelling study by Bintanja and van de Wal (2008) indicated that before (after) the MPT Eurasian ice sheets contained greater (less) ice volume than those in North America: despite the colder temperatures, the former could not expand further into the Atlantic Ocean and cold Siberian plains while the latter were able to expand until the Cordilleran and Laurentide Ice Sheets merged, on the western and northeastern regions of North America, respectively, and became the dominant Northern Hemisphere ice mass. Subsequently, the thicker, warm-based North American ice sheets were able to survive insolation maxima by rapid thinning through basal sliding, which initiated a glacial termination (Bintanja and van de Wal, 2008). Alternatively, Clarke (2005) suggested that changes to the basal properties of the Northern Hemisphere ice sheets altered their dynamics and led to increased volume. Thus, the climate of the Mid–Late Quaternary is dominated by a number of 'ice ages' with 100-kyr periodicity that, it is generally accepted, appear to be driven primarily by the ice dynamics and ice–climate interactions of the North American ice sheets.

4.2 The Arctic

4.2.1 The period before Termination II (~130 kyr BP)

In contrast to Antarctica, the ice core record in Greenland only reaches back as far the last interglacial (~120 kyr) and other sources of proxy data, such as marine sediment cores and glacial moraines, must be utilised to determine the climatological history of the Arctic before that time. Unfortunately, much of the record has been removed as a result of subsequent glaciations. Nevertheless, some very useful data may be obtained: for example, variations in sediment types within marine cores close to Greenland indicate significant variability in the extent of the Greenland Ice Sheet (Larsen *et al.*, 1994), the relative abundance of different plankton taxa gives proxy data on sea surface conditions (de Vernal *et al.*, 2005), while changes in the concentration and types of pollen provide information about climatic conditions (de Vernal and Hillaire-Marcel, 2008).

The latter authors examined a sediment core acquired off the southern tip of Greenland, the pollen content of which represents changes in vegetation close to the site. In general, low (high) pollen concentrations are recorded during glacial (interglacial) periods. During MIS 13 (~500 kyr BP) the record suggests shrub–tundra-type vegetation in southern Greenland, indicative of a substantially reduced Greenland Ice Sheet volume. This was also the case for the subsequent MIS 11 interglacial (~400 kyr BP) when pollen concentrations were an order of magnitude larger at the core site than during the Holocene. In addition, there was a high concentration of *Pinus* (spruce) pollen grains associated with this period, suggestive of forest vegetation, indicating milder regional conditions than at present, as do the proxy SST data (de Vernal and Hillaire-Marcel, 2008). Minimum values of pollen concentrations occurred in the glacial MIS 6 (190–130 kyr BP). The foraminifer abundance record from the northern Arctic Ocean provides evidence for the inflow of Atlantic Water to the Arctic Basin before MIS 6, which is reflected by strong ice-rafted debris (IRD) input to the deep sea (Spielhagen *et al.*, 2004).

During MIS 6 a huge ice sheet complex formed over northern Eurasia (Svendsen *et al.*, 2004), which must have blocked the Ob and the Yenisey Rivers (among others) draining into the Arctic Ocean and thus significantly influencing the freshwater budget of the Arctic Ocean (Mangerud *et al.*, 2004). Contemporaneously, a large ice shelf may have fringed the Barents–Kara Ice Sheet section of the ice sheet, perhaps reaching as far as the central Arctic Ocean (Jakobsson *et al.*, 2001). Dating of the IRD indicates that this so-called 'Late Saalian glaciation' lasted as long as 50 kyr.

The high IRD input ended during Termination II, at the transition to the last (Eemian) interglacial (MIS 5e) at ~130 kyr BP (Fig. 4.3) and corresponds to a strong meltwater signal, expressed by low oxygen and carbon isotope values of planktonic foraminifers. Svendsen *et al.* (2004) deduced that the meltwater spike originated from the drainage of a large ice-dammed lake south of the ice sheet margin in southwest Siberia. The meltwater may have even reached the Greenland–Icelandic–Norwegian (GIN) Seas and thus contributed to low-salinity events in the northern North Atlantic (Fronval and Jansen, 1996).

4.2.2 The last interglacial (130–115 kyr BP)

Most studies have suggested that during the last interglacial mean sea level was 4–6 m above current levels (e.g. Rohling *et al.*, 2008). A rapid increase in sea level at the start of MIS 5e,

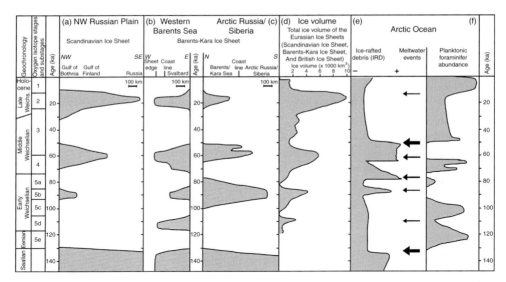

Fig. 4.3 Time–distance diagrams showing the growth and decay of the Eurasian ice sheets, together with their total ice volume and additional proxy information relating to their variability (Svendsen et al., 2004).

with levels 2–3 m above present has been attributed primarily to the melting of the Greenland Ice Sheet (Otto-Bliesner et al., 2006; Overpeck et al., 2006). A statistical analysis of local sea level indicators suggests it is highly likely (95% probability) that global sea level peaked at least 6.6 m higher than today and could have been several metres higher still (Kopp et al., 2009).

Oxygen-isotope data from the NGRIP core reveal that the Greenland climate was stable during the last interglacial and 5 °C warmer than present, based on a maximum value of −32‰ and a present value of −35‰ (assuming this is due solely to temperature changes) (North Greenland Ice Core Project members, 2004). In contrast, at Dye 3 in southern Greenland, there was a 5‰ decrease in $\delta^{18}O$ compared with the present, suggesting as much as 500 m lowering in the south while the northern and central ice sheet was similar to today.

This scenario is broadly matched by the modelling study of Otto-Bliesner et al. (2006), which indicated Greenland summer temperatures about 3 °C warmer than present during the early part of MIS 5 (slightly less than suggested from ice cores), which was a response to positive insolation anomalies (> 60 Wm^{-2}) in May–June for 130–127 kyr BP. After this time insolation anomalies were less than in the Holocene, during which the Greenland Ice Sheet changed little from its present configuration. For the Arctic as a whole, the model suggests temperatures were 2.4 °C warmer than today, with the greatest increases of more than 4 °C in the Hudson Bay–Baffin Island–Labrador Sea region. Arctic summer sea ice extent was also much reduced in the model, the 50% decrease being of similar magnitude to that seen in the past few years.

4.2 The Arctic

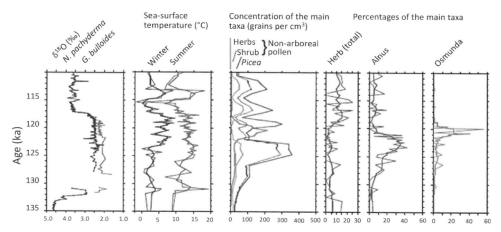

Fig. 4.4 The stratigraphy of MIS 5e based on a marine sediment core obtained south of Greenland. Shown are the isotope stratigraphy based on foraminifer shells, the sea surface temperatures estimated from dinocysts, and the concentration and percentages of the dominant pollen and spore taxa. (Adapted from de Vernal and Hillaire-Marcel, 2008.)

The vegetation of southern Greenland was dominated by *Alnus* (alder) in the middle of the interglacial (~125 kyr BP) and subsequently by *Osmunda* (fern) (Fig. 4.4). The latter appears to be unique in the last million years and corresponds to high winter SSTs and suggests dense fern vegetation under climatic conditions similar to those in which modern boreal forest, characterised by coniferous trees, exists. Towards the end of MIS 5e the concentration of *Osmunda* spores declined contemporaneously with an increase of herb tundra species, indicative of a cooling associated with the onset of the next glacial period (de Vernal and Hillaire-Marcel, 2008).

Marine cores taken in the North Atlantic and adjacent subpolar seas reveal very sharp changes at Termination II (~138 kyr BP), marked by increased biogenic productivity and replacement of sparse (Arctic-type) foraminifera assemblages with subpolar to cool-temperate assemblages (Fig. 4.3). These observations suggest that during MIS 5e ocean temperatures were on average similar to the present in the GIN Seas and warmer than at present in the Labrador Seas and Baffin Bay where sub-Arctic conditions were maintained for the entire duration of MIS 5e. From high concentrations of planktic foraminifers and the presence of coccoliths in Lomonosov Ridge cores, from the central Arctic Ocean, it is inferred that seasonally open waters existed in the eastern and central Arctic Ocean during MIS 5e and 5c (Svendsen *et al*., 2004). However, these proxy temperature records also display high-frequency variability indicating that the last interglacial was characterised by significant climate instability in northern latitudes and that marked variations in the northward heat flux may have existed in the absence of Northern Hemisphere ice sheet forcing. There is also qualitiative evidence (ice sheet variability) for climate change from deposition of coarse debris related to ice rafting (Fig. 4.3). Minimum IRD is seen in MIS 5e, indicative of a minimum in Northern Hemisphere terrestrial ice cover and suggesting that no pronounced ice advances occurred anywhere around the GIN Seas.

Ocean sediment cores from the North Atlantic indicate that both SSTs and salinity decreased gradually during the period of minimum ice volume at high latitudes, whereas at lower latitudes they remained stable or increased slightly (Cortijo *et al.*, 1999). This led to an increase in the meridional temperature and salinity gradients. In general, the influx of warm Atlantic water to the Nordic Seas was higher during MIS 5e than the Holocene (Knudsen *et al.*, 2002), indicating a strengthening of the North Atlantic Current. The Atlantic water masses, however, appear to have been deflected more to the west in the southern part of the Nordic Seas resulting in a much steeper meridional temperature gradient. Instead of today's stratification between Labrador Sea Water and NADW, a buoyant surface layer was present above a single water mass originating from the Nordic Seas (Hillaire-Marcel *et al.*, 2001).

The $\delta^{18}O$ record from the NGRIP Greenland ice core reveals a general, slow cooling from interglacial into glacial conditions from 122 to 115 kyr BP (MIS 5e to 5d). This was also a period of high-frequency climate change and includes the first defined DO event (DO 25), which follows a stadial (cold period) that occurred prior to full glaciation. Although weaker than following DO events, DO 25 had similar characteristics but occurred during ice sheet build-up and thus puts doubt on a large freshwater influx as a trigger. This climate signal encompassed the North Atlantic region as it is also observed in a marine core obtained near the Iberian coast (North Greenland Ice Core Project members, 2004). At this time (~119 kyr BP) the European icecaps were still building up and by 115 kyr BP IRD in the northwest Atlantic increased slightly, reflecting ice sheet growth in Greenland and northeast Canada. A contemporaneous drop in SST indicates a southward movement of the Arctic Front.

In Arctic Canada the onshore paleovegetational data and the offshore planktonic faunal and algal records demonstrate that both terrestrial and marine environments were much warmer than at present during MIS 5e. Optimal conditions persisted in the surface waters of the Labrador Sea during the early glacial inception when extensive ice masses developed over Arctic Canada.

The last interglacial maximum shoreline is well mapped around northern Alaska and eastern Siberia and is above the present level. Well-defined beach deposits indicate seasonal contraction of the Arctic Ocean pack ice. Over land, pollen assemblages and plant microfossils demonstrate that extensive closed boreal forest and a spruce tree-line existed north of the present latitude. Furthermore, soil profiles indicate permafrost degradation. In mainland North America there are relatively few such sites due to the subsequent formation of the Laurentide Ice Sheet, which caused intensive erosion during the last glaciation. However, some deposits do exist from the centre of the ice sheet where erosion was minimal. There are qualitative indications of warmer conditions in MIS 5e and the later interglacials in MIS 5 from microfloral and faunal evidence. The regional stratigraphy and aminochronology of the Hudson Bay lowland also suggests an early ice advance during a part of the last interglacial and full deglaciation of the region at least once and possibly twice after MIS 5e.

4.2.3 Later MIS 5: 107–75 kyr BP

The remaining (recent) part of MIS 5 (a–d) is termed the Early Weichselian in the European Arctic and the Early Wisconsinan in North America (cf. Fig. 4.3). MIS 5e sea level retreated from its maxima very rapidly indeed, implying the synchronous build-up of permanent Arctic ice cover in Laurentia and Fennoscandia together with those regions where the Greenland Ice Sheet and West Antarctic Ice Sheet (WAIS) had recently melted earlier in MIS 5e. Decreased $\delta^{18}O$ values from the Greenland Ice Sheet in 5d and 5b may be attributed to a cooling of air masses and/or increased continentality – due to oceanic changes and a southward shift of the polar front, accompanied by extensive sea ice. Lithological evidence suggests Canada and/or North Greenland as sources for the relatively low IRD deposits in the North Atlantic (Svendsen *et al.*, 2004). In the North Atlantic and GIN Seas recurrent warming in substages 5c and 5a are observed but not in the Labrador Sea and Baffin Bay. Recent evidence suggests that sea levels in MIS 5a were almost as high as in MIS 5e (Dorale *et al.*, 2010).

In Eurasia during late MIS 5c/early MIS 5b there was a major ice advance that terminated on the continent around ~90–80 kyr BP, as defined by moraines between southern Taimyr and the Pechora Lowland, in northern Russia and Siberia, respectively. Svendsen *et al.* (2004) suggested that highlands and islands fringing the Kara and Barents Seas acted as ice sheet nucleation areas, which subsequently confluenced to form a coherent Barents–Kara Ice Sheet. The ice sheet had a thickness exceeding 1000 m on the northwest coast of the Taimyr Peninsula and would have blocked northward drainage in this region (e.g. the Ob and Yennisey). At this time the Barents–Kara Ice Sheet terminated at or close to the northern margin of the continental shelf along the Arctic Ocean, a configuration supported by IRD records. Peak values in MIS 5 (cf. Fig. 4.3) indicate two advances of the ice sheet to the coast of Severnaya Zemlya (Knies *et al.*, 2001) while the western margin is difficult to define due to lack of evidence from the Barents Sea. However, there is evidence to indicate a restricted ice sheet across Scandinavia relative to the LGM (Svendsen *et al.*, 2004).

The IRD horizon of MIS 5b is weaker than in preceding glaciations, which may reflect a different routing of glacial meltwater, perhaps draining across the Barents Sea shelf and into the Norwegian Sea. Subsequently, during MIS 5a (85–72 kyr BP) a temporary return of planktonic forams and coccoliths indicates seasonally open waters. A significant IRD horizon in the central Arctic has a Kara Sea or western Laptev Sea provenance, and may correspond to the final retreat of the Barents–Kara Sea Ice Sheet from the northwest coast of the Taimyr Peninsula and the resultant emptying of large ice-dammed lakes in the southern Kara Sea Basin and West Siberian Lowland (Spielhagen *et al.*, 2004). Contemporaneous faunal assemblages in western Svalbard signify a seasonally sea ice free Norwegian Sea as at present.

During the latter part of MIS 5, climatic conditions similar to the present prevailed over southeast Canada and a recurrent warming is recorded in the south central Labrador Sea; fluctuating ice volumes in the Canadian Arctic led to episodic dilution of surface waters in Baffin Bay and the eastern Labrador Sea, and sub-Arctic conditions prevailed in nearshore

170   *The last million years*

environments of Baffin Island (de Vernal *et al.*, 1991). Lake-sediment cores from the island reveal that the vegetation was dominated by shrub–birch tundra. Its northward expansion may have been favoured by deeper snow cover associated with warmer winters and enhanced precipitation (Fréchette *et al.*, 2006).

4.2.4   *MIS4 and MIS 3: 75–25 kyr BP*

A readvance of the Barents–Kara Ice Sheet occurred during MIS 4, which terminated on the Russian mainland. The ice sheet was more (less) extensive in the west (east) than in the earlier advances during MIS 5. Dating of moraines associated with the North Taimyr ice-marginal zone have an age of 70–54 kyr BP implying a contemporaneous glacial advance with that in the European Arctic but no evidence suggesting ice advance onto the Pechora Lowland. In addition, there was a regrowth of the Scandinavian Ice Sheet: major glaciations in Svalbard and the northwest Barents Sea culminated at 65–60 kyr BP (MIS 4), while evidence from terrestrial lake cores indicates contemporaneous glaciations in Severnaya Zemlya (Raab *et al.*, 2003). Subsequently, there was a comprehensive deglaciation during MIS 3 (~50 kyr BP) as evidenced by raised marine sediments from islands across Svalbard and the Russian Arctic, indicative of strong isostatic rebound.

In the Arctic Ocean an increase in foraminifers and decrease in IRD after ~78 kyr BP is indicative of seasonally open water; Atlantic Water inflow into the GIN Seas is a convenient moisture source for ice sheet build-up. A very pronounced IRD layer, up to 1.5 m thick at 64–50 kyr BP (cf. Fig. 4.3) is believed to correspond to the Middle Weichselian (MIS 4 and 3) advance (Svendsen *et al.*, 2004) and has the highest sedimentation rates observed in the last 200 kyr. The rapid IRD deposition implies more icebergs and, hence, probably faster ice flow, which in turn suggests a marked increase in precipitation. A major meltwater spike was coincident with the end of the IRD input at ~50 kyr BP, as revealed by the $\delta^{18}O$ records of plankton foraminifers, and is likely to be associated with the drainage of ice-dammed lakes south of the Barents–Kara Ice Sheet. At a similar time a major freshwater event occurred in the GIN Seas (e.g. Baumann *et al.*, 1995). Subsequently IRD deposition in the Arctic Ocean was relatively limited indicating that ice was no longer present over the Barents–Kara sea shelf areas.

The inception and build-up of the Laurentide Ice Sheet has been estimated from numerical models utilising geomorphologic evidence to constrain them. The model described by Kleman *et al.* (2002) suggested that the inception of the Laurentide Ice Sheet occurred over Baffin Island and adjacent peninsulas. Subsequently, the modelled Laurentide Ice Sheet, which at this time the authors termed the Central Arctic Ice Sheet, expanded into the Keewatin region of Canada at 90–84 kyr BP where a large dome of ice formed. By 65 kyr BP this ice sheet had merged with the Cordilleran Ice Sheet in the west while to the east the Hudson Bay region was inundated by the Quebec–Labrador Ice Sheet (Kleman *et al.*, 2002). A minimum in the Laurentide Ice Sheet extent occurred in MIS 3 at 35–32 kyr BP when the ice margin followed the boundary of the Canadian Shield. Elsewhere across North America,

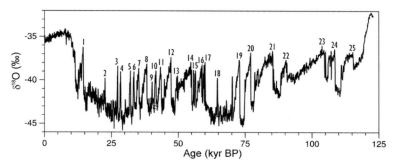

Fig. 4.5 Dansgaard–Oeschger events as observed in the NGRIP Greenland ice core $\delta^{18}O$ record.

to the west in the Cordillera and in the High Arctic the ice extent is believed to have been similar to the present.

During MIS 4 and MIS 3, the climate in Greenland was dominated by millennial-scale variability (DO events), as revealed by temperature-proxy records, principally derived from Greenland ice cores (see the review by Wolff et al., 2010; Fig. 4.5) but also observed in other proxy records across much of the Northern Hemisphere. They demonstrate a pattern of repeated significant warming of 8–15 °C in Greenland. Interestingly, DO events occurred sporadically in MIS 4, a period of rapid ice sheet growth, and throughout MIS 3, originally defined as an interglacial although sea level ranged from 60–90 m below present (Siddall et al., 2008). One of those that took place in MIS 4 is DO 19 (~75 kyr BP), which was studied in detail by Landais et al. (2004). These authors decomposed DO 19 into distinct sections. First, an initial cooling lasting ~300 years (stadial), corresponded to a weakened THC and was associated with the discharge of icebergs from the Laurentide or Fennoscandian Ice Sheets. Next, a slow ~5 °C temperature rise in ~1000 years: freshwater input into the North Atlantic was reduced due to the landward retreat of ice sheets following major calving, which led to a restart of the THC and associated atmospheric feedbacks. A cooling of ~1 °C in ~200 years reflects an instability in the THC 'off mode' (Fig. 4.6) or a triggering of the restart through sea ice formation. Subsequently the main rapid 16 °C warming occurred over ~100 years followed by a short maximum of ~200 years (interstadial). During this time the central Greenland accumulation rate increased from 0.07 m ice yr^{-1} to 0.21 m ice yr^{-1} (Lang et al., 1999). Finally, a cooling of ~16 °C took place over ~2000 years as the temperature relaxed back to cooler values. Contemporaneous changes in deuterium excess d (Section 2.5.2), in strong anti-phase with $\delta^{18}O$, reflect a significant warming of the oceanic source region during the cold phase of a DO, probably induced by a southward displacement of this evaporative region (Landais et al., 2004). In contrast, atmospheric CH_4 correlates strongly with $\delta^{18}O$ on millennial and submillenial timescales, although lagging temperature changes by 25–70 years, suggesting that Greenland temperature variability is broadly representative of the Northern Hemisphere (Huber et al., 2006).

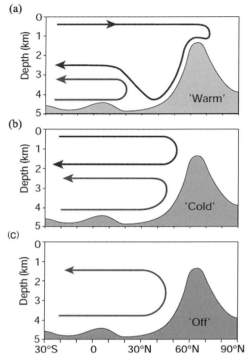

Fig. 4.6 Schematic of three possible modes of North Atlantic ocean circulation that prevailed during the Last Glacial Maximum, described as (a) warm, (b) cold, and (c) off, based on their physical characteristics in the North Atlantic. Shown is a section along the Atlantic with the topography symbolising the shallow sill between Greenland and Scotland. The upper lines represent the North Atlantic Meridional Overturning Circulation (NADW production) while the lower lines represent Antarctic Bottom Water. (Rahmstorf, 2002).

During this period Heinrich events H3–H6 (including H5a) occurred approximately every 7 kyr, in the cold period prior to the major DO events (Ahn and Brook, 2008). It has been postulated that very large tides in the Labrador Sea played a key role in liberating the armadas of icebergs into the North Atlantic at these times (Arbic et al., 2004).

4.2.5 The Last Glacial Maximum (LGM) and MIS 1: 24–11.5 kyr BP

During MIS 2 (Late Weichselian Glaciation: 25–10 kyr BP) the Barents–Kara Ice Sheet was actually smaller than in preceding glaciations, with a distinct Kara Sea advance not documented by the IRD record at this time (Fig. 4.3). In the Barents Sea, large quantities of glacially derived sediments were deposited by the ice sheet directly onto the upper continental slope at ~23 kyr BP (Kleiber et al., 2000). Moreover, sedimentary sequences on the Russian mainland indicate that an ice sheet did not reach this region during the LGM, except for the northern edge of the Taimyr Peninsula (Alexanderson et al., 2001). Svendsen *et al.* (2004) postulate that this event may have been very short lived and the northward flow of the Yennisey and Ob may have been blocked only briefly.

During the LGM the Scandinavian Ice Sheet had its maximum extension since the last interglacial: ice sheet margins on the northwest Russian Plain are indicated by prominent end-moraine systems that have been dated at ~20–18 kyr BP (Lunkka *et al.*, 2001), while in the Arkhangelesk region, to the north, the LGM has been dated as slightly later at ~17–15 kyr BP (Larsen *et al.*, 1999). In addition, there is evidence of large ice-dammed lakes that drained into the Volga River from the Russian Plain and then into the Caspian Sea, in contrast to those from Arkhangelesk that drained to the north and east. Radiocarbon dates from mammoth bones suggest that Finland was not glaciated during 33–25 kyr BP. This implies a rapid 1000 km advance of the Scandinavian Ice Sheet in 7000 years (Lunkka *et al.*, 2001), equivalent to ~140 m per year.

In North America the Laurentide Ice Sheet advanced to the northwest, south and northeast at 29–28 kyr BP while in the southwest and far north a maximum was reached later at 25–23 kyr BP (Dyke *et al.*, 2002). The main Laurentide Ice Sheet comprised three main sectors, each with its own system of ice domes and divides: these are termed the Quebec–Labrador, Keewatin and Foxe Basin sectors. The LGM ice thickness of the Laurentide Ice Sheet can only be determined directly near the margins where the ice terminated on land. Within the interior of the ice sheet the current topography is generally below the ice sheet level with the exception of a few overtopped mountain peaks that provide an estimate of minimum thickness. Hence, the best estimates of central ice thickness are derived from geophysical models that fit computed relative sea level histories to postglacial shorelines. Tarasov and Peltier (2004) estimated a maximum ice thickness of 3.3–4.3 km for the Keewatin dome using this methodology.

Striae on rock outcrops, glacial bedforms and postglacial depositional features are indicative of temporally shifting ice divides in the Labrador sector (e.g. Veillette *et al.*, 1999). The ice divide migrated northwestward from the initial build-up in the Quebec Highlands towards Hudson Bay. There followed two major changes in flow patterns (Dyke *et al.*, 2002): first, the main outflow centre shifted 900 km to the east, and second, a 'capturing' of the southwesterly outflow from the new centre by upstream propagation of a northward outflow into Ungava Bay, east of Hudson Bay. Veillette *et al.* (1999) link this capture to the major northward Gold Cove Advance of Labrador sector ice across Hudson Strait and suggest it was associated with the H1 Heinrich event of 16.8 kyr BP. The ice advance reached its maximum extent at ~12 kyr BP following the Younger Dryas, after which the Hudson Strait Ice Stream underwent major recession. However, there is not yet scientific consensus in the interpretation of the Labrador sector during the latter part of the LGM (e.g. Clark *et al.*, 2000).

Striation mapping and till stratigraphy also demonstrate the presence of a shifting ice divide in the Keewatin sector. A palaeo-ice-stream map of the northernmost part of this sector at the time of the LGM reveals ice streams, similar to those present on Greenland and Antarctica, flowing from within 200 km of the ice divide to close to the northern margin (de Angelis and Kleman, 2005). These were controlled by changes in the basal thermal

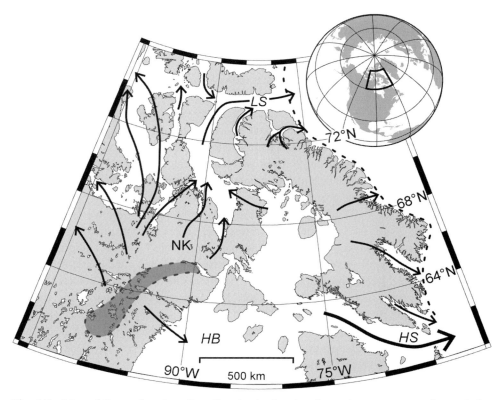

Fig. 4.7 Map of the northeastern Canadian Arctic showing the main export routes (arrows), ice margin (dotted line) and the Keewatin ice-divide zone (dark grey area) at the time of the LGM. Abbreviations: HB, Hudson Bay; HS, Hudson Strait; LS, Lancaster Sound; and NK, northern Keewatin. (After de Angelis and Kleman, 2005.)

configuration: that is whether or not the bed was frozen to the underlying rock or overlying a deformable bed of material. Elsewhere, warm-based ice streams were necessarily concentrated into topographic troughs. The evidence points to the Lancaster Sound trough being one of the major export regions of the Laurentide Ice Sheet (Fig. 4.7). Moreover, model data also indicate that geographically restricted fast flows due to subglacial till deformation are critical to obtaining a multi-domed late glacial Laurentide Ice Sheet structure (Tarasov and Peltier, 2004).

There is now a consensus for the existence of a so-called Innuitian Ice Sheet at the LGM that encompassed the northern part of the Canadian Arctic Archipelago (e.g. England *et al.*, 2006). This ice sheet advanced later than the Laurentide Ice Sheet, possibly as late as 22.5 kyr BP: the hypothesis proposed for this delay is that the height of the Laurentide Ice Sheet (maximum of ~3500 m a.s.l.) caused a split airstream around the ice sheet (Fig. 4.8) that led to increased precipitation over northern Canada allowing the subsequent build-up of

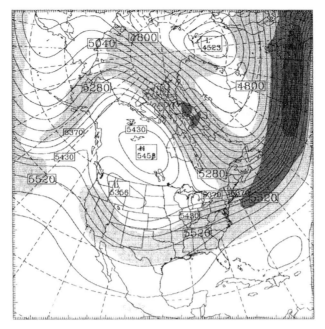

Fig. 4.8 Polar Mesoscale Model 5 LGM January monthly mean 500-hPa geopotential height and wind speed. Contour interval is 60 m and regions with wind speed exceeding 15, 30 and 45 ms^{-1} are shaded light, medium and dark grey, respectively (Bromwich et al., 2004).

the Innuitian Ice Sheet. The earlier maximum eustatic sea level lowering caused by the Laurentide Ice Sheet probably also aided the build-up. The Innuitian Ice Sheet comprised both lowland and alpine sectors from which several ice streams flowed, the largest being the Massey Sound Ice Stream, which flowed northwestward to the polar continental shelf.

The southwest margins of the Laurentide Ice Sheet lay next to the Cordilleran Ice Sheet that encompassed British Columbia, the southern Yukon Territory together with parts of Alaska, Washington, Idaho and Montana (Clague and James, 2002). Its history can be gleaned by dating glacial landforms and sediments. The Cordilleran Ice Sheet began to develop from the mountain ranges of the Canadian Cordillera at 35–30 kyr BP. However, its maximum extent was not reached until 18–17 kyr BP, that is several thousand years following the LGM. This lag relative to the Laurentide Ice Sheet may be due to the earlier decay of the latter altering the regional atmospheric circulation and allowing more moisture into the region (Clague and James, 2002). There is some contention regarding whether the Cordilleran Ice Sheet and Laurentide Ice Sheet coalesced along their mutual boundary with contrasting evidence in different regions.

To the southeast of the Laurentide Ice Sheet, in maritime eastern Canada, an Appalachian ice complex existed comprising a series of tidewater glaciers (Stea et al., 1998) that had the Gulf Stream as an accessible moisture source. At 22–19 kyr BP an ice centre formed over the

Magdalen Shelf, in the southern Gulf of St Lawrence (47 ° N, 62 ° W) with ice reaching as far as the outer shelf-slope margin. In ~18 kyr BP it retreated back to a tidewater margin at Sable Island with subsequent further retreats back to northern Nova Scotia and the Bay of Fundy. Haematite-stained quartz found in North Atlantic IRD deposits is indicative of the Appalachian ice complex and correlates well with periods of glacier retreat (Stea *et al.*, 1998).

During the LGM there were no pronounced IRD peaks in the central and eastern Arctic Ocean, but two do exist near the northern margin of the Barents Sea shelf (Knies *et al.*, 2001), dated as 15.4 and 13.6 kyr BP, indicating a two-stage collapse of the ice sheet. The delay may have been due to significant calving of icebergs and meltwater preventing further decay. Eventually, higher sea level and increased summer insolation would have become the dominant influence, triggering the second collapse.

IRD evidence from the North Atlantic Ocean suggests that until about 26 kyr BP open water in the GIN Seas provided sufficient water for the Fennoscandian–Barents Ice Sheets to grow significantly, after which they became increasingly glacial (Bauch *et al.*, 2001): sedimentary evidence suggests highly variable sea surface conditions and at least occasional advection of North Atlantic waters – analogous to the current Arctic Ocean (de Vernal *et al.*, 2006). The central North Atlantic and Norwegian Sea were seasonally ice free, with the southern winter sea ice limit located at ~55° N. Heinrich event H2 occurred at ~24 kyr BP, and may record the advance of the Hudson Strait Ice Stream beyond the sill at the mouth of the strait, culminating in the shift of the ice divide from the Quebec Highlands to the uplift centre of the Laurentide Ice Sheet (Dyke *et al.*, 2002). Extensive perennial sea ice cover existed along the eastern Canadian and Greenland coastal margins at this time while beyond this there was a narrow zone with sea ice for more than six months per year. The Labrador and Irminger Seas had seasonal sea ice for a few weeks to three months per year: in general the northwestern North Atlantic had much more extensive sea ice than at present (de Vernal and Hillaire-Marcel, 2000).

With regards to the strength of the North Atlantic MOC there are three conflicting theories to have emerged from both modelling and geological proxy-based studies, which are illustrated in Fig. 4.6. It should be noted that it is possible that more than one of these could have existed at different times during the LGM and certainly in the subsequent deglaciation. The first theory (Fig. 4.6a) identifies an enhanced THC, with NADW production at similar locations and with similar or stronger formation rates than at present (e.g. Hewitt *et al.*, 2003), but this result is difficult to reconcile with most available proxy data. The second theory (Fig. 4.6b) is characterised by shallower and weaker NADW production while there is enhanced northward flow of AABW such that the deep ocean basins are dominated by water formed in the Southern Ocean (e.g. Shin *et al.*, 2003). Finally, the third theory (Fig. 4.6c) suggests there was very little or no production of NADW, with much greater influence of penetration of AABW at all depths (e.g. Robinson *et al.*, 2005). An analysis of a kinematic proxy of the MOC suggests no more than 30–40% slowdown of the MOC during the LGM (McManus *et al.*, 2004), favouring the second theory, while a combined modelling and magnetic proxy-based study suggests that the third theory most closely matches iceberg trajectories derived from reconstructed IRD patterns (Watkins *et al.*, 2007). Iceberg trajectories and magnetic results imply that the LGM surface circulation was dominated by a cyclonic central North Atlantic gyre, which was separated from a southerly displaced North Atlantic Current.

In addition, the distribution patterns of surface temperatures and salinities for the LGM differ significantly from their modern counterparts. A larger seasonal temperature contrast is indicated with winter conditions colder than at present and relatively mild summers offshore. Data also suggest that a very strong salinity gradient existed between the surface and subsurface water layers, large enough to prevent deep/intermediate water mixing (as happens today) so a relatively dilute buoyant upper-water layer was present in the northwest North Atlantic during the LGM. Modelling studies also point to the importance of ocean feedbacks in driving the enhanced delivery of snow to high northern latitudes at this time (e.g. Khodri *et al.*, 2001).

Heinrich event H1, which occurred at ~16.8 kyr BP (Hemming, 2004), is thought to have resulted from a reorganisation of ice drainage in the Laurentide Ice Sheet interior: specifically the change of formerly southwestern outflow into northward flow into Ungava Bay (~59 $^\circ$ N, 67 $^\circ$ W) and a large eastward displacement of the outflow centre away from Hudson Bay. Dyke *et al.* (2002) suggest that this would have caused a drawdown of ice cover over the entire central part of the ice sheet by several hundred metres. Meltwater associated with H1 appears to have been particularly efficient at curtailing the MOC in the North Atlantic, perhaps owing to the delivery of icebergs to sensitive locations of NADW formation (McManus *et al.*, 2004). Proxy data suggest that the MOC was nearly eliminated at this time and for ~2000 years thereafter. Hence, this scenario appears analogous to the theory depicted in Fig. 4.6c. Further evidence for this is given by radiocarbon variability from North Atlantic corals and foraminifera, which demonstrates a high proportion of AABW in the water column (Robinson *et al.*, 2005), a finding that has been repeated in many model studies following a significant freshwater release (e.g. Ganopolski and Rahmsdorf, 2001).

Earlier estimates of Greenland temperatures at the LGM, as calculated from δ^{18}O values, have been revised downward following a better understanding of the changes in moisture sources at this time. Reconstructed temperatures from direct borehole temperature measurements have implied that LGM temperatures in central Greenland, from the GRIP/GISP2 ice cores, were ~22 °C colder than at present (Cuffey *et al.*, 1995; Dahl-Jensen *et al.*, 1998). LGM temperatures reconstructed at NGRIP, 300 km to the north, are several degrees colder still. GCM experiments suggest that during the LGM north Greenland precipitation was dominated by North Pacific moisture sources in contrast to southern Greenland, where precipitation was primarily derived from the North Atlantic (Charles *et al.*, 1994) A similar pattern is demonstrated in the (higher resolution) regional climate model experiment of Bromwich *et al.* (2004), particularly during winter when a blocking anticyclone exists over the Laurentide Ice Sheet splitting the upper-level flow (cf. Fig. 4.8). Another GCM study by Krinner and Genthon (1998) suggested that this circulation change was also partly responsible for the simulated reduction in precipitation across the Greenland Ice Sheet. The North

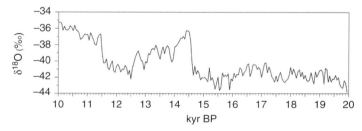

Fig. 4.9 Change in $\delta^{18}O$ from the Greenland GRIP ice core showing temperature changes between 20 and 10 kyr BP.

Greenland Ice Core Project members (2004) postulated that as a response to the Laurentide Ice Sheet expansion and coincident increase in sea ice extent and extensive North Atlantic ice shelves, NGRIP became further from the ocean. Thus, a higher fraction of air reaching this site came from the northern side of the Laurentide Ice Sheet, bringing colder and more isotopically depleted moisture than at GRIP.

Following the LGM, Greenland temperatures began warming at ~23 kyr BP, broadly following the Northern Hemisphere summer insolation curve (Labeyrie et al., 2003), but interrupted by a cooling associated with Heinrich event H1 at ~16.8 kyr BP (Fig. 4.9). North Atlantic SSTs also increased due to the northward retreat of the polar front. There then followed a colder period (stadial) that is generally known in Europe as the 'Oldest Dryas' (one of three Dryas periods named for a marker species of alpine plant, Dryas octopetala, which has been detected in ice cores and peat bogs). In the Greenland ice core record the end of the Oldest Dryas is clearly indicated. The new warmer period (interstadial) is known as the Bølling–Allerød. The (Bølling) transition from stadial to interstadial is observed as a marked warming – of 11 ± 3 °C (Severinghaus et al., 2003) – and an increase in accumulation (e.g. Kapsner et al., 1995) at 14.7 ± 400 kyr BP.

In addition, the Greenland summit ice cores show a marked increase in CH_4 concentrations. This may have been due primarily to there being a warmer and/or wetter tropical climate because potential Northern Hemisphere extratropical wetland sources were still largely covered by the Fennoscandian and Laurentide Ice Sheets (Severinghaus and Brook, 1999). Alternatively, a modelling study concluded there could have been an equal contribution from both latitudinal sources (Dällenbach et al., 2000). Another rapid change at the Bølling transition was a reduction in the concentration of dust and sea salt species (Mayewski et al., 1993; cf. Fig. 3.22). This is indicative of reduced cyclone activity in the Northern Hemisphere or possibly a reduction in Arctic sea ice extent during the Bølling/Allerød. At this time the THC intensified and 'modern' NADW production was initiated (Rahmstorf, 2002) and continued despite the impact of meltwater pulse 1A in ~14.2 kyr BP when sea level rose by about 20 m over 500 years (equivalent to a flux of 0.5 Sv). There is some debate as to whether this freshwater input was derived from melting Northern Hemisphere ice sheets (e.g. Peltier, 2005) or indeed from the Antarctic Ice Sheet (Clark et al., 2002).

The former study concluded that explicit evidence does exist for a contribution from the Fennoscandian and Laurentide Ice Sheets while the melting of the Antarctic Ice Sheet occurred some years later and could not therefore be the driver.

The disintegration of the northern part of the Barents–Kara Ice Sheet is indicated by a pulse of IRD at 15.4 kyr BP (cf. Fig. 4.3). This initial disintegration may reflect a reduction in precipitation in association with the cooling cycle predating Heinrich event H1, a decoupling of the glacier bed due to the relative sea level at the tidewater margin or a response to increased summer insolation. At 13.4 kyr BP the grounding line retreated from the shelf edge in this region (Franz Victoria Trough between Svalbard and Franz Josef Land). Subsequent stepwise deglaciation of the northern part of the ice sheet is suggested by distinct IRD pulses and by ~12 kyr BP most of the central Barents Sea was ice free (Landvik et al., 1998). Remnants existed on isolated icecaps on the Arctic archipelagos, which melted away during the Holocene.

The Younger Dryas cold event (stadial) saw a rapid return to glacial conditions in the Northern Hemisphere high latitudes. It can be considered as the second half of a DO event, the first part comprising the Bølling/Allerød interstadial. Using layer counting from the GISP2 core, Alley et al. (1993) dated the Younger Dryas as an event of 1300 ± 70 years that terminated at 11640 ± 250 BP (cf. Fig. 4.9). Some scientists have postulated changes in solar activity as the cause of the Younger Dryas, which interrupted the last deglaciation (Renssen et al., 2001a), partly based on the observation that a simultaneous cooling was also thought to have occurred in the Southern Hemisphere. However, recent studies have highlighted problems with the original dating of moraine sediments and, in contrast, demonstrated a warming in the south at this time (e.g. Barrows et al., 2007). Thus, it is generally considered that an Arctic freshwater flux that markedly reduced the MOC is responsible for the Younger Dryas (e.g. Tarasov and Peltier, 2005). This flux was originally derived from the large Keewatin Dome situated west of Hudson Bay. Model simulations revealed that the dominant drainage is to the northwest into the Beaufort Sea and into the Arctic Ocean.

The net retreat of the Laurentide Ice Sheet margin between 13 and 12 kyr BP was more rapid then previous rates everywhere. While not latitudinally asynchronous, as with earlier retreats, there was a strong east–west asynchrony. In the Keewatin sector there were changes in the pattern of ice flow in response to the relative stability of the eastern side, close to the Foxe and Labrador sectors, and significant retreat elsewhere. It is thought that the main north–south divide shifted 80 km to the east (Dyke and Prest, 1987). There remains some controversy as to the exact date of the formation of Lake Agassiz, the massive lake – maximum size $440\,000\,km^2$ – formed from glacial meltwater which encompassed much of central North America, but it is likely to have been 13–12 kyr BP. Many other large proglacial lakes, such as Algonquin and Iroquois–Vermont also existed at this time. Note that Lake Agassiz is also now thought to have drained predominantly into the Arctic Ocean through the Mackenzie River system to initiate the Younger Dryas (Teller et al., 2005; Murton et al., 2010). Models indicate that discharge peaked at 12.8 kyr BP at this time, with a 1σ range of 0.12–0.22 Sv, far greater than the 0.011 Sv of the Mackenzie River drainage today (Tarasov and Peltier, 2005). Once the fresh water reached the Arctic Ocean it subsequently flowed south, principally as sea ice, through the Fram Strait into the GIN Seas, the region of current NADW formation. Changes in Atlantic sediments (McManus *et al.*, 2004) and in the carbon cycle (Muscheler *et al.*, 2000) both imply a reduction in deep water formation during the Younger Dryas, although the former study reveals that it was not eliminated entirely as was the case during the H1 event. However, detailed regional sediment cores suggest that while reductions in deep water formation occurred at the beginning and end of the Younger Dryas, there was strong vertical water-mass convection in between (Bauch *et al.*, 2001).

The Laurentide Ice Sheet margin continued its substantial retreat between 12 and 11 kyr BP, particularly the dry terrestrial northwestern margin, between Melville Island and Prince of Wales Island to the Lake Agassiz Basin. The southeastern margin, which predominantly calved into lakes, retreated more slowly. However, within the general retreat there were significant but brief readvances (glacial surging). For example, at ~11.3 kyr BP such a surge formed a 60 000 km^2 ice shelf in Viscount Melville Sound (74 ° N, 100–115 ° W) (Hodgson and Vincent, 1984), which had disappeared 300 years later. Nonetheless, it must have evacuated a significant amount of ice from the northwestern ice sheet and enhanced the regional ablation. By 11 kyr BP eight glacial lakes were dammed along the south and west margins of the Laurentide Ice Sheet (Dyke and Prest, 1987). The Keewatin sector continued to experience changes in ice flow, with further migration of the north–south divide to the east as the western margin receded. In much of Prince of Wales Island (73 ° N, 99 ° W), the flow shifted from northwestward to northeastward as the divide migrated across the island.

The retreat of the Innuitian Ice Sheet occurred from west to east after about 13 kyr BP with much of it after the Younger Dryas in the Holocene. England *et al.* (2006) suggested that retreat was initiated by the ice sheet becoming destabilised by the decay of the Laurentide Ice Sheet. The Cordilleran Ice Sheet retreated rapidly and by 12 kyr BP much of the interior of British Columbia was free from ice and glaciers in the coastal mountains were of similar extent to today.

In Greenland the Younger Dryas was a period of reduced accumulation, by a factor of two from the early Bølling/Allerød (Alley *et al.*, 1993), and substantially elevated dust concentrations (cf. Fig. 3.22). Mayewski *et al.* (1993) postulated that the increased variability in dust concentrations at the start of the Younger Dryas resulted from significant atmospheric reorganisation produced by the rapid transition into this period. Additional source regions for the dust were probably outwash plains south of the North American ice sheets, whereas today Asia is the principal source for dust found within Greenland precipitation. Thus it seems likely that there was more meridional flow into Greenland and the negative phase of the NAO was dominant. Temperatures were approximately 15 °C colder than the previous interstadial but with much greater decadal frequency δ^{18}O variability, believed to be due to switches in the ratio between those air masses reaching Greenland from the northern and southern jetstreams around the Laurentide Ice Sheet (cf. Fig. 4.8). The end of the Younger Dryas, the beginning of the Holocene period, was marked by a very rapid increase in accumulation, with a doubling over just a few years. A step change in temperature of about 10 °C also occurred within a decade. This warming occurred ~30 years before a slower increase in CH_4, levels of which reduced from 700 to 500 ppbv during the Younger Dryas. The speed of the apparent link between the CH_4 source regions and Greenland supports an atmospheric, rather than a slower oceanic transmission of the climate signal (Severinghaus *et al.*, 1998) although an increase in NADW formation as the freshwater flux from the Laurentide Ice Sheet abated, and hence the MOC became more vigorous, is still generally believed to have ended the Younger Dryas cold period.

4.3   The Antarctic

4.3.1   *The period before Termination V (1000–430 kyr BP)*

The EPICA Dome C ice core extends back to ~800 kyr BP, spanning eight glacial cycles, back to MIS 20.2 (Jouzel *et al.*, 2007). The latest timescale (at the time of writing) for the ice core is the EDC3 chronology (Parrenin *et al.*, 2007), which was derived using a combination of snow accumulation and mechanical flow modelling but tuned and adjusted by independent age markers, especially in the more recent period. Variations in deuterium from the Dome C core plotted on the EDC3 timescale compare closely to benthic $\delta^{18}O$ from a stack (average) of 57 marine cores (Lisiecki and Raymo, 2005: Figure 4.1). Based on modelling results, Huybers and Denton (2008) proposed that at orbital timescales Antarctic temperature is primarily controlled by the duration of the Southern Hemisphere summer rather than the intensity of its Northern Hemisphere counterpart.

Both ice-core and marine-core data reveal that during the period prior to a major climatic reorganisation, the mid-Brunhes event (MBE), which took place at Termination V (~430 kyr BP) (e.g. Jansen *et al.*, 1986; Wang *et al.*, 2003), interglacials were both longer and cooler than during the last four glacial cycles. Summer SSTs, derived from a marine core drilled in the sub-Antarctic Zone of the Southern Ocean showed glacial/interglacial variations of 3–10 °C (Becquey and Gersonde, 2002). In addition, the partial pressure of CO_2 lies within a range of 170–260 ppmv in the earlier period, a range ~30% smaller than post-MBE (Siegenthaler *et al.*, 2005; Lüthi *et al.*, 2008). Although, generally, the sensitivity between temperature (as derived from δD) and CO_2 remains stable across the last six glacial cycles, Lüthi *et al.* (2008) demonstrated that atmospheric CO_2 concentrations during MIS 17 remained significantly below the levels of other interglacials (MIS 13, 15 and 19) while the lowest reported CO_2 levels of 172–180 ppmv during MIS 16 and 18 were lower than in subsequent glacial periods despite temperatures being similar. Thus, from 780–650 kyr BP (MIS 16–18) the temperature–CO_2 sensitivity is altered, with lower CO_2 values for a given temperature. Lüthi *et al.* (2008) suggest a long-term CO_2 increase from 800 to 400 kyr BP of ~25 ppmv, perhaps due to changes in weathering rates or ocean carbon storage. Spahni *et al.* (2005) and Loulergue *et al.* (2008) find similar results for CH_4. In the case of methane, it is MIS 19 that is most unusual, with the maximum concentration of ~740 ppbv being much higher than during MIS 13–17.

In addition to the longer orbitally driven changes in climate, the Dome C core also reveals significant millennial-scale variability along its length of similar magnitude to the Antarctic

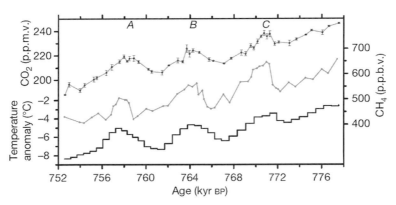

Fig. 4.10 Antarctic Isotope Maximum events during MIS 18. Variations of Dome C CO_2 (black solid circles; mean of four to six samples; error bars are 1σ of the mean (s.d.)), methane (middle line) and temperature anomaly based on the mean temperature of the past millennium (black step curve) between 778 and 752 kyr BP. Gases are plotted on the EDC3 age scale, temperature on the EDC3 timescale. The letters A to C denote the three events. (After Lüthi et al., 2008.)

Isotope Maximum (AIM) events observed in the last 50 kyr (MIS 3) (cf. Section 4.3.4). For example, during the glacial inception from MIS 19 to MIS 18 (778–752 kyr BP) there are three distinct millennial-scale variations observed in δD (temperature), CO_2 and CH_4 (Fig. 4.10).

Wolff et al. (2006) examined the Dome C record for signals of variability in sea ice extent, marine productivity and iron flux, and noted that, once again, no significant temporal changes in their relationship with temperature occurred. This study also demonstrated that the relative phase of changes in these variables was generally consistent across a termination, in that a decline in terrestrial dust (from Patagonia) preceded a reduction in sea ice by 2–3 kyr, and that the latter occurred over a similar time interval to an increase in CO_2. Thus, Fe fertilisation of the ocean can only impact CO_2 changes during the first part of a termination. In contrast, changes in sea ice are likely to become more important in the second part of a termination: Röthlisberger et al. (2008) showed that this is partly a threshold effect, with the sea ice proxy (sea salt) not being sensitive to temperature during very cold periods when sea ice was extensive. The temporal patterns of the variables is similar across other terminations, both prior to and following the MBE, suggesting that internal climate feedbacks are not responsible for the change in amplitude of the interglacial climate at this time (Wolff et al., 2006).

4.3.2 Termination V (~430 kyr BP)

Termination V is of particular interest because the orbital parameters at the time of MIS 11 – low eccentricity and a resultant damping of the precessional cycle – are similar to the present. MIS 11 is also an exceptionally long interglacial. The MBE followed a massive rise in sea level: in MIS 12 sea level was 139 m below present levels (Rohling *et al.*, 1998) to present levels in MIS 11 (e.g. Bowen, 2010) or possibly 20 m higher (Hearty *et al.*, 1999). Thus, the volume of global ice may have changed from having 15% more than in the LGM to significantly less than today: Hearty *et al.* (1999) suggest the disappearance of all of the Greenland and West Antarctic Ice Sheets together with 12 m (of sea level equivalent) from East Antarctica is necessary for the 20 m. This latter hypothesis is not supported by the modelling study of Raynaud *et al.* (2003), which indicates a volume only slightly smaller than at present and a maximum contribution of 5 m. Moreover, if sea levels were similar to that at present then melting of the Northern Hemisphere mid latitude ice sheets would be sufficient to cause the rise at Termination V (Bowen, 2010). Wang *et al.* (2003) showed that a change in low latitude oceanic carbon reservoir preceded the ice sheet expansion – from a maximum in planktonic $\delta^{13}C$ of 1.5‰ in MIS 13 to a minimum of 0.4‰ at the end of MIS 12 – and postulated this as a major control on global ice volume rather than Arctic summer insolation.

Concentrations of CO_2 from Dome C have a minimum at 200 ppmv at the end of MIS 12 and a maximum of ~275 ppmv at the start of MIS 11, mixing ratios that lie within the range observed during younger glacials and interglacials (EPICA Community Members, 2004), an observation also valid for CH_4 (Fig. 4.11). This appears to rule out previously proposed hypotheses of unusual greenhouse gas conditions during MIS 11 or a link between coral-reef growth and intense carbonate dissolution in MIS 11 through unusual CO_2 concentrations. The general temporal evolution is also similar to younger terminations, with an approximately linear increase in CO_2 across the termination and two-step increase for CH_4 (a slow rise prior to a rapid jump towards interglacial values). However, one significant difference in Termination V is in the apparent relative timing of the increases in CO_2 and CH_4 (EPICA Community Members, 2004). Unlike more recent terminations, where concentrations in both gases start to rise near simultaneously, the rise in CO_2 precedes the main increase in CH_4 by ~5 kyr, during which concentrations of the former had increased by 50 ppmv (more than half of the eventual total rise). The second (rapid) part of the CH_4 increase occurred when CO_2 values were near their maximum and continued for 2–3 kyr beyond the end of the termination (see Figure 4.11).

Diatom assemblages from marine sediment cores drilled in the South Atlantic allow the partial reconstruction of SSTs and sea ice limits during Termination V (Kunz-Pirrung *et al.*, 2002). During MIS 12 and 10 (glacial) there was seasonal sea ice cover north of the current ice extent, coincident with a 3–4 ° northward shift in the latitude of surface water isotherms. Diatoms associated with sea ice extent show periodicities of 3 and 1.85 kyr, similar to the DO events revealed in Greenland ice cores (cf. Section 3.4.3). During Termination V SSTs increased by 4–6 °C, reaching a maximum ~15 kyr before the minimum in global ice volume (~405 kyr BP). The warming was disrupted by two (cold) reversals that Kunz-Pirrung *et al.* (2002) postulate might be related to the collapse of the WAIS. In MIS 11 (interglacial) summer SSTs were ~2 °C higher than current values with a coincident 3 ° southward shift of the Polar Front. The summer SST record co-varies closely with the Vostok $\delta^{18}O$ proxy

184   *The last million years*

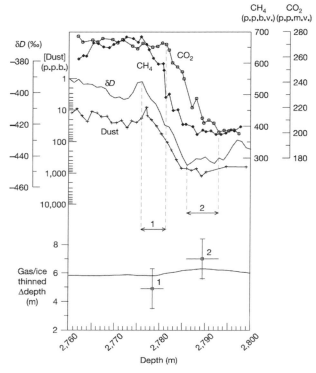

Fig. 4.11 Termination V in the EPICA Dome C ice core on an ice depth scale. The top panel shows the ice-core parameters: circles, CO_2; diamonds, CH_4; line with no symbols, δD; crosses, dust. The lower panel shows the modelled difference in depth between ice and air of the same age (line) along with estimates of the actual difference (error bars are based on uncertainty in aligning common events) for events considered roughly contemporaneous on the basis of their behaviour in later terminations at Vostok. Event 1, CO_2 peak/δD peak; event 2, CO_2 early increase/ δD early increase. (From EPICA Community Members, 2004.)

temperature record, indicating that the ACC region was the dominant source region for precipitation over East Antarctica.

4.3.3 *Terminations IV to II (~320 to ~138 kyr BP)*

There are three Antarctic ice core records that reach as far back as Termination IV: in addition to Dome C, there is the earlier Vostok core, on which much of the earlier work on ice ages was based, and the Dome Fuji core (Watanabe *et al.*, 2003), which is located in a different sector of East Antarctica. Records of temperature and greenhouse gases from the cores indicate a high level of homogeneity between the three records. However, an analysis using GCM data revealed that the relationship between temperature and the isotopic

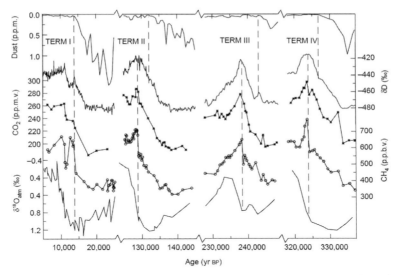

Fig. 4.12 Vostok time series during glacial terminations. Variations with respect to time of: a, dust; b, δD of ice (temperature proxy); c, CO_2; d, CH_4; and e, $\delta^{18}O_{atm}$ for glacial terminations I to IV and the subsequent interglacial periods (Holocene, stage 5.5, stage 7.5 and stage 9.3). (From Petit *et al.*, 1999.)

signature did vary between the three sites (Sime *et al.*, 2009) so the isotope data from the different sites are not directly comparable in terms of indicating temperature.

Terminations II, III (248 kyr BP) and IV follow a broadly similar pattern in the proxy records of Antarctic temperature and greenhouse gases: data from the Vostok core are illustrated in Fig. 4.12. Based on these data, Fischer *et al.* (1999) found that CO_2 concentrations increased by 80–100 ppmv 600 ± 400 years after the initial temperature increase from glacial conditions: using a more sophisticated technique Caillon *et al.* (2003) suggested 800 ± 200 years for Termination III. Typically, the increase in CO_2 is more regular while that for CH_4 occurs in two distinct stages, a low rise followed by a rapid jump to interglacial values. Termination III is the least similar of these three terminations: Fig. 4.12 reveals that the changes in the various parameters are much smaller than the preceding and following terminations. As a result, temperatures in the interglacial MIS 7.5 are cooler than MIS 5.5 or MIS 9.3. Also of interest is that the main rise in temperature is preceded by a cold reversal (Cheng *et al.*, 2009), similar to that prior to Termination I (the Antarctic Cold Reversal; cf. Section 4.3.4). Jouzel *et al.* (2007) postulated that the weaker interglacial results from the 65° N summer insolation (precession parameter) and obliquity being in antiphase rather than in phase. Following MIS 7.5 there are further periods of warmth (MIS 7.1 and 7.3) when temperatures were almost equal to that in MIS 7.5 and were actually higher than average Holocene temperatures. Temperatures in MIS 5.5, following Termination II, are the warmest recorded in the Dome C ice core (Jouzel *et al.*, 2007) and

186   *The last million years*

are thought to have been at least 6 °C higher than the present day (Sime *et al.*, 2009). In contrast, CH$_4$ values were highest in MIS 9.3.

Marine sediment records from the Atlantic and Indian Ocean sectors of the Southern Ocean have been used to infer summer SSTs and sea ice extent during the period through Termination II (Bianchi and Gersonde, 2002). Diatom-based summer SST estimates indicate that at the end of the glacial period (MIS 6) there was an intensification of cold water output from the Weddell Sea and a northward expansion of the summer sea ice extent by 3–5 ° latitude. At the start of the warming associated with Termination II (132–131 kyr BP) sea ice retreated rapidly to a minimum in the early part of MIS 5.5 (128–125 kyr BP) when summer SSTs were 2.5–3.5 °C warmer than at present. Significant centennial-scale variability existed at this period, with meltwater events a likely cause. During MIS 5.4 the sea ice expanded once again, although with SSTs only slightly cooler than today and a slightly weaker meridional temperature gradient the Southern Ocean could not be considered as being in a true glacial mode (Bianchi and Gersonde, 2002).

4.3.4   *MIS 3 and 4 and Termination I*

During MIS 3 and MIS 4 a number of significant warming events occurred in Antarctica, as first observed in the $\delta^{18}O$ record from the Byrd ice core in West Antarctica (Blunier and Brook, 2001) and subsequently also seen in the Vostok and Dome C records (Spahni *et al.*, 2005). These changes in climate are smaller than the DO events identified in Greenland ice cores, both in terms of their magnitude and speed of change. Comparisons of high-resolution Arctic and Antarctic ice core records have revealed a one-to-one coupling between Antarctic warm events and DO events, as discussed in Section 4.4. More recently, smaller-scale variations in the Antarctic $\delta^{18}O$ chronology have also been named (cf. Section 4.1.1): EPICA Community Members (2006) identified 12 AIM features in a core drilled in the Dronning Maud Land (Atlantic) sector of East Antarctica. AIM 12 corresponds to Antarctic warming event A2 at ~48 kyr BP. The authors suggested these features reflected reduced oceanic heat transport into the North Atlantic at these times.

Studies of Antarctic climate during the LGM (~18 kyr BP) and the subsequent retreat, based on marine-geological evidence, reveal that the West and East Antarctic Ice Sheets have not moved in concert (Anderson *et al.*, 2001a). During the LGM the West Antarctic Ice Sheet (WAIS) reached right to the continental shelf margin in the Ross Sea while also advancing significantly elsewhere along its margins. For example, in Pine Island Bay the grounding line was located at least 240 km seaward than at present (Lowe and Anderson, 2002): thicker ice was grounded on bedrock on the inner shelf while further offshore thinner ice slid on a deforming bed. In the Siple Dome region, in the interior of the WAIS, ice was 200–400 m thicker than at present (Waddington *et al.*, 2005). In contrast, advances of the East Antarctic Ice Sheet (EAIS) were smaller and in some regions, such as Queen Maud Land, there was little if any advance. Moreover, it is believed that drier conditions in East Antarctica may have actually led to a slight thinning of the interior of the ice sheet. In the

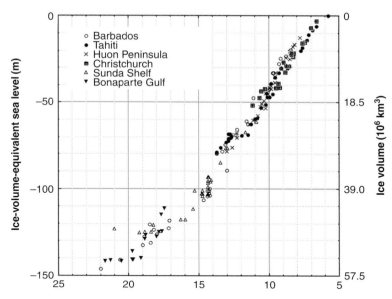

Fig. 4.13 Changes in sea level during the glacial–interglacial transition following the LGM as recorded at various locations. The x-axis shows time in kyr BP. (From Lambeck and Chappell, 2001.)

Antarctic Peninsula region the ice extent at the LGM was not as great as during other advances in the previous 100 kyr (Hjort et al., 2003). However, geomorphological data do indicate ice thicknesses a few hundred metres above present-day sea level in the north, increasing further south along the peninsula. The ice sheet over the peninsula merged with an expanded and thicker WAIS in the Weddell Sea region (Bentley et al., 2006).

During the transition from glacial to interglacial following the LGM, sea level rose by ~130 m (Lambeck and Chappell, 2001; Fig. 4.13). A significant jump in sea level occurred at ~14.2 kyr BP, when rates exceeded 40 mm yr^{-1} (giving a rise of 20 m over 500 years), which corresponds to a freshwater flux of ~0.5 Sv. This feature has been termed Meltwater Pulse 1A (MWP-1a) and there has been some debate as to its source. Clark et al. (2002) suggested a substantial contribution from the Antarctic Ice Sheet and, as a consequence, Weaver et al. (2003) postulated a Southern Hemisphere forcing for the Bølling–Allerød warm period in the north through a strengthening of NADW formation. However, Peltier (2005) believed that previous theories, based on a partial collapse of the Northern Hemisphere ice sheets, could not be ruled out based on the available evidence. Moreover, reconstructions of the Antarctic Ice Sheet indicate that its maximum possible contribution to MWP-1a could only have been ~14 m. A second meltwater pulse (MWP-1b) occurred at about 11 kyr BP (Peltier, 2005).

Initial retreat of the WAIS appears to have begun about 16 kyr BP while parts of the EAIS actually started retreat prior to the LGM itself (Anderson et al., 2001a). Marked differences are observed in the retreat from the Ross Sea continental shelf. Retreat from the western

188   *The last million years*

shelf (EAIS) was broadly continuous while that of the grounding line of the eastern shelf (fed by the WAIS) was episodic and asynchronous between different ice streams. For Pine Island Bay in West Antarctica, the ice had retreated from the outer shelf by ~16 kyr BP (Fig. 4.14) and subsequently grounded on a bedrock high and long enough to produce

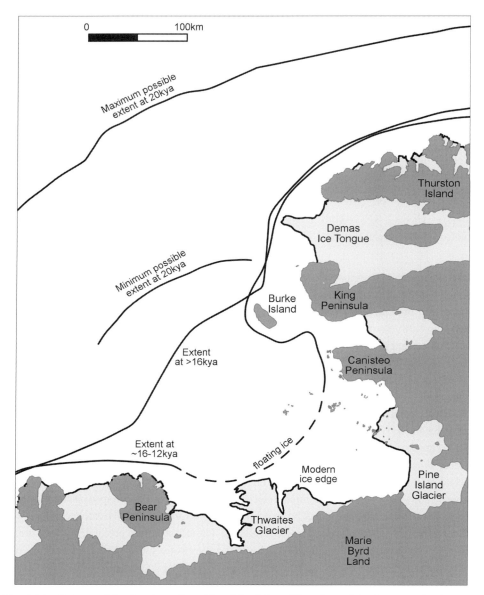

Fig. 4.14   Retreat of the ice sheet from Pine Island Bay. The edge of the ice cover is shown at approximately 20 kyr, 16 kyr and 16–12 kyr BP. (Modified from Lowe and Anderson, 2002.)

a 70-m-thick grounding wedge (subglacial material deposited at the grounding line). The second phase of deglaciation occurred from 16–12 kyr BP, coincident with the period encompassing the two meltwater pulses, so may be associated with increased calving and grounding-line retreat caused by the rise in sea level. Antarctic Peninsula deglaciation began in the southwestern region at ~15 kyr BP (Bentley *et al.*, 2006).

Monnin *et al.* (2001) subdivided the warming period following the LGM into four distinct periods based primarily on variations in CO_2 from Dome C, but which also correspond to discontinuities in δD and CH_4 (Fig. 4.15). The start of interval I is characterised by the synchronous start of increases in CO_2 and CH_4 at ~17 kyr BP, which appear to lag the start of warming temperatures by some 800 ± 600 kyr. Note that this corresponds with Heinrich event H1, as discussed later in this chapter in Section 4.4. During interval II the rate of CO_2 increase declines while CH_4 values remain constant. The δD data reveal a further warming (AIM 1) that occurs at the same time as MWP-1a (14.2 kyr BP). Following this, Antarctic

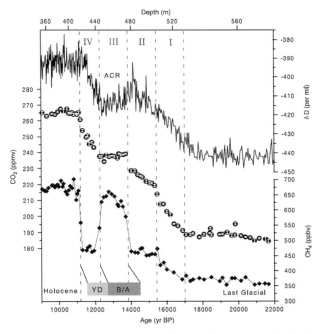

Fig. 4.15 The solid curve indicates δD in the ice of Dome C as a proxy for local temperature. Solid circles represent CO_2 data from Dome C (mean of six samples; error bars, 1σ of the mean). Diamonds show methane data from Dome C (the 1σ uncertainty is 10 ppbv). In the CO_2 and methane records, four intervals (I through IV) can be distinguished during the transition. The δD record is highly correlated with the CO_2 record, with the exception that the increased rates during intervals I and II are not significantly different in the deuterium record. The Younger Dryas (YD) and the Bølling/Allerød (B/A) events recorded in Greenland ice cores are indicated by shaded bars according to the GRIP timescale. Comparisons of the methane record with that of GRIP demonstrate that the YD corresponds to interval IV and the B/A event corresponds to interval III. (From Monnin *et al.* 2001.)

temperatures cooled by about 3 °C, a period known as the Antarctic Cold Reversal (ACR), which is observed in ice cores from across the continent (Watanabe *et al.*, 2003) (interval III in Fig. 4.15). During the ACR, which commenced at ~13.8 kyr BP and lasted for two millennia, CO_2 concentrations were stable but at slightly higher values than those in interval II. In contrast, the CH_4 concentration increased dramatically in this interval but then fell back to similar values to interval II in interval IV. This rapid increase and subsequent decrease in CH_4 concentrations can be linked to the expansion and contraction of northern latitude and tropical wetlands during the Bølling/Allerød and Younger Dryas, respectively (Section 4.2.5). The jump in CO_2 was probably related to changes in the THC. Through interval IV both CO_2 and Antarctic temperatures increased rapidly to levels observed in the Holocene.

During the transition from the LGM to Holocene encompassed by Fig. 4.15, there were also significant changes in the atmospheric chemistry: between 18–12 kyr BP there was a 24-fold decrease in dust deposition at Dome C. Although a smaller magnitude decrease in sea salt concentrations was observed, the temporal pattern of variability was markedly different, suggesting that the former record appears to primarily reflect climate changes in the source region of Patagonia rather than a reorganisation of the atmospheric circulation over the South Atlantic (Röthlisberger *et al.*, 2002).

While the δD record indicates the AIM 1 warming, the deuterium excess d (which can be related to SST changes in the moisture source region) has a similar maximum approximately 800 years later (Stenni *et al.*, 2001) followed by an oceanic cold reversal of about 0.8 °C. During the period between the two maxima the meridional temperature gradient between Dome C and its moisture source in the Indian Ocean would have increased significantly, suggesting an intensification of the high latitude atmospheric circulation (a more positive SAM). Note that the jump in CO_2 at the start of interval III occurs between the beginning of the ACR and the oceanic cold reversal, perhaps indicating a role for the Southern Ocean and/ or the circumpolar westerlies in controlling the carbon cycle. Such a mechanism has been proposed based on examination of recent data (e.g. Le Quéré *et al.*, 2007).

4.4 Linking high latitude climate change in the two hemispheres

Ice core and coral reef records indicate that low-frequency (10–100 kyr) late Pleistocene temperature and ice volume variations are roughly in phase – to within a few millennia – between both hemispheres and most studies imply that this variability is paced by Northern Hemisphere summer insolation forcing, itself controlled by the interplay between the obliquity (~43 kyr) and precession (~23 kyr) frequencies (e.g. Jouzel *et al.*, 2007; Kawamura *et al.*, 2007). Raymo *et al.* (2006) argued that following the MPT, ice sheet volume in both polar regions increased due to the establishment of positive globally synchronous feedbacks, such as CO_2 and albedo, at the precession frequency. Huybers and Denton (2008) believed that while the Northern Hemisphere climate did respond primarily to changes in the intensity of summer insolation (Milankovitch Theory) during the Pleistocene, in contrast the southern climate responded to the duration of the summer season in that hemisphere. The difference reflects the contrasting land–sea distributions, and hence how the ice sheets of the two hemispheres react to orbital forcing. The Northern Hemisphere ice sheets, in particular in North America, had extensive terrestrial margins and lost mass through surface ablation with ice-temperature feedbacks being of primary importance. In contrast the Antarctic Ice Sheet loses mass by calving and basal melting of ice shelves into the ocean: it is thus more stable and less affected by and able to influence atmospheric temperature. Instead, the temperature is controlled by the radiation balance and thus by the length of the summer season. Huybers and Denton (2008) also postulated that a long summer would reduce Antarctic sea ice and increase the outgassing of CO_2 from a less stratified Southern Ocean. This relationship is also found at millennial timescales (Ahn and Brook, 2008), where it takes place contemporaneously with a shallowing of NADW production (Fig. 4.6b).

Correlating high-resolution CH_4 records from Greenland and Antarctic ice cores has demonstrated that since the last interglacial the two hemispheres are generally asynchronous in terms of higher frequency millennial-scale variability, such that there is a so-called 'bipolar see-saw' in temperature. For example, Blunier and Brook (2001) showed that between 85–40 kyr BP seven Antarctic warm events (A1–A7) in the Byrd core preceded major Greenland interstadials (DO events 8, 12, 14, 16/17, 19, 20 and 21) by 1.5 to 3 kyr from the GISP2 core: warm events A1–A4 appear to have occurred immediately after Northern Hemisphere Heinrich events H4–H6 but chronological uncertainties remain as to their relative timing. Using the high-resolution EDML Antarctic core and NGRIP Greenland core, EPICA Community Members (2006) were able to reveal a one-to-one coupling between the majority of Antarctic warm events and DO events during MIS 3 (Fig. 4.16). This relationship has also been supported by independent dating using beryllium-10 variability (Raisbeck *et al.*, 2007). During MIS 3, sea level changes followed an 'Antarctic rhythm' (Siddall *et al.*, 2008) whereby a sea level rise occurred when the Antarctic was warm: if this were the case then Greenland temperatures would be out of step with the size of the Northern Hemisphere ice sheets. In fact, the out-of-phase relationship is actually about 90° with Antarctica leading, so the see-saw analogy is not strictly correct. Moreover, because the temperature increases in Antarctica are less rapid than those associated with DO events, the peak warmth in both hemispheres actually occurs at approximately the same time (Steig, 2006). Barker *et al.* (2009) suggested that Antarctic temperatures warm more slowly than those in the Arctic because the ACC, acting as a dynamical and thermal boundary in the Southern Ocean, slows the southward propagation of thermal anomalies from the South Atlantic.

EPICA Community Members (2006) noted that the amplitude of Antarctic warm events was linearly related to the duration of the concurrent Northern Hemisphere interstadial during MIS 3, suggesting the bipolar see-saw is achieved through an oceanographic teleconnection via changes in the strength of the MOC. They hypothesised that the duration of Northern Hemisphere stadials recorded in Greenland reflects the period of reduced MOC and hence the amount of heat accumulated within the Southern Ocean, acting as a heat

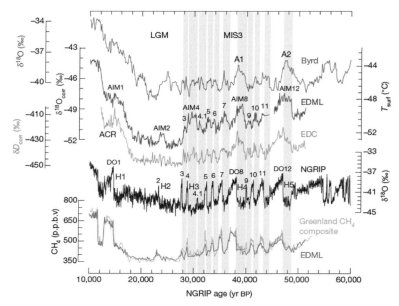

Fig. 4.16 Methane synchronisation of the EDML and the NGRIP records reveals a one-to-one assignment of each Antarctic warming with a corresponding stadial in Greenland. Displayed are 100-yr averages during MIS 3 in the EDML, EDC and Byrd ice core for the time interval 10–60 kyr BP in comparison with the NGRIP δ^{18}O record from northern Greenland. All records are CH$_4$ synchronised and given on the new GICC05 age scale for the NGRIP ice core, which has been derived by counting annual layers down to 41 kyr and by a flow model for older ages. Vertical bars indicate the Greenland stadial periods that we relate to respective Antarctic temperature increases. The approximate timing of Heinrich layers in North Atlantic sediments is also indicated. The y-axis on the right side indicates approximate temperature changes at EDML based on the modern spatial gradient between δ^{18}O and temperature. (From EPICA Community Members, 2006.)

reservoir. However, during the LGM, when conditions in the North Atlantic were very different from MIS 3, this relationship breaks down: the warming of AIM 2 at ~24 kyr BP is much less than would be expected. Nonetheless, the fact that AIM 2 lasts only 2 kyr suggests that the MOC was not significantly reduced throughout the entire LGM, and indeed this conclusion has been reached directly from North Atlantic sediment proxies (cf. Section 4.2.5). Based on model data and proxy records of temperature, Barker et al. (2009) proposed that during Northern Hemisphere stadials a southerly shift of the circumpolar trough and associated westerlies took place: the resultant stronger ACC meant more vertical mixing and greater release of CO$_2$ from the Southern Ocean. During the Oldest Dryas stadial (~18–15 kyr BP) this process, in combination with the build-up of heat in the Southern Ocean, was able to cause sufficient warming to start Southern Hemisphere deglaciation.

The trigger for the transition from interstadial to stadial remains uncertain but significant discharges of fresh water into the North Atlantic, primarily associated with the Laurentide

Ice Sheet, are likely to be a factor. A modelling study by Knutti *et al.* (2004), using a coupled ocean–atmosphere–sea ice model, revealed that such discharges have a marked effect on Southern Ocean temperatures due to anomalous southward oceanic heat transport. However, the origin of such water input has not been ascertained for all individual DO events; moreover, the largest inputs of North Atlantic fresh water, from Laurentide Ice Sheet iceberg discharge, do not systematically coincide with either the onset or the end of stadials. Knorr and Lohmann (2003) proposed that a warming Southern Ocean that reduced sea ice extent around Antarctica was the driver for increasing the MOC at the beginning of an interstadial, such as the Bølling/Allerød period: in their model the stronger THC, due to increased mass transport into the South Atlantic via warm (Indian Ocean) and cold (Pacific Ocean) currents, prevails over the impact of freshwater discharge into the North Atlantic. At higher frequencies, Robinson *et al.* (2005) demonstrated radiocarbon variations in intermediate depth (rather than deep ocean) water mass properties are associated with approximately synchronous but small climate changes in the two polar regions.

5

The Holocene

5.1 Introduction

The Holocene is the period of approximately the last 11.7 kyr and covers the time from the end of the last ice age up to the present. It therefore includes the so-called anthropocene, which is the period when humankind has influenced the climate system. For most of this latter period there are instrumental meteorological records, and this era is dealt with in the next chapter.

The Holocene marks the return of warmer and more humid conditions after the cold and dry period of the Last Glacial Maximum (LGM) (see Section 4.2.5). The start of the Holocene coincided approximately with the end of the Younger Dryas event (12.8–11.5 kyr BP) (see Section 4.2.5), an abrupt return to cold conditions (stade) during the gradual warming at the end of the last Pleistocene glaciation. Temperatures derived from ice cores collected on the Greenland icecap (see Section 2.4.2) suggest that the transition to the Holocene was a rapid switch of mode, with the Younger Dryas ending abruptly over a period of about 50 years. However, in other parts of the world the transition was not so rapid.

The Holocene can be split into a number of stages, and in this chapter we will divide it into the Early Holocene (11.7–5 kyr BP), the Mid Holocene (5–3 kyr BP) and the Late Holocene (3 kyr BP to present) (see Table 5.1).

Greenland ice core data suggest that temperatures during the Holocene have been about 12 °C higher than during the Pleistocene. In addition, there has generally been a reduction in climatic variability on all timescales from the decadal to the millennial (Dansgaard *et al.*, 1993). The Pleistocene–Holocene transition seems to have been triggered by orbital variations, but the rapid changes observed in the climatic conditions inferred from the proxy data point to the importance of feedbacks within the Earth's climate system that enhanced elements of the transition, so that rapid, synchronous changes took place in both polar regions. However, it took a considerable time for the major ice sheets to melt and it was several thousand years before global sea level reached its high level in the Mid Holocene.

Although climate variability has generally been reduced during the Holocene, with none of the major multimillennial-scale climatic shifts that have characterised the glacial periods, there have nevertheless been smaller climatic fluctuations. Realisation of this extends back to the early 1970s when Denton and Karlén (1973) found evidence of synchronous advances

Table 5.1   *Timeline of the Holocene*

	The Arctic		The Antarctic		Global	
	Date	Event	Date	Event	Date	Event
The Early Holocene *c.* 11.7–5 kyr BP	c. 11.5 kyr BP	Younger Dryas ends in Northern Hemisphere				
	11–9 kyr BP	More summer insolation in Northern Hemisphere coinciding with the Preboreal warming – the first phase of the Boreal	11.5–9 kyr BP	Early Holocene optimum		
	11.5–8.9 kyr BP	The Boreal cool dry episode of northern Europe				
	c. 10.75–10.2 kyr	The Erdalen cool event	*c.* 9.6 kyr BP	Collapse of King George VI Ice Shelf		
			c. 9 kyr BP	The maximum of Antarctic Holocene temperatures		
	8.9–5.7 kyr BP	The Atlantic warm, moist episode of northern Europe				
	5.5–5 kyr BP	The Piora Oscillation: a cold episode				
	c. 8.5 kyr BP	The European ice sheets had melted by this time				
	8.2 kyr BP	A major cooling event of the North Atlantic				
	c. 7 kyr BP	Start of the climatic optimum	5.4–5.2 kyr BP	Abrupt weakening of the westerlies		

Table 5.1 (cont.)

	The Arctic		The Antarctic		Global	
	Date	Event	Date	Event	Date	Event
The Mid Holocene c. 5–3kyr BP	c. 5.7–2.6kyr BP	The Sub-Boreal episode of northern Europe				
	c. 4kyr BP	End of the climatic optimum	4.5–2.8kyr BP	The Holocene climatic optimum		
	c. 4kyr BP	The peak of temperature in North America	6–3kyr BP in East Antarctica	Mid Holocene Hypsithermal		
The Late Holocene c. 3 kyr BP to present	c. 2.6kyr BP to present	The subatlantic period of northern Europe	1.2kyr BP	Intensification of the westerlies and the Amundsen Sea Low	End 17th century	The Maunder Minimum period of reduced solar irradiance
	AD 950–1300	The Medieval Warm Period	4–2kyr BP	The Antarctic Late Holocene climatic optimum		
			c. 1.9kyr BP	The ice in the Prince Gustav Channel may have reformed	AD 1350–1850	The Little Ice Age
			c. 1.4kyr BP	The Larsen A Ice Shelf reformed		

in mountain glaciers across North America and Europe suggesting the occurrence of broadscale climatic fluctuations. They proposed that these climate variations were part of a regular millennial-scale cycle, which coincided with climatic shifts during the last glaciation. Further evidence for climatic cycles came from proxy data, such as temperatures derived from Greenland ice cores and North Atlantic deep sea cores, which showed that there have been a number of abrupt climatic shifts. By contrast, the Antarctic climate record suggests that the Holocene has been the longest warm 'stable' period of the last 400 kyr (Petit *et al.*, 1999). But there have still been changes, although their amplitude is relatively small compared with the variations experienced during the Pleistocene. Many proxy records suggest that climate variability during the Holocene was greater than typically observed instrumentally during the twentieth century.

In the second half of the Holocene there were significant climatic fluctuations, such as the Little Ice Age (LIA) and the Medieval Warm Period (MWP) (see Section 5.4.4). This period is of particular interest in the study of climate because the orbital conditions, greenhouse gas levels and state of the cryosphere were similar to those in the era before humankind had a major influence on the environment. Hence natural climatic variability can be investigated immediately prior to the major increases in greenhouse gases that have taken place since the Industrial Revolution.

In this chapter we review the climatic conditions in the Holocene up to the start of the instrumental period, approximately 100 years ago. We examine the variations in forcing on the climate system during the period, the changes that took place in the atmospheric and oceanic circulations, sea ice extent, concentrations of various gases relevant to environmental change and the extent and thickness of the polar icecaps. Our knowledge of conditions during the Holocene has largely been obtained from proxy data, such as ice and sediment cores, and tree rings, and is therefore less detailed than for the last 100 years. The last 100 years is essentially the period of in-situ observational data and is dealt with in the next chapter.

5.2 Forcing of the climate system during the Holocene

5.2.1 Introduction

The climatic changes that occurred at high latitudes during the Holocene were a result of the combined effects of several different forcing factors operating on different timescales, coupled with feedbacks within the Earth system itself. On the longest, millennial and multimillennial timescale changes took place in the orbital characteristics of the Earth (eccentricity, precession and obliquity; see Section 3.4.1) relative to the Sun, resulting in variations in the seasonal and latitudinal receipt of solar radiation, although there was virtually no change in the total amount of radiation received by the Earth over the course of the year as a whole. However, changes in the solar radiation emitted by the Sun had an impact on the amount of energy received at the top of the atmosphere. Although this was felt proportionately across the whole of the Earth, the impact on surface conditions varied

198   *The Holocene*

across the globe because of feedbacks involving cloud, ice and other elements of the Earth system.

5.2.2   Orbital changes

Variations in the Earth's orbit around the Sun have the potential to force major changes in the climate system and work by the Cooperative Holocene Mapping Project (COHMAP) (COHMAP members, 1988) suggested that orbital changes could explain in broad terms the climatic variations since the LGM at about 18 kyr BP. Short-term climate events, which can be extremely important on a regional scale, were not considered.

During part of the Early Holocene, over 11–9 kyr BP, summer insolation in the Northern Hemisphere was about 8% greater than today due to a difference in orbital precession that aligned the boreal summer solstice with the perihelion (the closest approach of the Earth to the Sun). This gave greater levels of summer insolation at all latitudes of the Northern Hemisphere, with locations at $60°$ N receiving about $40°$ W m^{-2} more radiation than today (Bradley, 2003). Conversely, January insolation was 8% less than today. During the course of the first half of the Holocene the orbital conditions changed so that the present situation of perihelion occurring close to the winter solstice was reached by the Mid Holocene. In the Early Holocene insolation anomalies in the Southern Hemisphere summer were smaller than in the north and they mainly occurred at lower latitudes.

The fact that the greatest change in the Early Holocene occurred in the summer was especially important in the Arctic, when the ice-albedo feedback mechanism (Section 3.3.1) is most effective and changes in irradiance can be amplified and result in broadscale loss of sea ice.

Although we have a good understanding of how the changes in orbital characteristics resulted in variations in the solar radiation received at the top of the atmosphere during the Holocene, it is less clear how such changes to the solar input translated into modifications to the climatic parameters at high latitudes. There are no simple relationships between incoming solar radiation and parameters such as cloud cover or temperature since feedback mechanisms will play an important role, especially in the polar regions. In addition, model simulations have suggested that on the margins of the Arctic, vegetation feedbacks could have played a role in modifying the changes in solar forcing (TEMPO, 1996).

5.2.3   Solar output

As discussed later in this chapter, variations in solar output appear to have had an influence on the climate system throughout the Holocene, with signals being detected as changes in the atmospheric circulation and oceanic conditions. However, in order to establish relationships between solar activity and climate variables it has been necessary to make extensive use of proxy data, such as marine sediments, and this limits the robustness of the relationships that can be established.

5.2 Holocene forcing of the climate system

The variability of solar radiation

Solar irradiance varies most over the solar cycle of about 11 years, with output changes of about 0.08%, but there are also longer term variations superimposed on this cycle. Studies have suggested that the most marked solar variation cycles have periodicities of about 208, 350, 700 and 950 years (Lean, 2002). Obtaining data on past variations in solar irradiance is difficult, but some information can be obtained from production rates of the cosmogenic nuclides carbon-14 (^{14}C) and beryllium-10 (^{10}Be). Carbon-14 is produced in the upper atmosphere when cosmic rays convert nitrogen to this form of carbon. Since increased solar activity results in fewer cosmic rays, ^{14}C concentrations are *lower* at the time of the sunspot maximum.

Lean *et al.* (1992) examined the output of stars similar to our Sun and estimated that the Sun's output had increased by about 0.24% since the time of the Maunder Minimum at the end of the seventeenth century. Over this period, Northern Hemisphere near-surface air temperatures changed by about 0.2–0.4 °C (Lean *et al.*, 1995) and it has been suggested that much of the low frequency variability of temperature can be explained by changes in solar irradiance, along with the effects of volcanic forcing (Crowley, 2000). Periods of reduced solar irradiance, such as the Maunder Minimum, give generally cooler conditions, which are especially evident in the mid to high latitude interiors of North America and Eurasia. At the same time, warmer conditions are found over the mid to high latitude areas of the Atlantic – a pattern that has the characteristics of the North Atlantic Oscillation (NAO) in its negative index phase. This is reflected in less warm air advection across the Atlantic and colder conditions in Europe.

It is difficult to determine solar variability during the early part of the Holocene, but those estimates that have been made from production rates of ^{14}C suggest a large number of periods of anomalous solar activity (Stuiver and Braz;iunas, 1993) (Fig. 5.1) with irradiance levels possibly varying by as much as ± 0.4%. The extent to which such variations in solar activity are responsible for climatic changes in the Holocene is still being debated. However, in East Greenland the Holocene records show that the climatic proxies follow the broad trends indicated by changes in solar insolation. In addition, comparisons of proxy records of precipitation with the time series of solar variability have shown significant correlation at a number of locations, especially in the tropics. At higher latitudes, Stuiver *et al.* (1991) highlighted the similarity between an approximately 1470-year cycle in production rates of ^{14}C data and a similar cycle in the oxygen isotope temperature proxy data from the Greenland Ice Sheet Project (GISP2) ice core. But it is presently unclear how robust such relationships are and what mechanisms are responsible that could give rise to such direct climatological signals from changes in solar radiation.

In the Mid Holocene, the Antarctic climatic optimum occurred over *c.* 4.5–2.8 kyr BP (Bentley *et al.*, 2009) and coincided roughly with the solar maximum for these latitudes. For the more recent past, Beer *et al.* (1996) identified high levels of solar activity from about AD 1100–1250, with minima in the fifteenth century and at the end of the seventeenth century, followed by an increase up to the present day when values were comparable to the twelfth century (Fig. 5.2).

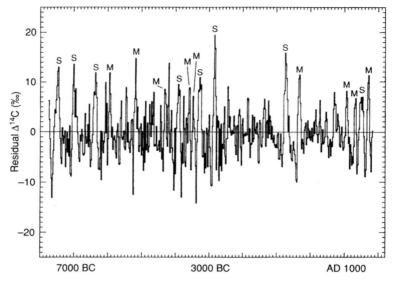

Fig. 5.1 Solar activity variations (lower values indicating higher solar activity/enhanced irradiance), as recorded by radiocarbon variations in tree rings of known calendar age. M, Maunder Minimum-like events; S Sporer Minimum-like events. (From Stuiver et al., 1991.)

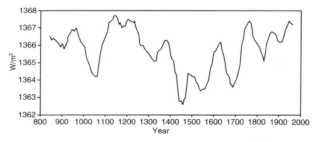

Fig. 5.2 Total solar irradiance from AD 843 to 1961, estimated from ^{10}Be variations recorded in an Antarctic ice core. (After Bard et al., 2000.)

The 1500-year cycle

A marked climatic cycle that has been observed in a number of Holocene proxy data sets, such as deep ocean cores and ice cores, has a period of about 1500 years. There are indications that this cycle persisted through the last glacial termination and well into the last ice age, suggesting that the cycle is a pervasive feature of the climate system. The causes of this cycle are not known with certainty (Campbell et al., 1998), but the fact that it has been observed across many parts of the Earth suggests that it could be forced by mechanisms external to the Earth, such as variations in solar output.

The events occurring roughly every 1500 years during the Holocene have been called Bond events after the work of Bond *et al.* (2001) who used marine sediment and ice core data to identify a correlation over the last 12 000 years between ice-rafted debris (IRD) in the North Atlantic and the production of ^{14}C and ^{10}Be in the atmosphere. The spatial distribution of IRD gives an indication of the surface winds and therefore the broadscale atmospheric circulation, while the two chemical species were taken as indicators of solar output. Bond's work suggested a close correlation between the solar output and millennial-scale variations in the surface winds and hydrography of the subpolar North Atlantic. In addition, the data indicated that every centennial timescale increase in drift from the north was linked to an interval of variable and overall reduced solar output. The most recent ice drift cycle was correlated with the LIA, which coincided with the Maunder solar minimum.

But the problem remains of how relatively small variations in solar irradiance can induce major changes in the climate system. Model experiments suggest that with reduced solar radiation there should be a cooling at high northern latitudes as a result of changes in stratospheric ozone concentration and distribution. In addition, there should also be a small southward shift of the northern subtropical jet and a decrease in the northern Hadley circulation. Together these effects could lead to the increased southward drift of North Atlantic sea ice, as well as the cooling of the ocean surface and atmosphere.

The cooling pattern of these cycles resembles those that resulted from reduced North Atlantic Deep Water (NADW) formation during the Younger Dryas, and some proxy data from ocean sediment records do suggest reduced NADW production during some of the peaks in drift ice. So the amplification mechanism may have been a reduction in the North Atlantic MOC coupled with increased southward sea ice transport into the Nordic Seas. Evidence to support these ideas comes from conditions observed in the North Atlantic during the 1960s and 1970s. At this time there was increased transport of sea ice through the Fram Strait resulting in surface water freshening in the Nordic Seas and the subpolar North Atlantic (the Great Salinity Anomaly). The freshening reduced convection in the Labrador Sea and there was basin-wide cooling. An atmospheric high-pressure anomaly over Greenland then resulted in strong northerly winds that spread sea ice into the subpolar North Atlantic. So amplification of the relatively small solar changes may have been the result of the Arctic Ocean's sensitivity to changes in surface-water salinity, so that greater freshening of the waters would reduce the NADW and thus allow the solar influence to spread beyond the Arctic.

While the 1500-year cycle has been linked to variations in solar output, it may also be partly a result of ocean variability. This possibility is discussed in Section 5.4.1.

5.2.4 *Volcanic aerosols*

Proxy records of volcanic eruptions during the Holocene have been created using data such as sulphate concentrations in the GISP2 ice core from Greenland (Zielinski *et al.*, 1994; Zielinski, 2000) and tephra layers in Irish peats (Pilcher *et al.*, 1995). Specific volcanic

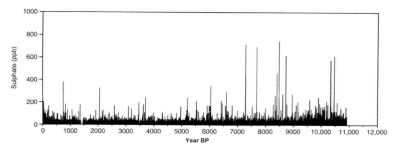

Fig. 5.3 The Holocene record of volcanic sulphate recorded in the GISP2 ice core from Summit, Greenland. (From Zielinski *et al.*, 1994.)

events identified via electrical conductivity measurements of the core include both local and intrahemispheric eruptions. Local events included the AD 1783 Laki and AD 1362 Oraefajokull eruptions from Iceland. More distant eruptions detected included Mount St Helen's, Washington (AD 1479) and Tambora (AD 1815). The data (Fig. 5.3) suggest that explosive volcanic eruptions were more frequent during certain parts of the Holocene, such as during 11.5–9.5 kyr BP. Three times as many volcanic events occurred from 9–7 kyr BP compared with the last 2 kyr. This volcanism may have contributed to volcanic cooling in the early Holocene. Figure 5.3 also shows that in the early part of the Holocene many of the eruptions were intense compared with the events in the later part of the record. Of course the GISP2 data are not giving a true global picture of volcanic activity and are biased towards registering the signals of eruptions closer to Greenland. However, modelling studies have shown the importance of volcanic eruptions in modulating the surface temperature and they have undoubtedly played a part in climate variability during the Holocene.

5.3 Atmospheric circulation

5.3.1 Introduction

The atmospheric circulation varied during the Holocene for several reasons, such as the blocking effects of the major ice sheets during the first half of this period and changes in insolation, which modified the distribution of surface temperatures.

For most of the Holocene the circulation of the atmosphere has to be estimated from various proxies, using data extracted from ice cores, driftwood records and marine sediment cores. Of course none of these can provide direct information on atmospheric circulation parameters, such as the locations of the major storm tracks. But analysis of driftwood at coastal stations, for example, can give some indication of the prevailing wind direction, provided that the origin of the different types of wood is known. To get some idea of the broadscale atmospheric circulation, empirical orthogonal function (EOF) analysis techniques have been applied to time series of the chemical species concentrations in ice cores and used to produce various atmospheric circulation indices.

5.3.2 The Arctic

Broadscale circulation

The polar vortex is a major feature of the high latitude atmospheric circulation and ice core records can provide some indications of its spatial extent and intensity. To try and estimate the state of the polar vortex, an EOF decomposition of various marine and terrestrial source species in the GISP2 ice core from Greenland summit was used to develop a polar circulation index (PCI) (Mayewski *et al.*, 1994). EOF 1 was found to reflect a well-mixed 'background' atmosphere representing 92% of the total variance in chemical species behaviour, with the time series of EOF 1 being referred to as the polar circulation index. This is thought to be a measure of the relative size of the polar vortex and gives an indication of the intensity of the polar atmospheric circulation.

EOF 1 was found to increase during a number of periods in the Holocene, and particularly during the period before 11.3 kyr BP, 610–0, 6100–5000 with smaller magnitude increases during 8.8–7.8 and 3.1–2.4 kyr BP (O'Brien *et al.*, 1995). These changes indicated that either the polar vortex expanded or the meridional (north–south) air flow intensified during these periods. Overall the EOF 1 analysis suggests significantly less atmospheric variance during the Holocene than in the previous 41 kyr (Mayewski *et al.*, 1994). In addition, analysis of the chemical species in the GISP2 core indicates that the variability in chemical records during the Holocene was more subdued than in the earlier glacial period but presented a more variable record of changes in source strength and atmospheric circulation (Mayewski *et al.*, 1997).

In the following sections we examine how the atmospheric circulation changed during the Holocene in various parts of the Arctic.

North America

At the start of the Holocene the Laurentide Ice Sheet was still present in central and eastern Canada and it exerted a major influence on the atmospheric circulation across the continent. Modelling experiments have shown that the ice sheet probably split the westerly jet stream into northern and southern branches across the North American continent prior to the start of the Holocene (see Section 4.2.5). However, by 12 kyr BP there was a single southern branch, although its strength was greater than today. This resulted in an increase in the number of storms. Over the ice sheet there was still a glacial anticyclone, but it was weaker than during the Pleistocene. The tree line moved northwards in northwestern Canada in the Early Holocene as a result of the increased summer insolation and there was reduced precipitation in the interior of Alaska. At boreal latitudes (60–65° N) there was a strong east–west difference in the climates in the Early to Mid Holocene that suggests an interaction between insolation forcing and the Laurentide Ice Sheet.

By 9 kyr BP the ice sheet had reduced in size but it still influenced the climate. The glacial anticyclone over the Laurentide Ice Sheet continued to weaken, while the Pacific subtropical high had strengthened off the west coast of North America. With the further retreat of the ice sheet there were large changes in the atmospheric circulation of the North American and

North Atlantic regions. The Dye 3 ice core from northwestern Greenland indicates that the climate here changed from an extremely continental to a more maritime character. By 6 kyr BP northwestern Canada began to cool and there was an increase in precipitation. By contrast, in central Canada there were changes in the tree line as a result of warming. So the available data suggest that across high latitudes in North America there was an increase in Mid to Late Holocene moisture and a Late Holocene cooling.

Northern Europe and the North Atlantic

The joint European Greenland Ice-core Project (GRIP) ice core provides valuable information on the changing atmospheric conditions across the North Atlantic region during the Holocene. Analysis of this core showed that following a period of low snow accumulation during the rapid return to glacial conditions of the Younger Dryas, accumulation increased in the Early Holocene. The nearby GISP2 core indicates that snow accumulation doubled rapidly from the end of the Younger Dryas event to the subsequent Pre-Boreal interval, in a period of about three years, with most of the change taking place in one year (Alley *et al.*, 1993). In addition, from the Younger Dryas to the Pre-Boreal the 25-year running mean snow accumulation increased by 80% in 25 years and doubled in 41 years. It appears likely that this greater accumulation was associated with an increase in the number and/or duration of active weather systems.

At the start of the Holocene the Fennoscandian Ice Sheet was still present and climatologically it had an anticyclone above it that resulted in more easterly winds along its southern boundary than occurs in the area today. However, by about 8.5 kyr BP the European ice sheets had melted and reconstructions of the 500-hPa circulation at this time (e.g. Lamb, 1995), suggest more ridging of high pressure across the USA than today with a trough across the North Atlantic. The polar vortex is thought to have been centred to the west of Greenland, with storms being steered towards both sides of Greenland and the Arctic Ocean. Iceland and Greenland were warm at this time as a result of the more southwesterly flow.

The GRIP and GISP2 ice core records indicate a marked period of environmental change about 8.4 kyr BP. This was at the same time that a glacial advance (the Cockburn Stade) took place in North America.

The Russian Arctic

In north-central Russia (68–70° N) the atmospheric situation was quite different from Europe or North America as there was no influence in the Early Holocene of the remnant ice sheet and the area therefore responded directly to increased summer insolation, with higher summer temperatures between 9 and 4 kyr BP (Macdonald *et al.*, 2000). Lake sediment data (Wolfe *et al.*, 2000) suggest that close to the Nordic Seas there were wet conditions at 9 kyr BP because of a deeper Siberian Low and the penetration of warm maritime air masses. More inland sites had a warm, dry continental climate. When sea surface temperatures (SSTs) declined and cyclonic activity weakened through the Mid to Late Holocene the strong moisture gradient weakened.

Lamb (1977) has suggested that the MWP and the LIA (see Section 5.4.4) in the Northern Hemisphere were related to the north–south movement of the mid latitude westerlies. Southward movement of the cyclonic storm track is associated with strong westerlies and advances of cold polar air masses towards the south. It is therefore possible that the meridional circulation was stronger during the LIA. Warm periods, such as the MWP, are associated with the northward movement of the storm track and weaker meridional to zonal (east–west) atmospheric circulation.

Some aspects of the atmospheric circulation over the Arctic Ocean during the Holocene can be inferred from computer simulations of recent years of extreme ice conditions, which are thought to be similar to ice conditions during early periods in the Holocene when the ice was more extensive. As discussed below in Section 5.6.2, Tremblay *et al.* (1997) used a sea ice model to reproduce sea ice conditions in 1968 and 1984 – years of extensive and little sea ice export respectively. This allowed them to consider the Holocene sea ice drift patterns proposed by Dyke *et al.* (1996) based on analysis of driftwood collected on the coast of the Canadian Arctic. Periods of extensive transport of ice out of the Arctic Ocean could be simulated using the atmospheric circulation data of 1968 (Fig. 5.4), which had a weak Icelandic Low and low pressure over northern Scandinavia. In this scenario the Transpolar Drift Stream (TDS) was broad and shifted to the east, resulting in very cold temperatures in the northern North Atlantic. In contrast, the atmospheric forcing for 1984, with a deep Icelandic Low, resulted in ice transport across the Arctic Ocean towards the Canadian Archipelago and anomalously warm temperatures across northern Europe. Overall, the results suggested that on the timescale of centuries to millennia the average atmospheric circulation over the Arctic Ocean during the Holocene may have resembled the current climatology with extremes similar to 1968 and 1984. In addition, there may have been abrupt changes between these two extreme states.

The North Atlantic Oscillation (NAO)

The NAO is a measure of the meridional flow across the North Atlantic and is computed from the surface pressure difference between Iceland and the Azores (see Section 3.4.5). The modelling of Tremblay *et al.* (1997) discussed above can provide some insight into how the NAO might have varied during the Holocene. Their simulation of 1968 had the characteristics of the NAO in its negative phase (weak Icelandic Low), while 1984 was a positive index year. They suggest that from their interpretation of the three modes of ice transport suggested by driftwood records (see Section 5.6.2) these can be equated to the positive and negative phases of the NAO and today's mean climatology. Their results indicate that over the last 10 kyr the atmospheric circulation of the Arctic at various times resembled the NAO in its positive and negative phases, with abrupt changes from one state to another.

The nature of the NAO can also be inferred from the locations of the subtropical high-pressure belt and the mid latitude low-pressure centre. Both were further south over 12–8 kyr BP and moved north around 8–6 kyr BP. But the zones moved south again over 4–3 kyr BP (see Lamb, 1995, p. 127).

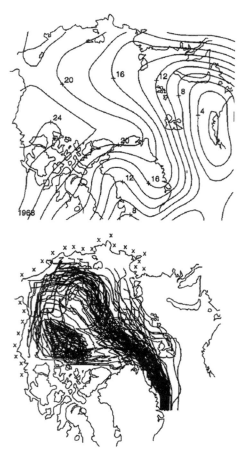

Fig. 5.4 Winter sea level pressure pattern (hPa–1000 hPa) and spaghetti plot including driftwood sources for 1968. The crosses indicate the source regions for northern Baffin Bay. (From Tremblay et al., 1997.)

5.3.3 The Antarctic

There has been much less research into the atmospheric circulation of the high southern latitudes during the Holocene than for the Arctic. However, in this section we will review the information that can been gained on circulation variability from the analysis of ice cores and lake and ocean sediments.

As discussed in the previous section, the presence of large ice sheets across North America, Greenland and Scandinavia in the Early Holocene resulted in considerable atmospheric circulation anomalies around the Arctic. In contrast, throughout the Holocene there was little zonal asymmetry around the Antarctic and the major changes therefore involved meridional shifts in the atmospheric circulation.

At the LGM some 20–18 kyr BP the land ice over the Antarctic and the sea ice around the continent were more extensive than at present, which resulted in the climatic atmospheric polar front being about 5–7° of latitude further north than today (Heusser, 1989). This resulted in stronger westerly winds over the Southern Ocean than we experience now. The easterly flow at the edge of the Antarctic Ice Sheet, to the south of the circumpolar trough, was also stronger at this time, so further strengthening the circumpolar vortex.

The ice core record has been particularly valuable in providing insight into changes in wind strength and direction over the Southern Ocean. For example, an increase in calcium at an ice core site indicates increasing transport of terrestrial dust while a corresponding increase in sodium reflects increasing transport of ocean sea salt. These two quantities increased together after AD 1000 in the Siple Dome core, suggesting strengthening of the Amundsen Sea Low (ASL) and the Southern Hemisphere westerlies.

The proxy records suggest a number of atmospheric circulation changes around the Antarctic during the Holocene. Ice core evidence indicates an abrupt weakening of the Southern Hemisphere westerlies at 5.4–5.2 kyr BP and the intensification of the westerlies and the ASL at about 1.2 kyr BP. In addition, the assessment of all the available proxy data for the Antarctic Peninsula region carried out by Bentley *et al.* (2009) points to abrupt shifts in the position of the belt of strong southern westerlies during the periods of the early Holocene warm period and the Mid Holocene Hypsithermal (see Section 5.4.3).

5.4   Temperature

The start of the Holocene was characterised by a major warming as the Earth came out of the last ice age. Gas isotope data from the GISP2 ice core from Greenland suggest that at the Pleistocene/Holocene transition the temperature change there was $10 \pm 4\,°C$ (Grachev and Severinghaus, 2005).

Over the Holocene temperatures are thought to have varied by about ± 1–$2\,°C$ from present-day values, which is much less than during the last ice age. The non-polar Holocene records suggest that lower latitudes had greater regional climatic variations than the polar regions. The Early Holocene was generally warmer than the twentieth century, but the period of highest temperatures varied regionally across the Earth. There seems to have been a south to north latitude pattern, with southern latitude areas having their maximum warming a few millennia before the Northern Hemisphere regions. Most ice cores from high latitude areas have the maximum at the beginning of the Holocene at about 11.5–9.5 kyr BP. This is the case with the cores from northwest Canada (Ritchie *et al.*, 1983) and the Antarctic Plateau (Ciais *et al.*, 1992; Masson *et al.*, 2000). But central Greenland (Dahl-Jensen *et al.*, 1998), and areas downstream of the Laurentide Ice Sheet, including Europe and eastern North America, did not warm up until after 8 kyr BP.

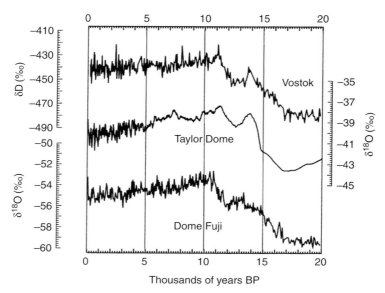

Fig. 5.5 Isotopic records from various Antarctic ice cores: Vostok, deuterium – δD; Taylor Dome, $\delta^{18}O$; and Dome Fuji, $\delta^{18}O$. (From Fisher and Koerner, 2003.)

In this section we consider first the overall degree of variability that took place, and then concentrate on specific periods of marked anomalous conditions that have received particular attention. As we are dealing here with the pre-instrumental period the discussion of temperatures is based on proxy data. These do not provide the temporal or spatial resolution that we now have with modern observing systems, and this is particularly the case for the polar regions. Nevertheless, they do provide us with data on temperatures at selected high latitude locations and, with recent advances in the analysis of ice cores in particular, are starting to provide temperature records with a resolution of years to decades, depending on the ice core accumulation rate (see Section 2.5.2). Throughout the Holocene it is possible to count annual layers in many ice cores, so providing high temporal resolution.

Many ice cores show an overall cooling from the Early Holocene until the LIA, but with some significant fluctuation during the Holocene, such as a Mid Holocene optimum (Masson et al., 2000). This is the case with the Antarctic cores collected at Vostok, Taylor Dome and Dome Fuji shown in Fig. 5.5. The cooling is also apparent in the Academii Nauk records from the Russian Arctic islands (Fig. 5.6). The trend is not present in all the Greenland cores, although it can be found in the GRIP record and the Arctic icecaps. The Agassiz core from Ellesmere Island, Northwest Territories, the most northerly island in the Canadian archipelago, is particularly valuable because it provides the only melt record extending back through the full length of the Holocene. The summer melt maximum in this core is between 11 and 9 kyr BP, peaking at 10 kyr BP. The record suggests that summers between 11.5 and 10.5 kyr BP were approximately 1–2 °C warmer than today, whereas the oxygen isotope records show

5.4 Temperature

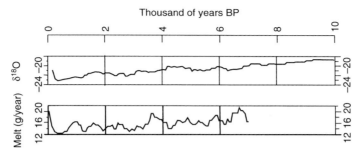

Fig. 5.6 The isotopic $\delta^{18}O$ record and melt (g cm^{-2} y^{-1}) from the Academii Nauk (Russian Arctic Islands) ice core. (From Fisher and Koerner, 2003.)

much cooler conditions in the immediate post glacial period. The ice core record from Dome C also shows higher than present temperatures over 11–9 kyr BP.

5.4.1 Temperature variability over the Holocene

Introduction

Proxy records show a number of periodic cycles in the temperature data extending back through the Holocene. In an analysis of a lake sediment core from western Canada extending back more than 4 kyr (Campbell et al., 1998) the major periods in decreasing power were 1540, 1440 1030, 590, 440, 330 and 280 years. When these cycles, along with the Milankovitch 23 708 years cycle (which dominates insolation variability at this latitude) were combined and a non-linear regression analysis carried out against the lake sediment core record, a correlation coefficient of 0.88 was obtained. This study showed that the main features of the proxy record from the lake, such as the LIA, MWP and the Neoglacial, can be related to these periodicities.

The Arctic

The proxy data that have been collected over recent decades provide a remarkable record of temperatures during the Holocene. For example, the GRIP ice core provides us with a detailed proxy of temperature changes over Greenland through the Holocene. Deep-ocean cores have provided us with further detailed records of temperature cycles during the period.

The various records present quite different pictures of temperature variability since the end of the last ice age. For example, the oxygen isotope and snow accumulation data from the Greenland ice cores show fewer millennial-scale fluctuations during the Holocene compared with the preceding glacial period, except for the cooling at 8.2 kyr BP discussed later. By contrast, the ocean cores provide much more information on millennial-scale climatic and oceanographic cycles.

Figure 5.7 shows the $\delta^{18}O$ record from the top 1500 m of the GRIP ice core, which spans approximately the last 10 kyr. This core suggests that the climate here was very stable during the Holocene, at least in comparison with the rapid changes inferred during the Pleistocene

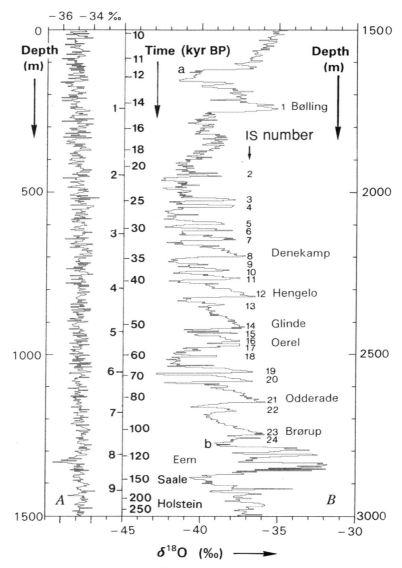

Fig. 5.7 The continuous GRIP Summit $\delta^{18}O$ record plotted in two sections on a linear scale: *A*, from the surface to 1500 m: *B*, from 1500 m to 3000 m depth. Each point represents 2.2 m of core increment. Glacial interstadials are numbered to the right of the *B* curve. The timescale in the middle is obtained by counting annual layers back to 14.5 kyr BP, and beyond that by ice flow modelling. The glacial interstadials of longest duration are reconciled with European pollen horizons. 'a' indicates the end of the Younger Dryas at 11.5 kyr BP and 'b' marine isotope stage 5d at 110 kyr BP. (From Dansgaard et al., 1993.)

and the late stages of the last glaciation. Greenland Summit borehole and isotopic data (Cuffey and Clow, 1997) indicate that the temperature range over the last 5000 years was about 3 °C. The only marked change in the GRIP data is a minimum of δ^{18}O at 8.21 kyr BP ± 30 yr (Dansgaard *et al.*, 1993). However, more recent studies, particularly those based on marine sediment cores, have suggested that the climate of the Holocene has been more variable than initially thought, and punctuated by a series of millennial-scale shifts (O'Brien *et al.*, 1995). An analysis of long-term trends has indicated an Early to Mid Holocene climatic optimum (the start of the Holocene until 6 kyr BP) with a cooling trend into the Late Holocene (deMenocal *et al.*, 2000).

Deep-sea cores from the North Atlantic provide an extremely valuable, century to decadal resolution sea surface temperature record for the Holocene that supplements the deep ice core data, which can only be collected at a very small number of locations. Cores collected on opposite sides of the Atlantic were analysed by Bond *et al.* (1997) and indicated that several ice-rafting episodes occurred during the Holocene. These suggested that the climate must have undergone a series of abrupt changes resulting in debris-bearing drift ice being deposited at sites more than 1000 km apart. The debris was then overlain by warm, largely ice free waters. The IRD events occurred at about 11.1, 10.3, 9.4, 8.1, 5.9, 4.2, 2.8 and 1.4 kyr BP. The ice-rafting events were interpreted as a series of ocean surface coolings, each of which appears to have been caused by substantial changes in the North Atlantic surface circulation. The magnitudes of the ocean surface coolings were probably no greater than 2 °C, at least across the eastern North Atlantic, implying that the coolings did not exceed 15–20% of the full Holocene to glacial temperature change.

The presence or absence of IRD suggests two modes of North Atlantic ocean surface circulation. Firstly, for conditions with little debris and warmer SSTs, the circulation was probably similar to today. Secondly, when large amounts of ice rafting occurred, there was advection of surface waters from the Greenland–Iceland Seas to much more southerly latitudes than occurs now. The circulation changes are thought to have been of large magnitude with ice-bearing surface waters shifted southwards by more than 5° of latitude, penetrating well into what is now the core of the warm North Atlantic Current.

As discussed in Section 5.2.3, the cooling events were found to take place every 1470 ± 500 years, and equate to the 1500-year cycle found in other proxy data. The temporal separation during the Holocene was about the same as the abrupt climate shifts during the last glaciations (Bond *et al.*, 1997). These authors suggest that these Holocene events were the most recent expression of a pervasive millennial-scale climate cycle that occurs during both glacial and interglacial periods. Bond *et al.* (1997) have called them mini Dansgaard–Oeschger events. This implies that Dansgaard–Oeschger cycles are not forced by internal instabilities of the large ice sheets during glaciations (Section 3.4.3), but instead originate through processes linked to the 1500-year climate cycle. The cooling events were abrupt, switching on and off within a century or two.

As noted in Section 5.2.3 the origins of the 1500-year cycle are not known with certainty at present. The fact that it maintains its periodicity across major stage boundaries, such as the last glacial termination, almost certainly means that it is not linked to ice oscillations.

212   The Holocene

Climate cycles may arise because of orbital periodicities, but most orbital cycles identified are longer than 1500 years.

Forcing of millennial-scale climate variability may also arise because of changes in solar output. Recent studies with climate models have suggested that a decrease of only ~0.1% in solar activity over the 11-year sunspot cycle can produce a change in surface climate via the atmosphere's dynamic response to changes in stratospheric ozone and temperature (Haigh, 1996). However, the mechanism is still controversial and there is currently no evidence of a solar cycle close to 1500 years.

The 1500-year climate shifts also appear to be linked to changes in the atmospheric circulation above Greenland. In particular, the deep-ocean core records correlate well with the flux of non-sea-salt potassium in the Greenland ice cores. As O'Brien et al. (1995) linked increases in soluble impurities in the Greenland cores with lower atmospheric temperatures, it is therefore likely that the North Atlantic Ocean surface and atmospheric conditions above Greenland were acting as a coupled system and changing on millennial timescales.

The O'Brien et al. analysis of the GRIP2 ice core data suggested that cooler climatic conditions occurred at quasi-2.6 kyr intervals during the Holocene. These were at approximately before 11.3 kyr BP, 6100–5000 and 610–0, and to a lesser extent during 8.8–7.8 and 301–204 kyr BP (all near multiples of 2.6 kyr) (O'Brien et al., 1995) (Fig. 5.8). These

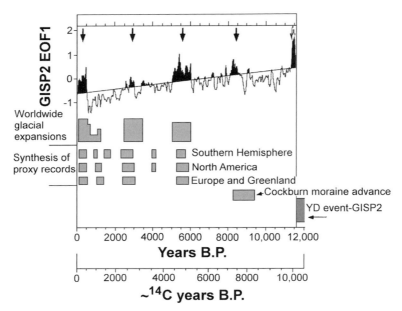

Fig. 5.8 Palaeoclimate cold events: GISP2 Holocene EOF 1; worldwide glacial expansions and their relative magnitude; synthesis of various climate proxy records from Europe, Greenland, North America, and the Southern Hemisphere showing cold periods; the Cockburn Stade, and the Younger Dryas (YD) event. (From O'Brien et al., 1995.)

5.4 Temperature

periods coincided with worldwide Holocene glacier advances (Denton and Karlén, 1973), with cold periods identified in palaeoclimate records from North America, Europe and the Southern Hemisphere (Harvey, 1980) and with periods of low solar output identified in tree ring data (Stuiver and Reimer, 1993). The 2.6 kyr quasi-periodicity is close to the roughly 2.5 kyr quasi-cycle identified via carbon-14 dating, but links between the carbon-14 record and climate are controversial. However, at least two sunspot minimum events identified in the carbon-14 record can be found in the GISP2 record. The earliest was more than 11.3 kyr BP at the same time as the Younger Dryas event. The most recent occurred at the same time as the LIA, which had the most abrupt onset of any of the events in the Holocene. Milder conditions, associated with contraction of the polar vortex or a weaker meridional circulation occurred between 10.6–9.3, 7.9–6.3, 2.7–1.5 and 0.96–0.61 kyr BP.

Data on sediment grain size collected from the Iceland Basin have also revealed a 1500 year signal. These data were used to reconstruct past changes in the speed of deep-water flow and the data suggested that flow changes had a periodicity of about 1500 years and coincided with the LIA and the MWP.

Tree ring data also provide high-resolution information on temperature variability over the last millennium. These show that although some of the major temperature fluctuations, such as the LIA and MWP, receive a lot of attention, there are many other temperature departures in these records. For example, the tree ring record from northern Fennoscandia indicates that alternating periods of generally cooler and milder conditions are found during the summers throughout the last millennium. The record also shows that the length and amplitude of the temperature fluctuations have varied dramatically over this period.

The Antarctic

Although there has been more focus on Holocene climate variability in the Arctic than the Antarctic, some studies have considered climatic fluctuations in the south. For example, Jones *et al.* (2000) created a multi-archive comparison of Antarctic temperatures during the last 7 kyr from lake, ice and ocean sediment cores. This is shown in Fig. 5.9 and illustrates the complexity of change that took place across the Antarctic and sub-Antarctic islands during most of the Holocene. Although deglaciation was still taking place on a number of the islands, the ice core record shows a general cooling over 7–5 kyr BP. However, South Georgia experienced a relatively warm period around 5 kyr BP.

Holocene climatic variability across the Antarctic Peninsula was considered in depth by Bentley *et al.* (2009), who utilized all the various forms of proxy data to gain insight into the mechanisms responsible for the changes. The two most pronounced warm episodes were found to be the Early Holocene warm period and the Mid Holocene Hypsithermal (See Table 5.1), and these events were linked to changes in the meridional location of the belt of strong westerlies.

One of the most marked climatic fluctuations in the Antarctic record is the 'climatic optimum' that occurred over 4–2 kyr BP. This is apparent in the ice core record, and can be identified in lake sediment records from the islands, as well as from some ocean sediments.

214   *The Holocene*

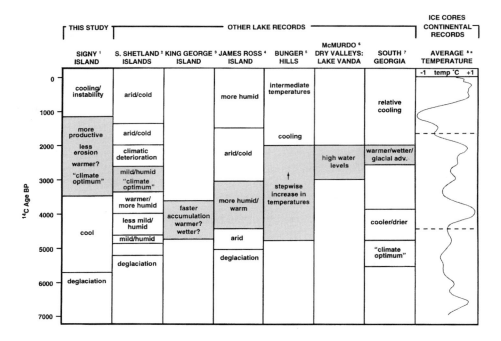

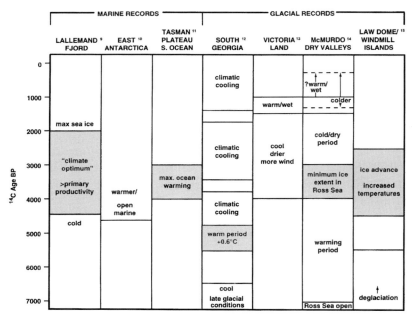

Fig. 5.9 Multi-archive comparison of Holocene palaeoclimates inferred from selected lake sediment, ice and ocean core records and glacier fluctuations. (From Jones *et al.*, 2000.)

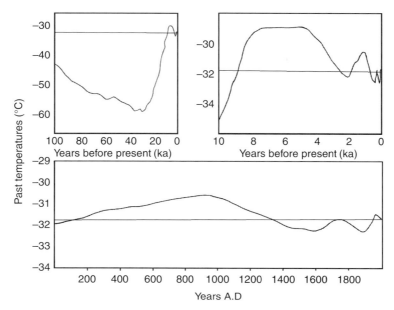

Fig. 5.10 The reconstructed temperature history for the GRIP ice core site in Greenland. (Top left) The last 100 kyr BP. (Top right) The last 10 kyr BP. (Bottom) The last 2000 years. (After Dahl-Jensen et al., 1998.)

5.4.2 The Early Holocene – 11.7–5 kyr BP

The Arctic

The ice core and borehole records from the Greenland Ice Sheet provide an excellent record of the transition from the Pleistocene to the Holocene. The GISP2 core indicates that the transition from the Younger Dryas occurred within a few decades and was of the magnitude of $10 \pm 4\,°C$. In parallel with the temperature rise there was also a substantial increase in snowfall.

The Holocene temperature time series retrieved by inverting the current temperature profiles measured down boreholes from the top of the Greenland Ice Sheet provide a particularly valuable record. Dahl-Jensen et al. (1998) analysed temperatures from the GRIP and Dye 3 boreholes to create 50 and 7 kyr records respectively for these sites, and the past temperatures for the Holocene segment of this record are shown in Fig. 5.10. It should be noted that the inversion method is not able to resolve rapid temperature fluctuations, such as the cold Younger Dryas and warm Bølling/Allerød events, but does provide an excellent indication of the broad temperature variations during the Holocene.

The present-day surface temperature at the GRIP site is $-31.70\,°C$ and this is indicated as the horizontal line in the parts of Fig. 5.10, so allowing easy identification of warmer and colder periods during the Holocene. After the end of the glaciation, temperatures increased steadily, reaching a maximum during the climatic optimum of 8–5 kyr BP when

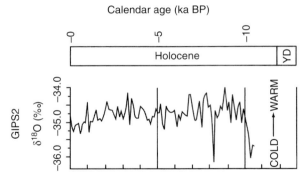

Fig. 5.11 The isotopic record from the GISP2 ice core since the Younger Dryas. (From Maslin et al., 2003.)

temperatures were 2.5 °C higher than today. A more recent study by Vinther et al. (2009) analysed the $\delta^{18}O$ record from four Greenland cores along with cores from small marginal icecaps and found a pronounced climatic optimum at this time with thinning at the ice sheet margins.

At the start of the Holocene the increased insolation resulted in higher summer temperatures across both Eurasia and western North America, where values were 2–4 °C warmer than today. However, it was still colder than at present just south of the Laurentide Ice Sheet in eastern North America.

A number of marked cold periods occurred during the Holocene. The first was the Pre-Boreal cold event (or the Pre-Boreal Oscillation) and occurred between 11.5 and 8.9 kyr BP. The peak of the event occurred at approximately 9.9 kyr BP and lasted about 200 years and was particularly pronounced in North Atlantic climate records (Bond et al., 1997), but is less marked in the GISP2 data (Fig. 5.11). Prior to this there was another cooling event, known as the Erdalen event at 10.75–10.2 kyr BP (see Table 5.1).

A number of records suggest an Early Holocene temperature maximum between 11.5 and 8 kyr BP. An indication of temperature change can also be obtained from tree fossils, which can be used to track changes in the latitude of the northern tree line. Such data from Siberia suggest that some species were already further north than at present by 10.6–8.8 kyr BP and spreading particularly fast at the start of the Holocene. They suggest that summer temperatures were about 4–5 °C, and possibly 7 °C warmer than the present in this area. The Early Holocene maximum in summer insolation was probably a major factor here, resulting in increased continentality and greater penetration of warm North Atlantic waters into the Arctic Seas, along with reduced sea ice cover. Other tree fossil data from northern Finland and the Kola Peninsula show a northward extension of the tree line over 7.7–4.3 kyr BP. As this is later than the Siberian data, it suggests regional climatic differences, probably related to the deglaciation history of the Fennoscandian Ice Sheet.

Studies of the Early to Mid Holocene undertaken with atmosphere-only climate models show a winter cooling in Europe at a time when there was reduced insolation in this season.

However, this conflicts with the northward movement of the tree line that implies a warming, which has been attributed to the taiga–tundra feedback. This is a feedback between migration of vegetation and the energy fluxes in the lowest layers of the atmosphere. It is based on the fact that the albedo for snow-covered low vegetation, such as tundra, is much higher than for forests. Snow-covered grass can have an albedo of 0.75, while that for snow-covered forest can be 0.2–0.4. Therefore the darker snow-covered taiga receives more solar radiation than the tundra, mainly in spring and early summer. Thus the warming in some parts of the high northern latitudes during the Mid Holocene has been called the biome paradox (Berger, 2001).

Across North America, at the edge of the Arctic, there was a strong west to east climate difference in the Early to Mid Holocene, suggesting an interaction between insolation and the Laurentide Ice Sheet. Over northwestern Canada the tree-line moved northwards in the Early Holocene at the time of maximum summer insolation and lake levels in the central part of Alaska indicate reduced precipitation and greater summer evaporation than occurs today. Data from Canada, including pollen and macrofossil data from the Mackenzie Delta region in the northwest, suggest that between 10 and 7.7 kyr BP some tree species grew about 60 km north of their present limit. The trees then retreated after 7.7 kyr BP, probably responding to climatic cooling as a result of decreased summer insolation.

In northern Alaska, lake sediment cores have suggested that between about 8.5 and 6 kyr BP there were alternating warm and cool intervals of 200 years duration (Anderson *et al.*, 2001b). Between about 6 and 4.6 kyr BP conditions were generally cooler than earlier.

The reconstruction of the temperature record from the GRIP ice core from Greenland shows an Early Holocene thermal maximum at 10 kyr BP and warmer temperatures than those measured at present until approximately 3–4 kyr BP. There is a pollen concentration maximum in the Canadian Agassiz ice core in the Early Holocene suggesting higher temperatures or different atmospheric circulation patterns, even though some of the sources were still ice covered. There are also higher concentrations between 11 and 7 kyr BP, with decreasing concentrations until 2 kyr BP.

A great deal of research has focused on the period around 6 kyr BP, which was approximately the time of maximum summer warmth in many Northern Hemisphere mid to high latitude regions – the so called climatic optimum. At this time summer temperatures were 2–4 °C higher than at present across the continental interiors of North America and Eurasia. Cheddadi *et al.* (1996) produced maps of temperature anomalies for this time using pollen and lake-level data. As can be seen in Fig. 5.12, they found that winter temperatures were 1–3 °C greater than at present in the far north and northeast of Europe, including over Fennoscandia, Svalbard and parts of the Russian Arctic. This implies a reduced east–west temperature gradient across northern Europe. Conditions during the summer were also warmer, with positive anomalies of growing degree days (temperatures above 5 °C) over Scandinavia and the European sector of the Arctic. These conditions were coupled with reduced net accumulation (precipitation minus evaporation) across northern Europe.

The most marked cooling event during the Holocene was an abrupt and short-lived temperature decrease at approximately 8.2 kyr BP, seen in a number of proxy records

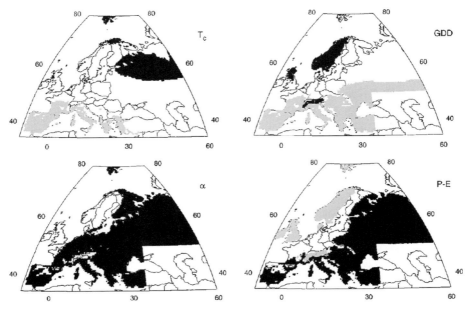

Fig. 5.12 Regions with significant climate anomalies at 6 kyr BP. T_c is the mean temperature of the coldest month; GDD is growing degree days; α is moisture availability; and P−E is annual precipitation minus evapotranspiration. Dark shading indicates regions with significant positive anomalies. Light shading indicates regions with significant negative anomalies. (From Cheddadi et al., 1996.)

from Greenland and around the North Atlantic. As well as introducing cooler temperatures to much of the North Atlantic region, the period was also characterised by dry, dusty and windy conditions.

Dating of the event varies slightly across the available ice cores with the best estimates being GRIP 8.21 kyr BP ± 30 yr (Johnsen et al., 1992), Dye 3 8.23 kyr BP ± 50 yr (Dansgaard, 1985), GISP2 8.25 kyr BP (Meese et al., 1997) and GRIP 8.19 kyr BP (Rasmussen et al., 2006). The only core that did not reflect this event was the Renland core from East Greenland. Muscheler et al. (2004) correlated tree ring ^{14}C records with cosmogenic isotopes in the GRIP ice core to determine that the most extreme $\delta^{18}O$ values corresponded to 8.15 kyr BP.

The magnitude of the cooling has been estimated using a number of techniques. Alley et al. (1997) estimated that in central Greenland the temperatures dropped by 4–8 °C and between 1.5–3 °C at terrestrial and marine sites around the northeastern North Atlantic Ocean.

The duration of the event has been estimated as a 160.5-year period when decadal-mean isotopic values were below average, and within which there was a central event of 69 years when values were consistently more than one standard deviation below the average (Thomas et al., 2007).

The 8.2-kyr event was experienced across a very wide area from the tropics to high northern latitudes. It is apparent in the Agassiz ice core record from Ellesmere Island,

Canada. The event led to cool and dry conditions lasting around 200 years, which has also been seen via multi-proxy studies from lakes in northern Sweden (Rosén *et al.*, 2001). It was followed by a rapid return to warmer and generally moister conditions than we have today. Alley *et al.* (1997) suggested that this event was also correlated with cool, fresh conditions across the North Atlantic and dry, windy conditions in the vicinity of the lakes around the Laurentian Ice Sheet.

The cause of the 8.2-kyr event has been debated for a number of years. O'Brien *et al.* (1995) linked it to periodicities in the climate system, and Bond *et al.* (1997) also felt that the event was linked to the climate system, with its large amplitude reflecting a mechanism that somehow amplified the climate signal at that time. The amplifying mechanism was not clear, but it was suggested that it might be linked to changes in the Atlantic MOC. This is also supported by the spatial pattern of change over the oceans and terrestrial areas being similar to that of the Younger Dryas event. In addition, the event certainly had many of the characteristics of the Younger Dryas, with cold, dry, windy conditions in Greenland, low temperatures in the North Atlantic Basin, and strong North Atlantic trade winds. This pattern is consistent with a reduction of North Atlantic heat transport as simulated in model experiments and the climate-anomaly patterns observed in the twentieth century in association with northerly flow from the Arctic into the North Atlantic, such as occurs during the negative phase of the NAO.

An increase in freshwater input to the North Atlantic triggering a slowdown of the thermohaline circulation was put forward as one possible reason for this cooling (Broecker *et al.*, 1990). It is now generally accepted that the 8.2-kyr event was caused by a massive injection of fresh water into the North Atlantic from the draining of glacial lakes Agassiz and Ojibway in North America. Before this event the ice in the Hudson Bay region and the water in the proglacial lakes had a combined volume estimated at $5 \times 10^{14}\,\mathrm{m}^3$ (Veillette, 1994), with Ojibway containing $1 \times 10^{14}\,\mathrm{m}^3$ of water and Agassiz possibly the same amount. So more than half of the water in the Hudson Bay area was impounded as ice that could not rapidly be released into the ocean.

These lakes were on the margin of the Laurentide Ice Sheet (Fig. 5.13) and the water was released via the Hudson Strait when the ice damming the lakes melted. Before the freshwater injection the surface of the lakes was more than 175 m above the sea level at that time, so there was a large hydraulic head to drive the flow of water into the ocean. Evidence for this scenario comes from the marine ^{14}C reservoir for Hudson Bay (Barber *et al.*, 1999), which indicated that the lakes drained catastrophically at around 8.47 kyr BP. Barber *et al.* estimate that around $2 \times 10^{14}\,\mathrm{m}^3$ of fresh water was released into the Labrador Sea at this time.

The injection of $2 \times 10^{14}\,\mathrm{m}^3$ of this fresh water into the North Atlantic resulted in a significant reduction in sea surface salinity and consequently the event altered the ocean's thermohaline circulation. Modelling of the deposits in the Hudson Strait suggests that the drainage occurred in less than one year and more recent modelling work of the subglacial drainage of Lake Agassiz by Clarke *et al.* (2004) has raised the possibility that the magnitude of the freshwater pulse was of the order of 5 Sv in less than 6 months.

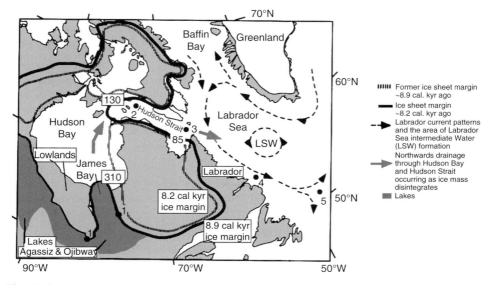

Fig. 5.13 Northeastern Canada and adjacent seas. Former ice sheet margins are shown for approximately 8.9 kyr ago (vertical-hatched line) and for ~8.2 cal kyr ago (thick grey line), before and after disintegration of ice in central Hudson Bay, respectively. The dark grey shading shows lakes Agassiz and Ojibway. The dark grey arrows show northward drainage through Hudson Bay and Hudson Strait. Arrows with dashed lines show Ladrador Sea current patterns and the area of Labrador Sea Intermediate Water (LSW) formation.

Experiments with ocean circulation models suggest that additional freshwater discharges of 0.06–0.12 Sv can reduce the formation of Labrador Sea Intermediate Water (LSW) and NADW, in turn affecting the ocean heat transport. However, the model experiments were concerned with freshwater injections on timescales of more than 500 years, not the much shorter period over which the 8.2-kyr event is believed to have taken place.

The injection of fresh water took place in the Hudson Strait, but the near-surface salinity anomaly propagated southeastwards into the region of LSW formation, where it had a greater impact than in the NADW formation region. Barber et al. (1999) noted that the present northward ocean heat transport associated with formation of LSW is 0.3 PW (1 PW = 10^{15} W), half that due to formation of NADW (0.6 PW). They noted that if the formation of both LSW and NADW ceased during the Younger Dryas, but that only LSW formation was disrupted during the 8.2-kyr event then we might expect regional atmospheric cooling of one third the magnitude at 8.2 kyr BP compared with the Younger Dryas and this is broadly in agreement with proxy records.

The Antarctic

Holocene temperatures across the Antarctic have been reconstructed from the ice core data collected at a number of sites on the continent. Most attention has been paid to Pleistocene conditions, but the records have also been analysed for the Holocene.

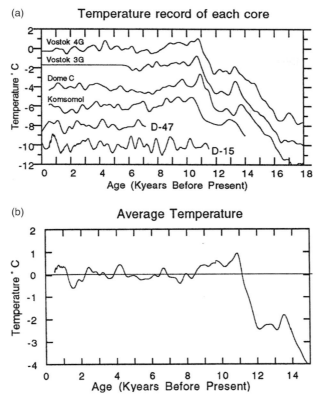

Fig. 5.14 (a) Temperature reconstructed from the isotope profile of six Antarctic ice cores. Temperature variations are expressed as the departure from the mean value over the past 5000 years. A constant offset has been added to each profile for display clarity. (b) Average temperature curve from the six Antarctic ice cores shown in (a). Temperature is expressed as the departure from the mean value over the past 5000 years. (From Ciais et al., 1994.)

For the Holocene itself, Ciais et al. (1994) examined the signals in six cores collected at plateau and coastal sites. The reconstructed temperatures from the cores are shown in Fig. 5.14a. It should be noted that the data have been filtered to remove all periods shorter than 500 years to minimise the noise. None of the temperature fluctuations in any of the cores exceeded 1–2 °C during the Holocene. A mean temperature from the six cores is shown in Fig. 5.14b. The Early Holocene experienced temperatures that were about 1 °C warmer than today, with conditions at about 9 kyr BP being the warmest in the whole record since the Last Glacial Maximum (Heusser, 1989). By about 8 kyr BP temperatures were very close to present-day values, and were followed by a cooling between 8 and 5.5 kyr BP.

The ice core records indicate a widespread Antarctic Early Holocene optimum between 11.5 and 9 kyr BP, centred on 10 kyr BP, which is observed in all cores from coastal and

222   *The Holocene*

continental sites (Steig *et al.*, 2000). The available data suggest that the Early Holocene warm period occurred almost synchronously across the whole Antarctic.

Across East Antarctica there were warm conditions in the Amery Oasis in the northern Prince Charles Mountains since the early Holocene, which lasted until about 6.7 kyr BP, and with an optimum between 8.6 and 8.4 kyr BP. But then cold conditions lasted from 6.7 kyr BP until 3.7 kyr BP. The complexity of change is reflected in the fact that a marine climate optimum is found in some areas, but it appears to be out of phase with the optimum recorded in terrestrial records and ice cores.

It should be noted that there is no evidence of the sharp 8.2 kyr cooling event found in the Greenland data and discussed earlier. But as recent research suggests that this was a result of fresh water release from the margins of the Laurentide Ice Sheet this is not surprising.

The smoothing applied to the data sets by Ciais *et al.* (1994) means that shorter term climatic fluctuations, such as any indications of the LIA and the MWP, cannot be seen, although, as discussed below, these events are seen in other palaeoclimatic records.

Other forms of data that give insight into Holocene conditions include sediments collected from lakes and the ocean. The lake sedimentary records have been particularly valuable in unravelling the climate record for the sub-Antarctic islands and at regional scales for all major ice-free areas of Antarctica. For example, Jones *et al.* (2000) analysed lake sediment cores from Heywood and Sombre Lakes on Signy Island in the South Orkney Islands and obtained a radiocarbon chronology of environmental change for the Holocene. The isolation and low latitude of Signy has made it very sensitive to recent climatic fluctuations and the island may well have been rather sensitive to climate changes during the Holocene. In the Early to Mid Holocene the authors identified a period of deglaciation and lake formation over about 5.9–3.8 kyr BP when the area experienced cool conditions and possible extended periods of winter lake ice cover. This is consistent with other evidence from ice cores of cool conditions at this time between 8 and 5.5 kyr BP (Ciais *et al.*, 1994).

The early Holocene climate optimum is not present in the terrestrial record since most of the area that is currently ice free was probably still ice covered before 9.5 kyr BP.

During the Early Holocene the King George VI Ice Shelf on the western side of the Antarctic Peninsula collapsed at about 9.6 kyr BP, while the Larsen B Ice Shelf on the Weddell Sea side survived throughout the Holocene. This suggests that the temperature gradient across the Antarctic Peninsula may have been even stronger during the Early Holocene than at present, when temperatures are of the order of 8 °C warmer on the west than the east.

5.4.3   *The Mid Holocene – 5–3 kyr BP*

Introduction

The Mid Holocene was the most stable warm period in postglacial times, when relatively high temperatures were experienced across much of the Earth. This period is often known as the Mid Holocene Optimum with temperatures 1–3 °C higher than those found today being experienced in many regions.

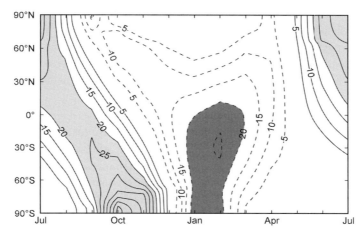

Fig. 5.15 The mean change in incoming solar radiation at the top of the atmosphere for the Mid Holocene (6000 yr BP) as a function of latitude and time of year. The contour interval is 5 W m^{-2} and negative values are dashed. Values greater than 20° W m^{-2} are lightly shaded and values less than −20° W m^{-2} are darkly shaded. Positive values (i.e. during northern hemisphere summer) indicate that the Holocene received more solar radiation than the present day. (After Valdes, 2003.)

The differences between the Mid Holocene climate and the present are mainly attributable to variations in solar radiation due to changes in the Earth's orbital parameters. Overall global changes in radiation received were very small. However, at this time there were significant seasonal differences: greater levels of solar radiation were received over the Antarctic in the austral spring and reduced irradiance in the summer. The changes were smaller in the Arctic with slightly less radiation in the winter and more in the summer (see Figs. 5.15 and 3.2). The insolation changes at the top of the atmosphere enhanced the seasonal cycle of incoming radiation in the Northern Hemisphere and to a lesser extent diminished it over the Southern Hemisphere.

During this period there is little evidence of synchronous climatic changes occurring in the Arctic and Antarctic. For example, the change to neoglacial conditions in the Antarctic Peninsula region lagged similar shifts in the Northern Hemisphere by several thousand years. The cooling trend in the Northern Hemisphere started around 4–5 kyr BP, while in the Antarctic the cooling started around 2.5 kyr BP and in some records as late as 1.4 kyr BP (Hodgson and Convey, 2005).

The Mid Holocene period has been studied intensively via model simulations, although most of the studies have been concerned with tropical and mid latitude features, such as ENSO variability, rather than polar conditions. One of the advantages of examining this period is that the orbital forcing is large, but the ice sheets had essentially the same configuration as today. It was the focus of a major international project called the Palaeoclimate Model Intercomparison Project (PMIP), which resulted in a number of modelling groups running their models for both this period and the present day

(Joussaume and Taylor, 1995). The Mid Holocene forcing for all these models was the same, and consisted of solar forcing set appropriately for the period, but with sea surface temperatures, vegetation cover, sea ice distribution and ice sheet extents set to present-day values. The models were run for at least 10 years and the differences computed between Mid Holocene and present conditions. Results from this project are discussed below.

The Arctic

The Mid Holocene period in the Arctic was characterised by relatively warm conditions when there was reduced amounts of sea ice and the high latitude boreal forests extended northwards of the modern tree line at the expense of tundra.

The warmer conditions experienced during the climatic optimum have been attributed to either a contraction of the polar vortex or a decrease in the northerly component of the atmospheric flow (O'Brien *et al.*, 1995). There were frequent westerlies in middle and sub-Arctic latitudes, and a shift in the location of the storm belt to relatively high latitudes.

The proxy records, such as the Greenland Summit ice core reconstructions, again provide valuable insights into the climatic conditions experienced. An overall view of the temperature changes can be gleaned from the GRIP Greenland borehole temperature record derived by Dahl-Jensen *et al.* (1998), which is shown in Fig. 5.10. Here the climatic optimum extended from about 7 to almost 4 kyr BP when temperatures were just over 2 °C warmer than those experienced today. From 4 kyr BP temperatures dropped steadily over the next 2000 years to reach a minimum of 0.5 °C colder than at present at about 2 kyr BP.

The relatively warm conditions can also be seen in other proxy records. Lake sediments from northern Norway provide a good temperature reconstruction for July (Fig. 5.16). This

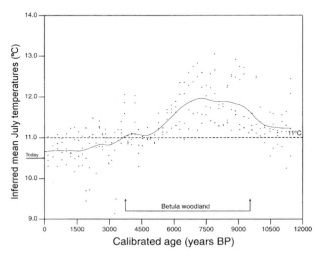

Fig. 5.16 Inferred mean July temperature for the Holocene at Björnfjelltjörn. The fitted line is a locally weighted least squares regression. (After Birks and Birks, 2003.)

time series was created using transfer functions from modern pollen–climate relationships in Scandinavia. Temperatures here were about 11 °C at the start of the Holocene and rose to around 12–12.5 °C between 9.5 and 6.5 kyr BP. The post-optimum decrease in temperatures began earlier here compared with the Greenland record, with temperatures decreasing to about 11 °C by 4.5 kyr BP. After this time the temperatures continued to decrease steadily to the present-day value of about 10.5 °C. Further evidence for an earlier decrease in temperatures comes from fossil tree data from Siberia. This suggests that the northern limit of some tree species began moving southwards about 6.7 kyr BP, reaching their present position about 4.8–3.1 kyr BP.

Across northern North America the warmest period was not until nearly 4 kyr BP, as residual ice from the Laurentide Ice Sheet was still melting until this time. But by this time the forests were further north across Canada and the permafrost zone was further north across Europe and Asia, as well as North America. This was the climax of the current interglacial with vegetation at high latitudes. There was probably also less cold swamp where the summer meltwater could not drain away, which limits the spread of trees in the Arctic fringe today. Further evidence for the warming in winter across eastern Canada comes from pollen records (Webb *et al.*, 1993) .

A cold climatic episode of the late Early Holocene that has received considerable attention is the Piora Oscillation, which occurred over 5.5–5 kyr BP and lasted about four centuries at the end of the Atlantic period (8.9–5.7 kyr BP, see Table 5.1). It was a more rapid shift towards cold climate conditions than any for several thousand years previously. It was first identified in the Val Piora in the European Alps, but evidence of the event has also been found in North and South America. A high latitude signal of the Piora Oscillation has been found in Alaska via changes in the vegetation. In addition, pollen data from the 60–70° N zone of Russia suggests a maximum July cooling close to 4.9 kyr.

Figure 5.17 shows the changes (from model experiments) in surface air temperature between the Mid Holocene warm period and the present day for December–February and June–August. The changes over the ocean are extremely small since present-day sea surface temperatures were used over all the oceans, but the temperature changes over the land are the result of external forcing. In the summer (JJA) temperatures over central North America and Eurasia were up to 2 °C warmer, although the differences at the edge of the Arctic were significantly less as a result of the present-day oceanic conditions used. During the winter (DJF) the differences were much less across northern Russia, although temperatures were more than 0.5 °C cooler over northeastern Canada. It should be noted that the greatest inter-model differences were at the edge of the Arctic, reflecting large uncertainties in the model runs in this area. Valdes (2003) points out that over northeast Europe the mean change was only about 0.1 °C, but that the range of predictions was from −4.0 °C to +7.3 °C. Model results do not suggest large differences in precipitation in either polar region, with the greatest changes being confined to the tropics.

The Antarctic

The most marked feature of the Antarctic Mid Holocene temperature record is the relatively warm conditions (the Holocene climatic optimum) that occurred over 4.5–2.5 kyr BP and

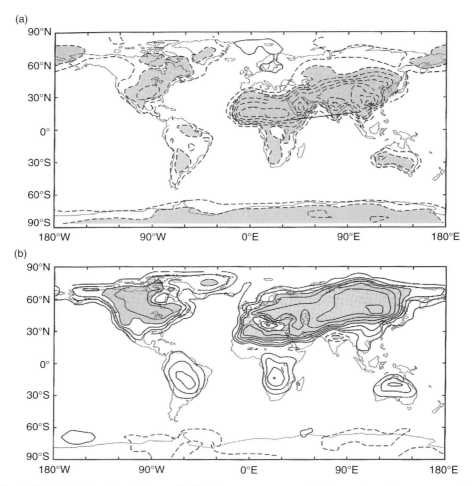

Fig. 5.17 Model simulated change (Mid Holocene to present day) in surface air temperature for (a) December to February) and (b) June to August for the average of all PMIP model simulations. The contour interval is 0.25 °C and negative contours are dashed. Positive values indicate that the Mid Holocene was warmer than the present day. (From Valdes, 2003.)

which coincided approximately with the solar maximum for these latitudes. This comparatively warm and humid period is recorded in a number of biological and chemical proxies in lake and ice cores, and in moss banks. However, the timing and strength of this event varied around the continent and it was more pronounced in the Antarctic Peninsula region than in the Ross Sea, although here there was a 'penguin optimum' (Baroni and Orombelli, 1994).

The event, which is also known as the Mid Holocene Hypsithermal, occurred at different times around the Antarctic. All East Antarctic proxy sites show a weak climate optimum between 6 and 3 kyr BP, but the period after deglaciation shows complex regional patterns.

The event was experienced between 4 and 2 kyr BP in the Larsemann Hills (69° 20'– 69° 30' S lat., 75° 55'–76° 30' E long.), but at the Amery Oasis of the northeastern Prince Charles Mountains (70–72° S, 64–69° E) it occurred between 3.2 and 2.3 kyr BP. In Wilkes Land enhanced biological production suggested a timing of between 4 and 1 kyr BP, coinciding with a local marine optimum and in the Bunger Oasis (66 ° 10' S, 101° E) warmer conditions were recorded in lake cores 3.5–2.5 kyr BP (Kulbe *et al.*, 2001).

The period coincided with an increase in the ice accumulation rate at Law Dome, which has been attributed to the warmer climatic conditions (Goodwin, 1998). During this general period of higher temperatures there were also shorter anomalous periods, such as the brief even warmer episode around 4 kyr BP identified by Ciais *et al.* (1994) from ice core records.

Across the Antarctic Peninsula the event took place over 3.5–3.2 to 2.9–2.7 kyr BP, and over 3.8–1.4 kyr BP on the sub-Antarctic islands to the north (Hodgson and Convey, 2005), which is considerably later than the Northern Hemisphere Holocene climatic optima. On the sub-Antarctic island of South Georgia warmer conditions than those experienced today occurred between about 5.6 and 4.8 kyr BP, although cooler conditions were also recorded over 4.8–3.8 and 3.4–1.8 kyr BP (Clapperton *et al.*, 1989). At Signy Island to the northeast of the tip of the Antarctic Peninsula lake sediments suggest that the optimum here started gradually from 3.8 kyr BP and lasted longer, coming to an abrupt end at about 1.4 kyr BP.

Conditions on the sub-Antarctic islands have been examined in several studies. Hodgson and Convey (2005) identified a late Holocene climatic optimum over about 3.8–1.4 kyr BP as suggested by biological evidence. This is consistent with the *c.* 3 kyr BP 'late Holocene climatic optimum' identified in other marine Antarctic records and based on high organic productivity in Midge Lake, South Shetland Islands (*c.* 3.3–2.7 kyr BP) and Lake Åsa (*c.* 4.3–2.7 kyr BP) on Livingston Island, South Shetland Islands. In this lake, which was deglaciated over 4300 years ago, a 1.5-m sediment sequence showed that the lake had undergone significant climatic changes over the last 4 kyr.

Warm conditions in the Late Holocene were also identified at Livingston Island (63° S, 60° W) (Björck *et al.*, 1991) via peak growth rates of a moss bank and at James Ross Island (64° 10'S, 57° 45'W) greater productivity in lake sediments (*c.* 4.4–3.1 kyr BP) (Björck *et al.*, 1996). Mäusbacher *et al.* (1989) and Schmidt *et al.* (1990) also found rapid sedimentation in lakes from King George Island from *c.* 5 to 3.2 kyr BP. In addition, radiocarbon dates indicate high water levels for Lake Vanda in the Dry Valleys at 2–3 kyr BP (Clapperton and Sugden, 1988). A warm period at this time was also detected in lake sediments from the Bunger Hills, East Antarctica, between 4.7 and 2 kyr BP (Melles *et al.*, 1997).

Pudsey and Evans (2001) investigated the Mid Holocene conditions in the north eastern sector of the Antarctic Peninsula via analysis of marine sediment cores collected in the Prince Gustav Channel. This location had been inaccessible by ship throughout much of the twentieth century, but could be reached following the disintegration of the ice shelf there in the 1990s. Ice-rafted debris consisted of local material in the Early and Late Holocene, but rocks from areas relatively far removed from the channel in the Mid Holocene (5–2 kyr BP). This indicated that the channel had open water at least in the summer season at this time and that the ice shelf was not present. Contemporaneously the

228   *The Holocene*

Larsen A Ice Shelf immediately to the south also showed fluctuations between 4 and 1.4 kyr BP (Brachfeld *et al.*, 2003); however, the Larsen B Ice Shelf, further south again, remained stable during this period.

Ciais *et al.* (1994) compared their East Antarctic record with the data from the Byrd site in Marie Byrd Land, West Antarctica. The Byrd record showed a rapid increase in δ^{18}O at 4.5 kyr BP, which was not present in the East Antarctic data. They suggested that this change in the isotope record indicated a regional warming of about 2 °C.

5.4.4   The Late Holocene – 3 kyr BP to present

Arctic

Records from the Arctic show a strong cooling in the second half of the Holocene, which is often called the Neoglacial period, and the colder conditions can be seen in ice core oxygen isotope and glacier records, which suggest advancing glaciers in Alaska and northern Norway. However, there were also shorter period temperature fluctuations superimposed on the general cooling trend. For example, North American tree ring data suggest that there were a number of rapid temperature drops and recoveries over this period, but these changes may not have occurred in Europe.

In northern Europe, following the Mid Holocene thermal optimum, temperatures dropped during the Sub-Boreal (5.7–2.6 kyr BP) and again at the onset of the colder, wetter Sub-Atlantic period (2.6 kyr BP to the present). By 2.5-2.2 kyr BP the temperatures in Europe were about 1 °C lower than at the peak of the Mid Holocene warm period.

The MWP, which has also been called the Little Climatic Optimum, was a relatively warm interlude that has been detected in many parts of the world. The timing and distribution of the event is still being debated, but many estimates have it covering the period of about AD 950–1300 in Europe, although other analyses have it starting as early as AD 400.

The event had a major influence on the high latitude human populations of the Arctic and it has been linked to the Norse expansion to Iceland in the ninth century and to Greenland and North America in the tenth and eleventh centuries, when favourable, relatively warm conditions occurred around the North Atlantic.

The MWP was a feature of many parts of the Arctic, with a clear signal being found in the records and proxy data for Iceland, Greenland and the Canadian Arctic. However, as discussed below, it is not so easily identifiable in the European sector.

The MWP has been investigated extensively in the Greenland/North Atlantic region via ice cores and ocean sediments. The record of Greenland surface temperature reconstructed from glacier borehole data (Dahl-Jensen *et al.*, 1998) shown in Fig. 5.10 suggests that temperatures during the MWP in Greenland were about 1 °C warmer than at present, with temperatures reaching a maximum around AD 1000. It therefore appears that the warm period reached its peak and ended earlier here than in Europe.

Ocean sediment data collected off the coast of Greenland have also provided very useful information on the MWP. Jennings and Weiner (1996) analysed sediments in two ocean

5.4 Temperature

cores from the Nansen Fjord, eastern Greenland, which reflected changing oceanographic and sea ice conditions from AD 730 to the present. Diamictons indicated periods of seasonally open water with iceberg rafting, while mud layers were deposited during periods of extensive sea ice cover. The data derived from the core revealed that the climate in the vicinity of Nansen Fjord was warmer and more stable than today indicating a clear MWP between about AD 730–1100. This coincides with the relatively warm period in the ice core record identified by Dansgaard *et al.* (1975).

In the Canadian Arctic a temperature time series has been created from a 1200-year sequence of glaciolacustrine varves (Moore *et al.*, 2001). This revealed that summer temperatures from AD 1200 to 1350 were up to 0.5 °C warmer than the mean for the last 1200 years and were followed by over 400 years of cooler conditions. The data also showed that although the late twentieth century temperatures in this area are higher than those over the last 400 years, they are not greater than those of the MWP. Lake sediments from western Canada also suggest that there were relatively dry conditions at this time.

The reconstruction of spring/summer (April–August) temperatures from tree ring data from Fennoscandia showed little evidence of a MWP (Briffa *et al.*, 1990). However, the data demonstrated that the second half of the twelfth century was one of the warmest 20 year periods in the entire record, while the first half of the century was very cold. Later work (Briffa, 2000) suggested that another warm period occurred over AD 950–1100 in high northern latitudes.

According to multiple proxy data sets the LIA was probably the most rapid and largest climate change in the North Atlantic area during the Holocene based on ice core and marine sediments along with other proxy data. However, although the LIA can be found in many polar climate records, its duration varied across Europe, and in northern Fennoscandia it lasted a much shorter time than in many other areas.

Evidence of a signal of the LIA comes from various sources, including ice cores, ocean sediment records and of course the historical records, although these are mainly limited to the extra-polar regions. For the Northern Hemisphere, Grove (1990) proposed that one of the most reliable sources of data comes from reconstructions of the elevations of mountain snowlines.

Over the period of the LIA, temperatures in many areas of both hemispheres are thought to have fallen to their lowest levels since the last glaciation. In the Northern Hemisphere it was the most recent of a series of millennial-scale cooling events that occurred during the Holocene.

The exact period that constitutes the LIA has frequently been debated in the scientific literature. For Europe, the LIA has been considered to start as early as the twelfth century and to have continued to as late as the early twentieth century. But it is often considered to span the period AD 1300 to 1850, although there is some consensus for the main European phase occurring between 1550 and 1700–1800 (Lamb, 1977).

In many ways the LIA can be considered as two distinct cold periods. The first followed the MWP, and is often referred to as the Medieval Cold Period (MCP) or LIA b. This began

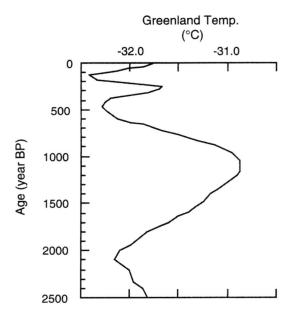

Fig. 5.18 A high-resolution record of the last millennial-scale climate cycle spanning the Medieval Warm Period and the Little Ice Age from Greenland. The record was constructed from glacier borehole temperature data and documents. (From deMenocal et al., 2000.)

around AD 1300 and ended about AD 1650, although some researchers have LIA b starting before AD 1200. The second cold period, covering AD 1650 to 1850, and spans what is often classically regarded as the LIA and is sometimes referred to as LIA a. This view of two separate events is supported by the Greenland surface temperature record (Dahl-Jensen et al., 1998), which shows two discrete cold periods at about AD 1500 and 1870 (Fig. 5.18).

The reasons for the LIA are still not entirely clear. Bradley (1988) ascribed it to an increase in volcanic activity, while Eddy (1976) linked it to a decrease in solar activity. Overpeck et al. (1997) reconstructed Arctic near-surface air temperatures for the last 400 years and attempted to explain the observed variability in terms of the relative roles of trace gases, solar insolation, aerosol loading associated with volcanic eruptions and changes in the atmospheric circulation. They concluded that the pronounced Arctic warming over 1820 to 1920 was primarily a result of reduced forcing by volcanic aerosols and increasing solar insolation. Since the early part of the twentieth century the high level of insolation and low aerosol loading continued to influence the Arctic climate, but the rapidly rising greenhouse gas concentrations have probably played an increasingly important role.

The LIA had a major influence on human populations surrounding the North Atlantic, causing the withdrawal of Norse colonies from Greenland, and a reduction in crop yields,

5.4 Temperature

resulting in famine in parts of Europe. It left large signals in the proxy records, such as the Greenland ice cores, the deep-sea sediment cores and lake sediments in Canada. When these data are combined with the historical records it is possible to create a fairly complete picture of this major climatic event.

The oxygen isotope records from many ice cores show a LIA, and an earlier warmer period that could be called a MWP. This LIA–MWP couplet can be seen in the Devon and Agassiz ice cores from northern Canada, and the Camp Century and North GRIP cores from Greenland, but the Summit and Dye 3 cores, also from Greenland, do not show any such signal.

The ocean sediment core from the Nansen Fjord off the east coast of Greenland (Jennings and Weiner, 1996) shows a relatively warm period peaking at about AD 1470 and then a decline towards cold conditions around AD 1600. The climate was variable with frequent periods of severe, cold weather over approximately AD 1630–1900. The ice core record suggests that temperatures fell by 0.5–1.0 °C (Dahl-Jensen *et al.*, 1998) during this period and the Nansen core shows abrupt switches to periods of cold, polar conditions around AD 1630, 1740, 1830 and 1905. These coincide well with the Icelandic historical records, but the Greenland ice core record suggests that the Greenland LIA had some differences of phasing compared with Europe.

The oxygen isotope data from the Renland (71° 18′ N, 26° 43′ W) ice core from a local icecap above Scoresby Sund, East Greenland, shows similar climatic conditions over this period. It indicates cooling after AD 1400. The available historical records generally support the picture provided by the proxy records, with settlements on the eastern side of Greenland failing by AD 1500, although it is not clear whether this was caused by climatic factors alone.

The records from Iceland are in broad agreement with those from Greenland. There is evidence of the LIA b period with an early cold interval culminating by about AD 1370. For the LIA a, the coldest decades from the Iceland historical record during the seventeenth and eighteenth centuries were the 1630s, 1690s, 1740s and 1750s (Ogilvie, 1984).

The LIA has been well documented for the well-populated areas of Europe (Grove, 1990); however, there is less information for Arctic latitudes. But in northern Fennoscandia tree ring data do provide an indication of climatic conditions (Briffa *et al.*, 1990). They show that the LIA was confined to the relatively short period over AD 1570–1650 when temperatures here fell to a level of about 0.5 °C below the long-term mean. Between AD 1600 and AD 1650 there were 13 summers with mean temperature below −1 °C, which is the largest number of such cold summers in any 50-year period. Following this, conditions were near normal from 1660 to 1750, but then the period between 1750 and 1770 was one of the warmest 20-year periods in the 1400-year record.

There is extensive evidence for a signal of the LIA in northern Canada and Alaska, with the Canadian winters being severe during this period. The widespread nature of the cold conditions is clear from the increased extent of the glaciers, which can be determined from the moraine deposits. Similar cold conditions were found in Alaska where the glaciers were also more extensive in the LIA period. LIA glaciation started at approximately AD 1200 and

232   *The Holocene*

continued through the nineteenth century (Porter, 1986). During this time, the majority of Alaskan glaciers reached their Holocene maximum extensions (Calkin *et al.*, 2001). Evidence that the LIA advances in the Kenai Mountains of Alaska were the most expansive events of the Holocene is also reported. Unusually cold conditions occurred in southern Alaska in the second half of the Maunder Minimum, approximately 1645–1715, when there was reduced solar insolation. This coincided with a period of exceptionally active volcanic eruptions.

Antarctic

Conditions varied markedly around the continent during the late Holocene and it is difficult to draw general conclusions about the climate. However, following the Mid Holocene warm period there was a marked cooling with glacier advance in some areas, although ice records suggest that at this time the Ross Sea sector experienced its warmest temperatures during the whole Holocene, exceeding those of the Mid Holocene.

The mean temperature of the Antarctic derived by Ciais *et al.* (1994) from six ice cores was described in an earlier section. Their time series (Fig. 5.14b) shows that the Late Holocene was characterised by generally stable mean temperatures across the continent, the only significant departure being a cooling over 2 to 1 kyr BP. However, there were obviously variations in conditions around the continent. For example, ice core records from the Newell Glacier in the Dry Valleys indicate warming around 1.5 kyr BP following an earlier cold, dry period (Mayewski *et al.*, 1995). In the Lambert Glacier region of East Antarctica there is evidence of cooling leading to dry conditions around 2 kyr BP, 700 yr BP and between 300 and 150 yr BP in some of the lakes. In the Amery Oasis cooling followed the MHH after 2.3 kyr BP, with a brief return to warmer conditions between 1.5 and 1 kyr BP.

On the eastern side of the Antarctic Peninsula there is some evidence that the ice in the Prince Gustav Channel may have reformed around 1.9 kyr BP, with the Larsen A Ice Shelf reforming around 1.4 kyr BP, although the dating of these events is uncertain.

The last two thousand years have been characterised by general cooling in a number of sections of the Antarctic. For example, Hodgson and Convey (2005), in their analysis of lake cores from Signy Island, found a period of cooling from about 1.4 kyr BP (roughly AD 700) to the present. Over the same period there was a decline of lake levels in the Dry Valleys near McMurdo, which fell to a low point at about 1.2 kyr BP with some lakes evaporating completely or being reduced to dense brine (Doran *et al.*, 1994). South Georgia also experienced cold episodes over the last 1.4 kyr BP with the most extensive glacial advance at about 2 kyr BP. This is consistent with the data for the Antarctic Peninsula where there is evidence of glacial expansion during the last 2.5 kyr suggesting cooler conditions. In addition, the lake sediment core from Livingston Island (Björck *et al.*, 1991) suggests that milder and more humid maritime conditions occurred about 3.2–2.7 kyr BP, when the summer climate, and possibly the winter conditions, were significantly milder and more humid than today. The reasons for the milder conditions are not fully understood, but could be a result of more maritime, northerly atmospheric

5.4 Temperature

flow, as implied by the increased amount of pollen found from South America. Since about 2.7 kyr BP the climate has gradually changed to that experienced today, with possibly warmer conditions being experienced around 2 kyr BP. The coldest and driest conditions seem to have occurred between about 1500 and 500 yr BP.

There has been an extensive debate regarding the region over which the MWP can be detected. Evidence for the event can certainly be found in the Alps, and in other parts of the Northern Hemisphere. But to date no clear signal of the MWP has been found in the Antarctic. Clow (personal communication to W. Broecker) has carried out a deconvolution of the temperature record at the Antarctic Taylor Dome site (Broecker, 2001) and his reconstruction showed that the air temperature was 3 °C colder during the time of the MWP than during that of the LIA. This record suggests that conditions at this location underwent an anti-phased oscillation compared with the north during the MWP–LIA period (see Section 4.4 for a discussion of the phasing of climate changes in the Arctic and Antarctic).

It is still debatable as to whether signals of the LIA, which are so clearly seen around the North Atlantic, can be found in the Antarctic. Evidence for several glacial advances during the last 750 years has been found in the Antarctic Peninsula, but Leventer *et al.* (1996) showed that the Antarctic Peninsula warmed during the time of the LIA. Ice cores collected across the region that extend back through the period of the LIA do not show a clear signal of cooler conditions over this period. Figure 5.19 shows the oxygen isotope record from six cores from the peninsula, with four of these extending back as far as AD 1500. While there are indications of slightly lower isotope ratios over the period 1500–1600 compared with later centuries, there is no statistically significant indication of a clear trend over this period.

There are some indications that other regions of the Antarctic were warmer at times during this period (Lamb, 1995), again revealing an anti-phase relationship between the Arctic and Antarctic climates. Evidence for a warming comes from the downhole temperature record at Taylor Dome (Broecker, 2000), which indicates that during the LIA the air temperature warmed by 3 °C. This is also suggested from the voyages of James Cook (in 1770) and James Weddell (in 1823), who were able to navigate a long way south in the pack ice. However, conditions in individual years, which may be anomalous, should not be taken to infer longer term climatic conditions. Nevertheless, radiocarbon dating of abandoned penguin rookeries at 77.5° S also indicates warmer periods there around 1250–1450 and 1670–1840 (Lamb, 1995).

There are extensive proxy climate records for many parts of the Antarctic and in these there is no indication of a major cooling at the time of the LIA. For example, there is no evidence of the events in the glacier records of the McMurdo Dry Valleys, nor in the many records for East Antarctica. Conversely, a number of outlet glaciers and ice shelves around the Antarctic Peninsula, such as Rotch Dome, Livingston Island and the Muller Ice Shelf, are believed to have advanced around the time of the Northern Hemisphere LIA, although this could be an independent climatic fluctuation.

234   The Holocene

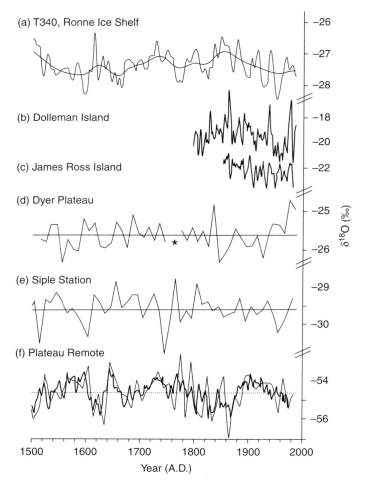

Fig. 5.19 The isotopic records from six Antarctic ice cores. (From Mosley-Thompson and Thompson, 2003.)

5.5   The ocean circulation

Details of the ocean circulation changes during the Holocene are only now being unravelled as proxy data become available from ocean sediment cores, and techniques are developed to infer water mass type and their changes during the period. However, we still have a very incomplete picture of how oceanic conditions changed during the Holocene and how the ocean, atmosphere and cryosphere interacted. Nevertheless, in this section we attempt to draw together the available observational data to examine oceanic conditions in both polar regions during the Holocene.

5.5.1 *The Arctic*

The Early Holocene

The oceanographic conditions of the North Atlantic have received considerable attention from researchers because of the importance that changes in the region could have on the THC and observation of events such as the Great Salinity Anomaly (see Section 6.4.2) (Dickson *et al.*, 1988). The area is also close to Greenland, which has yielded long palaeo-climatic records from ice cores that can be linked to oceanographic proxies. Ocean sediment cores have therefore been extracted from a number of locations in this region in order to understand ocean variability during the Holocene.

During the Younger Dryas event (see Section 4.2.5) there was ocean cooling in the North Atlantic, but this ended at about 10.5 kyr BP according to an ocean core collected off the coast of Norway (Lehman and Keigwin, 1992). Within 100 years there was then surface inflow to the Arctic of warmer water from the Atlantic. This resumption of circulation at the end of the cold Younger Dryas lasted for a few hundred years, and was followed by another reduction in circulation beginning around 9.7 kyr BP, which itself lasted only a few hundred years. This was followed by the establishment of a stable interglacial circulation by about 8.8 kyr BP.

The 8.2 kyr BP North Atlantic cooling event is seen in many palaeoclimate records and has been linked to a significant decrease in SSTs in the Norwegian Sea (Klitgaard-Kristensen *et al.*, 1998). The event is thought to have occurred because of the catastrophic proglacial lake drainage at the margins of the Laurentide Ice Sheet near to Hudson Bay (see Section 5.4.2). This event resulted in a release of fresh water and ice into the Labrador Sea and North Atlantic. NADW formation is sensitive to freshwater inflow (Alley *et al.*, 1997), so it is believed that the addition of fresh water to the marine surface waters resulted in a major slowdown of the THC. It has been suggested that during the 8.2-kyr event the deep-ocean state may still have been close to glacial conditions (Ganopolski and Rahmstorf, 2001).

The 8.2-kyr event has been simulated by Renssen *et al.* (2001b) using a model of intermediate complexity. They found that the response of the coupled system could be very sensitive to small changes in the initial conditions and that the state of the THC could be very different for the same freshwater input scenario. In some model runs the circulation recovered quickly, but in other cases the modelled THC remained weak for a millennium. If the model is producing a realistic simulation of the system, then this result has important consequences for modelling such events; implying a form of 'chaos' in the climate system, which may make it difficult to simulate the observed variability.

As discussed in Section 5.2.3 above, deep-sea cores collected in the North Atlantic suggest that during the Holocene cooling events occurred roughly every 1500 years.

During these periods there was southward advection of cooler and fresher surface waters into the core of the North Atlantic Current, implying links with the THC. Analysis of benthic material in the cores indicated that NADW production was reduced during some of the cooling events.

Determining water mass properties during the Holocene is very difficult, but Legrand *et al.* (1997) established a positive relationship between methane sulphonic acid (MSA) and salinity, allowing the variability of salinity to be estimated through the period – brine rejection under sea ice increases the salinity promoting a parallel positive relationship between MSA and sea ice extent. These workers found an Early Holocene MSA minimum, which, since the algae that produce dimethylsulphide (DMS) (see Section 2.5.5) prefer saltier water, they attributed to lower salinity in the Atlantic as a result of meltwater.

The Mid Holocene

Andrews *et al.* (2001) analysed three ocean sediment cores obtained off northern Iceland to investigate the variability of water masses in this area. This region is subject to extreme hydrographic variability on annual to decadal timescales, with the waters switching from cold low-salinity water arriving from the north, which is associated with extensive sea ice, to salty Atlantic Water. In the former conditions the marine productivity is low because the water column is well stratified, while in the latter scenario the vertical mixing results in nutrients being brought to the surface and high productivity. In the Andrews *et al.* (2001) study, these two types of conditions were investigated via the carbonate content of the sediments in the cores, which are a proxy for productivity. In one core, which had a resolution of 50 years, 13 oscillations were found over the last 4800 kyr with an average length of 370 years. These were superimposed on a long-term decrease of net carbonate accumulation. Peaks in carbonate were found at 3.8 and 2 kyr BP, with minima at 3 and 2.3 kyr BP. These carbonate records from the cores closely track the temperature records derived from the Greenland ice core proxy record covering the last 2000 years and show similar but slightly offset oscillations between 4 and 2 kyr BP. The offsets could indicate a real lag between marine conditions off Iceland and the temperatures at the summit of Greenland. The carbonate variations had a comparable periodicity in the Greenland Summit core of about 370 years.

Oceanographic conditions in the Nansen Trough on the East Greenland Shelf were investigated by Jennings *et al.* (2002) who analysed two ocean cores that yielded records of iceberg rafting and palaeohydrography. They found that the transition from distal glacial marine to postglacial sedimentation began in the Nansen Trough at about 10.9 kyr BP. Holocene conditions in the Nansen Trough between 10.9 and about 4.7 kyr BP were little influenced by glacial ice and indicate a dominant influence of Atlantic Intermediate Water (AIW) from the East Greenland Current. However, they found a Mid Holocene climatic shift from the low IRD of the Early Holocene to a Neoglacial interval with high IRD, of colder, lower salinity 'polar' conditions. This signal at about 4.7 kyr BP coincided with the onset of Neoglacial cooling in the Renland ice core $\delta^{18}O$ record from East Greenland.

Jennings *et al.* (2002) proposed that the carbonate flux peaks between 4.7 and 0.4 kyr BP were related to sea surface coolings associated with an increased flux of Polar Water (PW) and sea ice in the East Greenland current. These peaks were synchronous with sea surface coolings interpreted from North Atlantic deep-sea cores, but additional peaks identified around 3.8 and 2.4 kyr BP in another core imply that the shelf site captures higher frequency events. These data indicate that severe Arctic sea ice events began in the Neoglacial interval and that earlier Holocene cool events in deep-sea records were associated with other processes, such as release of meltwater from residual glacier ice and glacial lakes into the ocean. Jennings *et al.* also found that freshening of the East Greenland Current was pronounced after 2 kyr BP.

The Late Holocene

Ocean sediment cores from the Nansen Fjord (Jennings and Weiner, 1996) on the eastern side of Greenland reveal information on the ocean circulation during the MWP. During parts of the event the influence of PW on the area was sufficiently reduced so that the AIW impinged on the sea floor. Between AD 1110 and 1630 there were two fluctuations between colder and warmer conditions, with the cold intervals culminating at approximately AD 1150 and 1370. The earlier period was brief, lasting about 60 years, and was similar to present-day conditions in the fjord. The later interval may represent a time of frequent perennial sea ice cover and PW at depth in the fjord beginning about AD 1270 and ending about 1370. The two intervening warm intervals culminating at about AD 1170 and AD 1470 represent conditions relatively warmer than the present.

Nearby, in the Iceland Basin, ocean flow was reconstructed from sediment data, suggesting that there was faster flow of the Iceland–Scotland Overflow Water (ISOW) (see Section 3.9.1) during the MWP. The oceanic conditions around Iceland during the LIA have been investigated in several studies, with workers such as Jennings and Weiner (1996) finding that there were significant shifts in the ocean currents in this region. For example, a qualitative ocean flow in the Iceland Basin reconstructed from sediment data suggests that there was reduced flow of the ISOW during the LIA, around AD 1600. The Gulf Stream/Arctic waters boundary was located between Iceland and the Faeroe Islands at times during this period with waters in this area being about 5 °C colder than at present (Lamb, 1995).

Andrews *et al.* (2001) found that the largest decrease in carbonate content in the ocean sediment record occurred during the LIA. This suggests that the LIA was unusually severe at this location and may have been linked with the slowdown of the THC. The carbonate content did not show any change at the time of the MWP.

5.5.2 *The Antarctic*

The more zonally symmetric circulation around the Antarctic means that ocean changes during the Holocene were mainly in the meridional direction, and that the variability recorded around the hemisphere was less marked than in the Arctic.

238   *The Holocene*

Through the duration of the Holocene there was a steady reduction in the extent of the sea ice and a southward movement of the major oceanic frontal zones. There were also fluctuations in ocean temperatures with analysis of alkenone and $\delta^{18}O$ analysis of foraminiferan data from the Tasman Plateau in the Southern Ocean indicating a maximum in Holocene ocean warming at about 3–4 kyr BP (Ikehara *et al.*, 1997).

5.6   Sea ice and sea surface temperatures

5.6.1   *Introduction*

Reconstruction of sea ice extent during the Holocene presents a number of problems because of a lack of observations and the difficulty of finding reliable sea ice proxies in marine sediment and ice core data. However, diatom microfossils in marine sediment cores can be used as a proxy for past sea ice coverage.

Even when we have reliable temperature observations from sites on land these cannot always be simply translated into estimates of sea ice extent. For example, although the correlation between sea ice extent and land temperatures is generally quite strong (Ogilvie, 1984), we know that in the modern era severe cold weather in Iceland is not always correlated with heavy sea ice incidence. However, a strong inverse relationship exists between sea ice extent and temperatures on the Antarctic Peninsula (see Fig. 3.10).

There are no global reconstructions of SSTs available for the Holocene, which is a major handicap for climate simulations of the period. However, in the following we draw together the available data on sea ice distribution and SSTs from the isolated reconstructions.

5.6.2   *The Arctic*

Estimating the sea ice extent during the Holocene is difficult as there are clearly only direct observations for the last few hundred years and for earlier periods the ice extent has to be estimated from ice and sediment core data. Determining regional detail is therefore very difficult and most studies have concentrated on estimating the broader scale ice extent.

Estimates of ice extent for the North Atlantic and western Arctic area were used in the combined modelling/observational study carried out by the Cooperative Holocene Mapping Project (Kutzbach, 1988). Their fields of estimated surface type and modelled atmospheric circulation for 12, 9 and 6 kyr BP, along with data from the present, are shown in Fig. 5.20.

At the start of the Holocene the perennial sea ice cover is thought to have extended along the northern coast of Canada, down much of the Davis Strait and from the central part of the west coast of Greenland to Svalbard. At this time the winter sea ice maximum extended to Newfoundland, the northern tip of Scotland and to northern Denmark. Grumet *et al.* (2001) showed that the sea salt record from Penny Ice Cap was related to sea ice extent in nearby Baffin Bay, with Na^+ concentrations associated with more open water. The 12 000 year record suggests that in Baffin Bay the sea ice extent increased slowly through the Holocene

5.6 Sea ice and sea surface temperatures

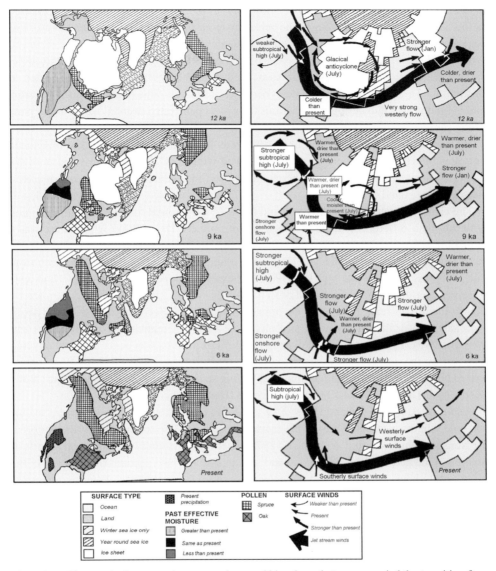

Fig. 5.20 Changes in the atmosphere, geosphere and biosphere that accompanied the transition from glacial to interglacial conditions in the North Atlantic and western Arctic during the past 18 000 years as illustrated by geologic and palaeoecologic evidence. (Adapted from COHMAP members, 1988.)

from very small amounts at 11.5–10.5 kyr BP; however, the nearby GRIP Na$^+$ record does not show such a strong trend over the Holocene.

The extent of the sea ice in the Atlantic was largely dictated by the cold, northwesterly winds from the North American ice sheet. Dansgaard *et al.* (1989) noted a rapid change in

240   *The Holocene*

deuterium in the Dye 3 core from the Younger Dryas to the Pre-Boreal, which they inferred as indicating a significant cooling of the oceanic vapour source at the start of the Holocene as sea ice retreated exposing a large region of cold ocean water.

It should be noted that there was a series of southward shifts in the sea ice edge and cooler surface waters in the North Atlantic associated with the 1500-year cycle (discussed in Section 5.2.3), and detected in the deep ocean core records. This happened in the Nordic and Labrador Seas and during these periods the waters penetrated deep into the warmer parts of the subpolar circulation.

By 9 kyr BP it is estimated that open water was found along the north coast of Canada during summer and that there was much less late winter ice along the west and northern coasts of Norway. In addition, the perennial sea ice had retreated from the east and west coasts of Greenland.

The changes in ice extent between 9 and 6 kyr BP are thought to have been rather small, with the main differences being confined to the southern Davis Strait where there was a reduction in late winter ice. However, there is thought to have been a marked change in Arctic sea ice variability in the Mid Holocene (Jennings *et al.*, 2002). As can be seen from Fig. 5.20, it is believed that by 6 kyr BP both the perennial ice and winter-only ice extents were very similar to those found today.

Information on sea ice cover for the time of the MWP has been obtained from the ocean sediment core drilled in Nansen Fjord along the eastern Greenland coast (Jennings and Weiner, 1996). These indicate that over AD 730–1110 the area was never or rarely covered with sea ice in summer. The later period of approximately AD 1270 to 1400 was characterised by cold intervals in the Jennings and Weiner sediment core, the Iceland sea ice index (Koch, 1945) and the Crête, Greenland ice core (Dansgaard *et al.*, 1975).

Historical data from Iceland (Figs. 2.19 and 5.21) indicate many years of severe sea ice and cold winters, especially during the fourteenth century. The coldest part of the period in the Nansen record occurred at about AD 1370, a period coinciding with historical reports of severe sea ice and cold winters on Iceland and an unconfirmed account of a change in the sailing route between Iceland and Greenland because of severe sea ice between these two areas (Ogilvie, 1991).

There is evidence that there was greater ice cover around Iceland during the LIA and Fig. 5.21 shows the number of months per year when ice was present around the island, along with the mean annual temperature available since 1850. The data show that from 1650 to 1890 Iceland was on average surrounded by sea ice for 2 months per year, but there is a large inter annual variability in the ice cover.

Recent proxy data, however, has cast doubt on some of these earlier estimates of ice extent. For example, the GISP2 ice core record from central Greenland shows low concentrations of MSA, indicating low sea ice concentrations between 11.5 and 10.5 kyr, increasing to high concentrations at approximately 2 kyr BP. So this suggests that sea ice concentrations around Greenland were at a minimum at the start of the Holocene, but then increased for 8 kyr BP (Legrand *et al.*, 1997). Further evidence for a minimum of sea ice between 11.5 and 10 kyr BP comes from the record of remains of bowhead whales in the

5.6 *Sea ice and sea surface temperatures*

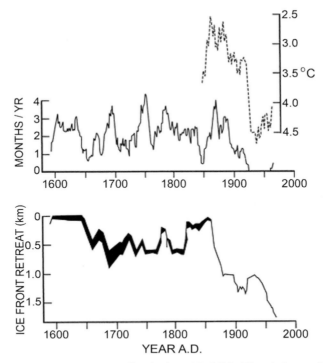

Fig. 5.21 Historic climate records extending back to AD 1600. (Upper) A record of the number of months per year sea ice was present around Iceland and its relationship to mean annual temperature available since AD 1850. (Lower) The extent of the front of the Grindelwald Glacier as reconstructed from oil paintings, drawings, prints, photos, maps and literature. (From Broecker, 2000.)

Canadian Arctic Archipelago (Dyke *et al.*, 1996), which suggests a maximum number at this time.

Sea ice drift patterns in the Arctic Ocean during the Holocene were inferred by Dyke *et al.* (1997) from radiometric analysis of driftwood collected in the Canadian Arctic Archipelago, with the samples covering the period 8.9 kyr BP to the present. Driftwood comes from North America and Eurasia, but the forest composition in these regions is quite different, allowing the origin of the fossil remains to be determined. There are suggestions that the driftwood record shows three modes of transport via the sea ice. Firstly, periods when the TDS was shifted to the west resulting in an expanded Beaufort Gyre and more wood delivery to the Canadian Arctic Archipelago. Secondly, occasions when the TDS was shifted to the east with all the wood being exported from the Arctic via the Fram Strait. In the third situation the TDS was split to the north of Greenland with wood being transported to both the Canadian sector and Svalbard.

The periods when the different forms of the TDS were dominant can be appreciated from Fig. 5.22 (Tremblay *et al.*, 1997). This figure shows that the position of the TDS shifted abruptly on timescales of 40–50 years, separating century to millennial periods of stable

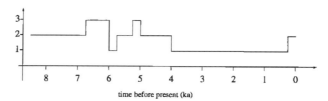

Fig. 5.22 Excursions of the Transpolar Drift Stream (TDS) during the last 8.5 kyr inferred from driftwood collected in the Canadian Arctic Archipelago. On the vertical axis the scale indicates TDS shifted to the west (3), split TDS (2) and TDS shifted to the east (1). (From Tremblay et al., 1997.)

conditions. Of particular note is the period of 6–5 kyr BP when there were a large number of shifts, and the long stable periods before 6.75 kyr BP and in the Neoglacial period of 4 kyr BP to AD 1750. The most recent shift at AD 1750 coincides with a sudden warming on Ellesmere Island (Beltrami and Taylor, 1995). The driftwood record suggests that each of these different transport modes was stable or at least dominant for centuries to millennia during the Holocene (Tremblay et al., 1997).

Sea surface temperatures in the Arctic during the Holocene essentially varied in phase with the changes in the sea ice extent, with SSTs being lower (higher) as sea ice became more (less) extensive. Analysis of ocean sediment cores again provides some insight into how SSTs have varied during the Holocene. High-resolution diatom data from the Vøring Plateau in the eastern Norwegian Sea (Birks and Koc, 2002) allowed the estimation of February and August SSTs, with maximum temperatures being found between about 9.7 and 6.7 kyr BP and lower temperatures in the second half of the Holocene. This SST cooling in recent millennia was also identified in a marine sediment core from 50 km off Iceland (Jiang et al., 2002). A diatom-inferred SST model was applied to the core, which indicated a distinct period of cooling from around 2.2 kyr BP up to the present.

5.6.3 The Antarctic

At the time of the LGM the Antarctic Ice Sheet was more extensive and the sea ice zone extended further north. Gersonde et al. (2005) used diatom assemblages from marine sediment cores to map the sea ice edge at this time and showed that the ice covered twice the area than at present (Fig. 5.23). Similarly, Allen et al. (2005) found that winter sea ice extent was at least 5° further north than at present until at least 19.0 ka BP, based on an ocean sediment core collected just south of the Polar Front in the southwest Atlantic.

In the Atlantic and Indian Ocean sectors the ice reached 47° S, which is in the modern Polar Frontal Zone and close to the Sub-Antarctic Front that today defines the northern edge of the ACC (Gersonde et al., 2005). The ice also had double the present extent during the summer as a result of the more northerly sea ice edge in the Weddell Sea, and possibly the Ross Sea. SSTs were therefore lower at more equatorward latitudes at the start of the Holocene. In fact at the start of the period the Antarctic sea ice was in

5.6 Sea ice and sea surface temperatures

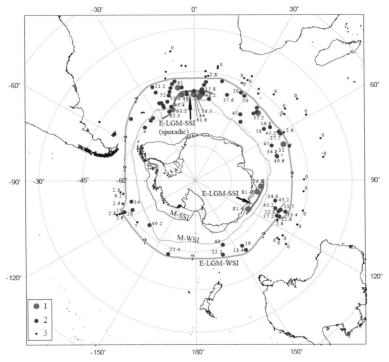

Fig 5.23 Winter and summer sea ice northern limits at the Southern Ocean Last Glacial Maximum (LGM) (northern line) and the present ice edges (southern line). The three sizes of dots indicate the presence of different diatoms. (From Gersonde et al., 2005.)

proximity to South America and retreated throughout the Holocene in parallel with increasing SSTs.

The regional extent of sea ice around the Antarctic during the Holocene has been investigated in a number of studies. Taylor and McMinn (2001) examined sea ice off East Antarctica during the Holocene using data derived from two ocean cores collected on the Mac. Robertson Shelf. These were analysed for diatom assemblages with the results being compared with present-day data from the same location. Open marine deposition commenced at core site GC 1 (Iceberg Alley, 67° S, 63° E) on the outer continental shelf less than 12.8 kyr BP and at about 8.6 kyr BP a sea ice diatom assemblage was deposited. This assemblage is similar to that being deposited in the surface sediments on the shelf today and suggests that perennial sea ice has persisted in the vicinity of GC 1 since that time. Interspersed within the sea ice assemblages, however, some deposits suggest Mid to Late Holocene high-productivity events associated with a climatic optimum. The diatom record from Nielsen Basin (67° S, 66° E) on the inner continental shelf is in contrast relatively uniform. Glacial ice was present over the region more recently than 5.6 kyr BP. Following ice retreat, an ice edge diatom assemblage was deposited briefly before sea ice conditions similar to that on the continental shelf today developed. At the site there is no evidence for the Mid to Late Holocene high-productivity events identified at GC 1. Diatom assemblages from the outer continental shelf on the Mac. Robertson Shelf contain evidence for a fluctuating Holocene climate. If an ice shelf had extended across the Mac. Robertson Shelf during the LGM it had retreated from the outer shelf by about 10.7 kyr BP (Taylor and McMinn, 2001).

The sediments from the inner shelf core started at about 5.6 kyr BP and the sandy sediment facies at the base of the core indicate that glacial ice was present over Nielsen Basin. Grounded ice on the outer shelf began to retreat less than 9.0 kyr BP (Sedwick *et al.*, 2001) and the front of a floating ice shelf had retreated to the inner shelf by about 5 kyr BP. Climatic cooling following the Mid Holocene climatic optimum (4.5–2.8 kyr) probably led to the development of conditions similar to those on the shelf today.

Ocean sediment cores provide a valuable source of data with which to investigate climatic conditions around the continent. However, it should be noted that there is a mismatch between the marine optimum and the optimum in the ice/terrestrial records. Nevertheless, the ocean sediment cores show that during the Early to Mid Holocene there was generally reduced sea ice around the Antarctic (Crosta *et al.*, 2007), which is consistent with a Holocene, marine-inferred climatic optimum. This event seems to have ended at different times around the continent. In the Atlantic sector the marine conditions were warm and ice free at 50° S until about 6–5 kyr BP. This warm event was followed by a cool episode that lasted until at least 2 kyr BP, with a subsequent late Holocene warming. Offshore of Adélie Land the diatom record suggests that the transition to greater sea ice extent in this marine-inferred cool period occurred at around 4 kyr BP and this lasted until at least 1 kyr BP. A core from off the western Antarctic Peninsula supports a transition to cool marine neoglacial conditions at about 3.6 kyr BP.

The SSTs across the Southern Ocean, as determined from the marine record, followed the broad pattern of climate change during the Holocene with, as shown in Fig. 5.14b, generally higher SSTs in the Early Holocene that were warmer than today (Labracherie *et al.*, 1989). SSTs then gradually dropped throughout the Holocene, although a Late Holocene SST maximum has been inferred just before 3 kyr BP from $\delta^{18}O$ records from the Indian sector of the Southern Ocean (Labracherie *et al.*, 1989).

In the Ross Sea sector an expansion in the extent of penguin occupation between 7 and 4.5 kyr BP indicates less sea ice than today, but enough ice for penguins to forage during spring (Emslie *et al.*, 2007). During the Mid Holocene the penguin and ocean data indicate that there was less sea ice than today, but during the Late Holocene the sea ice extent declined even more as the area experienced its warmest temperatures of the last 6 kyr over the periods 2.6–2.3 and 1.2–0.9 kyr BP.

The extent of sea ice during the Holocene can also be estimated from the amount of MSA in ice cores (Curran *et al.*, 2003), although the relationship between MSA concentration and the amount of ice off the coast appears to vary around the Antarctic. Due to post-depositional changes in MSA records at sites with low accumulation rates, it means that the use of this proxy is essentially limited to high accumulation records of the Holocene. Nevertheless, reconstructions of Antarctic sea ice extent based on proxy data from ice cores have the advantage of providing a regionally averaged view, rather than being sensitive to sea ice conditions over a single (sediment core) location in the Southern Ocean. Ice core records also have the potential to provide both long-term records of sea ice changes through the Holocene, and very detailed (i.e. annual resolution) reconstructions of sea ice changes during recent centuries. Early comparisons of high-resolution MSA reconstructions of sea ice extent from sites around the Antarctic continent reveal clear regional differences in the timing and speed of Antarctic sea ice decline (and growth) during the twentieth century (Curran *et al.*, 2003; Abram *et al.*, 2007).

5.7 Atmospheric gases and aerosols

5.7.1 Well-mixed gases

The ice core records give us a very valuable time series of carbon dioxide levels through the Holocene, and data from the Vostok core from the Antarctic plateau (Barnola *et al.*, 1991) and the Taylor Dome core (Indermuhle *et al.*, 1999) have been used to investigate CO_2 variability over this period. The Vostok data (Fig. 5.24) indicate that atmospheric CO_2 concentrations generally rose from the time of the LGM, when they had a value of about 180 ppmv, with only a slight dip at the start of the Holocene. In the Early Holocene, at about 9–8 kyr BP, CO_2 levels were approximately 20 ppmv lower than the pre-industrial value of 280 ppmv, but since that time have been increasing.

A comparison of high-resolution methane records for the Holocene from Greenland and the Antarctic are shown in Fig. 5.25. Methane levels peaked at 11 kyr BP with a decrease to around 5 kyr BP and a subsequent increase up to the present. This pattern is also apparent in the analysis of data from six ice cores from East Antarctica by Ciais *et al.* (1994) who showed that levels of CO_2 and CH_4 were higher during the Early Holocene, when local temperatures were at a maximum.

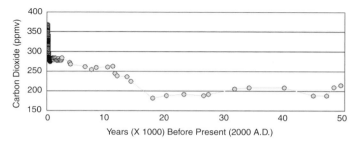

Fig. 5.24 Atmospheric CO_2 concentrations (ppmv) for the last 50 000 years as determined from Antarctic ice and air samples. (From www.geocraft.com/WVFossils/last_50k_yrs.html)

246   *The Holocene*

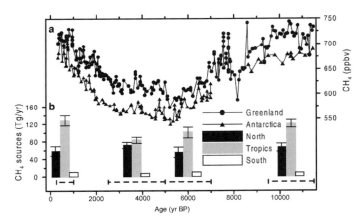

Fig. 5.25 Holocene methane records from Greenland and Antarctica. (From Raynaud *et al.*, 2000.)

The analysis of Raynaud *et al.* (2000) showed that prior to the period of anthropogenic influence, there was a CH_4 variability of 150 ppbv, and that Greenland concentrations were on average 6% higher than the Antarctic values. They point out that this interpolar gradient is about three times less than today's, with the current larger difference being a result of anthropogenic factors.

5.7.2 *Aerosols in the Arctic*

The various ice cores from Greenland have produced a remarkable record of aerosol variability during the Holocene, with one of the key Arctic records of gases and aerosols extending through the Holocene coming from the GISP2 deep ice core from the summit of Greenland. The air masses arriving at this site are dependent on several factors, including the state of the circumpolar vortex, which affects the zonal and meridional wind components. Dust amounts also increase with the frequency and intensity of storms and with the extension and dryness of the source areas of the dust. For example, levels of dust in the Greenland ice cores are sensitive to wind scouring of dust in the North American region.

Early work (Dansgaard *et al.*, 1989) suggested that the general Holocene background level of dust was about 0.05 mg of solid particles per kg of ice, with chemical variations and the aerosol loading of marine and terrestrial source species during the Holocene being lower than during the glacial period or subsequent Younger Dryas period. Ice cores have also given indications that the atmosphere over Greenland was more acidic during the Holocene (Mayewski *et al.*, 1994).

The GISP2 core from Greenland reveals that while there was more dust in the atmosphere during the Younger Dryas event, there was an abrupt transition (≤ 20 years) in wind-

5.8 The cryosphere, precipitation and sea level 247

blown continental dust concentrations at the end of this event (Dansgaard *et al.*, 1989; Taylor *et al.*, 1993).

The temperature changes during the Holocene (progressively milder winter and cooler summers) may have influenced the hydrological conditions, with subsequent impacts on the aerosol loadings. There are indications that conditions in the eastern Mediterranean were very arid around 9 and 6 kyr BP, increasing atmospheric dust loading over the Northern Hemisphere, which is apparent in the Greenland Summit ice cores. During the cooling event that occurred across the North Atlantic around 8.2 kyr BP the Greenland ice core record shows that dust levels increased in the region.

5.7.3   *Aerosols in the Antarctic*

The Antarctic ice cores provide a remarkable record of aerosols deposited onto the continent through dry deposition or within precipitation. Although atmospheric aerosol levels are much lower than for the Arctic as a result of the distance from other terrestrial land masses, the Antarctic still receives deposits from both marine and continental sources.

During the LGM there was a stronger meridional atmospheric circulation than is found now, and Antarctic ice cores suggest increased quantities of aerosols being transported into the continent from more northerly latitudes. However, during the Holocene aerosol amounts were fairly constant when examined in an analysis of ice cores from East Antarctica (Ciais *et al.*, 1994). Nevertheless, volcanic eruptions do result in increased aerosol deposition. The Siple Dome ice core has yielded a very detailed aerosol record for the Holocene via a 2–4 year resolution for the last 10 000 years. The volcanic glaciochemical record identified about 70 volcanic events during the Mid–Late Holocene with the largest sulphate signal (350 ppb) being measured at 2.242 kyr BP.

5.8   The cryosphere, precipitation and sea level

5.8.1   *Introduction*

The available cryospheric precipitation and sea level records for the Holocene provide a valuable record of climate variability and change in the polar regions, with the ice core records providing information on net surface mass balance, which is closely related to precipitation, especially in the centre of Greenland and the Antarctic where evaporation and sublimation are very limited. Information on the former marginal positions of glaciers can also be determined by examination of moraine geomorphology.

In this section we examine how the ice sheets, glaciers and ice shelves varied during the Holocene. Since these features are also strongly affected by the amount of snow falling on them we also here consider the available records of precipitation and precipitation minus evaporation.

248   *The Holocene*

5.8.2   *The major ice sheets*

The large ice sheets of North America, Greenland, Europe and Antarctica had a profound influence on the climates of the high latitude areas. They were so extensive that they influenced the atmospheric circulation and also resulted in significant modifications to the ocean circulation through the injection of meltwater. The LGM was ~20 kyr BP and the Holocene was a period of overall glacial retreat, although the decline of the various ice sheets occurred at different rates. In this section we review the decline of the ice sheets during the Holocene and also examine how the glaciers, and ice shelves fringing the ice sheets, changed over this period.

The Arctic

At the start of the Holocene, ice sheets covered extensive parts of central and eastern Canada (the Laurentide Ice Sheet), Greenland, northern Europe and the Russian Arctic. As discussed in Section 5.3.2, these ice sheets had a major influence on the atmospheric circulation of the Arctic and mid latitude areas of the Northern Hemisphere.

The great Northern Hemisphere icecaps took a long time to melt during the Holocene. Figure 5.26 shows the extent of the Laurentide Ice Sheet at three points during the Early Holocene – 10, 8.4 and 8 kyr BP. The dark shading indicates that near the start of the Holocene the ice sheet extended to almost the Great Lakes area on its southern boundary and to about 115° W on its western flank. At this location the ice sheet was very unstable as a result of the extent and depth of the water bodies at the margins of the ice sheet. By 8.4 kyr BP the area covered by the Laurentide Ice Sheet had reduced in areal extent, with a large lake (Lake Agassiz) forming adjacent to its western edge. Over the next 400 years there was a dramatic reduction in the extent of the ice sheet and by 8 kyr BP the remnants consisted of two separate ice sheets being located close to the present-day upper Quebec and extending from roughly Baffin Island westwards to the Northern Territories. The Laurentide Ice Sheet only finally disappeared around 6 kyr BP. Deglaciation of Hudson Bay and retreat of the mid latitude ice sheets from many maritime margins took place before the 8.2-kyr BP cooling event, but calving margins in Foxe Basin and along Ungava Bay, Canada, persisted through the event. At this time the remnants of the Laurentide Ice Sheet were smaller than the modern Greenland Ice Sheet in volume and area.

Analysis of sediment cores covering the last 8500 years from lakes in the Brooks Range, Alaska (Anderson *et al.*, 2001b), indicates that conditions here between about 8.5 and 6 kyr BP were drier than at present. After this time increasingly wet conditions were experienced. Between about 6 and 4.6 kyr BP the moisture balance (precipitation minus evaporation) increased to near-modern values. Since this time, it has been similar to or higher than modern values.

At the LGM the Fenoscandian Ice Sheet and the Barents Ice Sheet coalesced to form a continuous ice cover that extended from Western Europe eastwards to the Kara Sea. Melting began about 10 kyr BP with sea water filling isostatically depressed areas. The European ice sheets had essentially melted by about 8.5 kyr BP, although at this time there was still ice

5.8 The cryosphere, precipitation and sea level

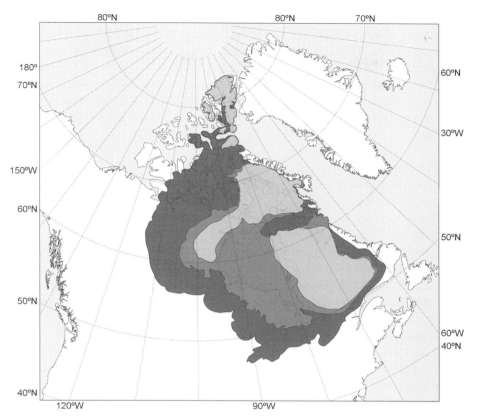

Fig. 5.26 The extent of the Laurentide Ice Sheet at three points during the Early Holocene: 10 (dark grey shading), 8.4 (mid grey shading) and 8 (light grey shading) kyr BP. (Adapted from Dyke and Prest, 1987.)

across North America. Eustatic sea level therefore took several millennia to reach its Mid Holocene maximum.

The Antarctic

Changes in the extent and volume of the Antarctic ice sheet since the LGM generally followed the mean temperature of the region, as illustrated in Fig. 5.14a. There was marked warming from the LGM to the start of the Holocene with a resultant decrease in the extent of the ice sheet. During the Holocene itself there was recession of the ice sheet until about 6.5 kyr BP, which was then followed by a small expansion.

Nevertheless, deglaciation took place at different times around the continent and data from the currently ice-free parts of the continent suggest that deglaciation occurred earlier in East Antarctica than the Antarctic Peninsula.

250 *The Holocene*

After the LGM the deglaciation around the Antarctic Peninsula began with a lowering of the ice surface on the southwestern part of the region about 14.5 kyr BP and with ice-front retreat before 14 kyr BP on the continental shelf west of the peninsula (Hjort *et al.*, 2003). Following the LGM, the retreat of the ice sheet continued throughout the Holocene with most of the currently ice-free areas probably still being covered by ice before about 9.5 kyr BP. On the western side of the Antarctic Peninsula the deglaciation occurred prior to the Holocene at about 13.3 kyr BP at the Palmer Deep drilling site and at 15.7 kyr BP at Lafond Trough. However, on the eastern side of the peninsula it occurred later, in the Holocene, at around 10.6 kyr BP.

Of the currently ice-free areas, there are differences in the timing of the deglaciation. Lakes in the north eastern part of the peninsula and on some sub-Antarctic islands show that sedimentation began at around 9.5 kyr BP. The Marguerite Bay area and parts of King George Island were ice free immediately after the Early Holocene climatic optimum (11.5–9 kyr BP), but other areas deglaciated as late as 5–3 kyr BP. Barsch and Mäusbacher (1986) concluded that King George Island was completely covered by ice until about 6 kyr BP. Further evidence for ice-free conditions around this time come from Tatur and del Valle (1986) who performed radiocarbon dating on lake sediments from King George Island and concluded that the area was deglaciated between 7 and 6 kyr BP.

The deglaciation of northern James Ross Island has now been dated to around 7.5 kyr BP (Hjort *et al.*, 1997) and a similar date has also been suggested for the deglaciation of the east coast of Alexander Island. The wide range of dates occurs because of factors local to the different sites, such as glacier ice extent, thickness and multiple local readvances.

In the Late Holocene there is evidence for two ice advances around the Antarctic Peninsula. The start of the neoglaciation has been put at about 3 or 2.5 kyr BP. The second has been suggested as around 1 kyr BP and is close to the dates put forward by Clapperton and Sugden (1988), who found evidence of ice advance at AD 1300–1500. An ice advance that was synchronous with the LIA ice during the last 750 years has been identified in the South Shetland Islands.

In the Ross Sea sector McMurdo Sound was filled by a grounded ice sheet at the LGM which only began to retreat at about 15.2–14.7 kyr BP and was then followed by a more rapid decline at 12.9–12.7 kyr BP. The ice sheet resulted in the formation of several proglacial lakes that started to dissipate over 14–8 kyr BP, with the final retreat of the Ross Ice Sheet and deglaciation estimated to have taken place around 6.3–5.9 kyr BP.

In East Antarctica some parts of the Larseman Hills escaped glaciation during the LGM, whereas other areas became ice free between about 13.5 and 4 kyr BP. Parts of the Bunger Hills in Wilkes Land were also ice free at the LGM, but the Windmill Islands were probably glaciated by an expanded Law Dome.

Increasing East Antarctic ice volume over the last 4 kyr has been interpreted by Goodwin (1998) on the basis of increasing surface elevation as determined from gas concentrations in ice cores, and from sedimentological and glaciological studies showing the extension over the ocean of the East Antarctic ice margin. The increase in ice volume in East Antarctica may have contributed as much as 1 m of equivalent eustatic sea level lowering during the Late Holocene. This is attributed to Holocene climate warming and a parallel increase in precipitation over the continent.

During the Late Holocene the maritime West Antarctic Ice Sheet may have continued to retreat, supplying melt water to the Southern Ocean as a result of a lag in ice dynamic response to Late Pleistocene climate change (Conway *et al.*, 1999). It is probable the cumulative Antarctic contribution to millennial-scale sea level changes during the Late Holocene was small, with the negative contribution from increased ice volume in East Antarctica equivalent to the positive contribution from melting in West Antarctica.

5.8.3 Glaciers

The Arctic

Section 5.8.2 dealt with changes in the major ice sheets during the Holocene and in this section we examine the advances and retreats of the glaciers, with particular emphasis on the Late Holocene.

With the relatively high temperatures experienced during the first half of the Holocene compared with today, it can be assumed that the first few thousand years of the Holocene must have been a time of rapid glacier retreat, not only of the Laurentide and Fennoscandian Ice Sheets, but also all the circumpolar icecaps and the Greenland Ice Sheet. Some of the ice cores from the Svalbard and Vavilov, Severnaya Zemlya icecaps show no evidence of pre-Holocene ice, suggesting that the icecaps melted completely. The cooling during the second half of the Holocene resulted in the regrowth of the various icecaps and they probably reached their maximum extent during the LIA, but have been retreating recently as temperatures have again risen. Across Europe cold conditions were experienced over 2.8–2.7 kyr BP when the glaciers of northern Norway advanced to their most extensive position for several centuries.

Considering regional variations, in the Brooks Range, northern Alaska, there were fluctuations in glacier extent during the Holocene. Advances have been documented between 6 and 4.6 kyr BP, during the Neoglacial of 3.5–3.3 kyr BP and since 2.2 kyr BP. Moraines also suggest multiple glacial expansions since 1.5 kyr BP and possibly since about 5 kyr BP (Ellis and Calkin, 1984). Lichenometry indicates glacial expansion between 3.5 and 2 kyr BP and during the last 1200 to 1500 years.

Consistent with the fact that many areas recorded lower temperatures during the LIA, glacier advance has been detected worldwide during this period. Such an advance has been quantified by Molnia (1986) who suggested that during the LIA the glaciers in Alaska covered a 10% greater area than they do today. In contrast, during the twentieth century, most Alaskan glaciers receded and, in some areas, disappeared. Across Alaska LIA glaciation started approximately at AD 1200, and continued through the nineteenth century. During this time the majority of Alaskan glaciers reached their Holocene maximum extensions (Calkin *et al.*, 2001). Evidence also exists that the LIA advances in the Kenai Mountains were the most expansive of the Holocene.

250

The Holocene

In northern Europe, most glaciers reached their Holocene maximum extent during the LIA. The impact of the LIA on Iceland was considerable, with agricultural land abandoned at the same time as glaciers advanced (Stotter *et al.*, 1999).

The Antarctic

Due to its relative ease of access, much of the research into changes in Antarctic glaciers has been concerned with the Antarctic Peninsula and the subpolar islands to the north and northwest. Here there was a complex pattern of change during the Holocene, with periods of glacier advance and retreat.

Signy Island (60.7° S, 45.6° W), to the northeast of the tip of the Antarctic Peninsula, currently has about one third of its land area covered by ice. But at the LGM it was thought to have been completely covered by ice, with the ice retreat starting about 7 kyr BP.

On King George Island glaciers were at or within their present limits by 7 kyr BP (Martinez-Macchiavello *et al.*, 1996) , and prior to 5.4 kyr BP on James Ross Island (Hjort *et al.*, 1997). Some parts of Byers Peninsula on Livingston Island in the South Shetland Islands seem to have been deglaciated as late as 5.8–3.5 kyr BP and on some south-facing parts of Elephant Island the deglaciation may have taken place around 7.2 kyr BP.

During the Mid Holocene there was a short period of glacial readvance in the Peninsula region from about 5.8 to 5.3 kyr BP. Evidence for this comes from James Ross Island, and King George Island. There was also glacial advance over 3.2–2.7 kyr BP on Livingston Island as a result of more humid, maritime conditions (Björck *et al.*, 1993), which presumably gave greater precipitation.

In more recent times there have been further glacial advances around the peninsula, with South Georgia experiencing the most extensive advance at about 2 kyr BP, and several regions showing advances during the last 750 years.

5.8.4 Ice shelves

The Antarctic

The ice shelves around the Antarctic Peninsula have received a great deal of attention in recent years because of the break-up of several in the second half of the twentieth century. This has prompted a number of studies that have investigated the evolution of the shelves during the Holocene.

The retreat of the ice shelves around the Antarctic Peninsula probably began around 14–13 kyr BP (Evans *et al.*, 2005) and continued throughout the Holocene. On the western side of the Antarctic Peninsula the ice shelves collapsed after the early Holocene optimum, with the Palmer Deep ocean core indicating warmer waters on the continental shelf at this time, implying that the collapse was a result of oceanic as well as atmospheric change (Domack *et al.*, 2005). The King George VI Ice Shelf collapsed around 9.6 kyr BP

5.8 The cryosphere, precipitation and sea level

Table 5.2 *Sea level around Alaska at various times since the Last Glacial Maximum*

Year before present	Sea level below the present value (m)
18000 (LGM)	120
13000	88
7900	22
5000	4

immediately following the Early Holocene optimum, but reformed at about 7.9 kyr BP (Bentley *et al.*, 2005). On the eastern side of the peninsula the ice shelves were still stable after the early Holocene optimum, and the Larsen B Ice Shelf remained stable throughout the Holocene until its collapse in 2002. Pudsey and Evans (2001) considered the Mid Holocene conditions in the northeastern sector of the Antarctic Peninsula through an analysis of marine sediment cores collected in the Prince Gustav Channel. As discussed earlier, an ice shelf was present in the channel for most of the twentieth century until its disintegration in the 1990s. The sediment data provided a record of the presence of an ice shelf for much of the Holocene via the IRD that was present. The record indicated that the ice shelf was established in the Early and Late Holocene, but was absent in the Mid Holocene period of 5–2 kyr BP.

5.8.5 Sea level

During the last ice age, when water became locked into the ice sheets, global sea level was drawn down, but with the end of the ice age deglaciation allowed sea level to rise again over a period of more than 10 kyr. The exact amount that sea level rose is not known, but has been estimated at 120 m (Peltier, 2002), but could be as much as 130–135 m (Yokoyama *et al.*, 2000). About 100 m of this came from the melting of the Northern Hemisphere ice sheets, with the remainder coming from the Antarctic Ice Sheet and smaller ice sheets. Models suggest that the Antarctic Ice Sheet contributed about 20 m to global sea level rise, but estimates vary from 6–13 m (Bentley, 1999) to 37 m (Nakada and Lambeck, 1988).

In the Northern Hemisphere, sea level variability around Alaska has been estimated for various times since the LGM, and these values are indicated in Table 5.2. Eighteen thousand years ago sea level was some 120 m lower than at present and just before the start of the Holocene at 13 kyr BP it was 88 m below present values. But sea level rose rapidly during the subsequent millennia, being 22 m lower by 7.9 kyr BP. As discussed earlier, there was a major cooling event in the North Atlantic at 8.2 kyr BP, but at this time there was only a slow rise in sea level compared with the several preceding millennia. It

254   *The Holocene*

has been estimated that the Antarctic Ice Sheet contributed about 21.8 m to the total eustatic sea level rise between 12 and 5 kyr BP (Peltier, 1994). It has also been reported that there was an abrupt rise of sea level starting at 7.6 ± 0.1 kyr (Blanchon and Shaw, 1995). By the Mid Holocene at 5 kyr BP most of the major ice sheets had melted and sea level was only 4 m below the current level.

The available data from global coastal and sea level research suggest that multi-decadal sea level changes of up to about 1 m, but typically ± 0.5 m or less, can be detected in the records covering the last few hundred years. The global data suggest that sea level was lower by 0.2–0.5 m than current levels over the period AD 1400–1850, a period that spans the LIA. There is also some evidence (Gibb, 1986) for sea level being 0.5–1.0 m higher than current levels for the period AD 1000–1300.

5.8.6 *Permafrost*

The permafrost that was present during the Holocene mostly formed during the earlier glacial periods, with some persisting during the warmer interglacials. However, some new, relatively shallow (30–70 m depth) permafrost did form during the Holocene.

Changes in permafrost extent during the Holocene to a large degree followed the variations in near-surface air temperature. Permafrost was therefore extensive at the start of the Holocene when temperatures were relatively low and there was still extensive ice across northern North America with permafrost on the fringes of the Laurentide Ice Sheet. The highest Holocene temperatures were not reached across North America until nearly 4 kyr BP, as ice was melting until this time. But by then the permafrost zone had retreated northwards across North America, as well as Europe and Asia.

As temperatures rose towards the Mid Holocene Climatic Optimum (4.5–2.8 kyr BP) there was a general retreat of permafrost. For example, Zoltai (1995) found that across west central Canada most permafrost had turned to peat by 6 kyr BP. At this time the permafrost distribution zone was about 300–500 km further north than at present, which is estimated to correspond to temperatures about 5 °C warmer.

During the second half of the Holocene surface temperatures were generally lower than during the Early Holocene so most new permafrost formed during this time. This was particularly the case during the LIA when temperatures were about 1 °C colder than today. Permafrost formed in a number of areas, such as central Canada where it occurred further south than today. Since the LIA much of this permafrost has degraded but it can still be found in some areas where it has been insulated by thick peat cover.

5.9 Concluding remarks

The Holocene was once thought of as a climatically very stable period, but we now know that there were millennial-scale climatic cycles throughout this period. Some of the most pronounced have periodicities of about 2500 and 950 years, possibly as a result of changes in solar flux. A further cycle at about 1500 years identified in the oxygen isotope record from ice cores may be a result of internal oscillations of the climate system. No clear signal of a 1500-year periodicity has been reported in ^{14}C or ^{10}Be from ice cores, implying that solar variation is probably not the forcing mechanism responsible and that an oceanic internal oscillation in the strength of the THC is more likely to drive this. The most recent event associated with the 1500-year cycle was the LIA. But debate is still ongoing as to the robustness of the 1500-year cycle and as to how similar the Holocene Dansgaard-Oeschger events are to those during the last glacial period.

Marked anomalous periods such as the LIA and the MWP receive a great deal of attention in the literature, but the spatial and temporal extent of these events is very variable and regionally equally anomalous periods with a duration of up to about 100 years can be found. Williams and Wigley (1983) have pointed out that the proxy evidence for spatially coherent century-timescale climatic fluctuations around the North Atlantic Basin is relatively weak and the historical evidence for these events is also patchy: therefore the significance of these events may have been overstated.

6

The instrumental period

6.1 Introduction

Analyses of the conventional surface meteorological observations indicate that the near-surface air temperature of the Earth as a whole has increased by about 0.6 °C over the last century (IPCC, 2007). However, the patterns of surface change across the Earth in the instrumental era are complex and sensitive to the period examined. Many studies highlight that some of the largest environmental changes have taken place at high latitudes.

In this chapter we are concerned with high latitude atmospheric, oceanic and cryospheric changes over the period for which there are a reasonable number of in-situ instrumental records. This is obviously shorter than for the more populated mid latitude regions and covers only about the last 100 to 150 years in the Arctic, and about 50 years in the Antarctic. The first long meteorological records started in Europe during the seventeenth century at locations such as Paris and London, but measurements from the Arctic generally began during the nineteenth century. However, as will be discussed later, there are several Arctic or near-Arctic temperature records that extend back to 1840–1860, such as those from Murmansk, Russia and Reykjavik, Iceland, and around a dozen starting from the second half of the nineteenth century. The greatest increase in the number of Arctic meteorological records came over 1930–40, and later in this chapter we discuss the temperature records from 59 stations in the high latitude areas of the Northern Hemisphere that provide reasonable longitudinal coverage over the areas around the Arctic Ocean.

In the Antarctic the data records start much later than in the north. Although many of the early expeditions made meteorological observations, these were for short periods and were mainly made during the summer season. The first data collected over an Antarctic winter were made by the ship the *Belgica* when it was trapped in sea ice during 1898. Then the first land station was established at Cape Adare by the British Antarctic Expedition in 1899. However, the longest continuous record from the Antarctic comes from Laurie Island in the South Orkney Islands. This station was established in 1903 by the Scottish National Antarctic Expedition and handed over to the Argentine government in 1904. It is still in operation today as Orcadas Station and provides the longest climatological record on any of the sub-Antarctic islands (Zazulie *et al.*, 2010).

256

After the Second World War a number of stations were established on the Antarctic Peninsula, some of which are still in operation today. However, the greatest advance in in-situ observing came with the International Geophysical Year (IGY) of 1957–58 when over 60 research stations were established by 12 nations. These not only made surface meteorological observations, but also established radiosonde programmes, many of which are continuous up to today. In addition to the meteorological programmes, many of these stations make a wide variety of other measurements, such as snow accumulation via stake arrays, air sampling, ozone measurements and, recently, observations such as cloud properties via LIDAR.

A further great advance in observing came with the launch of the first polar-orbiting satellites. The first data available comprised visible and infrared imagery with a relatively coarse horizontal resolution. This was valuable in showing the location of weather systems in the remote areas of the polar regions, but also provided the first broadscale view of the extent and concentration of sea ice, albeit in cloud-free conditions. Sea ice charts were produced regularly based on the satellite data, ship reports and reconnaissance observations from aircraft. The extent of snow cover could also be determined via its high albedo if there was some sunlight. The introduction of passive microwave sensors in the 1970s provided a further significant advance.

Of course many important parameters cannot be determined from satellite data. Although we have long data sets of many atmospheric parameters, the records for the oceans of water mass type and property are unfortunately rather short and sparse. Exciting new autonomous observing systems are being deployed in many of the world's oceans, although there are problems in using many of these in the harsh environment of the sea ice zone.

6.2   The main meteorological elements

6.2.1   Introduction

From the earliest days of meteorological observing in the polar regions the primary quantities of near-surface air temperature and atmospheric pressure have been measured on ships and land stations (see Section 2.2). These could be fairly easily measured using standard instruments, and provide the longest climatological records that we have for the polar regions. The near-surface wind speed and direction is also easily measured from ships, although on land stations this quantity can be strongly influenced by local orographic factors, is difficult to compare even with nearby stations and can change markedly when a station is relocated, even a short distance.

The number of stations making upper air measurements is much smaller than those measuring surface conditions. However, analysis of the data from selected radiosonde stations can provide valuable insight into changes at particular locations.

In this section we will examine some of the changes in conditions across the Arctic and Antarctic that have been determined via in-situ observations and analyses created from these measurements. We will concentrate in particular on near-surface air temperature since this is a primary quantity used in trying to detect anthropogenic impacts on the climate system.

258 *The instrumental period*

6.2.2 The Arctic

Temperature

The palaeoclimate data suggest that the Arctic warmed from about 1840 to the mid twentieth century, when it reached the highest temperatures in the past four centuries (Overpeck *et al.*, 1997). The temperature records from stations with long records for December–January and April are shown in Fig. 6.1, along with a map indicating their locations. Data are shown for 59 stations in all with the temperature data being presented eastwards from the Greenwich meridian. Temperatures are available for several islands in the Arctic Ocean, with the remaining stations being located in the northern parts of North America, Scandinavia, Russia and around the coast of Greenland.

The overriding impression of Fig. 6.1 is the complexity of temperature changes that take place on a range of timescales. Decades of warmer or colder conditions can be found throughout the records, followed by switches back to more average conditions. The more extreme conditions often occur across a range of longitudes, but are very rarely found all the way around the Arctic. So the trends in temperature depend on the season, region and duration examined. For the central Arctic, the largest temperature trends over 1979–95 were for the spring followed by winter (Fig. 6.2). Trends for summer and winter were much smaller (Serreze *et al.*, 2000). Analysis of satellite data also shows that the surface temperature trends are greater for inland regions compared with ocean areas (Comiso, 2003).

The Arctic-wide temperature changes since 1890 can be appreciated from Fig. 6.1b. This shows that one of the most pronounced warm periods in the instrumental record occurred over 1930–40. The increase in temperature was seen in the in-situ observational records, by the lack of sea ice around Iceland and via the Greenland ice core boreholes (Dahl-Jensen *et al.*, 1998). The 1930s was also the warmest decade of the twentieth century in Iceland (Hanna *et al.*, 2004). The warming at this time may well have occurred because of natural climate interdecadal variability since climate models forced by increasing levels of greenhouse gases are not able to reproduce this warm period. After the 1940 warm period, Iceland experienced a cooling until the early 1980s, although the region has experienced an overall warming since the mid 1900s (Hanna *et al.*, 2004).

Another pronounced warm period apparent in these figures occurred during April in the 1950s. The higher temperatures were recorded across northern Russia, Alaska and northern Canada, although were less apparent close to the Greenwich Meridian. Overall, Fig. 6.1c shows generally warm temperatures in the 1990s, which is consistent with the global trend, whereas over the 1930s to the 1950s there were more regional/temporal episodic warm events.

In terms of annual mean surface temperature, north western North America and central Siberia have experienced the greatest temperature rises over the last 50 years, which is of the order of 2–3 °C. The increase in temperature across Alaska and northern Canada is mainly the result of a sudden warming in the mid 1970s. Climatic conditions across Alaska are strongly influenced by the Aleutian Low, which is a climatological atmospheric feature of the North Pacific. A deep Aleutian Low results in predominantly southerly air masses affecting the region and higher temperatures. Cold anomalies across the state are associated with a weak

Aleutian Low and more cold, northerly air masses. The depth of the Aleutian Low is strongly influenced by the phase of the Pacific Decadal Oscillation (PDO), which is a major climate cycle of the Pacific region. In the mid 1970s there was a shift in the nature of the PDO, which resulted in a deepening of the Aleutian Low and warmer conditions across Alaska. The higher

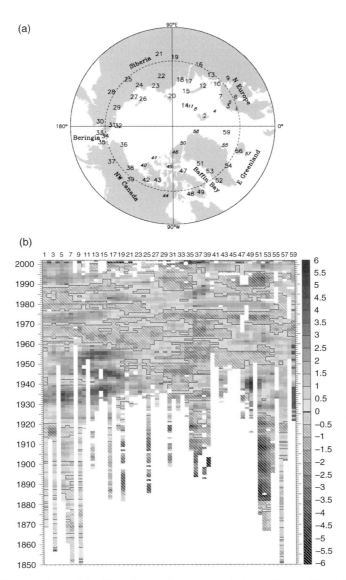

Fig. 6.1 (a) The locations of the 59 stations with long records in the Arctic. The station names are provided in Overland *et al.* (2004). Surface temperature anomaly values for the 59 stations are shown in (b) December–January and (c) April. The anomalies are relative to the 1961–90 mean for each station. (The illustrations are updated from Overland *et al.*, 2004.)

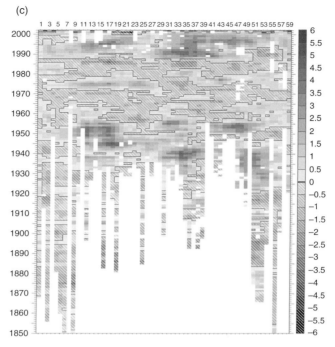

Fig. 6.1 (cont.)

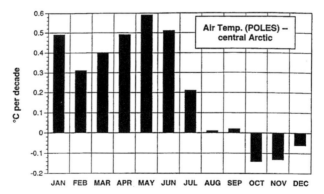

Fig 6.2 Seasonal variation of linear trends (1979–95) of the central Arctic Ocean surface air temperature in degrees centigrade per decade. (From Serreze *et al.*, 2000.)

temperatures in the region have affected many aspects of the environment and resulted in extensive melting of permafrost. In addition, there has been the loss of an estimated three million acres of white spruce on the Kenai Peninsula by spruce bark beetle as a result of the greater survival rate of the larvae in the warmer conditions during the winter.

Temperatures have increased markedly across Siberia in recent decades. Observations from 52 stations across the eastern part of the region indicate that the annual mean temperature increased by 0.06–0.59 °C per decade over the period 1956–90, with an average increase of 0.29 °C per decade (Romanovsky et al., 2007). The most significant trends were in the southern part of the region between 55 and 65° N.

The warming across Siberia seems to be associated with changes in the North Atlantic Oscillation (NAO)/Arctic Oscillation (AO). Changes in the NAO result in a strengthening or weakening of the westerly winds and the advection of relatively warm air masses into Siberia. From the 1970s until about 2000 the NAO was more in its positive phase (see Section 6.3.2). This has resulted in greater transport of warm Atlantic air across Europe and into central Asia.

Conversely, a positive NAO also results in some regional cooling. With a deep Icelandic Low there is strong northerly flow down the Davis Strait, which can result in cooler temperatures in parts of Greenland. Data from eight stations in coastal southern Greenland for 1958–2001 showed a significant cooling of −1.29 °C over the 44-year period (Hanna and Cappelen, 2003).

Since about 1970 there has been a steady rise in the annual mean surface temperature for the Arctic (see Fig. 6.1 for the time series of Arctic temperatures for December–January and April). In 2006 the mean annual surface temperature for land areas north of 60° N was 1.0 °C above the mean value for the twentieth century (Arguez et al., 2007). By 2008 the Arctic 'scorecard' produced by the US National Atmospheric and Oceanic Administration (www.arctic.noaa.gov/reportcard/) reported that this year had the fourth warmest annual mean Arctic temperature, which continued the trend of temperature anomalies of more than 1 °C above the 1961–90 reference point.

Models predict that the greatest impact of increasing greenhouse gas concentrations will be felt in the Arctic (Räisänen, 2001), and the pattern of recent change agrees with this. However, determining the reasons for the recent temperature rise across the Arctic is not trivial, but studies with climate models and the in-situ observations can provide insight into the possible causes. During the twentieth century there were two anomalously warm periods in the 1930s/1940s and starting in the 1990s (Fig. 6.3). Both periods have been linked to sea ice variability

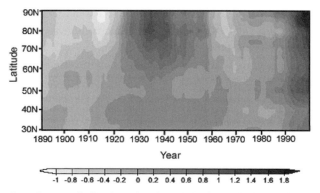

Fig. 6.3 Hovmöller diagram indicating the time–latitude variability of observed surface air temperature anomalies north of 30° N, 1891–1999. (From Johannessen et al., 2004.)

262 *The instrumental period*

(Johannessen *et al.*, 2004), but climate models suggest that the warm period during the 1930s and 1940s was a result of intrinsic variability in high latitude circulation patterns. Conversely, the models reveal that the warming in recent decades is more likely a result of anthropogenic forcing and cannot be explained by natural processes alone. The more recent study of Wang *et al.* (2007) found a similar result using 63 realisations from 20 coupled climate models from the IPCC fourth assessment. When run through the twentieth century all ensemble members reproduced the warm Arctic anomalies of the last two decades, but only eight models generated warm anomalies of at least two thirds of the observed mid-century warm event, and here the timing was not correct. This is further powerful evidence that the recent warming across the Arctic has a strong anthropogenic component.

While there has been a great deal of research into changes in surface temperature across the Arctic there is much less knowledge about how upper air conditions have changed. There are obviously no long time series of radiosonde observations from across the Arctic Ocean, but there are several sites fringing the ocean that have made ascents for several decades and these data have been assimilated into the reanalysis data sets. Graversen *et al.* (2008) considered the temperature trends at upper levels over the Arctic derived from the reanalysis fields. They found a positive temperature trend which, during the summer half year, they attributed to changes in the atmospheric heat transport. However, it was been pointed out that the quality of the ECMWF reanalysis fields is poorer in high northern latitudes (Thorne, 2008), which may cast doubts on the reliability of these results. Grant *et al.* (2008) examined only the radiosonde data, which revealed that the greatest warming is near the ground.

6.2.3 The Antarctic

Surface temperature

The in-situ observational record of Antarctic surface temperatures is poor and sporadic before the International Geophysical Year (IGY 1957–58), however, the Orcadas series of observations from Signy Island begins in 1903 and the Faraday/Vernadsky Station record begins in 1947. There are 16 stations on the Antarctic continent or on the sub-Antarctic islands that have reported on a near-continuous basis since the IGY. In addition, a further six stations started reporting during the 1960s, so that there are approximately 20 time series that can be used to investigate temperature trends. However, the vast majority of the stations are in the Antarctic coastal region or on the islands of the Southern Ocean, with only two providing long-term meteorological data for the interior of the continent.

The Reference Antarctic Data for Environmental Research (READER) data base of mean Antarctic temperature observations was created by the Scientific Committee on Antarctic Research to provide a quality controlled set of data with which to investigate climate change (Turner *et al.*, 2004). Table 6.1 gives the annual and seasonal surface temperature trends for selected Antarctic stations that have long, reliable records.

Table 6.1 *Annual and seasonal surface temperature trends ($^{o}C\ yr^{-1}$) at selected Antarctic stations*

Station	Annual	Spring	Summer	Autumn	Winter	Period
Novolazarevskya	$+0.0187 \pm 0.0116^{a}$	$+0.0138 \pm 0.0208$	$+0.0089 \pm 0.0214$	$+0.0117 \pm 0.0275$	$+0.0376^{a} \pm 0.0253$	1962–2009
Syowa	$+0.0089 \pm 0.0167$	$+0.0021 \pm 0.0189$	-0.0001 ± 0.0121	-0.0013 ± 0.0350	$+0.0293^{b} \pm 0.0241$	1957, 1959–61 1966–2009
Mawson	-0.0026 ± 0.0110	0.0000 ± 0.0144	-0.0076 ± 0.0144	-0.0083 ± 0.0238	$+0.0072 \pm 0.0245$	1954–2009
Davis	$+0.0106 \pm 0.0163$	$+0.0125 \pm 0.0244$	$+0.0101 \pm 0.0157$	$+0.0043 \pm 0.0321$	$+0.0064 \pm 0.0321$	1957–64, 1969–2009
Mirny	$+0.0100 \pm 0.0142$	$+0.0158 \pm 0.0210$	-0.0049 ± 0.0171	$+0.0025 \pm 0.0325$	$+0.0287^{b} \pm 0.0250$	1956–2009
Vostok	$+0.0121 \pm 0.0186$	$+0.0096 \pm 0.0274$	$+0.0155 \pm 0.0217$	$+0.0100 \pm 0.0338$	$+0.0272 \pm 0.0384$	1958–2009
Casey	$+0.0139 \pm 0.0196$	$+0.0169 \pm 0.0201$	-0.0043 ± 0.0158	-0.0075 ± 0.0417	$+0.0392 \pm 0.0358^{b}$	1957–2009
Dumont d'Urville	$+0.0001 \pm 0.0144$	$+0.0144^{c} \pm 0.0165$	$+0.0003 \pm 0.0170$	$-0.0270^{a} \pm 0.35^{c}$	$+0.0124 \pm 0.0366$	1956–2009
Scott Base	$+0.0177 \pm 0.0223$	$+0.0361^{c} \pm 0.0371$	$+0.0033 \pm 0.0207$	$+0.0062 \pm 0.0343$	$+0.0170 \pm 0.0410$	1957–2009
Rothera	$+0.0618^{c} \pm 0.0629$	$+0.0850^{a} \pm 0.0591$	$+0.0183 \pm 0.0237$	$+0.0721^{c} \pm 0.0718$	$+0.0992 \pm 0.1239$	1977–2009
Faraday/Vernadsky	$+0.0540^{a} \pm 0.0286$	$+0.0342^{b} \pm 0.0272$	$+0.0222^{a} \pm 0.0097$	$+0.0529 \pm 0.0324^{a}$	$+0.1035 \pm 0.0565^{a}$	1950–2009
Bellingshausen	$+0.0252 \pm 0.0181^{a}$	$+0.0050 \pm 0.0218$	$+0.0168 \pm 0.0100^{a}$	$+0.0350 \pm 0.0405$	$+0.0376 \pm 0.0466$	1968–2009
Esperanza	$+0.0296 \pm 0.0140^{a}$	$+0.0215 \pm 0.0164^{b}$	$+0.0417 \pm 0.0100^{a}$	$+0.0414 \pm 0.0351^{b}$	$+0.0203 \pm 0.0286$	1945–2009
Marambio	$+0.0503 \pm 0.0298^{a}$	$+0.0332 \pm 0.0448$	$+0.0607 \pm 0.0158^{a}$	$+0.0818 \pm 0.1072$	$+0.0207 \pm 0.0689$	1971–2009
Orcadas	$+0.0200 \pm 0.0073^{a}$	$+0.0186 \pm 0.0071^{a}$	$+0.0158 \pm 0.0040^{a}$	$+0.0205 \pm 0.0098^{a}$	$+0.0270 \pm 0.0240^{b}$	1904–2009
Halley	-0.0144 ± 0.0222	-0.0040 ± 0.0329	-0.0023 ± 0.0191	-0.0492 ± 0.0347^{a}	-0.0025 ± 0.0434	1957–2009
Neumayer	-0.0060 ± 0.0248	-0.0209 ± 0.0573	-0.0157 ± 0.0486	-0.0269 ± 0.0625	$+0.0079 \pm 0.0641$	1981–2009
Amundsen–Scott	-0.0014 ± 0.0117	$+0.0073 \pm 0.0254$	-0.0046 ± 0.0276	$+0.0073 \pm 0.0289$	-0.0199 ± 0.0313	1957–2009

^{a}, Significant at the 1% level. ^{b}, Significant at the 5% level. ^{c}, Significant at the 10% level.

This table of trends is kept up to date at www.antarctica.ac.uk/met/gjma/temps.html

264   *The instrumental period*

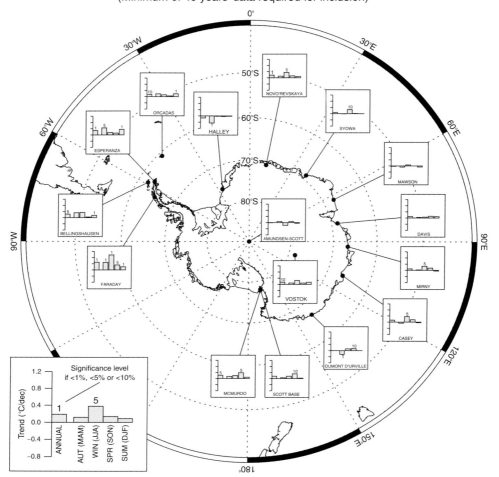

Fig. 6.4 Antarctic near-surface temperature trends 1951–2009.

Temperature trends across the Antarctic since the early 1950s illustrate a strong dipole of change, with significant warming across the Antarctic Peninsula, but with little change across the rest of the continent (Fig. 6.4). The largest warming trends in the annual mean data are found on the western and northern parts of the Antarctic Peninsula. Here Faraday/Vernadsky Station has experienced the largest statistically significant (< 1% level) trend in annual mean temperature of +0.54 °C per decade for the period 1950–2009. Rothera Station, some 300 km to the south of Faraday/Vernadsky, has experienced a larger annual mean warming trend, but the shortness of the record and the large interannual variability of the temperatures mean that the trend is only statistically significant at the 10% level. Although

The warming on the western side of the Antarctic Peninsula has been largest during the winter season, with the temperatures at Faraday/Vernadsky increasing by +1.03 °C decade^{-1} over 1950–2009. In this area there is a high anticorrelation during the winter between the sea ice extent and the surface temperatures (cf. Fig. 3.10), suggesting more sea ice during the 1950s and 1960s and a progressive reduction since that time. King and Harangozo (1998) found a number of ship reports from the Bellingshausen Sea in the 1950s and 1960s when sea ice was well north of the locations found in the more recent period of satellite data availability, suggesting some periods of greater sea ice extent than found in recent decades. However, there is very limited sea ice extent data before the late 1970s so we have largely circumstantial evidence of a mid-century sea ice maximum at this time (de la Mare, 1997). At the moment it is not known whether the warming on the western side of the peninsula has occurred because of natural climate variability or as a result of anthropogenic factors.

Temperatures on the eastern side of the peninsula have risen most during the summer and autumn months, with Esperanza having experienced a summer increase in annual mean temperature of +0.42 °C decade^{-1} over 1945–2009. This temperature rise has been linked to a strengthening of the westerlies that has taken place as the Southern Annular Mode (SAM) has shifted into a more frequent positive phase (Marshall *et al.*, 2006). Stronger winds have resulted in more relatively warm, maritime air masses crossing the peninsula and reaching the low-lying ice shelves on the eastern side.

Estimating temperature trends across the remote interior of the Antarctic is difficult because of the lack of staffed stations, with only Vostok and Amundsen–Scott having long climate records. Over 1957–2009 Amundsen–Scott Station at the South Pole has only small trends in all seasons, none of which are statistically significant. The other plateau station, Vostok, has also not experienced any statistically significant change in temperatures, either in the annual or seasonal data, since the station was established in 1958.

However, a number of attempts have been made to determine temperature trends in the interior by extrapolating coastal data across the continent. Chapman and Walsh (2007) used such an approach and found that for the period 1958–2002 there had been a modest warming over much of the 60°–90° S region, although the largest warming trends were over the Antarctic Peninsula. They also identified a zone of cooling stretching from Halley Station to the South Pole. They found overall warming in all seasons, with winter trends being the largest at +0.172 °C decade^{-1} while summer warming rates were only +0.045 °C decade^{-1}. For the 45-year period the temperature trend in the annual means was +0.082 °C decade^{-1}. Interestingly the trends computed were very sensitive to start and end dates, with trends calculated using start dates prior to 1965 showing overall warming, while those using start dates from 1966 to 1982 show net cooling over the region. Due to the large interannual

266 *The instrumental period*

variability of temperatures over the continental Antarctic, most of the trends are not statistically significant.

A statistical climate-field-reconstruction technique was used by Steig *et al.* (2009) to produce similar fields of trends for the four seasons and the year as a whole. They found an annual mean warming in excess of 0.1 °C decade^{-1} over the last 50 years across most of West Antarctica, with the greatest temperature increases during the winter and spring.

Around the rest of the Antarctic coastal region there have been few statistically significant changes in surface temperature over the instrumental period (Table 6.1). The largest warming outside the peninsula region is at Novolazarevskya, where temperatures have risen at a rate of +0.19 °C decade^{-1}. The high spatial variability of the changes is apparent from the data for Novolazarevskya and Syowa, which are 1000 km apart. Although relatively close to Novolazarevskya, Syowa has only experienced a temperature rise of +0.09 °C decade^{-1} since the record started in 1957.

One area of the Antarctic where marked cooling has been noted is the McMurdo Dry Valleys. Here automatic weather station (AWS) data for 1986–2000 show that there has been a cooling of 0.7 °C decade^{-1}, with the most pronounced cooling being in the summer (December–February = 1.2 °C decade^{-1}, $P = 0.02$) (Doran *et al.*, 2002). The trends in the other seasons are not statistically significant.

Research has shown that changes in the major modes of variability strongly affect the temperatures across the Antarctic. Perhaps the largest change in climatic conditions across the high southern latitudes has been due to the shift in the SAM into its positive phase (see Section 6.3.3). The SAM has changed largely because of the increase in greenhouse gases and the development of the Antarctic ozone hole, with models revealing the latter to have had the greatest influence (Arblaster and Meehl, 2006). Thompson and Solomon (2002) considered the Antarctic surface temperature trends over 1969–2000 and showed that the contribution of the SAM was a warming over the Antarctic Peninsula and a cooling along the coast of East Antarctica (Fig. 6.5). They only considered the months of December to May, which is when the largest change in the SAM has taken place. Due to the much higher regression coefficient between the SAM and temperatures in autumn than in summer, it is in autumn when changes in the SAM have had the greatest impact on Antarctic temperatures (Marshall, 2007). Thompson and Solomon (2002) attributed the trends primarily to changes in the polar vortex as a result of the development of the Antarctic ozone hole. While the major loss of stratospheric ozone occurs in the spring, the greatest changes in the tropospheric circulation, such as the strengthening of the westerlies, is found in the summer and autumn. As discussed earlier, the summer–autumn warming on the eastern side of the Antarctic Peninsula has been linked to the stronger westerlies associated with the changes in the SAM. However, the large winter-season warming on the western side of the peninsula appears to be largely independent of changes in the SAM.

It is difficult to determine the surface temperature trends across the high Antarctic Plateau because there are only two stations with long temperature records. Since the mid 1980s many AWSs have been deployed in the interior, filling important gaps in the observational network. These can provide valuable indications of temperature

6.2 *The main meteorological elements* 267

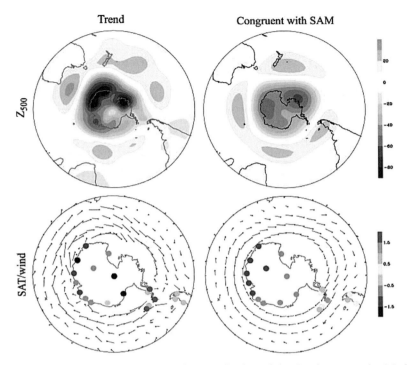

Fig. 6.5 December–May trends (left) and the contribution of the Southern Annular Mode to the trends (right). (Top) 22-year (1979–2000) linear trends in 500-hPa geopotential height. (Bottom) 32-year (1969–2000) linear trends in surface temperature and 22-year (1979–2000) linear trends in 925-hPa winds. Shading is drawn at 10 m per 30 years for 500-hPa height and at increments of 0.5 K per 30 years for surface temperature. The longest vector corresponds to about 4 m s^{-1}. (From Thompson and Solomon, 2002.)

trends at remote locations, although few AWS systems have been maintained at the same locations since the 1980s and there can be gaps in the data when systems fail during the winter.

Another means of examining temperature trends is via the infrared imagery from polar-orbiting satellites. Such imagery can only be used under cloud-free conditions, and provides data on the snow surface rather than at the standard meteorological level of 1.5 m above the surface, but with such high spatial coverage it provides a very valuable supplement to the in-situ observations. Comiso (2000) used the NOAA AVHRR imagery to investigate the trends in skin temperature across the Antarctic over the period 1979–98 (Fig. 6.6). The satellite-derived temperatures were compared with the in-situ observations from 21 stations and found to be in good agreement with a correlation coefficient of 0.98. The trends showed a cooling across much of the high plateau of East Antarctica and across Marie Byrd Land, with the former being consistent with the trends derived by Thompson and Solomon (2002), although the trends in Fig. 6.6 are for annual data and the Thompson and Solomon study is

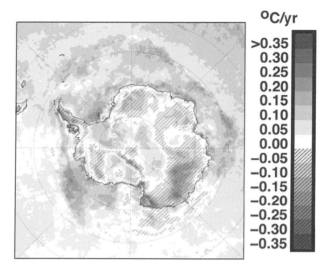

Fig. 6.6 Temperature trends determined from satellite imagery. The trends were derived using the January and July monthly average data from 1979 to 1998. (From Comiso, 2000.)

concerned only with the summer and autumn seasons. However, the apparent cooling across the Antarctic Peninsula in Fig. 6.6 is surprising since this area has experienced a marked warming in recent decades, although the in-situ observations are all from low-level coastal stations.

In recent decades many relatively short ice cores have been drilled across the Antarctic by initiatives such as the International Trans Antarctic Science Expedition (ITASE) (Mayewski et al., 2006). These provide data over approximately the last 200 years and therefore provide a good overlap with the instrumental data. Large-scale calibrations have been carried out between satellite-derived surface temperature and ITASE ice core proxies (Schneider et al., 2006). Their reconstruction of Antarctic mean surface temperatures over the past two centuries was based on water stable isotope records from high-resolution, precisely dated ice cores. The reconstructed temperatures indicated large interannual to decadal scale variability, with the dominant pattern being anti-phase anomalies between the main Antarctic continent and the Antarctic Peninsula region, which is the classic signature of the SAM. The reconstruction suggested that Antarctic temperatures had increased by about 0.2 °C since the late nineteenth century. Schneider et al. (2006) found that the SAM was a major factor in modulating the variability and the long-term trends in the atmospheric circulation of the Antarctic.

Upper air temperatures

Analysis of Antarctic radiosonde temperature profiles indicate that there has been a warming of the troposphere and cooling of the stratosphere over the last 30 years. This

6.2 The main meteorological elements

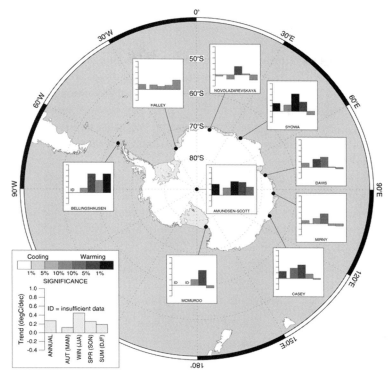

Fig. 6.7 Annual and seasonal 500-hPa temperature trends for 1971–2003. (After Turner et al., 2006a.)

is the pattern of change that would be expected from increasing greenhouse gases; however, the mid-tropospheric warming in winter is the largest on Earth at this level. The data show that regional mid-tropospheric temperatures have increased most around the 500 hPa level with statistically significant changes of 0.5–0.7 °C decade^{-1} (Fig. 6.7) (Turner et al., 2006a). Figure 6.7 indicates warming at many of the radiosonde stations around the continent, but a clear pattern of predominantly winter warming is apparent around the coast of East Antarctica and at the pole.

The exact reason for such a large mid-tropospheric warming is not known at present. However, it has recently been suggested that it may, at least in part, be a result of greater amounts of polar stratospheric cloud (PSC) during the winter (Lachlan-Cope et al., 2009). PSCs are a feature of the cold Antarctic winter, forming at temperatures below about 195 K. However, the Antarctic stratosphere has cooled in recent decades because of the Antarctic ozone hole and rising levels of greenhouse gases. Analysis of stratospheric temperatures in the reanalysis data sets has suggested that over the last 30 years the area where PSCs might form in winter has increased in size, so promoting the formation of more PSCs. Once present, PSCs act like any other cloud, giving a warming below their level and cooling above. We have little data on the optical properties of PSCs, but modelling suggests that if

270 *The instrumental period*

the optical depth is around 0.5 then a greater amount of PSCs could give a mid-tropospheric warming. PSCs are not currently represented explicitly in climate models, but if further research shows that they are responsible for the large winter season mid-tropospheric warming they need to be represented more realistically in the models.

Great advances have been made in our understanding of recent temperature changes across the Antarctic in the last few years. We now know that anthropogenic activity, and particularly the presence of the Antarctic ozone hole, has played a large part in the near-surface warming on the eastern side of the Antarctic Peninsula through its role in driving stronger westerlies. However, we still do not know the reasons for the large winter season warming on the western side.

At upper levels the anthropogenically induced stratospheric ozone hole has had a major impact on temperatures. It is clearly responsible for the stratospheric cooling, especially during the austral spring, but there are also indications that it has indirectly played a part in the recently discovered large mid-tropospheric warming above the continent in winter (Turner *et al.*, 2007).

6.3 Changes in the atmospheric circulation

6.3.1 *Introduction*

In this section we are concerned with the broadscale changes in the atmospheric circulation that have taken place across the two polar regions in recent decades. Much of our understanding of atmospheric variability and change has come from assessments of the reanalysis data sets that have been prepared recently by a number of NWP centres (see Section 2.3.3). The fields allow the investigation of how the major modes of atmospheric variability, such as the SAM, the NAM and the NAO, have changed in recent decades. They can also be used to track individual weather systems, so allowing the examination of changes in cyclone tracks, and regions of cyclogenesis and cyclolysis.

In this section we examine how the major modes of variability of the two polar regions have changed in recent decades, and also look at changes in cyclone frequency and distribution.

6.3.2 *The Arctic*

Mean sea level pressure values across the Arctic and in mid latitudes are strongly influenced by the NAO. Figure 6.8 (Polyakov *et al.*, 2003) shows the annual MSLP for the area north of $62°$ N since 1880. Overall there has been a small negative trend of -0.05 hPa/100 years, although the MSLP values show large fluctuations over the period. The largest drop in pressure was from the early 1980s to the early 1990s. However since the pressure minimum in the early 1990s MSLP values have been increasing over the time that record high temperatures and minima in sea ice extent were being recorded.

6.3 *Changes in the atmospheric circulation* 271

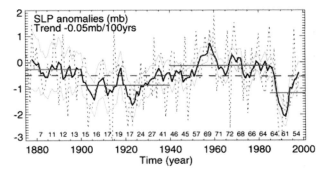

Fig. 6.8 Composite time series of the annual sea level pressure (hPa) anomalies for the region poleward of 62° N. The dashed line shows annual means, solid, dark line the 6-yr running mean and the solid, light line the 95% confidence intervals. The horizontal dashed line shows the trend, and solid horizontal line shows the means for positive and negative low-frequency oscillation phases (see Polyakov *et al.*, 2003, for details, from which the figure was taken). Numbers in the bottom part of the panel show the number of stations used for averaging.

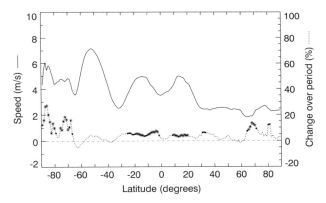

Fig. 6.9 The zonal, annual mean 10-m wind speed for January 1979 to December 2009 between 90° N and 90° S (solid line). The trend over this period (percentage) is shown by the broken line. The circles on the broken line indicate trends that are significant at the 5% level.

The pole-to-pole cross-section of zonal, annual mean near-surface wind speed for the period 1979–2009, along with the trend, is shown in Fig. 6.9. The figure shows that the latitude with the strongest mean wind speed is close to 50° S, and that poleward of this location the speeds generally drop towards the coast of the Antarctic continent. Since 1979 wind speeds have increased close to 50° S, but only by about 5%. The greatest increase in speed of up to 20% has been over 65–70° S, where the increases have been statistically significant at the 1% level. This latitude band encompasses the Antarctic coastal zone around East Antarctica and the open ocean across the Weddell, Bellingshausen and Amundsen Seas.

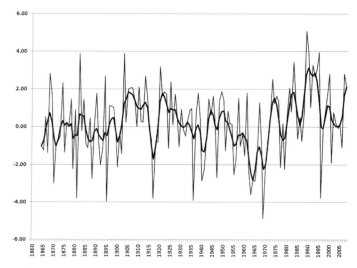

Fig. 6.10 The winter season North Atlantic Oscillation index based on station data. The thick line is the 5-year running mean. (Data from the Climate and Global Dynamics group, National Center for Atmospheric Research, Boulder, CO, USA.)

The NAO/NAM

In the Northern Hemisphere we have a long record of the variations in the NAO that extends back to the mid nineteenth century (Fig. 6.10). In the late nineteenth and early twentieth centuries there were fluctuations in the NAO index, and periods of positive anomaly, such as around 1910, but no overall trend. The time series of the winter NAO shows generally positive values from approximately 1900 to 1935 and negative values for 1935–75 (Fig. 6.10). This index is standardized with respect to 1950–99. There was a period of predominantly positive values over 1989–95, followed by a 10-year period of more neutral albeit variable values (Overland and Wang, 2005). The most marked increase in the NAO index has been from the mid-1960s to the mid 1990s when the index shifted from strongly negative to strongly positive. This change resulted in a strengthening of the mid latitude westerlies and advection of warm air from the North Atlantic into the Arctic Ocean and northern Russia, the consequences of which are discussed in the relevant sections on sea ice and ocean conditions.

The positive NAM index reflects strengthening of the polar vortex and cold stratospheric temperatures; therefore the 1980s were a period of warm stratospheric temperatures followed by a cold, strong polar vortex in the positive NAM years. Until a few years ago, the upward trend in the NAM from the 1960s to the mid 1990s (Fig. 6.11) was considered to be a statistically significant (and continuing) climate change relative to earlier values of the NAM (Feldstein, 2002). Increasing concentrations of greenhouse gases can shift the NAM into its positive phase and there was a great deal of speculation during the 1990s that the NAO increase was a result of anthropogenic activity. However, since the mid 1990s, at a time of

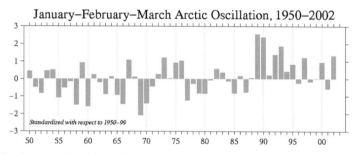

Fig. 6.11 The January–March Arctic Oscillation/Northern Annular Mode index for 1950–2002. (From http://jisao.washington.edu/ao/)

record concentrations of greenhouse gases, the NAO has reverted to more neutral conditions, indicating the complex nature of climate change and the danger in making assumptions about the reasons behind short-term trends.

Despite the return of the NAM to more neutral conditions since the mid 1990s, some modelling studies suggest that external forcing, including increased greenhouse gas concentrations and stratospheric ozone loss, may favour a higher frequency of its positive state (e.g. Kuzmina et al., 2005). However, the NAM/NAO record is also consistent with a red (lower frequency) noise time series model of atmospheric variability (Wunsch, 1999). In the Antarctic the marked shift of the SAM into its positive phase has been linked to the springtime decrease of stratospheric ozone (Arblaster and Meehl, 2006). However, the Arctic has not experienced the same level of stratospheric ozone loss that has been observed in the Antarctic since there is more meridional heat transport in the north and stratospheric temperatures are not cold enough for large-scale ozone destruction. Nevertheless, there has been some ozone loss and a modelling study by Volodin and Galin (1999) showed a link between ozone depletion and NAM trends, but this finding has not been reproduced in other models. Moreover, simulations forced by observed ozone trends did not show a significant Northern Hemisphere circulation response, suggesting that the much weaker ozone depletion observed in the Northern Hemisphere has not been an important driver of circulation changes there.

The influence of the tropics

High northern areas of the Pacific Ocean are affected by tropical atmospheric and oceanic changes via the Pacific North American Association (PNA) and its oceanic equivalent the Pacific Decadal Oscillation (PDO) (Zhang et al., 1997). The PNA is a series of quasi-stationary MSLP anomalies extending from the west-central equatorial Pacific into higher latitudes. EOF analysis of the region north of 20° N shows the NAM as the first component of atmospheric circulation variability, but the second EOF (Fig. 6.12) is the Pacific-oriented, PNA-like pattern, indicating the importance of this mode of

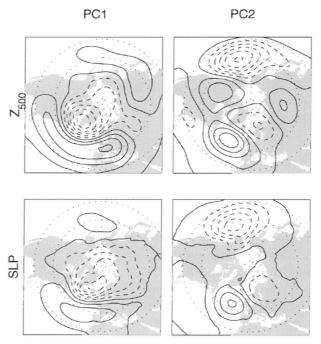

Fig. 6.12 Monthly mean 500-hPa height and sea level pressure fields regressed on the first two standardized principal components based on data for 1958–99. (From Quadrelli and Wallace, 2004.)

variability in influencing the Arctic climate (Fig. 6.12) (Quadrelli and Wallace, 2004; Wu and Straus, 2004).

The PDO high–low latitude link (teleconnection) varies on longer timescales than the PNA, with, as its name implies, a strong decadal influence. The PDO time series based on sea surface temperature (SST) (Fig. 6.13) represents the low-frequency variability of the PNA in the North Pacific sector and has an interdecadal regime-like structure. The PDO is important in influencing the climate of Alaska via the Aleutian Low, which is one of the main features of atmospheric circulation in the Northern Hemisphere winter. It has a controlling influence on North Pacific Ocean circulation, as well as Bering Strait sea ice extent and the surface climate of western North America. There has been a statistically significant twentieth century shift to a deeper and more poleward winter Aleutian Low, which may have been a result, at least in part, of anthropogenic forcings. In addition, there was a marked shift in the nature of the PDO in the mid 1970s (Fig. 6.13), with the index shifting from negative to positive. This resulted in a deepening of the Aleutian Low and greater warm air advection into Alaska, which was the main contributing factor to the warming across the state.

On an Arctic-wide scale, the EOF analysis of sea level pressure carried out by Wu *et al.* (2006) showed that there is a western/eastern hemisphere Arctic dipole for EOF 2. This

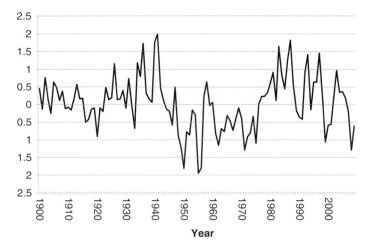

Fig. 6.13 Time series of the annual mean Pacific Decadal Oscillation (PDO). (From data published by the University of Washington at http://jisao.washington.edu/pdo/PDO.latest)

dipole is associated with a strong pressure gradient over the central Arctic with winds blowing toward or out of the North Atlantic. Much of the recent multi-decadal decrease in pressure and strengthening of the Icelandic and Aleutian lows is consistent with positive trends in the NAM and PNA, that is the early 1990s minus the 1960s for the NAM and early 1980s minus the 1950s for the PNA. For the recent period (2000–05), their contribution to Arctic change is reduced.

The interannual variability of the PNA component has a strong contribution from the El Niño–Southern Oscillation (ENSO) and there have been changes in the PNA in recent decades as El Niño events have become more frequent and of greater intensity.

Synoptic-scale weather systems

Changes in cyclonic activity across the Arctic have been investigated using the long series of analyses that are available. Zhang *et al.* (2004) used an automated cyclone identification and tracking algorithm to examine depression distribution and variability over the 55-year period 1948–2002. They used a cyclone activity index, which integrated information on cyclone intensity, frequency and duration into a comprehensive index of cyclone activity.

The authors found that Arctic cyclone activity had increased during the second half of the twentieth century (Fig. 6.14), while mid latitude activity generally decreased from 1960 to the early 1990s (Fig. 6.15). Zhang *et al.* (2004) also determined that the number and intensity of cyclones entering the Arctic from mid latitudes had increased, suggesting a shift of storm tracks into the Arctic, particularly in summer. Arctic cyclone activity displays significant low-frequency variability, with a negative phase in the 1960s and a positive phase in the 1990s, upon which 7.8- and 4.1-yr oscillations are superimposed. The 7.8-yr signal generally corresponds to the alternation of the cyclonic and anticyclonic regimes of the

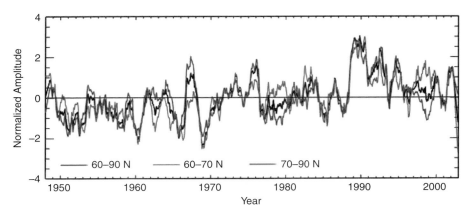

Fig. 6.14 The cyclone activity index anomalies for the Arctic region (60–90° N) and its two subregions, the Arctic Ocean (70–90° N) and the Arctic marginal zone (60–70° N), to which the 13-point smoother has been applied. (From Zhang et al., 2004.)

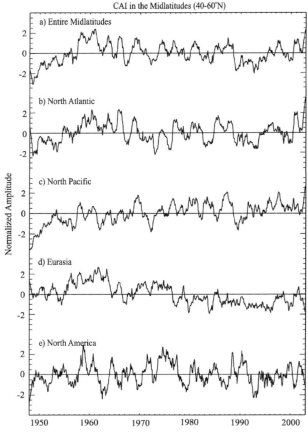

Fig. 6.15 The cyclone activity index anomalies in (a) the entire mid latitudes and its four sectors, (b) the North Atlantic, (c) the North Pacific, (d) Eurasia, and (e) North America. (From Zhang et al., 2004.)

6.3 Changes in the atmospheric circulation 277

Arctic sea ice and ocean motions. For the period 1970–99 relative to 1945–69, there has been an increase in the intensity of major weather systems (centres of action) for sub-Arctic regions that influence the Arctic. Mean pressures have decreased in the Icelandic and Aleutian centres, while pressures in the Siberian high have increased. Thus the pressure gradient, and therefore wind forcing, in the regions between these centres has increased.

6.3.3 Antarctic

Surface pressure and the wind field

There are few continuous records of MSLP for the Antarctic before 1957, when many of the research stations were established. However, the records for all the continental stations for which MSLP data are available show negative trends in the annual mean MSLP over the full length of their records (Turner *et al.*, 2005). This can be attributed to the trend in recent decades towards the SAM being in its positive phase. The MSLP trends for all stations on the Antarctic continent for 1971–2000 were more negative than for 1961–90, except for Halley, where the trends were similar.

The trend in near-surface wind strength is positive across the Southern Ocean, with winds having increased by about 15% since 1979. On the continent, all but two of the coastal stations have recorded increasing mean wind speeds over recent decades.

The Southern Annular Mode

The SAM is the principal mode of variability of the atmospheric circulation of the high and mid latitude areas of the Southern Hemisphere, and its variations have a profound effect on many aspects of the climate of the Antarctic and Southern Ocean (see Section 3.4.5). The SAM changes as a result of variability in the internal dynamics of the atmosphere, as well as a result of anthropogenic factors, such as increasing levels of greenhouse gases and the loss of stratospheric ozone.

Marshall (2003) produced an index of the SAM based on station data that starts in 1957, so allowing investigation of changes in the SAM in the pre-satellite era. The index shows a general increase beginning in the mid 1960s (Fig. 6.16), which is consistent with a strengthening of the circumpolar vortex and intensification of the circumpolar westerlies, as observed in northern Antarctic Peninsula radiosonde data.

The positive trend in the SAM is statistically significant annually and in summer and autumn (Marshall, 2007), with the largest change being during the summer and autumn seasons. This shift of the SAM into its positive phase has been linked to increases of greenhouse gas concentrations (Hartmann *et al.*, 2000; Kushner *et al.*, 2001), but the work of Arblaster and Meehl (2006) suggests that ozone depletion has made the greatest contribution to recent change. Longer records of SAM variability have been reconstructed using principal component regression techniques (Jones *et al.*, 2009), which reveal that the magnitude of the recent SAM trend in summer is unprecedented during the twentieth century, while that in autumn is not.

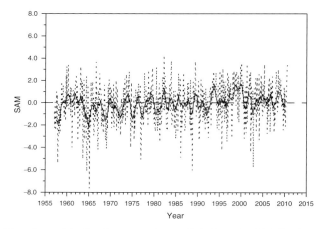

Fig. 6.16 Monthly values of the Southern Annular Mode index. The thick line is a 12-month running mean. (From www.antarctica.ac.uk/met/gjma/sam.html)

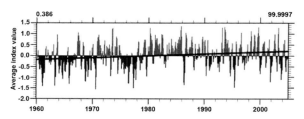

Fig. 6.17 Time series of the index of zonal wave number 3. Time increments are three-month running means. (From Raphael, 2007.)

Changes in the wave number 3 pattern

Mean surface and upper air analyses of the Southern Hemisphere show a climatological wave number 3 pattern throughout the troposphere with troughs in the area of the Ross Sea, close to the Greenwich Meridian and near 90° E. This has a major influence on the atmospheric circulation around the Antarctic, giving, for example, the mean northwesterly flow on the western side of the Antarctic Peninsula.

It is possible to quantify the strength and variability of the wave number 3 pattern using indices, such as that developed by Raphael (2007). This index is a measure of the strength of that part of the large-scale atmospheric circulation associated with the wave number 3 pattern and is calculated as the normalized deviation of the pressure field at the centre of the mean annual positions of the three ridges that define the pattern (49° S, 50° E; 49° S, 166 E; 49° S, 76° W) from its climatological mean at those locations. Figure 6.17 shows the time series of the index, illustrating that the wave number 3 pattern has negative and positive phases. When the index is positive the geopotential height field has a pronounced meridional component while a negative index indicates that the geopotential height field is strongly zonal.

6.3 Changes in the atmospheric circulation

Figure 6.17 shows that there has been an apparent shift of the index to more positive values beginning in the middle to late 1970s. This positive trend has a statistically significant slope and represents an increase in the amplitude of the wave number 3 pattern and therefore a shift toward a more meridional pattern of circulation. This increase in amplitude of the wave number 3 pattern has been found in other atmospheric circulation data (van Loon *et al.*, 1993) and may be part of an overall change in the circulation that is not limited to the Southern Hemisphere and has been noted in other studies (Trenberth and Hurrell, 1994).

The influence of the tropics

The southern equivalent of the PNA is the Pacific South American Association (PSA), which is a comparable series of MSLP anomalies extending from the central, tropical Pacific Ocean towards the tip of South America and the Antarctic Peninsula. Since the late 1970s El Niño events have become more frequent and stronger; however, it has been difficult to find high latitude climate changes than can be attributed to this tropical variability. Around the Antarctic a shift to more El Niño events would tend to promote a stronger PSA pattern with more high-pressure/blocking events to the west of the Antarctic Peninsula. However, the available atmospheric analyses would suggest that there had been the reverse of this trend with more storms in the area, again indicating the highly non-linear links between the high/low latitude areas (Lachlan-Cope and Connolley, 2006).

Synoptic-scale weather systems

As for the Arctic, various series of meteorological analyses have been used to investigate changes in cyclonic activity in high southern latitudes. Simmonds and Keay (2000b) used a depression identification and tracking scheme to examine changes in cyclone occurrence and distribution over 1958–97. Their work was based on the NCEP reanalysis fields, which were available every 6 hours. They found that over the 40-year period the annual and seasonal numbers of cyclones had decreased at most locations south of $40°$ S, with the greatest reduction close to $60°$ S, and an increase in numbers to the north. Since MSLP pressure values have dropped around the Antarctic in recent decades, it is concluded that there has been a trend to fewer but more intense cyclones in the circumpolar trough. One exception is the Amundsen–Bellingshausen Sea region (Simmonds *et al.*, 2003) where, as discussed earlier, there has been greater cyclonic activity.

The annual average number of cyclones per Southern Hemisphere analysis increased from the start of the period to a maximum of about 39 in 1972, followed by an overall decline, with the numbers of systems in the 1990s being particularly low. The study concluded that the decrease in cyclone numbers was associated with a warming Southern Hemisphere. The study also found that the mean radius of extratropical cyclones over the Southern Ocean displayed a significant positive trend, which had taken place in parallel with the drop in central pressure.

280   *The instrumental period*

6.4 The ocean environment

6.4.1 Introduction

Determining recent changes in the polar oceans is difficult, since there are far fewer time series of ocean observations than for the atmosphere. In the open ocean research vessels have made subsurface measurements of temperature and salinity for many years using instruments such as expendable bathythermographs (XBTs). However, the spatial distribution of such measurements is quite variable. Some areas, such as the Drake Passage south of South America and the northern North Atlantic, have been studied intensively for many years and here we have a reasonable knowledge of changes in the ocean currents and water masses. Other areas have far fewer observations and determining change is very difficult. Advances have been made with some newly released data sets, such as military archives of submarine data provided by the USA and Russia (Dickson, 1999). Perhaps the most challenging area is the sea ice zone. Here measurements can be made by vessels during the summer and autumn months, but there are very few observations during the other seasons when there is extensive sea ice.

In the future new observing systems offer great promise for autonomous measurement in the polar oceans. The Argo floats (see Section 2.2.4), which automatically collect ocean profiles and transmit the observations to data centres when they surface, are providing excellent coverage of the ice-free ocean. Several thousand such systems have been deployed, which provide an almost synoptic view of the world's oceans. Work is under way to develop systems that can withstand the very much more challenging environment of the sea ice zone. All these systems will provide excellent data for determining oceanographic change in the future.

In this section we will review our understanding of high latitude oceanographic change based on the available observations. For the surface, we have satellite data sets that start in the 1970s, and these can indicate changes of sea surface temperature, ocean colour and sea ice conditions. For the subsurface layers we are dependent on more sporadic in-situ measurements.

6.4.2 The Arctic

The Arctic Ocean is very sensitive to environmental changes above the ocean surface itself, over the surrounding land masses and within the northern Pacific and Atlantic Oceans. Weather systems can easily penetrate to high latitudes bringing relatively warm air masses into the heart of the region and, as discussed below, changes in atmospheric circulation on a range of timescales can radically alter the advection of air and water masses into the Arctic Basin. Changes in the atmospheric circulation over the surrounding continents can result in precipitation changes and alterations to the freshwater runoff into the Arctic Ocean. There can also be changes in the oceanic circulation at mid latitudes that result in modifications to the water masses that enter or exit the Arctic Ocean. For example, the subsurface layers

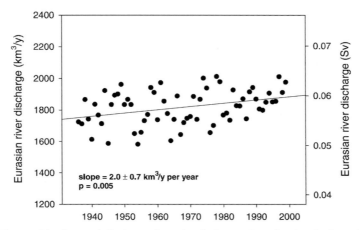

Fig. 6.18 The combined annual discharge from the six largest Eurasian Arctic rivers for the period 1936–99. (From Peterson *et al.*, 2002.)

within the Arctic Ocean are sensitive to the penetration of heat from Atlantic Water and Pacific Water via the Fram Strait and the Bering Strait.

The Arctic Ocean has experienced remarkable changes in recent years, with perhaps the most high profile event being the marked reduction in summer/autumn sea ice extent (see Section 6.5.2). However, there have also been large changes in the circulation of the ocean and the nature and distribution of water masses.

Over the last half century there have been a number of alterations in the freshwater cycle of the Arctic Ocean. These changes are associated with an increase in freshwater input from river discharge, and the consequent freshening of the Nordic Seas and Subpolar Basins. River discharge reflects net precipitation (precipitation − evaporation (P–E)) over a drainage basin, and is therefore a sensitive indicator of broadscale climatic conditions. The discharge records from the North American rivers are quite short, but long records are available for many Eurasian rivers, which dominate inflow to the Arctic Ocean. In the late 1960s/early 1970s there was a sharp increase in P–E over much of the Arctic and Eurasian river discharge began to increase. The six largest Eurasian rivers are the Yenisey, Lena, Ob', Pechora, Kolyma and Severnaya Dvina and the average annual discharge of these into the Arctic Ocean has increased at a significant rate of 7% over the period 1936 to 1999, with the increase being 2.0 ± 0.7 km^3 per year (Fig. 6.18). This suggests that the average annual discharge from these six rivers is about 128 km^3 per year greater than when routine discharge measurements began in the 1930s. Over the period that Eurasian river discharge increased the pan-Arctic mean surface air temperature rose by 0.7 °C. However, a positive correlation has also been found between discharge and global mean surface temperature, which is perhaps not surprising since conditions across a large part of the globe affect poleward moisture transport.

Many of the broadscale changes observed have been linked to alterations to the NAO/NAM, which is the primary mode of variability at high latitudes. Over 1965–95 the NAO

282   *The instrumental period*

index increased at the same time as global air temperatures rose, while freshwater input to the Arctic Ocean paralleled these changes. However, after 1995 the NAO and the NAM became more neutral, while temperatures across the Arctic continued to rise.

The positive phase of the NAO gave a greater flux of warm Atlantic Water from the Atlantic to the Siberian Arctic. The water was at a level of 100–400 m below the surface. The upward flux of heat from this layer may be contributing to the maintenance of the open water there.

Other changes occurred during the 1990s when the NAM was in its positive phase. The cold halocline layer (CHL) of the Arctic Ocean lies between the cold upper mixed layer and the underlying Atlantic Water layer. It is close to the freezing temperature and vertically stratified in salinity, and its density gradient suppresses upward heat flux to the surface from the warmer underlying water. During the first half of the 1990s, and simultaneously with the strong NAM years, the CHL disappeared in the Nansen, Amundsen and Markov Basins for the first time in the historical record (Steele *et al.*, 2004). Following 1998 the CHL has partly recovered, associated with a shift of the flow of low-salinity shelf waters back to the Laptev Sea.

The shift in the NAO into its positive state since the 1960s is thought to have contributed to the decreasing deep convection in the Greenland Sea and the increasing convection in the Labrador Sea (Dickson *et al.*, 1996). This has led to warming of the deep water in the Greenland Sea and cooling of the water of the Labrador Sea.

The greatest change in freshwater discharge has been during the low flow period of November to April, and the increase in discharge has been linked to the shift of the NAO into its positive phase, which has been greatest during the winter months (cf. Fig. 6.10). The positive phase of the NAO is associated with positive precipitation anomalies across Scandinavia, but they also extend across Siberia to the Lena River (Dickson *et al.*, 2000).

Outflow of fresh water from the Arctic Ocean into the Nordic Seas is extremely variable and can be very large in some years. There are long salinity records for the Faroe–Shetland Channel and these indicate that there were low salinity values at the start of the twentieth century, before values rose through the middle part of the century (Fig. 6.19). However, the largest salinity change began in the late 1960s and has been described as the Great Salinity Anomaly by Dickson *et al.* (1988) (see Section 6.4.2). This was an increase in freshwater flux from the Arctic Ocean over the late 1960s to the early 1980s, but it also resulted in changes across much of the North Atlantic. The Great Salinity Anomaly accounted for about half of the total freshening of the Nordic Seas and the Subpolar Basins that has been observed in recent decades. The event involved advection of Polar Water beyond its normal bounds in the East Greenland Current to the north and east coasts of Iceland and a freshening of the upper 500–800 m of the northern North Atlantic. It could be traced around the Atlantic subpolar gyre for over 14 years from its origins north of Iceland in the mid to late 1960s to its return to the Greenland Sea in 1981–82. Its propagation speed has been estimated at about $3 \, \mathrm{cm \, s^{-1}}$.

A number of mechanisms for the freshwater event have been put forward, including changes in the surface moisture flux; however, it is now generally accepted that it was an

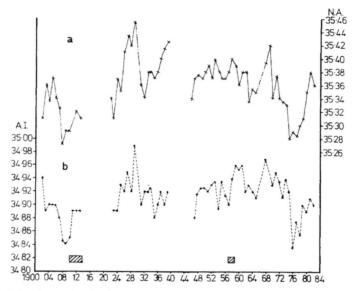

Fig. 6.19 Salinity time series for (a) North Atlantic and (b) Arctic Intermediate water masses in the Faroe–Shetland Channel, 1902–82. (From Dickson et al., 1988.)

advective feature that developed in the Greenland and Icelandic Seas during the 1960s as a result of a persistent anticyclonic anomaly over Greenland.

Winter season high atmospheric pressure over Greenland, and therefore northerly winds over the GIN seas, is associated with the negative phase of the NAO, which was more common in the middle of the twentieth century. In the late 1980s and early 1990s the NAO was in its positive phase with strong south to southwesterly flow to the east of Greenland. This resulted in warm, fresh Norwegian Atlantic Water being carried northwards in the lower layers into the Barents Sea. So warm Atlantic Water resulted in a warming of up to 2 °C in the Eurasian Basin and a shift towards the Pacific Ocean of the front separating waters of Atlantic and Pacific origins from the Lomonosov to the Alpha–Mendeleev Ridge. Subsequently the freshwater cap near the surface that insulated the sea ice from the warm water below dissipated in the Eurasian Basin, which had a major impact on the sea ice in the region.

Although the GSA, and particularly the period in the early 1970s, gave the greatest change in salinity, the freshening continued as a lesser rate until the late 1990s. This has been a result of increasing river discharge into the Arctic Ocean coupled with the greater net precipitation across the region (see Section 6.6.2). It has been estimated that these factors contributed around 20 000 km^3 of fresh water to the Arctic and high latitude North Atlantic between the 1960s and the 1990s (Peterson et al., 2006). In addition, the smaller sea ice extent added another 15 000 km^3 through a reduction in brine rejection, with glacial melt contributing a further 2000 km^3. These inputs to the Arctic Ocean are the same as the amount of fresh water that accumulated in the North Atlantic over the period 1965–95.

Despite the greater freshwater flux into the Arctic Ocean in recent decades the near-surface layers have experienced an increase in salinity (Swift *et al.*, 2005). This study identified salinity increases in the upper 175 m of the Arctic Ocean between the periods 1949–75 and 1976–93, with the salinity changes being found in all areas of the Arctic Ocean except parts of the Makarov Basin and the northeasternmost part of the Canada Basin. The increase in salinity suggests that additional fresh water was exported from the Arctic Ocean.

Near-surface waters in the Nordic Seas and the subpolar basins also became more saline as P–E over the ocean declined, despite river discharge increasing. An additional factor was that the Greenland Ice Sheet, along with high latitude glaciers, continued to melt, contributing further freshwater input to the Arctic Ocean. The greater river discharge and melting of sea ice and glaciers appear to be linked with rising temperatures, while the ocean P–E anomaly is tied closely to the NAM and NAO (Peterson *et al.*, 2006).

Field experiments in the Arctic Ocean have highlighted changes in ocean heat content at high latitudes. Over the period 1995–99 the US Navy made available a Sturgeon-class nuclear-powered submarine for unclassified science cruises to the Arctic Ocean. This programme, called SCICEX (Scientific Ice Expeditions), identified a peak warming near the North Pole in 1995. During the first decade of this century the North Pole Environmental Observatory, an experiment established at high northern latitudes, suggested that there was a minimum of temperature at the same location in 2005, followed by a pulse of warming thereafter.

An analysis of 80 years of summer-mean sea surface and upper 100 m temperatures in the surface layer of the Arctic Ocean indicated control by both ocean and sea ice advection. During 1965–95, as the NAM index rose, warming was observed from Atlantic Water advection into the Barents/Kara Seas and from Pacific Water advection into the Chukchi and western Beaufort Seas. This period coincided with greater winter sea ice transport away from the eastern Siberian shelves, leading to thinner ice that melted quickly during the summer, and leading to a longer summer open water period that allowed warming of the ocean through the absorption of greater amounts of solar radiation. Recent ocean surface warming in the Beaufort/Chukchi/East Siberian Seas since 2002 seems forced more by shortwave energy absorption than by northward-flowing ocean advection, although more in-situ data are needed to confirm this.

Conditions in the Arctic Ocean are very sensitive to water masses passing through the Bering Strait. Over the period 2001–04 there was an increase in northward ocean heat transport through the Bering Strait which is thought to have had an impact on sea ice decrease (Richter-Menge *et al.*, 2006). In particular, the amount of warm Pacific Water entering the Arctic Ocean via the Chukchi Sea has been very important in the rapid reduction of Arctic sea ice in recent years, which has not been uniform but disproportionately large in the Pacific sector of the Arctic Ocean. The spatial pattern of ice reduction has been similar to the spatial distribution of warm Pacific Summer Water that crosses the upper portion of the halocline in the southern Canada Basin north of the Chukchi Sea. The reduction in the amount of sea ice near the coast has enhanced the ice motion in the more central parts of the ocean, along with the upper ocean circulation. As a result, heat transport into the basin has increased. Ocean warming has resulted in less ice formation and faster ice motion, leading to outflow of sea ice through the Fram Strait. This has the potential to give a positive feedback, involving ocean warming, activation of sea ice motion and ice extent decrease.

6.4.3   The Antarctic

Ocean currents and water masses

One of the most studied features of the Southern Ocean is the Antarctic Circumpolar Current (ACC), which is the strongest current system on Earth, linking the Atlantic, Pacific and Indian Ocean Basins. Recent observational studies have suggested that the waters of the ACC have warmed more rapidly than the global ocean as a whole. Gille (2002) examined data from Autonomous Lagrangian Circulation Explorer floats, which measured temperatures in the Southern Ocean at depths between 700 m and 1100 m. She compared these measurements with earlier hydrographic temperature observations and found that between the 1950s and 1980s the mid-depth Southern Ocean temperatures had risen 0.17 °C. The warming was found to be concentrated within the ACC, with the temperature increase being comparable to Southern Ocean atmospheric temperature increases. A recent update to this work (Gille, 2008), showed that there has been substantial warming throughout the upper 1000 m of the ocean since the 1930s. The warming has been greatest in the uppermost layers, with summer temperature increases being of the order of 1 °C in the top 100 m. Again the warming has been concentrated in the region of the ACC and seems to be associated primarily with a marked temperature increase during the 1960s.

The mechanisms behind the recent Southern Ocean warming are not fully understood at present. However, it has been suggested that it is a result, at least in part, of the shift of the SAM into its positive phase. The change in the SAM has given stronger winds across the Southern Ocean and greater eastward wind stress. There are also theories, supported by climate model experiments, that the ACC may have moved poleward in response to the changes in the SAM (e.g. Fyfe and Saenko, 2006), so bringing warmer water further south and giving a signal of warming. However, there are only limited data at present to support this theory, although the work of Gille (2008) does provide some evidence.

Climate models suggest that the stronger winds over the Southern Ocean may be leading to an acceleration of the ACC (Hall and Visbeck, 2002; Fyfe and Saenko, 2006). However, observational data shows that the strength of the ACC depends on the phase of the SAM on a range of timescales from days and weeks (Aoki, 2002) to years (Meredith *et al.*, 2004). But there is no evidence yet for a long-term increase in transport. This may be a result of the lack of suitable monitoring systems, although the indications are that any change in transport over recent decades will have been small, consistent with the theory that the ACC is 'eddy saturated', with excess energy from the accelerating winds being cascaded from the large (circumpolar) scales to smaller (eddy) scales on interannual and longer timescales (e.g. Hallberg and Gnanadesikan, 2006). This idea is supported by satellite altimeter measurements of Southern Ocean eddy activity (Meredith and Hogg, 2006). It has also been demonstrated in both parameterised and eddy-resolving models that the poleward eddy heat flux associated with this process can be a significant contributor to the observed Southern Ocean warming (Hogg *et al.*, 2008).

It is also likely that a further contribution to the South Ocean warming has come from increased atmosphere–ocean heat flux as a result of increasing levels of atmospheric greenhouse gases. Modelling studies have suggested that the rate of Southern Ocean warming would have been even greater but for the masking effects of volcanic and other aerosols (Fyfe, 2006).

The subpolar gyres are important oceanographic features south of the ACC, and these have also undergone very significant change in recent decades (e.g. Boyer *et al.*, 2005). For example, there has been a large freshening in the coastal region between the Amundsen Sea and the Adélie coast (Jacobs *et al.*, 2002). This has been of the same magnitude, and perhaps larger, than the freshening that was observed in the North Atlantic from the 1960s to the 1990s (Dickson *et al.*, 2002) (see Section 6.5.2). Such changes indicate alterations in the hydrological cycle affecting both the northern and southern limbs of the global overturning circulation.

The freshening of the Southern Ocean is occurring at a range of depths and is not just confined to the surface layers. For example, the freshening is found in the area of AABW formation in the Ross Sea, where it has been shown that this water mass is getting progressively fresher in recent years (Rintoul, 2007; Aoki *et al.*, 2005). The reasons for the freshening are not entirely clear at present, but it may be occurring because of greater melt of glacial ice at the margins of the ice sheet (Jacobs *et al.*, 2002), with the ice melting possibly because of higher ocean temperatures.

Changes have taken place in the Weddell Sea, which is also a major site of AABW formation. Here the Warm Deep Water (WDW) layer has warmed by 0.3 °C since the 1970s (Robertson *et al.*, 2002). One possibility is that this may be a recovery from the Weddell Polynya of the 1970s when the deep layers of the Weddell Sea were directly ventilated. WDW is the deep-ocean precursor of AABW, so the changes in its properties are likely to influence AABW formation. Possibly consistent with this, Weddell Sea Bottom Water (WSBW, the densest form of AABW in the Weddell Sea) has recently been observed to be warming (Fahrbach *et al.*, 2004). WSBW is topographically constrained to lie within the Weddell Sea and is therefore too dense to be directly exported to the global overturning circulation. However, Weddell Sea Deep Water (WSDW) can readily escape from the Weddell Sea and this form of AABW has been observed to be warming downstream in the South Atlantic, with the warming signal reaching as far north as the Equator (Zenk and Morozov, 2007). It is unlikely that the WDW warming in the Weddell Sea can explain all of the WSDW warming observed at lower latitudes. This is relevant since a reduction in the export of the very densest WSDW would be manifest as a warming of WSDW, with potential long-term impacts on the meridional overturning circulation in the ocean. Current research is concerned with relating the export of WSDW from the Weddell Sea to cyclonicity of the wind stress over the Weddell Gyre, and possibly a link to El Niño.

There have also been some large regional changes in the surface and near-surface layers of the Southern Ocean. One of the most pronounced has been a large warming to the west of the Antarctic Peninsula. Here summertime surface ocean temperatures have increased by more than 1 °C since the 1950s, with an accompanying increase in salinity (Meredith and King, 2005). As discussed in Section 6.2.3, near-surface air temperatures on the western side of the Antarctic Peninsula have risen significantly over the last 50 years. There is strong evidence that the warming, which is largest during the winter season, has been accompanied by a decrease in sea ice over the eastern Bellingshausen Sea, although for a long time it was not clear whether the reduction of sea ice was a result of atmospheric or oceanographic factors. The strong salinification of the surface ocean over this period is due to mixed-layer processes as a result of decreasing sea ice formation. It is believed that the ocean is not just responding to the higher air temperatures, but that the oceanographic changes are a result of a positive feedback, acting to promote a decrease in ice production and further atmospheric warming.

There has also been a large warming downstream of the Antarctic Peninsula, close to South Georgia in the Scotia Sea. The data record for this location starts with the *Discovery* voyages in the 1920s and recent analyses indicate a subsequent strong summertime warming in excess of 1 °C and an even stronger wintertime warming of more than 2 °C. The reasons for this warming may well be linked to the location of South Georgia at the latitude of the ACC where it comes under the influence of the general circumpolar warming of the Southern Ocean. In addition, the Scotia Sea is a region where the increases in wind speed as a result of the shift of the SAM into its positive phase induce instantaneous warming rather than cooling, therefore the more positive trend of the SAM may have an anomalously large effect on ocean warming here.

Such changes in ocean temperature may have important implications for the Antarctic biota. Species that exist at depth (benthic) are well evolved to deal with low temperatures; however, they are much less able to adapt to increases in temperature. An increase of just 2 °C could have major deleterious effects on populations and species here (Peck, 2005). Over the coming decades rising ocean temperatures could be a major threat to the biota of the Southern Ocean.

As discussed in Section 6.9.2 there has been a marked loss of ice in the Amundsen Sea Embayment region of West Antarctic as the glaciers that drain the region have accelerated and thinned. The ice loss has been attributed to oceanographic changes, with Circumpolar Deep Water melting the undersides of the floating glaciers (Thoma *et al.*, 2008). Walker *et al.* (2007) used oceanographic data to calculate that 2.8 terawatts of heat had been transferred onto the continental shelf and towards the glaciers via a submarine glacial trough. They estimated that this was enough heat to account for most of the basal melting in the region.

The Southern Ocean carbon cycle

The global carbon cycle is the biogeochemical cycle by which carbon moves between reservoirs in the ocean, on the land and in the atmosphere. It therefore involves the biosphere, atmosphere, geosphere and hydrosphere, with complex exchanges taking place on a range of timescales from days to millennia.

The amount of carbon in the four reservoirs varies with time, changing most rapidly with the terrestrial biosphere, and on long geological timescales with the carbon deposits in the Earth. However, the current carbon content is about 750 Gt in the atmosphere, 2200 Gt in the terrestrial biosphere, 40 000 Gt in the oceans and 4000 Gt in sediments, including fossil fuels.

Carbon is mainly held in the ocean in the form of dissolved carbon dioxide and in marine organisms, such as plankton. Phytoplankton play a very important part in the carbon cycle as they are involved in about half the Earth's photosynthesis.

The oceanic carbon is split between 38 000 Gt in the deeper layers, about 1000 Gt in the surface layers, 700 Gt in dissolved organic carbon and 3 Gt in marine biota. On timescales up to decades and centuries the carbon cycle is dominated by the physical and biological exchanges of carbon within the system, but on longer timescales of several centuries or longer the deep ocean plays a more significant role.

The ocean is particularly important in the carbon cycle because of its ability to take CO_2 from the atmosphere. The concentration of CO_2 in the atmosphere increased rapidly throughout the twentieth century, reaching an annual mean in 2009 of nearly 387 ppm. However, almost half of the carbon currently emitted to the atmosphere by fossil fuel burning is sequestered into the ocean. Without this CO_2 drawdown atmospheric levels would be much greater and air temperatures far higher than they are today.

The flux of CO_2 between the atmosphere and the ocean is affected by many factors, including the surface mixing as a result of the surface wind and the difference in concentration of the gas in the air and water. The concentration of the gas in the ocean is dependent on the atmospheric and ocean CO_2 partial pressures, which themselves are a function of temperature, salinity, photosynthesis and respiration.

The Southern Ocean is a huge reservoir for CO_2 and is especially important in the carbon cycle because the cold ocean temperatures favour the uptake of CO_2 from the atmosphere. The high latitude areas of the Southern Ocean are characterised by oceanic descent as the waters are cooled and become more saline as a result of brine rejection as sea ice forms. The sinking water carries CO_2 into the deep ocean where the gas can remain for many centuries.

The Southern Ocean is one of the world's largest carbon sinks, absorbing about 15% of all CO_2 emissions. At a time of rapidly increasing anthropogenic CO_2 emission there is a great deal of interest in the exchanges of carbon between the different reservoirs and whether the different components are acting as more or less efficient sources or sinks of carbon dioxide. Over recent decades the surface wind speeds across the Southern Ocean have increased (see Section 6.3.3) as the SAM has shifted into its positive phase. This has resulted in the ocean being unable to absorb CO_2 at the rate it did previously (Le Quéré *et al.*, 2007). This has implications for future climate change and suggests that global temperatures may rise even faster than currently expected.

In a more recent study, Le Quéré *et al*. (2009) reconfirmed the importance of the Southern Ocean as a carbon sink. They noted that over the past 50 years the fraction of CO_2 emissions that remains in the atmosphere has increased from about 40% to 45%, with models implying that this is a result of a decrease of CO_2 uptake by the carbon sinks in response to climate variability and change.

6.5 Sea ice

6.5.1 Introduction

Prior to the 1970s our knowledge of sea ice extent and concentration was based primarily on ship and aircraft observations, together with some information from coastal stations. The majority of these data are from the Arctic, where it has proved possible to prepare ice analyses for the Eurasian Arctic that extend back to 1930. Some Norwegian ice charts also exist for the North Atlantic going back to the 1500s, although clearly the data on which these are based are extremely sparse in the early years.

Antarctic sea ice data for the pre-satellite era are very sparse, and based on coastal observations and some whaling ship records, most of which are from the summer season. Nevertheless, as discussed below, these data have been used to infer changes in sea ice extent during the early to mid twentieth century. Recently sea ice extent has started to be estimated from chemical signals in ice cores collected in the coastal regions of Antarctica. Such estimates appear to agree well with the satellite data from recent decades, although such techniques may not be applicable to all sectors of the continent.

From the 1960s it was possible to obtain a broader-scale view of the distribution of sea ice from visible and infrared satellite imagery. However, the imagery was only of value in cloud-free or partly cloudy conditions, which was a major handicap as both polar regions are characterised by extensive low cloud. However, with the rapid advances in satellite technology during the 1970s and 1980s, new microwave instrumentation allowed improved monitoring of sea ice regardless of the cloud cover. Although the data sets derived from these observations only start in the late 1970s, they are sufficient to provide an indication of variability and change in sea ice extent and concentration.

6.5.2 Sea ice extent and concentration

The Arctic

The pre-satellite era The best records of sea ice extent in the pre-satellite era come from coastal observations. For example, Iceland has an excellent record of sea ice around the island that extends back to 1600. Here there was a dramatic decrease in sea ice around the coast in the period from the end of the Little Ice Age (approximately AD 1300 to 1850) to the twentieth century. These data, which record the number of months that the island was

290 The instrumental period

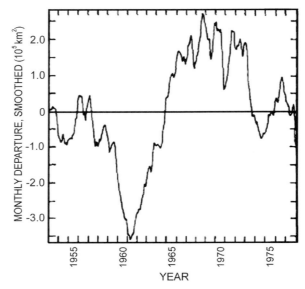

Fig. 6.20 A 24-month running mean of the departure from the monthly mean of the Arctic sea ice area. (From Walsh and Johnson, 1979.)

surrounded by sea ice, are shown in Fig. 5.21 (Broecker, 2000). This record indicates that Iceland was surrounded by sea ice for an average of 2 months a year through much of the period up to the late nineteenth century. However, there was a steady decline in ice from about 1870, and by about 1920 the seas around Iceland were free of sea ice for the whole year. Not surprisingly, there is a close correspondence between the rising air temperatures recorded on Iceland and the reduction in sea ice off the coast.

Other indications of extensive sea ice in the Atlantic sector during the early part of the twentieth century come from spring and summer ice-limit charts from the Danish Meteorological Institute covering the period 1901–56 (Mysak *et al.*, 1990). These showed positive ice anomalies in the Greenland Sea during 1902–20 and in the late 1940s. The data also suggested generally negative ice anomalies during the 1920s and 1930s.

More sea ice observations became available in the second half of the twentieth century, although the density of these data varies widely on a regional basis. Many of the observations were brought together by Walsh and Johnson (1979), who digitised data covering 1953–77 and performed a trend analysis. As can be seen in Fig. 6.20, there are large variations in ice cover over the period they examined, including a marked decrease of ice extent around 1960, followed by a positive anomaly in the late 1960s/early 1970s. This latter period coincided with a sea ice maximum noted around the coast of Iceland. In addition, 1968 was a year of very large ice export from the Arctic Ocean via the Fram Strait. For the period as a whole, Walsh and Johnson (1979) found a slight, but statistically significant positive trend in sea ice extent of $3140 \, \text{km}^2 \, \text{yr}^{-1}$.

These authors highlighted the anticorrelation between atmospheric temperature and sea ice extent, with this relationship being found for the Arctic as a whole. In a more recent study Chapman and Walsh (1993) used data from 1961 to 1990 and found that Arctic sea ice changes were consistent with the temperature variations. This study found the greatest warming in the spring and winter, but the greatest decreases of sea ice in the spring and summer.

The large amount of sea ice in the East Greenland Sea in the mid 1960s (Mysak *et al.*, 1990) has attracted a lot of attention because of the accompanying oceanographic consequences. The large ice extent was accompanied by low salinity water (the Great Salinity Anomaly; see Section 6.4.2) that migrated into the western Atlantic in subsequent years and then reached Iceland after 14 years after following a clockwise track (Dickson *et al.*, 1988). Over the next 3 to 5 years the anomaly propagated into the Labrador Sea and then across to the Labrador and east Newfoundland coast.

Conditions in the tropical Pacific, and especially the ENSO cycle, clearly affected the sea ice extent in certain sectors of the Arctic in recent decades. The heavy ice years in Hudson Bay, Baffin Bay and the Labrador Sea in each of the periods 1972/73, 1982/83 and 1991/92 coincided with strong El Niño events and a positive NAO (Mysak *et al.*, 1996). This is consistent with the earlier study by Mysak (1986) which found low air temperatures over the northwest North Atlantic during El Niño events.

Arctic sea ice variability and trends from the satellite record Reliable, high horizontal resolution sea ice extent and concentration data are available for the polar regions from 1978, when the Scanning Multichannel Microwave Radiometer (SMMR) instrument was first flown on the Nimbus 7 polar-orbiting satellite. Parkinson *et al.* (1999) analysed ice extent variability over the period 1978–96 using these data and observations from subsequent satellites. Over this period the yearly ice extents were at a maximum in 1982 and a minimum in 1995. In this study they found the overall decrease of Arctic sea ice was $34\,300 \pm 3700\,\mathrm{km}^2\,\mathrm{yr}^{-1}$ or -2.8% per decade over the 18-year period. Figure 6.21 shows that the ice extent was still very variable over this period and that the summer of 1996 had near record amounts of ice for this season. Nevertheless, the trend has been towards less ice throughout the 1990s. Evidence of widespread sea ice melting is corroborated by the substantial recent increase in freshwater content of the Arctic Ocean – of the order of three times more fresh water. This study also found that ice decreases were found in all seasons and over the year as a whole, but the greatest losses were found in the spring and the smallest in the autumn.

The more recent study of Comiso and Nishio (2008) examined sea ice variability and trends for the period 1978–2006 and confirmed the ongoing and accelerating loss of sea ice from the Arctic. This can be seen in Fig. 6.22, which shows the monthly anomalies of sea ice extent for the whole Northern Hemisphere. Over the period considered the ice extent had decreased at a rate of $-3.4 \pm 0.2\%\,\mathrm{decade}^{-1}$ ($-39\,600 \pm 2170\,\mathrm{km}^2\,\mathrm{yr}^{-1}$) with, as can be seen in Fig. 6.22, an increasingly rapid decrease over the last few years.

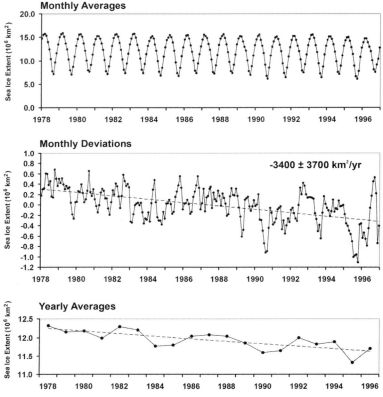

Fig. 6.21 (Top) Monthly averages and (middle) deviations of Arctic sea ice extent for the period 1978–96 along with the least squares fit. (Bottom) Yearly average Arctic sea ice extents for 1978–96. (From Parkinson *et al.*, 1999.)

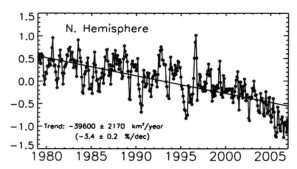

Fig. 6.22 Monthly anomalies of sea ice extent and trend for the Northern Hemisphere for 1978–2006. (From Comiso and Nishio, 2008.)

6.5 Sea ice   293

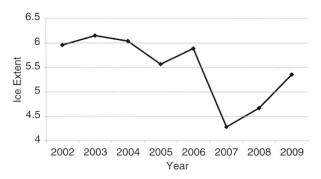

Fig. 6.23 September average Arctic sea ice extent for 2002–09 (10^6 km^2). (Data from NSIDC.)

Regionally, the greatest percentage loss has been in the Gulf of St Lawrence, where the decrease has been $-10.7 \pm 3.6\%$ decade^{-1}, followed by the Sea of Okhotsk/Sea of Japan ($-8.7 \pm 2.0\%$ decade^{-1}), Baffin Bay/Labrador Sea ($-8.6 \pm 1.1\%$ decade^{-1}), the Greenland Sea ($-8.0 \pm 1.0\%$ decade^{-1}), Kara/Barents Sea ($-7.2 \pm 0.8\%$ decade^{-1}), Hudson bay ($-5.1 \pm 0.7\%$ decade^{-1}), the Canadian Archipelago ($-1.4 \pm 0.3\%$ decade^{-1}) and the Arctic Ocean ($-1.4 \pm 0.2\%$ decade^{-1}). Only one sector showed an overall increase in ice extent over the period, which was the Bering Sea, where the ice had increased by $1.7 \pm 2.0\%$ decade^{-1}.

In recent years there has been a succession of minima in the extent of mean September sea ice (Fig. 6.23). September 2005 saw a new record minimum in Arctic sea ice extent surpassing the previous record low in 2002. All four years 2002–05 had ice extents approximately 20% less than the 1978–2000 mean, with the loss of sea ice amounting to approximately 1.3 million km^2.

At the time of writing the lowest observed mean September ice extent was in 2007 when it covered an area of 4.28×10^6 km^2 (Fig. 6.24), which was 39.2% or (2.76×10^6 km^2) below the 1979–2000 average and some 23% below the 2005 value (Stroeve et al., 2008). The second and third lowest September average ice extents were in 2008 and 2009 respectively. The sea ice extent anomalies with respect to the average from 1979 to 2000 indicate that the September extent is decreasing at a rate of -8.9% per decade over the period 1979–2009.

Based on the frequency of near-record low ice extents the US National Snow and Ice Data Center has concluded that Arctic Sea is likely to be on an accelerating, long-term decline. In addition, the recent ice minimum is believed to be lower than the Arctic ice minima of the 1930s and 1940s. Regionally, the largest loss of ice has been in the Pacific sector of the Arctic (Fig. 6.24), with smaller retreats north of Russia.

There have been a number of studies into why the extent of Arctic sea ice has been decreasing at such a large rate. The ice loss has been linked to a 'flushing' of much of the ice out of the Arctic Basin in the early 1990s, which has resulted in a decrease in the age of ice across the area. A large volume of thick multi-year ice is thought to have exited the Fram Strait in the 1990s, leaving the Arctic with thin, first-year ice more prone to melt in summer (Kwok et al., 2004). The recent extreme minima may in part represent a response to this

294   *The instrumental period*

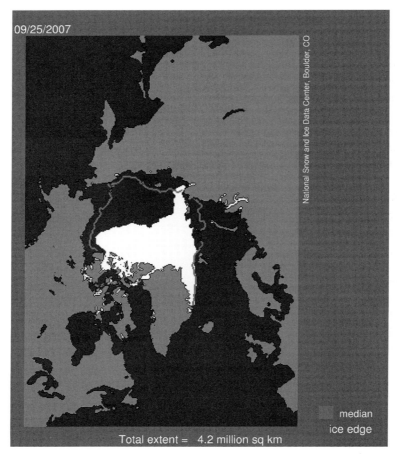

Fig. 6.24   Arctic sea ice extent on 25 September 2007. (From the US National Snow and Ice Data Center.)

effect (Rigor and Wallace, 2004). More recent work indicates that while the impacts of altered wind fields on ice circulation are important, the overall downward trend is more clearly allied with a general warming on at least the regional scale. There appear to be feedbacks at work in the sense that increased open water and thin ice absorb more solar energy in summer, leading to less ice growth the following winter. It has been proposed that the Arctic may be near a 'tipping point' where there is a new equilibrium state between increased solar absorption in the ocean during summer and the amount of first-year sea ice that can grow during the following winter.

Links with the broadscale circulation The NAO (see Section 3.4.5) is very important in determining the sea ice extent in particular years in the North Atlantic sector of the Arctic.

The NAO was in its positive phase over 1980–2000 (cf. Fig. 6.10) and the associated deep Icelandic Low had an especially strong impact on sea ice extent in this sector of the Arctic. The anomalously deep Icelandic Low resulted in strong southwesterly winds across the North Atlantic and advection of warm air masses across Europe and into the Barents Sea, generating negative ice extent anomalies.

Greater cyclonic flow in the central North Atlantic also results in more northerly flow down the Labrador Sea, giving a greater amount of ice in this area and greater export of ice out of the Arctic Ocean. It is believed that the positive NAO in the mid 1990s may have contributed to the sea ice reduction at this time. However, the NAO has become more neutral since 2000 suggesting that other factors may have contributed to the ice loss in the last few years.

Data from drifting buoys in the Arctic Ocean have indicated that surface pressure values have been decreasing in recent decades (Walsh *et al.*, 1996). This has resulted in a decline in the anticyclonic wind forcing on the Arctic Ocean sea ice and decreased the wind forcing on the Transpolar Drift Stream (TDS) across the Arctic Basin, through the Fram Strait and into the Greenland Sea.

Examination of years of extensive or very limited ice can provide insight into the links between the ice cover and the broadscale atmospheric circulation. Extensive ice export took place in 1968, while 1984 had very low ice export. These two years were characterised by very different winter MSLP patterns and with different phases of the NAO. In model simulations (Tremblay *et al.*, 1997) 1968 had a weak Beaufort Gyre with a broad TDS shifted to the east, leading to a large ice export from the Arctic. Conversely, the 1984 simulation led to an expanded Beaufort Gyre with a weak TDS shifted to the west and a low ice export.

The Antarctic

The pre-satellite era It is much more difficult to estimate the sea ice extent around the Antarctic in the pre-satellite era than for the Arctic, since the ice edge is over the Southern Ocean for much of the year, well removed from most coastal observatories. Some island stations, such as Signy, have provided information on sea ice variability, and revealed details of the Antarctic Circumpolar Wave (see Section 3.4.7). But the only type of data that provide any form of spatial coverage with which to assess the sea ice cover are the observations of the ice edge made by whaling vessels. The region close to the ice edge is rich in food and many whales congregated here, attracting the whaling fleets. Most of these observations are for the summer season. De la Mare (1997) examined the whaling records, which provide the location of every whale caught since 1931. He suggested that there had been a big change in the location of the whaling vessels between the 1940s and the 1970s, with the summer sea ice edge having moved southward by 2.8° of latitude between the mid 1950s and the early 1970s. He inferred this as meaning that there had been a decrease in the area covered by sea ice of 25%. However, there is a great deal of debate over how the locations of whale catches can be translated into ice edge estimates that are comparable to those from satellite observations. It is particularly unfortunate that there is no overlap between the period

296 *The instrumental period*

covered by the whale catch data and the modern satellite observations of the ice edge. At the moment the de la Mare results are questioned in many quarters and cannot be regarded as proof of a major sea ice extent decrease between the 1940s and 1970s.

Nevertheless, there is additional evidence for more extensive Antarctic sea ice in some sectors of the Antarctic in the first half of the twentieth century. An analysis of the Law Dome MSA record has provided a proxy record of the sea ice extent in the 80° E to 140° E sector extending back to 1841 (Curran *et al.*, 2003). This showed variability in the ice extent from the start of the record until about 1950, but no overall trend. However, the data revealed a 20% decline in extent between 1950 and 1995. The good correlation ($P < 0.002$) between MSA and satellite-derived ice extent in the 22-year overlap period indicates the reliability of this result. Interestingly, the record also shows a large cyclical variation in the extent with a period of about 11 years, demonstrating the care that needs to be exercised in interpreting short sea ice records.

Unfortunately, in other sectors of the Antarctic the relationship between MSA and sea ice extent is not so clear. Abram *et al.* (2007) examined MSA from three sites around the Weddell Sea and found that the amount of MSA reaching the ice core sites was reduced after years of increased winter sea ice in the Weddell Sea, which is the opposite to the expected relationship if MSA is to be used as a sea ice proxy. They suggested that in this sector atmospheric transport strength, rather than sea ice conditions, was the dominant factor that determines the MSA signal in near-coastal ice cores.

Another investigation into sea ice extent in the pre-satellite era was undertaken by King and Harangozo (1998). They were concerned with understanding the major winter season warming that has taken place on the western side of the Antarctic Peninsula since the 1950s. Here there is a strong inverse relationship between sea ice extent over the Amundsen–Bellingshausen Sea (ABS) and temperature at coastal sites on the western side of the peninsula, suggesting more extensive sea ice in the middle of the twentieth century. Very few ship observations of the winter sea ice edge over the ABS exist for the first half of the twentieth century. However, several ships reported an ice edge location to the north of anything found during the satellite era, providing some evidence of more ice at this time.

Antarctic sea ice variability from the satellite record The extent and concentration of Antarctic sea ice is only known with confidence since the early 1970s when reliable satellite passive microwave observations became available. These data have been analysed by several authors.

Interannual variability with periods of 3–5 years in the regional sea ice extent has been noted (Zwally *et al.*, 2002a). However, anomalies around the Antarctic tend to cancel each other out so that such variability is not seen in the total ice extent. The interannual variability of the annual mean extent for the Antarctic is only 1.6% compared with 6–9% for each of the five regional sectors examined. Cavalieri *et al.* (2003) carried out a study of Antarctic sea ice for the period of 1973–2002. They found that the total Antarctic sea ice extent ($> 15\%$) decreased dramatically over the period 1973–77, then gradually increased through the period to 2002. The trend reversal is attributed to a large positive anomaly in the early 1970s, an anomaly that apparently began in the late 1960s.

The study by Comiso and Nishio (2008) analysed the sea ice extent data for 1978–2006. They found that across the whole Southern Hemisphere sea ice extent had increased at a rate of $1.0 \pm 0.2\%$ decade^{-1} (11200 ± 2680 km^2 yr^{-1}); however, this figure masks large regional differences. There have been contrasting trends between the Antarctic Peninsula and Ross Ice Shelf regions, with the Ross Sea sector experiencing an increase of $4.2 \pm 0.7\%$ decade^{-1} while the ABS has lost ice at a rate of $-5.7 \pm 1.0\%$ decade^{-1}. This pattern of change has been attributed to a deepening of the Amundsen Sea Low since around 1980 as a result of the ozone hole (Turner *et al.*, 2009b).

Around the rest of the continent there have been small increases in ice extent since the late 1970s, with the trends being: the Weddell Sea ($0.7 \pm 0.6\%$ decade^{-1}), the Indian Ocean ($1.9 \pm 0.6\%$ decade^{-1}) and the West Pacific Ocean ($1.2 \pm 0.8\%$ decade^{-1}).

6.5.3 Sea ice thickness

There is only limited data on sea ice thickness, but the available observations have been used to examine changes in recent decades. As discussed in Section 2.4, our knowledge of sea ice thickness changes comes from in-situ ice core measurements, upward looking sonar (ULS), and more recently, in the Arctic, satellite radar altimeter measurements.

The Arctic

With the reduction in Arctic sea ice extent over recent decades it is intuitively expected that there will have been a corresponding decrease in sea ice thickness. However, the data with which to investigate this are sparse. But the data from a 1000 km transect made during two summer cruises in 1958 and 1970 suggested that the draft (thickness of ice below the water) had decreased by 0.2 m in the Transpolar Drift Stream and the Eurasian Basin, and by 0.7 m in the Canadian Basin (McLaren, 1989). More recent reports of thinning ice have been made by Wadhams (1990, 1992). He found thinning over 1976–87 of 0.8 m between Fram Strait and the pole and of 2.2 m north of Greenland. A summary of earlier reports by Wadhams (1994) suggested that there had been a reduction in thickness between the 1970s and the 1980s of 0.6 m in the area bounded by the Fram Strait, Greenland and the North Pole.

ULS measurements of sea ice draft from submarines provide a record extending back into the 1950s. These have provided the first persuasive evidence of large-scale thinning over the entire Arctic Basin. This time series was used by Rothrock *et al.* (1999) to examine changes between the two periods 1958–76 and 1993–97. They found that the mean ice draft at the end of the melt season had decreased by about 1.3 m in most of the deep water parts of the Arctic Ocean, from 3.1 m in the early period to 1.8 m in the 1990s, or about 40%. Figure 6.25 shows a map of the changes in draft over the same period. They observed that the greatest decreases had taken place in the central and eastern part of the Arctic compared with the Beaufort and Chukchi Seas. The recent cruises also found continued thinning at a rate of

298   *The instrumental period*

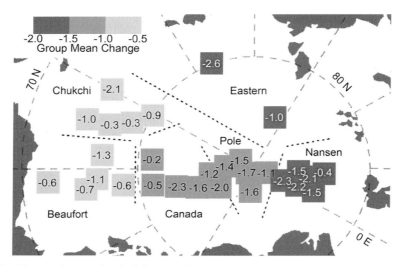

Fig. 6.25 Changes in mean ice draft from 1958–76 to 1993–97. (From Rothrock *et al.*, 1999.)

10 cm per year from 1993 to 1997. A number of numerical experiments also demonstrated a thinning of the Arctic sea ice over the last few decades (Rothrock *et al.*, 2003).

A study by Rothrock and Zhang (2005) examined the reasons for the decrease in ice thickness. Part of the general downward trend in ice thickness and extent was attributed to altered wind fields associated with the upward trend in the NAM (see Fig. 6.11) up to the mid 1990s (Rigor *et al.*, 2002).

With the short and sparse data set of Arctic sea ice thickness it is difficult to investigate the variability of thickness using the in-situ data. However, Laxon *et al.* (2003) used an eight-year time series of Arctic ice thickness derived from satellite altimeter measurements of ice freeboard to determine the mean thickness and variability over 65–81.5° N. The data revealed a high frequency interannual variability in the thickness that was dominated by changes in the amount of summer melt. The study suggested that a continued increase in melt season length would lead to further thinning of the sea ice.

The Antarctic

We have far fewer observations of Antarctic sea ice thickness than for the Arctic, because of the lack of the valuable submarine ULS measurements. Our knowledge of Antarctic sea ice thickness is therefore limited to point measurements ranging from in-situ sampling, ship-based observations, some electromagnetic soundings from aircraft- and ship-based instruments. The observations from ships operating in the Antarctic sea ice zone since 1980 have been assembled by the SCAR Antarctic Sea Ice Processes and Climate (ASPeCt) project. This data set comprises 23 373 observations of sea ice thickness collected over a period of more than 20 years (Worby *et al.*, 2008) allowing the investigation of regional and seasonal variability in the thickness distribution of the Antarctic sea ice zone. The data indicated that the long-term mean and standard deviation of the total ice thickness (including ridges) was 0.87 ± 0.91 m.

6.6 Snow cover

6.6.1 Introduction

Snow cover plays an extremely important part in the climate of the Arctic, being involved in a number of key feedback processes (Section 3.3.1). As the snow usually contains a large amount of air, it is also very effective at insulating the underlying land or sea ice. The snow cover is also very important in storing water, which is eventually released into the oceans and the timing of this release can impact the high latitude ocean circulation.

The extent and thickness of the snow on land is influenced by many factors, including the near-surface air temperature, aerosols falling on the snow and changes in the albedo, the local orography, the state of the vegetation, the tracks of depressions and the extent of sea ice over nearby ocean areas. It is therefore highly sensitive to many climatic factors – local, regional and global – as well as changes in land use and vegetation. Due to this sensitivity, it has often been suggested that snow extent is an excellent measure of climate change and may provide an indication of human influence on the Earth system.

Although snow cover has its most visible influences over land areas, it also plays a very important role in the thermodynamics of sea ice. A layer of snow on top of the sea ice can provide insulation for the sea ice, but the weight of the snow can also push the ice down into the water forming snow ice (see Section 3.2.5).

For many years snow extent information was only available from the limited number of in-situ meteorological observing stations. However, it is now routinely determined using data from instruments on the polar-orbiting satellites. In fact a number of websites provide daily analyses of Northern Hemisphere snow cover. Such data, which start in the 1960s, have been assembled into climatologies of snow cover extent, which can be used to investigate trends. The early satellite instruments provided visible imagery that allowed the extent of the snow cover to be estimated. However, data from the more recent passive microwave instruments can now be used to determine the important snow/water equivalent.

In this section we will use the available data to examine the long-term mean snow cover extent and consider the trends over recent decades. The data will be interpreted in conjunction with our knowledge of temperature and other changes described in earlier sections.

6.6.2 The Arctic

The annual cycle of snow cover

The US National Oceanographic and Atmospheric Administration (NOAA) have been producing maps of Northern Hemisphere snow cover based on satellite imagery since

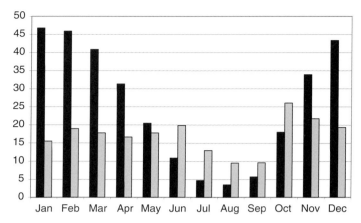

Fig. 6.26 Monthly snow cover over the Northern Hemisphere (black) (millions of km^2) and the standard deviation multiplied by 10 (grey). (From the monthly mean fields of snow cover produced by the Rutgers University Global Snow Laboratory for 1979–2005.)

November 1966, when the early visible satellite imagery became available. Unfortunately, the early imagery was rather poor and there was little expertise at that time in estimating snow cover from satellite pictures, especially when the snow cover was rather patchy. The snow cover therefore tended to be underestimated in the early years. Since 1972, however, the data are felt to be reliable provided that there is a reasonable amount of solar radiation so that the areas of snow can be discriminated. Since 1978 comparable snow extent data have been derived from passive microwave instruments flown on the polar-orbiting satellites, and a comparison of these two forms of data has been carried out by Armstrong and Brodzik (2001). They considered Northern Hemisphere snow cover for a 20-year period and found that the interannual variability and hemispheric-scale long-term trends were similar in the two forms of data. The passive microwave data gave less snow cover than the visible imagery in the autumn and early winter when the snow was shallow. However, since the data are of comparable quality we will here examine the mean snow cover from the NOAA data set, which is made available at the Rutgers University Global Snow Laboratory.

The annual cycle of Northern Hemisphere snow cover extent, including Greenland, is shown in Fig. 6.26. This indicates that the maximum snow extent occurs in January (46.8×10^6 km^2) and the minimum in August (3.5×10^6 km^2). Between February and July there is a fairly linear decrease in the extent of snow, followed by a slower reduction into August and a rapid increase in extent from September. This annual cycle means that most of the Arctic is covered by snow for 6–8 months of the year. The average snow extent for the year is 25.5×10^6 km^2, which is split approximately between 14.8×10^6 km^2 (58%) in Eurasia and 10.7×10^6 km^2 (42%) in North America. The mean annual snow cover on Greenland covers 2.2×10^6 km^2, most of which is present throughout the year. Greenland therefore represents about 8.6% of the Northern Hemisphere snow cover.

Snow cover extent

Maps of the mean January and August snow cover fraction as derived from the visible satellite imagery by the Rutgers University Global Snow Laboratory are shown in Fig. 6.27. In January almost continuous snow cover with an average concentration of 91–100% is found across Canada, the north central states of the USA, Russia and northern Fennoscandia. Iceland and the islands of the Arctic Ocean are also characterised by extensive snow. The comparable map for April (not shown) indicates that snow is still present at this time of year across much of Canada, with lower snow cover fractions only being found in the central part of the country and close to the border with the USA. Near-continuous snow cover is also found across virtually the whole of Russia. However, by August (Fig. 6.27b) the only region with extensive snow cover is Greenland, and even here parts of the coastal region now have fractional snow cover values of less than 80%. At this time of year the only other high latitude areas with extensive snow cover are along the Coast Mountains of western Canada, the Arctic islands of northern Canada, the mountains of Norway, the higher parts of Iceland and some of the islands in the Arctic Ocean. By November (map not shown) snow has returned to many of the locations described above for April, indicating the shortness of the Arctic summer.

Variability of snow cover

Figure 6.26 also shows the monthly standard deviation of the snow cover for the 34-year period. The largest variability is during the transitional months and early summer, with the autumn values being slightly larger than those for the spring. The smallest variability is during the late summer. During the Arctic winter temperatures are generally very low, so at this time of year the greatest variability in snow cover is in the sub-Arctic. However, as can be seen in Fig. 6.26, there is still a large variability in snow cover across the Northern Hemisphere as a whole.

Trends in snow cover

The time series of Northern Hemisphere annual mean snow cover extent is shown in Fig. 6.28. Over the 34-year period the greatest snow cover occurred in 1978 ($27.6 \times 10^6 \, km^2$) and the smallest in 1990 ($23.4 \times 10^6 \, km^2$), which represents a range of $4.2 \times 10^6 \, km^2$ or just over 16% of the annual mean. The linear regression line fitted to the annual snow cover indicates that since 1972 the annual mean snow extent has been decreasing at a rate of 0.43 million square km per decade, which amounts to 1.7% per decade. Annual snow cover extent can vary markedly from year to year, but there are periods of several years with consistent positive or negative anomalies. For example, the late 1980s and early 1990s had especially low snow extent.

The annual cycle of the trends for Northern Hemisphere snow cover extent are shown in Fig. 6.29. Computing the changes during the summer and autumn months is very difficult since the actual areas of snow cover are very small, so Fig. 6.29 is restricted to the months outside this period. However, the figure shows a clear pattern of decreasing snow cover from the start of the year until May, followed by trends towards greater snow cover at the end of

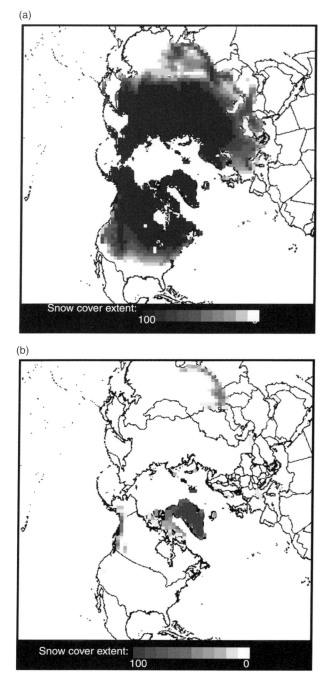

Fig. 6.27 Charts of mean snow cover fraction for 1967–1999. (a) January and (b) August. (From the Rutgers University Global Snow Lab, http://climate.rutgers.edu/snowcover/index.php)

6.6 Snow cover

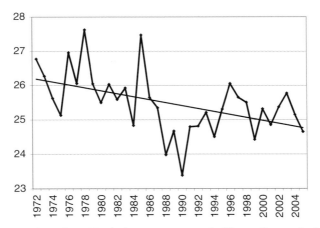

Fig. 6.28 Mean annual Northern Hemisphere snow extent (millions of square km) for 1972–2005.

Fig. 6.29 The monthly trends in snow cover (% per decade) for the period 1972–2005.

the year. The greatest decrease has been during the spring (March–May), where the snow cover has reduced by 3% per decade over the 34 years.

The snow cover has been considered separately for America and Eurasia with Fig. 6.30 showing the annual mean extents and Fig. 6.31 the annual and monthly trends. There is only a moderate correlation of 0.5 between the extent in these two areas because local atmospheric circulation anomalies can influence the snow cover in particular areas. However, some features are common to both time series. There was a marked decrease in the snow extent from the mid 1980s to the early 1990s, consistent with the observation that negative snow cover anomalies have been a feature of Canada and northern Russia since the late 1980s. The early 1990s snow extent minimum was a feature of both regions, as was the recovery in the following years.

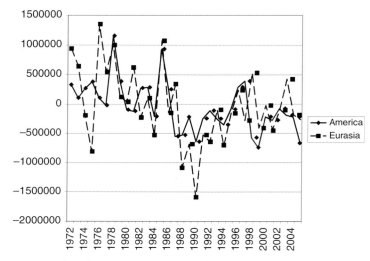

Fig. 6.30 Mean annual northern snow extent anomalies (square km) for America and Eurasia.

Fig. 6.31 The monthly trends in snow cover (% per decade) for the period 1972–2005 for Eurasia (dark grey) and America (grey).

The variability of annual mean snow extent is slightly larger in Eurasia compared with America but the percentage trend in annual cover has been slightly larger in America (Fig. 6.31). In the spring the loss of snow cover across Eurasia is larger than in America, but the annual trend does not reflect this as the positive trend in extent at the end of the year is much greater across Eurasia than over America.

Although in this section we have concentrated on the snow cover extent information derived from satellite imagery, since this provides consistent, hemisphere-wide data, observations at the meteorological stations support the trends discussed earlier. For example, at Barrow, Alaska, the snow melt onset was about 10 days earlier in the mid 1990s compared with the 1970s (Stone *et al.*, 2002). Barrow is influenced by advection of air both from the Arctic and the North Pacific, so snow cover characteristics in this region are likely related to shifts in the predominant flow pattern.

The decrease in Northern Hemisphere snow cover is consistent with the rise in near-surface air temperatures discussed in Section 6.2.2 and other indications of climate change noted in high northern latitudes, such as the decrease in sea ice extent and the loss of permafrost. The loss of snow cover has major implications for the climate system and has the potential to influence conditions far removed from the Arctic. For example, the coupled warming and loss of snow cover in the spring across northern Russia has resulted in an overall increase in the runoff and seasonality from Siberian rivers from the 1950s to the 1990s (Peterson *et al.*, 2002; ACIA, 2005). There is debate about the relative contributions of warmer temperatures, changes in precipitation amount, the role of melting permafrost and human impacts (dams, diversions) to this increase, although dams and diversions appear to affect the seasonality more than the total annual runoff. The result has been a greater freshwater input to the Arctic Ocean that has the potential to significantly alter ocean circulation.

Where snow cover is disappearing earlier in the spring, the large amounts of energy that would have melted the snow can now directly warm the soil. This has an impact on the terrestrial biosphere. In addition it can result in changes to the atmospheric circulation. For example, Groisman *et al.* (1994) showed that a warming during late winter and early spring in parts of northern North America could be correlated to earlier snow retreat in these areas.

In addition to a loss of snow cover, there have been other changes associated with precipitation, such as an acceleration of the Arctic hydrologic cycle that is suggested by some indicators, such as an increase in terrestrial precipitation. Observations demonstrate that most mid and high latitude areas of the Northern Hemisphere have had an increase in precipitation of 5–10% over the last 100 years (IPCC, 2007).

6.6.3 The Antarctic

Introduction

Most of the Antarctic has temperatures that are perpetually below freezing so the majority of the surface is covered by snow. However, the small areas that are snow and ice free are important habitats for terrestrial plant life and are often referred to as oases. These rocky areas have a low albedo and warm rapidly in the summer months, proving excellent environments in which plants can flourish. They have often been compared to the oases in deserts of lower latitudes.

The seasonal cycle of snow cover is not a major factor across the Antarctic, as it is in the Northern Hemisphere. The amount and variability of snowfall in recent decades is dealt with in Section 6.9.2. so here we will only consider snow cover on the sub-Antarctic islands and in the parts of the continent where the ground becomes snow free for part of the year.

Fig. 6.32   The Dry Valleys of Victoria Land.

Snow cover extent

The central part of the Antarctic has perpetual snow or ice cover, but several parts of the coastal region of the continent are snow free for at least part of the year. Overall between 2% and 4% of the Antarctic is snow free. These areas include:

- the peaks of mountains that protrude through the ice sheet (nunataks)
- the northern and western parts of the Antarctic Peninsula; these are the warmest parts of the Antarctic and many of the coastal areas are snow free in the summer
- a small number of permanently snow free areas in East Antarctica where there is little precipitation; the largest of these are the Dry Valleys of Victoria Land, which get virtually no precipitation at all (Fig. 6.32)
- several areas around the coast of East Antarctica; some of the main oases are the Bunger Hills, the Vestfold Hills, the Prince Charles Mountains and the Lasserman Hills.

Seasonal snow cover is much more common on the sub-Antarctic islands. These islands are characterised by strongly maritime climates and the temperatures and the snow cover can vary markedly in all seasons with the passage of active depressions.

Trends in snow cover

Variations in snow cover over the relatively small areas that are snow free in the Antarctic can only be determined using observations from the stations or fairly high resolution satellite imagery. The western side of the Antarctic Peninsula has warmed most during the winter season, but there has also been a summer warming of 0.2 °C per decade (significant at the 1% level) during the summer. Early aerial photographs of this part of the Antarctic, when compared with modern images, have shown that summer warming has reduced seasonal snow cover (Fox and Cooper, 1998).

6.7 Permafrost

6.7.1 Introduction

Permafrost is soil or rock that stays at or below the freezing point of water for two or more years. It is therefore defined only on the basis of temperature and not on the composition of the frozen ground. Permafrost may comprise frozen rock, mineral soils or organic material, such as peat.

It can also be continuous or discontinuous, with the latter being the case when the nature of the soil allows permafrost to exist in an otherwise permafrost-free region. For example, the presence of peat favours the maintenance of permafrost since its thermal conductivity changes during the year, having low conductivity in the summer when it is dry, and higher conductivity in the autumn and winter when it is wet.

Permafrost is usually overlaid by a relatively thin layer of ground between the surface and the permafrost called the active layer, which undergoes seasonal freezing and thawing. The active layer can vary in depth from a few tens of centimetres to several metres and in many areas it experiences a large interannual variability.

It is found under approximately 25% of the land surface of the Northern Hemisphere, with most occurring at high latitudes, although some does occur at high altitudes in alpine regions.

The spatial distribution, temperature and thickness of permafrost is strongly dependent on the temperature at the ground surface, but is also influenced by other factors, such as soil type, drainage, the nature of the overlying vegetation and snow cover. The actual permafrost layer can be hundreds of metres in depth, depending on the annual cycle of temperatures at the location. For example, across Alaska the permafrost has a depth of 400 m at Barrow (71° 18′ N, 156° 47′ W) and 90 m at Fairbanks (64° 49′ N, 147° 52′ W) in the discontinuous zone of permafrost in central Alaska.

The depth of permafrost is also dependent on the temperature of the rocks at greater depth and the variability of temperature with depth, which is known as the geothermal gradient. The ground temperature generally decreases with increasing latitude, usually being several degrees warmer than the air temperature due to factors such as the extent of lakes and rivers, along with vegetation that maintains a deep snow cover.

A feature of the permafrost zone is the presence of ground ice, which is a body of clear ice in frozen ground. Ground ice is found extensively in Antarctica, often in the form of rock glaciers, which form on slopes when there is a plentiful supply of rock debris. As with ordinary glaciers, there is a large amount of ice present that allows the glacier to flow slowly downhill, but in a rock glacier the ice is hidden. Ground ice is extremely important and its

Fig. 6.33 A photograph of an ice wedge.

presence influences vegetation, the topography and the response of the landscape to environmental changes whether natural or anthropogenic. Thawing of ice-rich soils can result in weakening, collapse or settlement of the soil.

The temperature within the permafrost layer varies over the year, but at greater depths the seasonal difference in temperature decreases. The level at which there is no significant change in temperature over the year is called the 'depth of zero annual amplitude'. At Yellowknife (62° 28′ N, 114° 27′ W) in the Northwest Territories of Canada this depth has been measured at 15 m.

The annual cycle of freezing and thawing of the soil results in cracking and buckling that creates ice wedges, pingos (ice-cored hills), polygons and thermokarst lakes. A photograph of an ice wedge is shown in Fig. 6.33. Seasonally frozen ground is soil that freezes and thaws over the annual cycle. The US National Snow and Ice Data Center has defined it as the near-surface soil that experiences freeze for more than 15 days per year. By contrast, intermittently frozen ground experiences fewer than 15 days of freeze per year.

Permafrost has a profound effect on hydrology, erosion, vegetation and human activities. It limits movement of ground water and the rooting depth of plants. On slopes, it allows fluid-like movement of surface soil and deposits. Seasonal thawing over continuous permafrost creates a saturated surface layer in which pools of meltwater accumulate, conducive to marsh and tundra ecosystems and peat formation. When thawing extends below the active layer it can melt ice in the upper parts of the permafrost, creating an underground lake called a thermokarst. This can result in settling of the ground, giving uneven surface topography that includes pits, troughs, mounds and depressions, which can fill with water. Thermokarst damages agricultural fields and ecosystems such as forests by drying in mounded areas and flooding in low-lying zones. Further, it can contribute to erosion and increased sedimentation and siltation of rivers, which pose additional environmental concerns.

Permafrost is extremely important in the climate system because of the methane that is locked within it and the water that it contains. Methane is an extremely important greenhouse gas and its release from permafrost will contribute to increasing global temperatures. The thawing of permafrost is an irreversible process as the meltwater generally escapes. It also increases ground-water mobility, increases susceptibility to erosion and landslides, and it can affect soil storage of CO_2. If the thawed soil drains and dries it can release CO_2 to the atmosphere; conversely the storage of CO_2 in the soil could increase if the soil remains flooded.

The thickness of permafrost can change as a result of changes in the climate or disturbance of the surface. Permafrost becomes thinner and the active layer thicker when the temperature of the ground rises and the soil layer that was unfrozen during the summer does not refreeze completely. The residual unfrozen layer is called talik. Talik is formed when the climate warms and/or when snow accumulation increases. Increased thawing of the permafrost layer can also occur because of changes in the ground surface above the permafrost as a result of anthropogenic changes in land use or floods or forest fires.

In this section we will examine the occurrence of permafrost in the Arctic and on the sub-Antarctic islands and limited areas on the Antarctic continent that are not permanently covered by the Antarctic Ice Sheet. We will also consider the changes that have taken place in permafrost extent in recent decades.

6.7.2 The Arctic

Permafrost underlies approximately $23.4 \times 10^6 \, km^2$ or 24.5% of the exposed land area in the Northern Hemisphere. As shown in Fig. 6.34, the region of continuous permafrost (coverage of 90–100%) extends across eastern Russia and northern Alaska and Canada. Approximately half of the land surface of Canada has permafrost and nearly two thirds of Russia and one fifth of China. Smaller areas of permafrost are also found in many of the nations of northern Europe and Asia, for example in northern Japan. A small region of continuous permafrost occurs in northern Scandinavia, although sporadic permafrost (covering 10–50% of the area) is more common here.

South of the main continuous permafrost zones in Russia and North America there are narrow belts of discontinuous permafrost (50–90% coverage), sporadic permafrost and a somewhat wider belt of isolated permafrost with less than 10% of the land having permafrost.

In addition to the permafrost zone, another 57% of the land area of the Northern Hemisphere ($54.4 \times 10^6 \, km^2$) experiences seasonally frozen ground. In North America this extends down through Canada into parts of the USA, and is also found in parts of northern and central Europe.

As discussed in Section 6.2.2, in recent decades parts of the Arctic have experienced marked surface warming, which has had a major impact on the permafrost. In a number

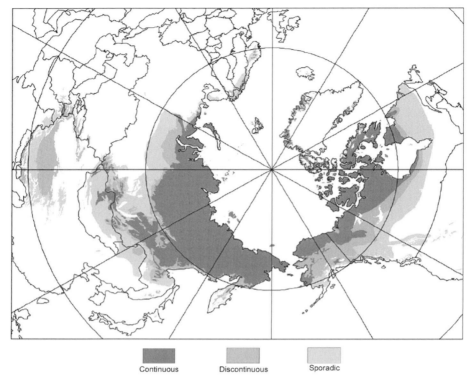

Fig. 6.34 The extent of permafrost in the high latitude areas of the Northern Hemisphere. (From Turner et al., 2007.)

of areas summer soil temperatures are approaching 0 °C, with a consequent impact on the permafrost. Surface vegetation represented by tundra is now increasingly changing to shrubs or wetlands, as monthly summer temperatures exceed 10 °C. Tundra covers approximately the same regions as permafrost and its area, as estimated from satellite measurements, has decreased by about 17% over the last 25 years (Overland and Wang, 2005). Photosynthetic activity as measured by the Normalised Difference Vegetation Index (NDVI) has also increased, with a peak over 45–70° N (Myneni et al., 1997). Counterintuitively, an increased area covered by shrubs traps snow, which insulates ground temperatures, leading to an early snow melt (Sturm et al., 2005). Although permafrost temperatures continue to rise, data on warming of the active layer are less conclusive and there is little evidence of a recent thickening of the active layer (Richter-Menge et al., 2006).

Permafrost underlies about 85% of Alaska; everywhere in fact except for a narrow belt along the southern coast. It varies widely, in depth, continuity and ice content (Fig. 6.34). Approximately the northernmost one third of the state has continuous permafrost, with a broad discontinuous zone across the central region and sporadic permafrost in the south.

6.7 Permafrost

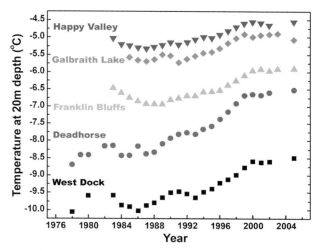

Fig 6.35 Time series of permafrost temperatures at a depth of 20 m for five sites in Alaska. (From Richter-Menge *et al.*, 2006.)

There is a good record of permafrost temperatures across Alaska and the data from the Arctic show a substantial warming over 1985–2005 (Fig. 6.35). The rate of warming varies with location, but has been typically 0.5–2.0 °C at the depth of zero seasonal temperature change (Richter-Menge *et al.*, 2006). However, this report noted that there has not been continuous warming of the permafrost over the period, but there were several periods of cooling, which at times could be attributed to a shallower than normal winter snow layer. During 2005 soil temperatures in interior Alaska reached the temperatures of the mid 1990s, which were the highest during the last 70 years. Continuous permafrost on the North Slope of Alaska has warmed 2–4 °C over the last century.

In Alaska there are many indications that the loss of permafrost started several hundred years ago. For example, Jorgenson *et al.* (2001) found that the extent of totally degraded permafrost in a landscape with rapidly degraded permafrost had increased from 39% to 47% in 46 years, with the degradation starting about 250 years ago.

The discontinuous permafrost to the south is warmer, usually above −2 °C and increased warming at multiple sites (1–2 °C since the late 1980s) suggests that much of the discontinuous permafrost south of the Yukon River and on the south side of the Seward Peninsula must already be thawing. At some sites, the discontinuous permafrost is thawing at the top and the bottom.

There is much less information on changes in the active layer, but at present there is no indication of any long-term trend in its thickness, possibly because of the shortness of the records (Brown *et al.*, 2000). However, the active layer around Fairbanks in 2005 was the deepest observed for the previous 10 years.

During historical and recent geological times the Arctic tundra was a net sink for carbon dioxide and large amounts of carbon are stored in the soils at high northern latitudes.

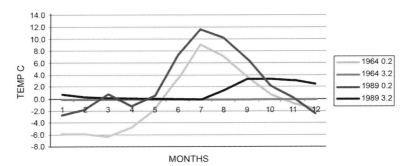

Fig. 6.36 The annual cycle of 0.2 and 3.2 m ground temperatures at Khoseda-Khard in northeast Russia. (From Richter-Menge et al., 2006.)

However, there is evidence that the rise in Arctic temperatures over recent decades has resulted in a lowering of the water table, accelerating the rate of soil decomposition and changing the soil from a sink to a source of CO_2. Oechel et al. (1993) showed that the tundra on the North Slope of Alaska had indeed become a source of CO_2 to the atmosphere and that this was linked to recent warming across the region. This suggests that tundra ecosystems can play a part in a positive feedback involving atmospheric CO_2 and greenhouse warming.

Permafrost temperatures have been rising in northwest Canada for the last few decades. Thie (1974) estimated that 75% of the permafrost at the southern limit of discontinuous permafrost (thought to originally cover 60% of the area) in Manitoba had degraded by 1967, with the extensive thawing starting some 100–200 years ago. More recently, Beilman and Robinson (2003) estimated that 30–65% of the permafrost along the southern margin of the discontinuous zone in Canada had degraded over the last 100–150 years. In the central Canadian Arctic, a general northward retreat of the southernmost margin of discontinuous permafrost by about 60 miles over the twentieth century has been reported. Continued climate warming is highly likely to bring accelerated thawing of warm discontinuous permafrost.

The widespread increase in near-surface air temperatures across parts of Russia discussed earlier in Section 6.2.2 has also resulted in some thawing of the permafrost and talik formation. For example, the annual cycles of 0.2 m and 3.2 m ground temperatures at Khoseda-Khard in northeast Russia in 1964 and 1989 (Fig. 6.36) show a warming of several degrees throughout the year at the upper level and a comparable amount at the greater depth during the winter.

Changes have also been observed in the depth of the active layer from the data collected at the Russian meteorological stations (Frauenfeld et al., 2004). These data suggest an increased depth of more than 20 cm between the mid 1950s and 1990 for the continuous permafrost regions of the Russian Arctic. However, some specialised permafrost research stations found no significant changes in the active layer thickness, indicating the problems in getting a clear picture of change. Observations also show rising permafrost temperatures across northern Europe.

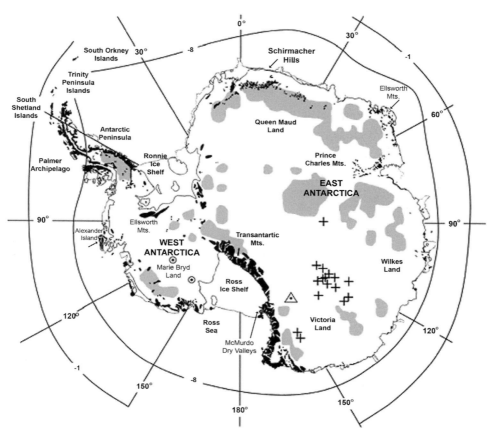

Fig. 6.37 The distribution of permafrost in continental Antarctica. Permafrost is found throughout the ice-free areas, which are shown in black. Subglacial permafrost beneath the Antarctic ice sheet may be restricted to the shaded areas. Crosses indicate the locations of subglacial lakes. (From Bockheim and Hall, 2002.)

6.7.3 The Antarctic

The orography of the high latitude areas of the Southern Hemisphere means that permafrost distribution is quite different from in the north. Less than 1% (55 000 km^2) of the Antarctic continent is ice free, the rest being covered by the Antarctic Ice Sheet. However, the very low temperatures across the continent mean that permafrost is found across all the ice-free areas (Bockheim and Hall, 2002).

Permafrost is found on the Antarctic continent (Fig. 6.37) and on the sub-Antarctic islands. The main exposed permafrost areas are down the length of the Transantarctic Mountains, across many coastal areas and some interior regions of the Antarctic Peninsula and in certain coastal areas of West and East Antarctica.

Estimating the total area of Antarctic permafrost is difficult. French (1996) estimated that 37% of the world's permafrost was found in the Antarctic. This included the ice-free area

314 *The instrumental period*

Table 6.2 *Permafrost characteristics in Antarctica*

Region	Permafrost distribution	Permafrost		Active layer	
		Form	Moisture content (%)	Thickness (cm)	Moisture content (%)
Antarctic Peninsula and Islands	Discontinuous	Wet	16–20	40–150	5.8–21
Maritime East Antarctica	Continuous	Wet, dry (?)	6–22	60–150	1.5–28
Transantarctic Mountains coastal	Continuous	Wet	6.1–77	30–60	1.0–10
Inland valley floors	Discontinuous	Dry	1.6–9.5	20–40	0.3–3.0
Inland valley sides	Continuous	Dry, wet		20–30	0.3–2.0
Upland valleys	Continuous	Dry, wet		20–25	0.3–3.0
Plateau fringe	Continuous	Dry, wet	1.7–20	0.3–4.0	0.3–4.0

From Bockheim and Hall (2002).

and regions under the ice sheet where subglacial permafrost may be located (Fig. 6.37). However, much of the land below the East Antarctic Ice Sheet is above the pressure melting point and is unfrozen (Herterich, 1988), so the actual coverage may be around 25%.

Most of Antarctica, with the exception of the inland valley floors and sides of the McMurdo Dry Valleys, has ice-cemented permafrost. In the Dry Valley the permafrost is predominantly of the 'dry' class, with ice-cemented permafrost being restricted to some very limited areas.

Compared with the Arctic, very little is known about the thickness, properties and age of permafrost in Antarctica. The most comprehensive data on permafrost have been collected by the Dry Valley Drilling Project, when 15 boreholes were drilled with depths ranging from 4 m to 381 m. Beyond the Dry Valleys boreholes have been drilled in North Victoria Land and permafrost characteristics were also studied on Seymour Island and King George Island.

The Transantarctic Mountains contain the largest ice-free area (55%) in the Antarctic, which extends across 30 400 km^2 (Bockheim and Hall, 2002). In the interior of the continent the active layer and the near-surface permafrost are characterised by exceptionally low moisture content, which is less than 3%, so that the permafrost is 'dry'. Along the mountains the active layer thickness depends on the distance from the coast and the elevation, with it being 30–60 cm close to the coast and 15–20 cm on the high plateau.

In the Dry Valleys there is little snow cover and the surface is covered with loose gravelly material where ice-wedge polygons may be observed. Parts of the coastal Transantarctic Mountains have the highest recorded permafrost moisture content in the Antarctic (Table 6.2).

Marie Byrd Land in West Antarctica contains 12% of the ice-free area (6600 km^2), mainly in the Ford Ranges and the Executive Committee Range. Around the rest of the continent there are smaller areas of ice-free ground that contain permafrost. These include Queen Maud Land (4.8% or 2600 km^2), coastal East Antarctica (4.6% or 2500 km^2), the Prince Charles Mountains (4.4% or 2400 km^2) and the Ellsworth Mountains (3.1% or 1700 km^2).

The Antarctic Peninsula and adjacent offshore islands contain about 14% of the ice-free area of the Antarctic (8000 km^2). Islands off the Peninsula where permafrost is found consist of the South Orkney Islands, the South Shetland Islands, Adelaide Island, James Ross Island and Alexander Island. Many of these locations have an active layer, such as Signy Island (60° 43′ S, 45° 36′ W) where there is permafrost present below an active layer of 0.3–0.7 m (Chambers, 1966). Permafrost is discontinuous across the Antarctic Peninsula, predominantly wet with a moisture content of 16–20% (Bockheim and Hall, 2002). The active layer in this relatively mild part of the Antarctic is quite deep, ranging from 40 cm to 150 cm.

It will be clear from the above that there are far fewer data on Antarctic permafrost than for the Arctic. For many areas there have been no boreholes drilled and there is little knowledge on the properties of the ice. So we have no data sets of the changes in permafrost over recent decades. However, from the information we have on climate change across the continent in recent decades it is possible to produce broad estimates on the likely changes in permafrost in the recent past. Much of coastal East Antarctica has experienced little change in the annual mean temperatures, and there has even been a slight cooling during the summer. Doran *et al.* (2002) has pointed to a cooling in the Dry Valleys based on an analysis of automatic weather station observations. We may therefore expect that there has been little change, or a slight increase, in the amount of permafrost in recent decades.

A contrasting situation exists across the Antarctic Peninsula and the sub-Antarctic islands. Here there has been an increase in both winter and summer temperatures since the first records were taken in the early 1950s. In parallel with the marked retreat of glaciers and loss of ice shelves we can expect a degradation of some of the permafrost. However, the extent of this is not known, and observations are mainly limited to the areas around the research stations.

6.8 Atmospheric gases and aerosols

6.8.1 Introduction

Monitoring the emissions of greenhouse gases is critical when assessing the extent of climate change in a world where industrialisation is progressing at a rapid pace. It is relatively easy to determine the amount of fossil fuel that is burnt each year. It is known that about 50% of the carbon put into the atmosphere remains there, with a further 25% being absorbed by the ocean. The remaining 25% enters the biosphere.

6.8.2 Well-mixed gases

The concentration of carbon dioxide has increased from a level of 280 ppm in pre-industrial times to about 388 ppm at the start of 2010. This level was higher than the natural range over the last 650 000 years (180 to 300 ppm), as determined from ice cores. In recent decades the annual CO_2 concentration has been increasing at a faster rate. Over 1960–2005 it increased by 1.4 ppm per year, which rose to 1.96 ppm per year for 2005–09.

Carbon dioxide levels have increased primarily because of the greater use of fossil fuels, but also, to a smaller extent, because of land-use changes.

Methane is another important greenhouse gas and its atmospheric concentration has increased from a pre-industrial level of 715 ppb to 1788 ppb in 2008. As with CO_2, the current concentration is outside the range found in ice cores spanning the last 650 000 years, which recorded levels of 320 to 790 ppb. The IPCC noted that it was very likely that the observed increase in methane concentration was a result of anthropogenic activities, and especially agriculture and fossil fuel use.

A third important atmospheric greenhouse gas is nitrous oxide, which has increased from a pre-industrial level of 270 ppb to 319 ppb in 2005. Since the 1980s the growth rate of this gas has been almost constant. Anthropogenic processes, primarily from agriculture, are responsible for more than a third of all emissions.

6.8.3 *Pollution*

The Arctic

The Arctic is often thought of as a pristine wilderness, but over the twentieth century there was a huge increase in the amount of industrial pollution reaching the area. This came from densely populated parts of the Northern Hemisphere, but also from some high latitude areas, such as the Kola Peninsula and the industrial complexes of north-central Siberia.

The climatological west to southwesterly flow of mid latitudes picks up pollution at low levels and this can be transported polewards at various levels of the troposphere, before descending at high latitudes. The main routes by which airborne pollutants reach the Arctic are over northern Europe and Asia, and then across the Arctic Ocean towards Alaska and northern Canada.

Mercury and other heavy metals, PCBs (polychlorinated biphenyls), furans, and even radioactive fallout, have reached previously unpolluted areas of the Arctic, with a consequent serious impact on the wildlife. Many of these contaminants become locked into ice and snow, and any melting of the snow in the future will release these toxins into the water systems and have the potential to enter the food chain.

The particles that constitute 'Arctic haze' (see Section 3.4.4) are similar to those found in smog at lower latitudes and consist of sulphates, along with carbon. Ice cores indicate that sulphate pollution is a comparatively recent phenomenon, starting after the Industrial Revolution and reaching a peak in the middle of the twentieth century. At high northern latitudes the reduced amounts of sunlight received lengthen the degradation process of many types of pollution and increase the probability that toxic substances will become locked into ice or ground or enter the food chain.

6.8 Atmospheric gases and aerosols 317

On a more positive note, in many parts of the world the emission of pollutants has become increasingly regulated so that some forms of pollution are decreasing in the Arctic. For example, the Arctic Monitoring and Assessment Programme has determined that persistent organic pollutants (POPs) in Arctic biota have been decreasing for the past 2–3 decades.

The Antarctic

The Antarctic is far more remote from human influence than the Arctic and clear evidence of pollution in the north, such as Arctic haze, is not present. However, human influence on the continent can still be detected.

In the vicinity of the research stations there is local contamination of the air, ice, soil and marine environment through fuel combustion for transportation and energy production, along with waste incineration and the release of sewage. However, the impact is quite local and the nations operating the stations have introduced stringent controls on emissions.

The Antarctic is well removed from most of the other land masses in the Southern Hemisphere, but the closeness of the Antarctic Peninsula to South America means that this sector of the continent does receive some pollutants. A variety of elements have been measured at the stations in this area. Black carbon has been carried to the northern part of the peninsula when the winds are from the north, and the source region has been identified as areas of burning biomass in Brazil (Pereira *et al.*, 2006). An observing programme at the Korean 'King Sejong' station on King George Island recorded various major metals and ions in suspended particulate matters, including cadmium, aluminium, barium, chromium, nickel and bismuth. The record showed that aluminosilicate dust deposition had doubled during the twentieth century, which was thought to be consistent with increasing air temperatures, decreasing relative humidity, and widespread desertification in Patagonia and northern Argentina.

Ice cores collected across the continent provide an outstanding record of the background levels of many chemical species, aerosols and even biological material. As described in Section 2.5.2, concentrations of oxygen isotopes have been used to determine past temperatures changes across the Antarctic on decadal to multi-millennial timescales. However, the ice core data have also been used to investigate changes in many chemical species in the Antarctic, such as lead (Wolff and Suttie, 1994), sulphur (Boutron and Wolff, 1989), and copper, zinc and cadmium (Batifol *et al.*, 1989).

6.8.4 Stratospheric ozone

Introduction

Concentrations of stratospheric ozone vary across the Earth with the total column ozone amount above the polar regions being usually greater than at lower latitudes (Fig. 3.20). Such data show a clear annual cycle, with a peak in the spring and a decrease in late summer and through the autumn. As discussed in Section 3.4.4, since the early 1980s there has been

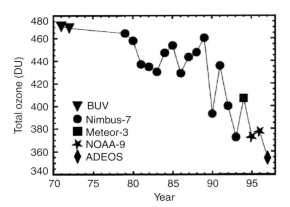

Fig. 6.38 March average from a variety of satellite-borne instruments of total ozone from 63° N to 90° N. (From Newman *et al.* 1997.)

a springtime loss of stratospheric ozone in both polar regions, although it has been most marked in the Antarctic, where it is known as the Antarctic ozone hole. The physical and chemical processes responsible for ozone loss are the same in both polar regions, but the 'hole' is more marked above Antarctica because stratospheric temperatures are lower.

Section 3.4.4 describes our current knowledge of the mechanisms of how chlorofluorocarbons destroy stratospheric ozone each spring. Here we consider the variability and trends in stratospheric ozone concentration, which at high southern latitudes have been responsible for large changes in the Antarctic climate since the early 1980s.

The Arctic

Although the greatest springtime loss of stratospheric ozone has occurred in the Antarctic, the low stratospheric temperatures above the Arctic can provide the necessary conditions for substantial destruction of ozone. When the dynamic conditions at high northern latitudes give particularly cold stratospheric temperatures, significant ozone loss is observed (Newman *et al.*, 1997).

Figure 6.38 shows the mean total ozone over 63–90° N as determined from various satellite instruments. From the start of the record in 1970 until around 1980 values were fairly constant at around 460–470 DU, although data were sparse during this period. However, even though total ozone amount is rather variable on a year-to-year basis, there has been an overall decline since the 1980s. The greatest decrease of up to 30% has been in the winter and spring, when the stratosphere is colder.

The more recent study by Jiang *et al.* (2008) examined trends in total column ozone from the combined ozone data product and the European Centre for Medium-range Weather Forecasts assimilated ozone for January 1979 to August 2002. They found that the largest ozone loss had been at the North Pole, with a decline of 1 DU yr^{-1}.

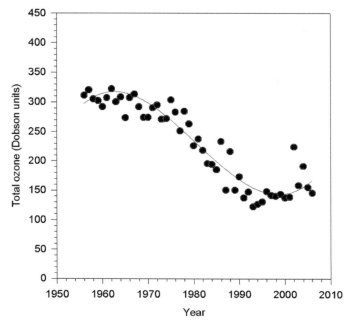

Fig. 6.39 The mean October total column ozone amount (DU) measured by surface-based instruments at Halley Station over 1958–2009.

The Antarctic

The loss of stratospheric ozone above the Antarctic during the austral spring was first detected using surface-based instruments at Halley station on the eastern side of the Weddell Sea (Farman et al., 1985), but was subsequently confirmed using satellite observations (Stolarski et al., 1986).

The loss of ozone can be appreciated from the time series of October mean total column ozone measurements made at Halley over the period 1958–2009 (Fig. 6.39). These observations show modest interannual variability in the ozone total up to about 1980, but then a rapid decline into the late 1990s. The total ozone amount stopped declining after this time since there was almost complete destruction of stratospheric ozone during each spring. There has been some speculation that springtime stratospheric ozone concentrations are beginning to recover, and while there is interannual variability in the depth of the ozone hole, at the time of writing there is no clear indication of a sustained recovery.

The springtime ozone loss is more marked in the Antarctic than the Arctic since stratospheric temperatures are lower as a result of the polar vortex being stronger. The vortex is also more annular than in the north since the orography of the Southern Hemisphere is well north from the Antarctic.

The form of the 'ozone hole' can vary considerably on a range of timescales. It can extend northwards almost into mid latitudes and move over southern South America where the UV-B radiation measured at the surface can increase dramatically over the space of a few day (Jones *et al.*, 1998). The ozone hole can also 'split' as happened in 2002 (Roscoe *et al.*, 2010; cf. Fig. 3.44).

6.9   Terrestrial ice and sea level

6.9.1   Introduction

Understanding the recent changes in the volume of ice locked into the ice sheets (ice masses that cover areas of more than $50\,000\,\text{km}^2$), icecaps (domed masses of ice covering less than $50\,000\,\text{km}^2$) and glaciers is of major importance because of the implications of ice loss in the future on sea level. Globally there are just over 66 m of sea level rise contained in the land ice, which splits between the Antarctic ice sheet (60 m), Greenland (6 m), and glaciers worldwide (0.38 m).

The ice shelves around Antarctica and Greenland are also important with regards to sea level rise. Even though they are floating, and any melting would not contribute directly to sea level rise, they buttress many of the glaciers that flow down from the icecaps, and there is evidence that if the ice shelves are removed then these glaciers will begin to flow faster and discharge more ice into the oceans.

Determining whether the ice sheets are growing or shrinking (the question of mass balance) is not trivial, since the ice volume is affected by climatic factors in complex ways. Net mass balance is defined as the difference between the mass gained by snowfall onto the surface and the loss by evaporation/sublimation, iceberg calving, runoff and snow blown off the ice sheet at the coast. The most efficient way of measuring mass balance is by instruments flown on the polar-orbiting satellites, such as altimeters, which measure their height above the surface. But field observations of ice elevation are also of great importance. However, the satellite systems can make frequent, repeat measurements so that errors can be reduced and they have near-global coverage, with only some of the most high latitude areas being beyond their range. The first altimeters used microwave radiation to determine their height above the ice surface (Zwally, 1989), but the most recent instruments carry lasers (Schutz *et al.*, 2005). Within the last few years other techniques have been developed to estimate mass balance from satellites, such as the use of gravity data (Velicogna and Wahr, 2005).

The field of mass balance studies is advancing very rapidly as new forms of satellite data become available, but in this section we will present the current best estimates of how the ice sheets have changed in recent decades. We will start by considering the large ice sheets of Greenland and Antarctica, before turning to the high latitude glaciers and then the state of the ice shelves. Finally we will examine how sea level has changed in recent decades and the contribution from the high latitude cryosphere.

6.9 Terrestrial ice and sea level

6.9.2 The ice sheets

Greenland

The ice sheets, icecaps and glaciers of the Arctic contain just over 3 million km³ of ice, the bulk of which is locked into the Greenland Ice Sheet, with the remainder being in the icecaps and glaciers of the Canadian Arctic, Iceland, Svalbard, Alaska, Franz Josef Land, northern Scandinavia, Severnaya Zemlya and Novaya Zemlya.

A great deal of attention has been paid to the mass balance of Greenland because of the possible effects on the North Atlantic MOC of greater freshwater input. The fear is that a freshening of the North Atlantic may slow down the THC resulting in colder conditions across northern Europe. In this section we assess our current understanding of the mass balance of the Greenland Ice Sheet.

Precipitation The precipitation falling on the Greenland Ice Sheet and the net accumulation (P–E) and their change in recent decades can be estimated in several ways. The ECMWF operational atmospheric analysis fields for 1992–98 were used by Hanna *et al.* (2002), who estimated that the mean accumulation for the ice sheet was 0.287 or 0.307 m yr^{-1}, depending on the land mask used. These figures were in good agreement with earlier studies, such as Ohmura *et al.* (1999), who found a mean value of 0.297 m yr^{-1}. They noted the large interannual variability of the accumulation and found a standard deviation of 0.039 m yr^{-1} and a range from 0.232 m yr^{-1} in 1995 to 0.333 m yr^{-1} in 1992.

Modelling studies have suggested that there has been a trend towards greater accumulation over the higher parts of Greenland (Bugnion, 2000; van de Wal *et al.*, 2006). These results are supported by laser altimeter measurements from aircraft (Krabill *et al.*, 2004). However, the changes in elevation that were detected were small (less than 20 cm yr^{-1}).

Loss of ice In contrast to the centre of Greenland, there has been significant thinning at the margins. This has been attributed to increases in melting, as well as dynamic thinning (Abdalati *et al.*, 2001; Krabill *et al.*, 2004). Warmer summers may have caused melting of the order of a few tens of centimetres per year, but the nature and magnitude of the elevation changes suggest that glacier dynamics and creep thinning have also played a role (Abdalati *et al.*, 2001). The thinning at the margins exceeded 1 m yr^{-1} close to the coast. Thinning of ice was found in most coastal areas, except in the southeast, where the ice was found to have thickened by more than 1 m between May 2002 and May 2003. The rate of the thinning generally has exceeded that expected from increased melting during recent warmer summers.

Krabill *et al.* (2004) estimated the total ice-sheet melt (runoff) during 1993–98 and 1997–2003 using ECMWF data. They found net ice loss associated with melting/snowfall anomalies of 35 ± 5 km³ for the earlier period and 46 ± 7 km³ for the latter, indicating that melt losses had increased over recent years. They estimated that ice loss from Greenland since 1997 was 35% higher than for 1993–94 to 1998–99, with over half from increased runoff and the rest from increased losses by dynamic thinning.

322   *The instrumental period*

Outlet glaciers are important to mass balance since their discharge accounts for nearly half of the ice sheet mass loss. Krabill *et al.* (2004) found that the velocity of some glaciers exceeded those needed to balance upstream snow accumulation. They reported a doubling of the velocity of the Jakobshavn Isbræ glacier between 1997 and 2003, giving net loss from its drainage basin of about 20 km^3 of ice between 2002 and 2003. In addition, Abdalati *et al.* (2001) reported thinning rates as high as 10 m yr^{-1} near the terminus of the Kangerdlugssuaq Glacier in southeast Greenland.

Overall mass balance Several estimates of the mass balance of the Greenland Ice Sheet are available. Krabill *et al.* (2004) used altimeter data and modelled snowfall to derive figures that showed a loss of 80 ± 12 km^3 yr^{-1} between 1997 and 2003. This is greater than the figure of 60 km^3 yr^{-1} for 1993–9 to 1998–99.

The ten estimates of Greenland mass balance examined by Shepherd and Wingham (2007) had a range of −227 to 11 Gt yr^{-1}, although only one study had a positive mass balance. The current best estimate is that Greenland is losing mass through increased melting at the edge of the ice sheet, and that the central area above 2000 m is approximately in balance.

The Antarctic

Precipitation The input to the Antarctic Ice Sheet in terms of snowfall and hoar frost is difficult to estimate because of the high variability of the snowfall in some parts of the continent. However, a number of studies have used the available ice core records and the output from atmospheric models to estimate the net accumulation or surface mass balance.

Monaghan *et al.* (2006a) combined ice core data and model output to produce a 50-year record of snow accumulation for the continent. They found that since the 1950s there had been no statistically significant change in snowfall. This is consistent with recent climate model experiments, which suggest no statistically significant increase in snowfall since 1980 over the grounded ice sheet as a whole, the West Antarctic Ice Sheet or the East Antarctic Ice Sheet (van de Berg *et al.*, 2005, 2006; Monaghan *et al.*, 2006b). However, it is in disagreement with the results of Davis *et al.* (2005) who used satellite altimetry data to infer an increase in snow accumulation over East Antarctica for the period 1992–2003. They calculated that the interior of the Antarctic north of 81.6° S (70% of the total ice sheet area) increased in mass by 45 ± 7 billion tons per year from 1992 to 2003. Davis *et al.* (2005) found that overall the interior of East Antarctica was thickening at a rate of 1.8 ± 0.3 cm yr^{-1}, while West Antarctica exhibited bimodal behaviour with thinning and thickening in different parts of the region. The greatest thinning was found in Marie Byrd Land and especially the Pine Island and Thwaites Glaciers drainage basins. Conversely, marked thickening of the ice of 8–19 cm yr^{-1} was found across basins adjacent to the Antarctic Peninsula and the Ronne Ice Shelf. Overall West Antarctica showed a slight thinning of 0.9 ± 0.3 cm yr^{-1}. The authors suggest that much of the change in elevation across East Antarctica is a result of greater snowfall. However, there is less agreement between elevation and snowfall changes across West Antarctica, where changes due to ice dynamics are thought to play a greater role. They pointed out that a mass gain of the

6.9 Terrestrial ice and sea level

magnitude found across the whole Antarctic was enough to slow sea level rise by $0.12 \pm 0.02 \, \text{mm yr}^{-1}$.

Monaghan *et al.* (2006a) discuss in depth the possible reasons for the discrepancy between these two studies; however, it does indicate the problems in determining the changes in snowfall across the continent with such sparse data and only recently developed tools such as satellite altimeter instruments.

Loss of ice The loss of ice from iceberg calving at the edges of the continent and the discharge of ice into the Southern Ocean are extremely important because of the possible impact on sea level rise. The calving of icebergs from the edges of the Antarctic is part of the natural hydrological cycle of the continent. However, determining changes in the amount of ice that leaves the continent by this route cannot be determined with confidence at this time because of the difficulties in detecting many smaller pieces of ice. The US National Ice Center uses satellite data to track the larger icebergs and their records covering the last 25 years have reported a greater number of icebergs. However, Ballantyne and Long (2002) determined that this increase was largely a result of improvements in iceberg tracking, rather than any increase in discharge of ice from the continent.

Loss of ice locked into the Antarctic Ice Sheet can also be detected by changes in surface elevation of the continent. Here certain sectors of the Antarctic have received particular attention because of rapid changes of elevation noted in repeat satellite altimeter data, especially around the coast of West Antarctica. The discharge of ice from glaciers in the Amundsen Sea sector of West Antarctica has been estimated at $250 \, \text{km}^3 \, \text{yr}^{-1}$ using satellite and aircraft data, which is about 60% more than is accumulated in their catchment basins (Thomas *et al.*, 2004).

Satellite altimeter measurements of the West Antarctic Ice Sheet, along with satellite-derived velocity data, indicate that parts of this region have been thinning since the early 1990s. The glaciers in this region, including the Pine Island Glacier (PIG) and Thwaites Glacier (Fig. 6.40), are accelerating and at the time of writing contributing about 10% of the observed rise in global sea level (Rignot *et al.*, 2008). Theoretical considerations have suggested that a retreat of the PIG may accelerate ice discharge from the interior of the West Antarctic Ice Sheet, since there is no extensive ice shelf at the coast and the bedrock topography deepens inland.

Satellite radar interferometry data have shown that the grounding line of the PIG retreated 5 km inland between 1992 and 1996 (Rignot, 1998). More recently Shepherd *et al.* (2001) used satellite altimetry and interferometry to investigate thinning of the glacier. They found that over 1992–99 the interior of the drainage basin lowered at a rate of about $0.1 \, \text{m yr}^{-1}$, which is comparable to the expected variability of snowfall. However, near the grounding line the PIG had thinned by up to $1.6 \, \text{m yr}^{-1}$ (Fig. 6.41), which is consistent with the grounding line retreat. The thinning was reduced further inland, but could be identified 150 km upstream of the grounding line. Shepherd *et al.* concluded that the thinning could not be explained by short-term variability in snowfall and therefore must be a result of glacier dynamics. They estimated that the PIG is currently thinning and is reducing the mass of the glacier by $3.9 \, \text{Gt} \pm 0.9 \, \text{Gt yr}^{-1}$, or about 5% of the mass flux across the grounding line,

324   *The instrumental period*

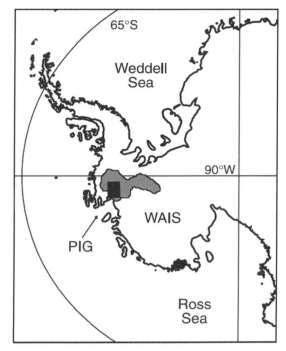

Fig. 6.40   Map of West Antarctica showing the Pine Island Glacier (PIG) drainage basin (grey). (WAIS), West Antarctic Ice Sheet. (From Shepherd *et al.*, 2001.)

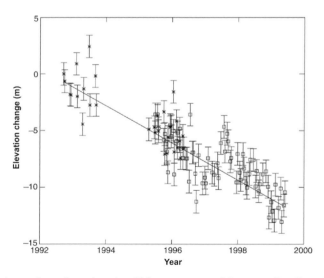

Fig. 6.41   The change in surface elevation 13 km upstream of the grounding line of the Pine Island Glacier between 1992 and 1999. (From Shepherd *et al.*, 2001.)

which is equivalent to 0.01 mm yr^{-1} of eustatic sea level rise. If the entire grounded section of the PIG discharged into the ocean the net contribution to eustatic sea level rise is estimated to be 6 mm. They also estimated that if the trunk of the glacier continued to lose mass at the present rate it would be entirely afloat within 600 years.

Since the 1990s the glacier thinning rates have increased, with data from aircraft surveys during 2002–03 showing thinning along the coast of the Amundsen Sea from the PIG to the Thwaites and Pope Glaciers compared with earlier elevation data from satellite laser altimeters. These data suggest an average thinning of 0.4 m yr^{-1} mainly over the PIG to 1.8 m yr^{-1} for the areas over the seaward ends of the Thwaites, Haynes, Pope, Smith and Kohler Glaciers. Thinning rates near the grounding lines of all the glaciers surveyed reached local maxima in excess of 5 m yr^{-1}. The mean thinning indicated by the four flights described in Thomas $et\ al.$ (2004) was 1 m yr^{-1}. They pointed out that if the 1 m yr^{-1} thinning rate was typical across the 60 000 km^2 area surveyed by their aircraft, then this represents a loss of about 60 km^3 yr^{-1} from only 15% of the total catchment area. For individual glaciers, the results of Thomas $et\ al.$ (2004) indicated thinning along the main trunk of the PIG of an average of 1.2 m yr^{-1} between 100 km and 300 km inland of the grounding line.

Overall mass balance Over the last decade at least seven estimates have been made of the mass balance of the Antarctic Ice Sheet (Shepherd and Wingham, 2007), which ranged from −139 to 42 Gt yr^{-1}. These estimates were published between 1998 and 2006, with the data spanning periods of four to 11 years. There is also no clear trend in the estimates towards a particular figure for the mass balance, with both negative and slightly positive estimates being published in 2006. However, this is in line with the difficulty in determining the current status of Antarctic mass balance from the limited data available. In summary, the recently published papers suggest that the Antarctic has an overall slight negative mass balance, which is a result of basal melting along the coast of the Amundsen Sea.

The satellite data have also been used to estimate the current mass balance of East and West Antarctica separately. East Antarctic is increasing in mass, with the six studies examined by Shepherd and Wingham having values ranging from −1 to 67 Gt yr^{-1}. The smaller range compared with that for the Antarctic as a whole reflects the more uniform topographic nature of East Antarctica and the lack of major weather systems penetrating into this section of the continent.

Five studies have examined the mass balance of West Antarctica, with all of them concluding that there has been a loss of mass in recent years. The values had a range of −136 to −47 Gt yr^{-1}, reflecting the difficulty of using the satellite data in this region with variable orography.

6.9.3 Glaciers

Glaciers are very sensitive to climate change, and particularly near-surface temperatures. However, the volume of ice within them varies constantly as a result of changes in input

326   *The instrumental period*

due to snowfall and mass loss through melting and evaporation. Nevertheless, the records of glacier extent are a powerful tool for investigating climate change at high altitude and high latitude locations where there are relatively few conventional meteorological observations. The length of the glacier records varies considerably according to location. In the Alps and parts of Scandinavia the records extend back several hundred years, but they are much shorter in the remote high latitude areas. However, from the middle of the twentieth century systematic studies of glacier extent and volume have been carried out, with the task gradually becoming easier as aerial survey data and satellite imagery became available.

In this section we will consider the changes that have been observed in the high latitude glaciers over approximately the last 100 years. We will not consider glaciers that form part of the major ice sheets of Antarctica and Greenland as these are dealt with elsewhere.

The Arctic

Northern Hemisphere glaciers had been losing volume more or less continuously since the nineteenth century (Dyurgerov and Meier, 2000). This general pattern of melting and retreat with a tendency to negative mass balance in recent decades is not surprising considering the high latitude warming that has been noted in Section 6.2.2.

The first global analysis of glacier extent and change was carried out by Thorarinsson (1940) who considered data extending back into the eighteenth century. He also computed the volume change of glaciers since 1850, which is often considered to be the time of maximum glacier extent since the last ice age in some areas. He found that glaciers had been shrinking globally, but with periods of greater retreat separated by stagnation and advance.

A powerful tool in considering changes in glacier extent has been glaciometeorological modelling based on precipitation and temperature data, with the estimates calibrated by recent direct observations (Tangborn, 1980). Such estimates for several geographical areas are shown in Fig. 6.42. These indicate that glaciers have been in retreat across the whole Northern Hemisphere since at least the late nineteenth century and that the retreat has accelerated in recent decades.

Volume data on small glaciers covering the period since the mid 1940s has been assembled by Dyurgerov and Meier (2000). Before the International Geophysical Year data were only collected for ten glaciers, but the number observed has increased steadily with data being available for 70–90 in the mid 1970s. During the period, mass balance has been measured for a total of 260 glaciers. Figure 6.43 presents volume changes for selected glaciers for the period 1961–97.

Glaciers in Alaska, as throughout the Arctic, have generally retreated through most of the twentieth century. Estimated losses in Alaskan glaciers are of the order of 10 m in thickness over the past 40 years, even while some have gained thickness in their upper regions. The Wolverine glacier experienced a period of mass gain between about 1975 and 1988, but has experienced a steady loss since that time. The mass balance of four Canadian Arctic glaciers studied all showed a steady decline since 1960, with only short periods of gain. All the

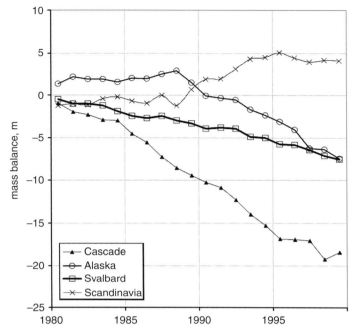

Fig 6.42 Time series of cumulative volume changes for glaciers in different geographical regions from 1980 to 1999. The units are mass (water-equivalent) changes per glacier area per year in metres. (From ACIA, 2005.)

glaciers show a greater rate of mass loss since the late 1980s, although the overall rate of loss is less for Devon Island.

Scandinavia presents a more complex picture of change in recent decades. Glaciers here have shown both steady decreases and marked increases since 1970. Some of the increases may be a result of increased precipitation associated with the NAM (ACIA, 2005). However, there are indications that the mass balance of Scandinavian glaciers has turned negative since the late 1990s.

The Antarctic

In this section we are primarily concerned with the glaciers on the sub-Antarctic islands, since the vast majority of the Antarctic continent is covered by the large Antarctic Ice Sheet, changes in which are dealt with in Section 6.9.2. However, we will make some reference to the glaciers on the Antarctic Peninsula, since this is a region of major change in recent decades.

There has been a large increase in near-surface temperatures across the Antarctic Peninsula since in-situ meteorological records were first collected on a routine basis in the early 1950s. The warming extends to the sub-Antarctic islands north of the tip of the

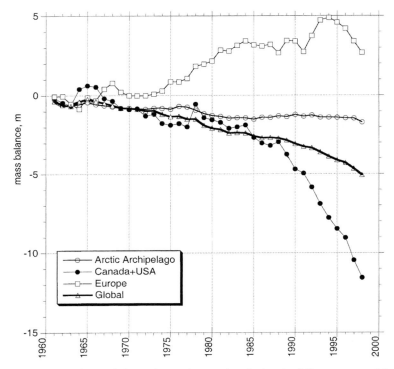

Fig 6.43 Time series of cumulative volume changes for glaciers in different geographical regions from 1961 to 1997. The units are mass (water-equivalent) changes per glacier area per year in metres. (From ACIA, 2005.)

peninsula and can be detected at Signy Island, South Orkney Islands, which has a record starting in 1903 (Zazulie *et al.*, 2010). Here there has been a widespread retreat of glaciers, snowfields and permanent ice, with a resultant increase in bare ground. Between 1949 and 1989 there was a 35% reduction in the ice-covered area. And between the mid 1960s and mid 1980s ice margins have receded by over 100 m at many locations. In addition, the thickness of the island's glaciers has been reduced by between 7 m and 8 m (Smith, 1990).

As temperatures have risen across the Antarctic Peninsula, so there has been a parallel retreat of glaciers. In a recent study of the glaciers around the coast of the peninsula it was found that over the last 50 years 87% of the 244 glaciers studied had retreated, and that the average retreat rates have accelerated (Cook *et al.*, 2005).

6.9.4 *Ice shelves*

Ice shelves form when ice streams flow down from the major icecaps to the coast and extend out over the ocean. They are important in the climates of the polar regions because of their

6.9 Terrestrial ice and sea level

possible role in controlling the flow of the ice streams, the cap that they provide on the upper part of the ocean and the production of bottom water under the shelves (see Section 6.4.3).

There is still some controversy as to the extent to which ice shelves affect the dynamics of glaciers flowing down towards the shelves. For many years it was believed that the loss of an ice shelf would result in an increase in flow from the inland ice sheet. However, the conditions at the base of the ice streams have also been put forward as determining the flow speed of the ice. But the acceleration of tributary glaciers along the eastern side of the Antarctic Peninsula soon after the break-up of ice shelves has suggested that the presence of a buttressing ice shelf is important, at least with some ice streams.

The Arctic

As temperatures have risen markedly across the Arctic in recent decades, so there has been the loss of some major ice shelves. In 2003 the Ward Hunt Ice Shelf on the north coast of Ellesmere Island, which at 443 km^2 was the largest ice shelf in the Arctic, split into two, releasing all the water from the freshwater lake that it dammed. It is believed that this ice shelf had been present for at least 3000 years (Mueller *et al.*, 2003).

As discussed earlier, there has recently been extensive melting at coastal sites around Greenland and this has resulted in the loss of a number of ice shelves. For example, the ice shelf at the head of the Jakobshavn Glacier thinned and broke up. Since then the glacier, which is the fastest flowing in Greenland, has experienced rapid acceleration.

The Antarctic

The warming across the Antarctic Peninsula over the last 50 years has resulted in the disintegration of a number of floating ice shelves on both sides of this major topographic barrier. This has prompted a great deal of discussion over whether these events have occurred as a result of natural climate variability or because of anthropogenic factors (Marshall *et al.*, 2006). In the northern part of the Antarctic Peninsula the ice shelves are near to the climatic limit where they can just survive through the summer. This makes them particularly sensitive to climatic fluctuations (Vaughan and Doake, 1996).

Many of the break-out events have been observed via satellite imagery, but historical records suggest that some of the small ice shelves on the eastern side of the peninsula have been retreating since at least the middle of the nineteenth century. For example, Cooper (1997) examined the historical observations for the Prince Gustav Ice Shelf between James Ross Island and Trinity Peninsula and constructed a map of the extent of the shelf from the mid nineteenth century (Fig. 6.44). This suggests that the ice shelf has been retreating for most of the period since 1843, but with a rapid retreat between 1957 and 1961, prior to the recent retreat of this ice shelf between 1992 and 1995.

The rate of ice shelf retreat around the peninsula increased in the late 1980s (Skvarca *et al.*, 1999) and in 1995 there was a major break-out of the Larsen A and the Prince Gustav Ice Shelves (Rott *et al.*, 1996). Over this period a number of ice shelves disintegrated on the western side of the Antarctic Peninsula (Vaughan and Doake, 1996). Ice shelf disintegration

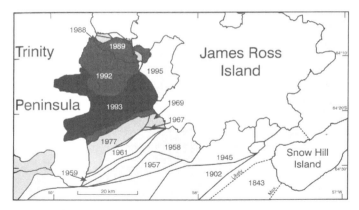

Fig. 6.44 The extent of the Prince Gustav Ice Shelf between 1843 and 1995. Two ice fronts are shown for 1843, one at the maximum possible extent, and the other indicating a likely extent. (From Cooper, 1997.)

does not seem to have been a frequent occurrence in the area, since Pudsey and Evans (2001) showed via analysis of marine sediments that the ice shelf in the Prince Gustav Channel had been present since the Mid Holocene (see Section 5.4.3).

The most publicised disintegration was the loss of the Larsen B Ice Shelf. In February 2002 the 3250 km² ice shelf on the eastern side of the peninsula disintegrated into many small fragments over a period of months, releasing 500 billion tonnes of ice into the Southern Ocean.

6.9.5 Sea level

It is well known that sea level is not uniform across the globe and that there are regional variations. For over a century tide gauges have allowed the investigation of sea level in coastal regions and more recently various satellite-borne radar altimeter missions have provided data across the world's oceans. By combining these observations it is possible to examine changes in mean global sea level.

Contributions to global sea level rise

Over the last century global sea level has increased by an average of 1.7 mm yr^{-1} (Church and White, 2006). However, available altimeter data for 1993–2007 suggests that sea level has been rising by 3.36 ± 0.41 mm yr^{-1} over the 14-year period (Beckley et al., 2007).

Greenland is an important contributor to global sea level. The modelling study of Mitrovica et al. (2001) indicated that ~0.6 mm yr^{-1}, or approximately one third, of the observed sea level rise over the last century could be explained by increased ablation and meltwater contributions from the Greenland Ice Sheet with rising air temperatures. Recent observations suggest that the loss of ice from the coastal fringes of Greenland are contributing about $0.13 \, \text{mm yr}^{-1}$ to global sea level rise (Krabill *et al.*, 2004). This figure was produced using data from aircraft laser altimeter measurements in 1993–94 and 1998–99, along with volume-budget comparison of snow accumulation with ice discharge.

Although there has been a major warming across the Antarctic Peninsula, much of the continent has experienced little change in surface temperature, or has had a small cooling. The contribution of the continent to global sea level rise is therefore thought to be rather modest. However, the loss of ice from the West Antarctic Ice Sheet primarily via the Amundsen Sea Embayment has been estimated at 0.13 to $0.16 \, \text{mm yr}^{-1}$ over recent decades (Rignot and Thomas, 2002; Zwally *et al.*, 2005). The more recent report of Bindoff *et al.* (2007) arrived at a broadly similar contribution of 0.1–$0.2 \, \text{mm yr}^{-1}$ over the last few decades, with some evidence for a slightly larger value in the 1990s.

Contributions from high latitude and non-polar glaciers

Although the Antarctic and Greenland Ice Sheets contain a huge amount of ice, their contributions to twentieth century sea level rise are thought to be rather modest, with the greatest rise coming from ocean thermal expansion and the melting of glaciers and icecaps. Estimating the contribution from glaciers and icecaps is difficult, but various studies have addressed this problem. Dowdeswell *et al.* (1997) estimated that Arctic ice masses (excluding Greenland) contributed about $0.13 \, \text{mm yr}^{-1}$ to global sea level rise. For the 1990s, Dyurgerov (2001) estimated that non-polar glaciers accounted for around $0.4 \, \text{mm yr}^{-1}$ to the overall rise.

Regional sea level changes

Not surprisingly the change in mean, global sea level gets the most publicity when considering climate change; however, there have been large regional changes. For example, radar altimeter data for the period 1995–2003 have revealed that Arctic sea level has been falling by just over 2 mm a year in recent decades. This may be a result of changes in temperature or salinity of the ocean and more research is needed to explain this observation. But confirmation of this change has also come from in-situ observations, with Russian tide gauge data suggesting a fall in Arctic sea level during the 1990s.

6.10   Attribution of recent changes

Attribution of changes in the Arctic and Antarctic climate systems is difficult since the observational records for the high latitude areas are relatively short and the natural variability is large because of feedbacks, especially involving the cryosphere. Detecting the impact of increasing greenhouse gas concentrations and other anthropogenic factors presents a number of problems. Attribution of change is further complicated since the Arctic and Antarctic are intimately coupled to the global climate system.

Climate change in the Arctic and Antarctic is complicated by feedbacks that may affect not only the high latitudes but their interactions with the global system. Among the

332 *The instrumental period*

processes involved in these feedbacks are (a) changes in albedo, going from highly reflective ice and snow surfaces to highly absorptive open ocean and uncovered vegetation, causing additional solar energy to be absorbed and accelerating the warming trend, (b) the interaction of changes in the wind field and surface heat fluxes with changes in cloud cover, (c) the potential release to the atmosphere of large stores of carbon from land and shallow coastal ocean areas, and (d) release to the North Atlantic of additional fresh water from melting sea and land ice and increased river runoff, altering the THC. All of these feedbacks have the potential to operate on decadal time scales, and possibly to lead to abrupt change.

Nevertheless, there are three general approaches to attribution: investigation of long time series records, understanding processes, and the use of general circulation models (GCMs). We can divide causes of decadal and longer variability into external forcing, intrinsic (internal) variability of the atmosphere and ocean, and feedbacks/amplifications through land and oceanic cryospheric processes. External forcing can further be divided into natural (solar variability and volcanoes) and anthropogenic (CO_2 and SO_4).

On the scale of the Northern Hemisphere, Crowley (2000) considered the twentieth century warming within an historical context using simulations from an energy balance climate model. He stressed the importance of changes in solar irradiance and volcanism and found that 41–64% of pre-1850 decadal-scale temperature variations were due to these factors. It was found that the impact of volcanic activity was large but short lived with maximum activity in the early and late 1800s and late 1900s, and minimums in the 1700s and mid 1900s. The effect of CO_2 in the late nineteenth century was found to be an order of magnitude larger than the variability related to the solar signal.

Attempts are now being made to try and use the available climate data to identify anthropogenic influences in the two polar regions. Gillett *et al.* (2009) carried out a formal attribution study to determine whether the observed changes seen in the data were within the range of natural climate variability or whether they were a result of anthropogenic forcing. They found that recent changes were not consistent with internal climate variability or natural climate drivers alone, and were directly attributable to human influence.

6.10.1   The Arctic

Arctic climate is influenced by external forcing, intrinsic variability of the atmospheric circulation and internal feedbacks. External forcing (solar, volcanoes and CO_2) affects the absorption of radiation and is particularly effective in the subtropic stratosphere. Increased absorption raises the temperature and can increase the meridional temperature gradient, and thus increase the forcing to the atmospheric general circulation. Although the polar vortex has at least two positive feedbacks, one through dynamic cooling of the stratosphere and ozone chemistry and one through modification of vertical wind shear and gravity wave propagation (Baldwin and Dunkerton, 2001), these are difficult to discriminate with confidence because of the large amount of atmospheric variability affecting the polar regions.

Ozone loss has been both an anthropogenic influence and a dynamic response to Arctic stratospheric cooling. Decreasing trends (1979–93) in March total column ozone are congruent with the NAM trend. There are positive temperature feedbacks from surface processes in the Arctic: that is, albedo and surface flux changes through loss of sea ice and tundra. Warmer surface temperatures influence the geopotential height field in the troposphere over the Arctic and affect northward advection of heat.

It is important to consider the early/mid twentieth century warm event. A surface temperature times series, averaged over the Arctic and an extended winter period, shows a 0.7 °C warm anomaly during the 1930s–1940s, which was of the same magnitude as those in the 1990s. There is no early twentieth century warm anomaly at mid latitudes. One interpretation of this event is a low frequency oscillation with maxima in the 1930s and 1990s. Another interpretation comes from comparing the meteorological events that contribute to the anomalies. The composite signal for the earlier period is made up of a number of individual events that vary greatly in region and season with local amplitudes of 3–4 °C. Some years during the 1930s were unusual in that West Greenland and Eurasia were simultaneously warm. The temperature anomaly patterns in the 1990s were Arctic-wide and more associated with NAM-type sea level pressure patterns, in contrast to the earlier period (Johannessen *et al.*, 2004; Overland *et al.*, 2004). Temperature anomaly patterns in the 1990s appear unique in the twentieth century (Przybylak, 2003). While there is evidence that the initiation of these earlier warm periods was based on intrinsic variability, feedbacks may have had a major role in developing multi-year anomalies. There may have been a connection between stronger winds and reduction of sea ice in the Barents Sea (Bengtsson *et al.*, 2004). Sea ice losses and changes from tundra to shrub vegetation may have a 'memory' effect in the current Arctic system not unlike the 1930s–1940s.

Over approximately the second half of the twentieth century the positive trend in the NAM (Fig. 6.11) contributed to the positive surface temperature trend over Eurasia (Hurrell, 1995); likewise the positive PNA trend contributed to the positive temperature trends over the northern Rocky Mountains (Wu and Straus, 2004). Residual temperature trends not explained by the AO and NAO are present from eastern Siberia through northern Alaska to northeastern Canada (Quadrelli and Wallace, 2004). These are primary regions of sea ice and tundra changes, which suggest the operation of Arctic feedbacks or other unknown processes.

6.10.2 *The Antarctic*

For the Antarctic the major climate changes have taken place in the Antarctic Peninsula region. The summer warming that has been observed on the eastern side has been linked to the changes in the SAM, and the strengthening of the westerlies, bringing warmer air masses into the region. The changes in the SAM observed over the last 50 years are greater than any changes in the SAM found in a 1000-year control run of the Hadley Centre climate model. The SAM is known to vary for a number of reasons, including changes in the amount of stratospheric ozone and increases in greenhouse gases. From an investigation of the changes

334 *The instrumental period*

in the SAM in models forced with greenhouse gases and volcanic aerosols it has been suggested that the summer season warming in the peninsula is the first evidence of anthropogenic activity impacting the Antarctic climate.

The winter season warming has been linked to a decrease of sea ice over the Bellingshausen Sea at this time of year. However, it is still unclear exactly what has caused the reduction of sea ice, although research has suggested that the changes are driven by the atmosphere (Meredith and King, 2005).

The climate of the Antarctic Peninsula is strongly affected by tropical conditions via PSA teleconnections. Deep convection near the Equator results in upper level divergence generating a Rossby Wave train extending toward the southeastern South Pacific. During El Niño (La Niña) events there tends to be higher (lower) MSLP values over the Bellingshausen Sea giving colder (warmer) conditions over the Antarctic Peninsula. In recent decades El Niño events have been deeper and more frequent, which would have suggested colder conditions in the region, but as discussed earlier, the trend has been to warmer conditions. Clearly further research is required into tropical–Antarctic links and the reasons for change in the Antarctic Peninsula region.

Investigation of the mechanisms behind the climate changes observed in the Antarctic Peninsula region are hampered by the inability of some models to correctly reproduce the complex air–sea–ice interactions that are clearly so crucial over the eastern Bellingshausen Sea. Runs of some coupled atmosphere–ocean climate models forced by increasing greenhouse gases fail to reproduce the winter season warming over the peninsula, suggesting a failing in the representation of air–sea interactions.

Research has shown that the development of the Antarctic ozone hole has had a profound impact on the Antarctic environment, including being a major contributing factor to the slight cooling around the coast of East Antarctica, promoting the increase of sea ice around the continent and warming the eastern side of the Antarctic Peninsula through strengthening the winds over the Southern Ocean. In many ways the ozone hole has shielded the Antarctic from the effects of increasing greenhouse gas concentrations since 1980. The loss of stratospheric ozone, which has a clear anthropogenic origin, has therefore been one of the clearest indications of human impact on the Antarctic environment.

6.11 Concluding remarks

In recent decades there have been increases in near-surface temperatures, losses of sea ice and tundra, and disintegration of permafrost over large parts of the Arctic. In contrast, increases in Antarctic surface temperatures have generally been concentrated in the Antarctic Peninsula, and associated in winter with sea ice loss in the Bellingshausen–Amundsen Seas sector and with changes in the SAM during summer.

For recent decades – the 1950s to date in the Arctic and 1957 to date in the Antarctic – positive trends in large-scale atmospheric circulation represented by the NAM, SAM and the PNA patterns contribute to the long-term temperature trends. But the continuing Arctic warming in recent years of near-neutral NAM cannot be fully explained at present.

6.11 Concluding remarks 335

A synthesis of observational evidence and our present understanding leads to the conclusion that the Arctic changes of the past several decades are attributable to a combination of circulation-driven changes, probably augmented by the effect of increased greenhouse gas concentrations, and local feedback processes. While many modelling studies suggest that the increases in greenhouse gases favour shifts in the atmospheric circulation, in particular a higher frequency of occurrence for the positive NAM, we find no compelling evidence that recent variations of the circulation are solely greenhouse driven.

7

Predictions for the next 100 years

7.1 Introduction

Many of the high latitude climatic changes discussed in earlier chapters occurred because of natural climate variability, associated with fluctuations in the orbit of the Earth around the Sun, changes in the amount of solar radiation emitted by the Sun, volcanic eruptions that injected large amounts of dust into the atmosphere, changes in the ocean circulation and exchanges of heat between the ocean and atmosphere. However, from about the middle of the eighteenth century, at the start of the Industrial Revolution, humankind began to influence the climate system through the emission of increasing amounts of greenhouse gases. At first the impact was very small, but in the last decade of the nineteenth century Svante Arrhenius (Arrhenius, 1896) suggested that the increasing blanket of greenhouse gases above the Earth could raise temperatures in the troposphere. However, it was only in the second half of the twentieth century that there was widespread interest in climate change as decade on decade the mean temperature of the Earth started to increase and record high temperatures were registered with increasing frequency.

The occurrence in recent decades of high profile severe weather events, such as droughts, severe hurricanes and heat waves, resulted in major debates on the role of humans in these events. In the early years of the twenty-first century it was hard to open a newspaper or switch on a television without encountering discussions on the reasons for recent climate change, with environmentalists and global warming 'sceptics' presenting opposing views. While the scientific literature contained many articles that presented a compelling picture of a world that was progressively warming as greenhouse gases increased, governments, industry, policy makers and the general public were looking for definitive statements on the reasons for recent change and estimates of how the Earth might evolve in the future. The response of the international community was the establishment of the Intergovernmental Panel on Climate Change (IPCC) under the World Meteorological Organization and the United Nations Environment Programme. The IPCC was established in 1988 and tasked with assessing the scientific literature and other available information to understand the scientific basis of risk of human-induced climate change, its potential impacts and options for adaptation and mitigation.

The activities of IPCC were organised into three working groups concerned with (1) assessing the scientific aspects of the climate system and climate change, (2) determining the impact of climate change and adapting to it, and (3) options for mitigating climate change.

The first Assessment Report of IPCC was published in 1990, with subsequent reports coming in 1995 and 2001. The latest report, Assessment Report 4 (AR4), was issued in 2007 and its release was widely reported in the media. It provided compelling evidence that humankind *has* influenced the climate system during the last century and that our impact is increasing.

IPCC used the output from 23 state-of-the-art coupled atmosphere–ocean climate models. These were run through the twentieth century with the measured levels of greenhouse gases, volcanic aerosols and solar variability. Such experiments allowed the skill of the models in simulating past conditions to be assessed. The models were also run through the twenty-first century under different greenhouse gas emission scenarios to provide a range of climate projects that would aid policy makers when considering possible mitigation strategies. The projections were created by simply averaging the output from these models. However, not all the models have the same degree of skill at high latitudes, and a number of studies have weighted the output from these models, with the weights being based on the performance of the models in simulating the climatic variations over recent decades, the rationale being that if the models can realistically simulate past conditions then they will produce better projections of future conditions. This may not necessarily be the case, but in the following sections we will present results from the IPCC study, as well as weighted projections in order to try and provide the best estimates of how the Arctic and Antarctic climates will evolve over the next century.

In this chapter we will also refer to the climate predictions from the Arctic Climate Impact Assessment (ACIA). This was a major study concerned with investigating the possible impact of climate change on the fragile system of the Arctic. The climate projections were based on model output from the IPCC Third Assessment Report (TAR), which inevitably will not be as good as those from the AR4, considering the rapid pace of model development. However, the ACIA dealt in detail with many aspects of the Arctic climate system that could not be considered by the IPCC, which has a global remit. We therefore consider it important to deal with some of the climate projections from the ACIA in this chapter. The full ACIA report is available online at www.acia.uaf.edu/.

A third initiative that we have drawn from is the Antarctic Climate Change and the Environment (ACCE) report. This was an initiative of the Scientific Committee on Antarctic Research that aimed to provide a similar, thorough assessment of past, and potential future, climate change for the Antarctic as was carried out by ACIA for the north. The full ACCE report is available online at www.scar.org/publications/occasionals/acce.html.

7.2 Possible future greenhouse gas emission scenarios and the IPCC models

7.2.1 *Introduction*

The degree to which the climate of the Earth will change over the next century is heavily dependent on whether the recent rapid increase in greenhouse gas emissions continues or whether efforts are successful in reducing the rate of emission or actually reducing greenhouse gas levels to earlier values. Whatever happens it will take a long time for the levels of greenhouse gases to drop. For example, carbon dioxide once in the air will remain there for more than a century.

Future levels of greenhouse gas emissions will be determined by many complex factors, such as social and economic developments in different parts of the world, along with technological changes. How greenhouse gas levels will change is therefore highly uncertain. Nevertheless, in order to have emission levels for runs of general circulation models through the twenty-first century it is necessary to have a range of possible scenarios that can be prescribed. The IPCC developed emission scenarios in 1990 and 1992 that were used for the model runs that contributed towards the first two assessment reports. Following a review in 1995, it was decided to develop a new series of scenarios in light of the greater knowledge that was available on the contribution of different forms of energy to gas emissions and the rapid industrialisation that was taking place in some developing countries. A new set of scenarios was therefore developed and published in 2000.

7.2.2 The IPCC greenhouse gas emission scenarios

The scenarios developed in 2000 were based, in IPCC terminology, on four storylines, which are narrative descriptions of families of scenarios that highlight the main characteristics and dynamics, and the relationships between key driving forces. The four storyline and scenario families as described in detail in Nakicenovic *et al.* (2000) (available online at: www.grida.no/climate/ipcc/emission/index.htm) are:

- A1: a future world of very rapid economic growth, global population that peaks in the middle of the twenty-first century and declines thereafter, and rapid introduction of new and more efficient technologies.
- A2: a very heterogeneous world with continuously increasing global population and regionally oriented economic growth that is more fragmented and slower than in other storylines.
- B1: a convergent world with the same global population as in the A1 storyline but with rapid changes in economic structures towards a service and information economy, with reductions in material intensity, and the introduction of clean and resource-efficient technologies.
- B2: a world in which the emphasis is on local solutions to economic, social, and environmental sustainability, with continuously increasing population (lower than A2) and intermediate economic development.

Each storyline was then developed to include quantitative projections of major driving variables, such as possible population change and economic development. This resulted in a total of 40 scenarios, each of which was considered equally valid with no assigned probability. Six groups of scenarios were drawn from the four families; one group each in

7.2 Future greenhouse gas scenarios and models

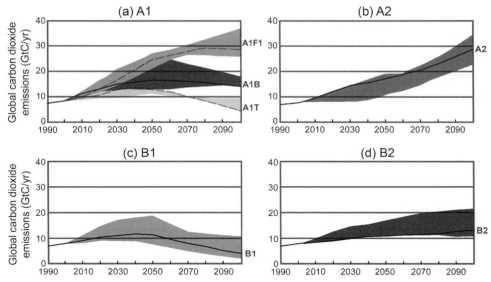

Fig. 7.1 Total global annual CO_2 emissions from all sources (energy, industry, and land-use change) from 1990 to 2100 (in gigatonnes of carbon per year) for the families and six scenario groups. The 40 SRES scenarios are represented by the four families (A1, A2, B1, and B2) and six scenario groups: (a) the fossil-intensive A1FI (comprising the high-coal and high-oil-and-gas scenarios), the predominantly non-fossil fuel A1T, and the balanced A1B; (b) A2; (c) B1, and (d) B2. Each grey-shaded emission band shows the range of harmonised and non-harmonised scenarios within each group. For each of the six scenario groups an illustrative scenario is provided, including the four illustrative marker scenarios (A1, A2, B1, B2, solid lines) and two illustrative scenarios for A1FI and A1T (dashed lines).

the A2, B1 and B2 families, and three groups in the A1 family. These characterised alternative developments of energy technology as A1FI (fossil intensive), A1T (predominantly non-fossil) and A1B (balanced across energy sources). Scenario A1B (often referred to as SRESA1B) is frequently taken as a likely emission scenario for the coming century, based on the observed increase in greenhouse gases over the first decade of the twenty-first century. Climate projections based on this scenario of greenhouse gas increases will therefore be presented frequently in the following. The A1B group is a representative scenario that postulates a balance between fossil-intensive and non-fossil energy sources.

Figure 7.1 illustrates how total global annual CO_2 emissions from all sources would change during the twenty-first century for the four emission families and the six scenario groups. The figure shows that the greatest emissions would be with the A1F1 scenario group, which includes high coal and high oil and gas scenarios. With this group the CO_2 emissions by the end of the century would have risen from the 1990 value of about 8 Gt yr^{-1} to be in the range

340 *Predictions for the next 100 years*

28–37 Gt yr^{-1}. The frequently used A1B scenario has emissions increasing to a peak around the middle of the century and then decreasing to about 14 Gt yr^{-1} by 2100.

With the other three scenario groups, A2 anticipates a steady increase of CO_2 emissions during the century, with values peaking around 30 Gt yr^{-1} in 2100. B1 suggests increasing emissions until around 2050, and then a decrease to about 4 Gt yr^{-1}, which is less than the 1990 level. B2 has a modest increase throughout the century to about 12 Gt yr^{-1} by 2100.

The ACIA made extensive use of model projections from the IPCC TAR, with the focus being on output from the runs with the A2 and B2 emission scenarios.

7.2.3 *The climate models used to produce the twenty-first century projections*

The IPCC AR4 was a major initiative to understand past climatic change and to produce the best possible projections of how the climate of the Earth may change over the next century. Output from the various climate models are available online and form the World Climate Research Programme's Coupled Model Inter-comparison Project phase 3 (CMIP3) multi model data set. While this is an extremely valuable resource for the study of climate it should be noted that not all models were run for all the scenarios and not all parameters were made available. When considering various aspects of future climate we will therefore use a subset of the CMIP3 models.

The basic set of experiments carried out as part of AR4 consisted of pre-industrial control runs, simulations of the present-day climate, climate of the twentieth century runs and twenty-first century runs with various increases of greenhouse gases. In the rest of this chapter we present mean simulations from the AR4 models, along with weighted averages for the Antarctic based on their ability to simulate recent climate change and variability (Connolley and Bracegirdle, 2007).

Studies have suggested that by weighting individual members of a model ensemble of future predictions according to their performance in reproducing the climate of recent decades, the spread of projections can be reduced when compared with statistics without model weighting (Murphy *et al.*, 2004).

The ACIA published their report in 2005 and based their findings on the output of a subset of five models from the IPCC TAR.

7.3 Changes in the atmospheric circulation and the modes of climate variability

7.3.1 *Introduction*

As discussed in the following sections, over the next century, if greenhouse gas concentrations continue to rise, the Earth as a whole is expected to experience a near-surface warming of several degrees, with the greatest temperature rises being in the two polar regions. The Equator to pole temperature difference will therefore decrease, although the greater amount of energy in the lower atmosphere will result in more vigorous weather systems. Climate models indicate that there will be changes to the broadscale atmospheric circulation with a poleward shift in the extratropical storm tracks, with subsequent changes in precipitation, temperature and wind speed and direction.

In recent decades there has been a great deal of research into the modes of variability of the climate system (see Section 3.4.5). We now know that the primary mode of variability in the Northern and Southern Hemispheres is an oscillation of mass between the high and mid latitudes, called the Northern Annular Mode (NAM) and the Southern Annular Mode (SAM). Both these modes can vary from the positive phase, where there is a decrease (increase) of atmospheric mass at high (mid) latitudes, to the negative phase where the mass change is in the opposite direction. In recent decades the NAM and the SAM have both shifted into their positive phases, which has resulted in a strengthening of the mid latitude westerlies. As discussed elsewhere, a number of natural and anthropogenic factors can cause the annular modes to move into their positive phase, but one important factor is increasing levels of greenhouse gases. Modelling studies suggest that the shift in the annular modes will largely be a result of increasing CO_2 concentrations (Rauthe *et al.*, 2004; Kuzmina *et al.*, 2005). If concentrations of greenhouse gases continue to increase in the future we can therefore expect to see the NAM and SAM be more in their positive phases, although because of natural factors there will be periods when the indices will decrease.

7.3.2 The Arctic

As discussed later, climate models suggest that the largest warming on Earth over the next century will be at high northern latitudes. This, and the accompanying large cryospheric changes, will result in major alterations to the atmospheric circulation.

The IPCC AR4 models when run with the mid-range greenhouse gas emission scenario SRESA1B, suggest that there will be a large relative decrease in the occurrence of strong high-pressure patterns, which are a feature of the Arctic Ocean. This was identified in a study that used the technique of Self Organising Maps, which is a neural network algorithm that uses an unsupervised learning process to find generalised patterns in data (Cassano *et al.*, 2006). The ACIA also projected a small drop in MSLP (mean sea level pressure) throughout the year across the Arctic over the next century. The greatest drop in pressure was predicted to be in the winter, but even then it was expected to be only of the order of 4 hPa over the central Arctic.

The fall in pressure at high northern latitudes is expected to be accompanied by a large increase in strong Icelandic Low patterns, and an extended storm track into the eastern Arctic Basin giving lower pressures across the Arctic. This scenario is predicted for both summer and winter seasons.

This circulation change is consistent with the expected shift of the NAM and the North Atlantic Oscillation (NAO) into their positive phases. Miller *et al.* (2006) carried out a very comprehensive study of the simulation of the NAM and SAM in 14 of the IPCC AR4 models and considered how the modes would change over the coming century if greenhouse gas concentrations increase according to the SRESA1B emission scenario. Figure 7.2 shows the multimodel average of the annular principal component for winter during the period

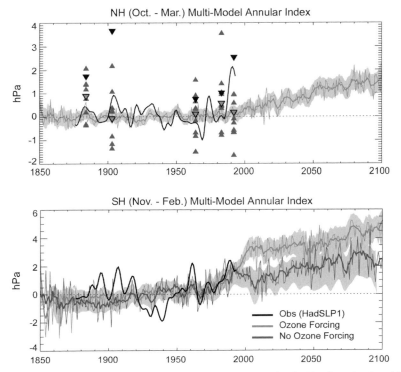

Fig. 7.2 Multimodel average of the annular principal component for the Northern (top) and Southern (bottom) Hemispheres from the IPCC AR4 models. A 10-year low-pass filtered version is shown by the thick, light grey line. (From Miller *et al.*, 2006; see paper for more details on the symbols.)

from 1850 to 2100. In the mean of the model simulation the mode began to increase about 1980 and is predicted to increase linearly until about 2080, after which it is almost constant. The grey shading indicates the variability within the multimodel ensemble, and shows that the positive trend is a robust feature. This result is consistent with the Cassano *et al.* (2006) study, which found that for both summer and winter the largest changes in the synoptic circulation patterns were found to occur during the first half of the twenty-first century, with similar signs but smaller magnitude trends during the second half of the century.

7.3.3 The Antarctic

Over the last couple of decades the SAM has shifted predominantly into its positive phase in austral summer and autumn, with the decrease of stratospheric ozone since the late 1970s being the most important factor in driving this change (Arblaster and Meehl, 2006). Most AR4 GCMs predict the SAM to be even more predominantly positive in future

7.3 Modes of atmospheric circulation variability

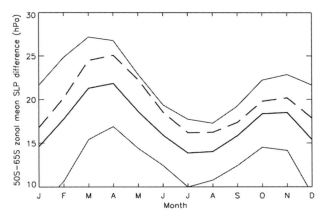

Fig. 7.3 Difference between zonal mean sea level pressure (SLP) at 50° S and 65° S. (a) Mean for 2004–23 (solid line) and 2080–99 (dashed line). (From Bracegirdle *et al.*, 2008.)

across all four seasons in response to increases of greenhouse gas concentrations (Shindell and Schmidt, 2004). However, the Montreal Protocol outlawing CFCs is now in force and it is expected that stratospheric ozone will return to pre ozone hole levels before the end of this century. Chemistry climate models, with fully interactive stratospheric chemistry, have shown that this predicted ozone recovery will lead to a less positive SAM in austral summer despite increasing greenhouse gases (Perlwitz *et al.*, 2008; Son *et al.*, 2008).

Attempts have also been made to quantify the expected changes in the SAM. The study of Bracegirdle *et al.* (2008), using weighted AR4 model data, found that during the twenty-first century the seasonal model ensemble mean changes of the SAM (defined as the zonally averaged difference between MSLP at 40° S and 65° S) were predicted to be 2.7, 3.8, 4.8 and 3.1 hPA for austral summer, autumn, winter and spring, respectively. However, given the discussion above, the summer value is actually likely to be negative.

The semi-annual oscillation is also a very important climatic cycle at high southern latitudes that affects many aspects of the climate in the Antarctic coastal zone (see Section 3.5.2). Various changes in the SAO have been noted during the instrumental period, with these being strongly linked to variations in the meridional gradients of MSLP and 500 hPa temperature.

During the twenty-first century changes can also be expected in the SAO as a result of alterations in pressure across the Southern Hemisphere. Bracegirdle *et al.* (2008) examined the weighted ensemble average MSLP and 500 hPa temperature changes in the IPCC AR4 models and considered how the SAO indices might change through the twenty-first century. Figure 7.3 shows the difference between the zonal mean SLP at 50° S and 65° S for 2004–23 and 2080–99. It can be seen that the pressure difference between these latitudes is greater during the second half of the twenty-first century compared with the first and that the largest increase is during April. The magnitude of the changes of the

500 hPa temperature SAO index in autumn is similar to the changes that were associated with a decreased amplitude of the SAO in the late 1970s (Meehl *et al.*, 1998), but with a different annual variation. The annual cycle of the changes of both the 500 hPa and MSLP gradients show a more pronounced strengthening of the autumn peak of the SAO compared with the spring peak. This is reflected in the near-surface winds as a strong autumn contraction and intensification of the circumpolar westerlies around Antarctica.

7.4 The main meteorological elements

7.4.1 Introduction

In the absence of any anthropogenic factors, we could expect climatic fluctuations over the coming century based on our knowledge of climate cycles observed during the Pleistocene and Holocene. For example, Campbell *et al.* (1998) used our understanding of periodicities, such as the 1500-year and 23 000-year cycles, to predict that even in the absence of anthropogenic climate forcing, there would be a continued warming until AD 2400. However, the increases in greenhouse gases as quantified in the IPCC emission scenarios will result in climatic forcing that is far in excess of that from any natural factors.

7.4.2 Projected global changes

The confidence in the projections of future global changes is higher than the regional changes, so we will start by considering the mean, global changes that we expect over the next century.

The IPCC AR4 has produced projections of global near-surface temperature change and changes in precipitation for a number of greenhouse gas emission scenarios. Across the range of emission scenarios, the temperatures in 2090–99 compared with 1980–99 are projected to increase by a range covering 0.3–6.4 °C. In terms of the best estimates, the range is 0.6–4.0 °C. If greenhouse gas concentrations are kept at 2000 levels the likely range of increases is 0.3–0.9 °C. However, the range of increases for the frequently quoted SRESA1B scenario is 1.7–4.4 °C, increasing to 2.4–6.4 °C for SRESA1F1.

Globally, AR4 suggests that the warming over the next century will not be uniform across the Earth, but greatest over land and at most high northern latitudes. It is also projected that there will be an increase in precipitation in both polar regions, which is consistent with the air masses being able to hold more water vapour at higher temperatures and subsequently releasing it as they move polewards and are cooled.

On longer timescales, if radiative forcing were to be stabilised in 2100 at SRESA1B levels there would still be a further increase in global average temperature of about 0.5 °C, mostly by 2200.

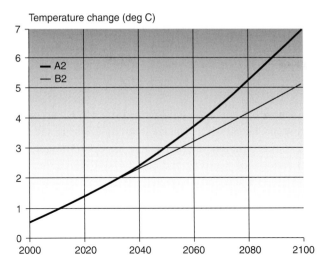

Fig. 7.4 Projected Arctic temperature increases for the A2 and B2 emission scenarios. (From ACIA, 2005.)

7.4.3 The Arctic

Temperature

In the early decades of the twenty-first century it may be difficult to separate natural climate variability from an anthropogenic signal in some parts of the Arctic because of the large, natural climatic fluctuations (ACIA, 2005). For example, an analysis of past warming trends based on tree ring data from Fennoscandia suggests that any summer warming that occurs as a result of increasing levels of greenhouse gases may not be detectable in this region until after 2030 (Briffa et al., 1990). However, even at the start of this period there are major changes taking place across the Arctic, and particularly the Arctic Ocean, that suggest that there will be a strong signal of anthropogenic climate change in high northern latitudes.

The ACIA estimated that the region north of 60° N will experience a temperature increase by the middle of the twenty-first century of 2.5 °C (with both the A2 and B2 scenarios), increasing to 7 °C (A2) and 5 °C (B2) by the end of the century (Fig. 7.4).

The temperature increases with the two different emission scenarios are very similar up to about 2040, but then diverge later in the century. For the period 2071–90 the five models used by ACIA suggested that Arctic temperatures would have risen by 3.7 °C (with a range of 2.8–4.6 °C for the five models) when compared with a 1981–2000 baseline. This is twice the temperature increase expected for the Earth as a whole.

The central Arctic is expected to increase in temperature by 5 °C by 2071–90. By the same time, the models run with the B2 scenario predicted temperature increases of about 3 °C for Scandinavia and East Greenland, around 2 °C for Iceland and up to 5 °C for the Canadian Archipelago and Russian Arctic.

346 *Predictions for the next 100 years*

Across the Arctic Ocean the greatest warming is expected to be in the autumn and winter, with a temperature increase of up to 9 °C by 2071–90 with the B2 scenario. The magnitude of the warming at this time of year is primarily because of the decrease in sea ice extent and thickness, coupled with the delay in the formation of sea ice and the initiation of snow cover. The albedo feedback associated with retreating sea ice dominates the signal of high latitude climate warming.

An apparent paradox is that while the long-range projections favour changes in autumn and winter over the oceans, recent Arctic change is strong in spring and larger over land. For this 'emerging greenhouse' period, there is large model-to-model and intra-ensemble variability in spatial patterns of warming and cooling, not unlike the spatial variability of recent temperature trends (Wang and Overland, 2005; Serreze and Francis, 2006). The output for these more proximate periods suggests both NAM/Pacific North American (PNA) pattern and a residual response of unknown origin. For the next decades, it is unclear in what proportion future changes will be driven by natural/intrinsic variations, unique Arctic feedbacks, or anthropogenic sources.

In contrast to the other seasons, the summer temperatures across the Arctic Ocean are not expected to increase by more than 1 °C at this time of year because of the presence of melting ice. This pattern of greatest warming during the autumn/winter and a smaller increase during the summer is also expected across the surrounding land areas, but to a lesser degree than at higher latitudes. In fact the summer warming over northern Eurasia and northern North America is expected to be greater than over the Arctic Ocean, with the reverse being the case in winter.

Although large temperature increases are expected in the Arctic, both the ACIA and AR4 predicted that parts of the North Atlantic Ocean will experience some of the smallest temperature increases. This is because the area forms the sinking branch of the Atlantic meridional overturning circulation (MOC) and the ocean is well mixed to a great depth, so that much of the surface heat is carried to deeper layers. Even though the Atlantic MOC is expected to weaken over the next century, with less transport of heat polewards, there is still expected to be a warming across the North Atlantic–Nordic Seas.

The changes in atmospheric circulation discussed earlier imply an extended North Atlantic storm track in summer, which acts to transport warm air into Eurasia, and a strong Aleutian Low that advects warm air into northwestern North America. We can therefore expect warmer conditions in these two areas. In winter we can expect warm anomalies north of Scandinavia, extending into the eastern Arctic Basin and cold anomalies over much of Alaska and portions of northwestern Canada. This tendency towards colder conditions in Alaska is in contrast to the large warming observed there in recent decades. This pattern of temperature changes is consistent with the temperature advection patterns associated with an MSLP pattern dominated by low pressure over the Arctic Ocean.

Precipitation

With the greater penetration of cyclonic systems into the Arctic and the higher air temperatures, we can expect an increase in precipitation across the high latitude areas of the Northern Hemisphere.

7.4 The main meteorological elements

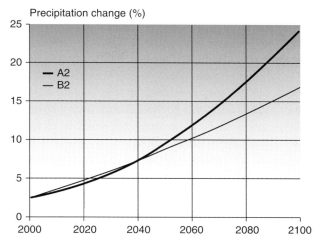

Fig. 7.5 Projected percentage changes in mean annual precipitation in the Arctic (60–90° N) for the A2 and B2 emission scenarios, relative to 1981–2000. (From ACIA, 2005.)

The ACIA five-model mean projected change in mean annual precipitation through the twenty-first century with the A2 and B2 emission scenarios (Fig. 7.5) has a similar form to that of temperature. The differences between precipitation with the two emission scenarios being small during the first half of the century and diverging after that time. Relative to the 1981–2000 reference period, by 2050 the Arctic is expected to receive about 8% more precipitation, increasing to 17% and 24% by the end of the century with the B2 and A2 emission scenarios respectively. As with temperature, the greatest increase of precipitation is expected to be during the autumn and winter, with the least in the summer.

On a more regional scale, the circulation patterns associated with a more pronounced Aleutian Low are expected to give positive precipitation anomalies during summer along the Canadian west coast and in southeast Alaska (Cassano *et al.*, 2006). Further east, the circulation changes associated with a strong and/or extended Icelandic Low are expected to give greater precipitation along the east coast of Greenland, and from the North Atlantic extending into Scandinavia. ACIA also estimated that the Arctic Ocean and the terrestrial Arctic regions of North America and Eurasia will experience the greatest increase in precipitation. The five models used by ACIA suggest that by 2071–90 the annual mean precipitation could increase from 5–10% in the Atlantic sector of the Arctic to 35% locally in parts of the high Arctic.

The Self Organising Maps study of Cassano *et al.* (2006) based on AR4 fields considered changes in winter season precipitation across the Arctic. Their work suggested positive precipitation anomalies over Alaska, the Greenland Ice Sheet, and western Scandinavia during the twenty-first century. However, the precipitation anomaly patterns were found to be less spatially coherent than the temperature anomalies and more sensitive to small changes in the details of the circulation patterns.

348 *Predictions for the next 100 years*

Due to the cold temperatures, evaporation rates are rather small across the Arctic, but are nevertheless very important for the terrestrial environment. Changes in evaporation over the next century are difficult to estimate; however, with higher air temperatures the atmosphere has a greater capacity to hold moisture and the general expectation is that evaporation will increase, although the rates will remain relatively small. But with the models examined by the ACIA, for every region that they considered at least one model projected a decrease. However, overall most models indicate a positive change over the coming century.

Precipitation − evaporation (P–E) is an important quantity at high latitudes since it constitutes the net moisture input to the terrestrial system, which is important for ecosystems. In the ACIA study all the models used bar one had an increase of P–E over the next century. The greatest increase of 14% was found over the Arctic Ocean, with smaller increases of 6–12% over the terrestrial watersheds. These changes were interpreted as a likely increase in the frequency and/or duration of wet periods. However, it was noted that the changes in P–E were expected to be smaller over the river basins during the warm season.

Cloud cover

The Arctic is a very cloudy environment as a result of moist air masses being advected poleward and gradually cooled through contact with the cold snow and ice surfaces. As discussed above, air temperatures are expected to be higher over the coming decades, which may suppress the cooling of air masses. Yet there will also be less sea ice, so increasing the flux of moisture into the atmosphere. Such opposing influences on cloud formation make it difficult to predict how cloud amounts will change over the coming century. However, the ACIA estimated that there will be a small increase in cloud cover across the Arctic by the end of the twenty-first century. Their five-model study indicated that by 2071–90 mean annual cloud cover would have increased, with a decrease in shortwave radiation received at the ground. By 2071–90 the flux was projected to have decreased by $10 \, \mathrm{W \, m^{-2}}$ compared with 1981–2000.

The IPCC Fourth Assessment also suggested that there would be an increase in Arctic cloud. Their multimodel assessment of future climate change with the SRESA1B scenario predicts that there will be an increase in annual mean Arctic cloud cover of up to 5% by 2100. The largest change was projected to be over the central Arctic Ocean, with the reduction in sea ice during the summer and autumn months allowing greater fluxes of moisture into the atmosphere. The increases are also expected to be large over the land areas, with a small reduction over the northern North Atlantic.

7.4.4   *The Antarctic*

Temperature

Over the last 50 years the largest near-surface temperature changes have taken place over the Antarctic Peninsula, where annual mean temperatures on the western side have increased by

7.4 The main meteorological elements

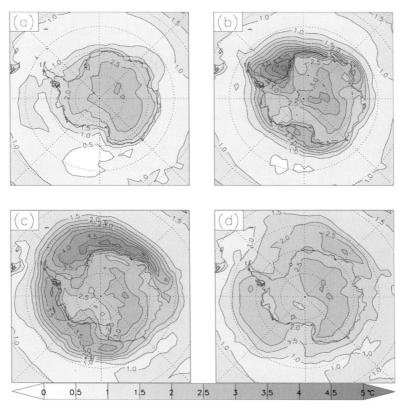

Fig. 7.6 Predicted twenty-first century near-surface (10 m) temperature change for (a) DJF, (b) MAM, (c) JJA and (d) SON using a weighted selection of the IPCC AR4 models and the SRESA1B emission scenario. Difference between 2080–99 mean and 2004–23 mean. (From Bracegirdle et al., 2008.)

about 3 °C (Turner et al., 2005). However, over the twenty-first century a much more general warming across the continent and Southern Ocean is expected (Carril et al., 2005). Chapman and Walsh (2007) found this to be the case in an analysis of the output of 11 of the GCMs used in the AR4 study. Bracegirdle et al. (2008), using a weighted selection of the IPCC AR4 models, showed that the surface warming averaged over the continent was projected to be 0.34 °C decade^{-1} with an intermodel standard deviation of 0.10 °C decade^{-1}. This warming is approximately the same magnitude as that over other land regions around the globe.

The temperature trends for the four seasons (Fig. 7.6) suggest that the largest warming will be across the high latitude areas of the Southern Ocean in winter as a result of the loss of sea ice and the greater fluxes of heat into the atmosphere. Here the models suggest a temperature increase in excess of 0.5 °C decade^{-1} close to the Greenwich Meridian, although great care must be taken when interpreting the model results at the regional level.

However, temperature increases of more than 0.3 °C decade^{-1} are predicted around much of the continent during winter. Further to the north over the ice-free Southern Ocean the IPCC AR4 predicted that there would be some of the smallest surface temperature increases over the twenty-first century, indicating the complexity of change that is anticipated over the coming decades.

A feature of the projections for all four seasons is the increase in temperature over the interior of the continent. Here temperatures are projected to rise by more than 0.25 °C decade^{-1} throughout the year. The reasons for this are not fully understood at present, but may be related to greater downward longwave radiation or cloud cover.

Some doubt must be cast on the regional-scale predictions for climate change over the coming century since the models have difficulty in reproducing the observed changes over the last 50 years in some parts of the continent. For example, the AR4 models are not able to replicate the large warming on the western side of the Antarctic Peninsula during winter. Here models cannot help us in understanding how temperatures will change over the coming decades. However, it is difficult to imagine that the very rapid increases of temperature observed across the peninsula during the second half of the twentieth century will continue. The rapid rises of winter temperature have been associated with a dramatic decrease of sea ice just to the west of the observing stations. Once the sea ice has retreated to a more southerly location, then the winter season temperatures should stabilise at a higher level. But until we understand whether the changes here have been a result of natural climate variability or have an anthropogenic origin it is not possible to predict how conditions here will change with any degree of confidence.

The large summer warming that has occurred on the eastern side of the Antarctic Peninsula in recent decades is better understood (see Section 6.2.3), and has been linked to the shift of the SAM into its positive phase. As discussed in Section 7.3.3, chemistry climate models indicate that the summer SAM will become less positive in response to ozone recovery so this regional warming may reverse during the twenty-first century.

In terms of upper air conditions, we expect the temperature increases over the next century at the 500 hPa level to be slightly less than those at the surface, with the model ensemble annual mean increase being 0.28 °C decade^{-1} (Bracegirdle *et al.*, 2008).

When run over the second half of the twentieth century the IPCC models showed no evidence of a mid-tropospheric peak in warming, such as that reported by Turner *et al.* (2006a) based on the last 30 years of radiosonde data. However, this warming has been attributed, at least in part, to an increase in polar stratospheric cloud, as a result of a cooling stratosphere in winter. This is turn has been a result of increasing levels of greenhouse gases and the ozone hole. While we can expect a recovery of stratospheric ozone levels during the coming century, we anticipate that greenhouse gas levels will continue to increase. Until we can realistically reproduce the recent mid-tropospheric increase in winter-season temperatures above the Antarctic we cannot predict how they will evolve over the next century.

7.4 The main meteorological elements 351

Precipitation

As temperatures rise across the Antarctic over the next century so the air will be able to hold a greater amount of moisture. Since air masses are advected southwards and forced up on to the Antarctic Plateau by depressions over the Southern Ocean, we can expect greater amounts of precipitation, especially in the coastal region. The expected poleward shift of the mid latitude storm track in the Southern Hemisphere will also contribute to greater precipitation across the Antarctic over the twenty-first century (Lambert and Fyfe, 2006; Lynch *et al.*, 2006).

The model projections can help us in quantifying the expected precipitation changes, and the analysis of the weighted IPCC projections (Bracegirdle *et al.*, 2008) predict that there will be an increase of net precipitation averaged over the continent of $2.9 \pm 1.2\,\mathrm{mm}\,\mathrm{yr}^{-1}$ decade^{-1} by 2100. This represents an increase of 20% over current amounts.

Predicting the spatial and seasonal variation of precipitation is very difficult since the IPCC models have a wide range of trends. However, the IPCC models predict that the largest absolute increases of precipitation will be near to the coast and increases relative to current climate are largest (up to 25%) over the interior region of East Antarctica.

Over the Southern Ocean south of $60°\,$S the annual precipitation accumulation is estimated to increase at $10.2 \pm 3.8\,\mathrm{mm}\,\mathrm{yr}^{-1}$ decade^{-1} based on the weighted output of the AR4 models. The largest trend is in the autumn, which is consistent with the largest expected increase in the SAO during that season and the associated contraction and strengthening of the storm track.

The wind field

The changes expected in the near surface winds over the Antarctic and the Southern Ocean will be considered via the differences in the monthly-mean wind vector magnitude between 2080–99 and 2004–23. It should be noted that changes of magnitude are nearly equivalent to changes of the mean wind speed if the directional constancy (ratio of the average wind vector to the average wind speed) is large. Over the steep Antarctic coastal region the directional constancy is very large, but over the Southern Ocean the directional constancy is small and changes of the mean wind speed will in general differ from monthly-mean wind vector magnitude.

Figure 7.7 shows these differences for summer and winter as determined from 15 weighted IPCC AR4 projections with the SRESA1B greenhouse gas emission scenario (Bracegirdle *et al.*, 2008). The simulations reveal an increase in the near-surface wind vector magnitude with a peak near $60°\,$S, where the annual zonal mean circumpolar westerlies are predicted to increase by $0.73 \pm 0.36\,\mathrm{m}\,\mathrm{s}^{-1}$ ($21 \pm 10\%$) over the twenty-first century. The increase is greatest in the autumn, with a projected zonal mean increase of 27%. Note that the results of the chemistry–climate model runs indicate that the annual value is probably too high as the summer westerlies (Fig. 7.7a) are likely to actually decrease in response to greater stratospheric ozone concentrations.

352 *Predictions for the next 100 years*

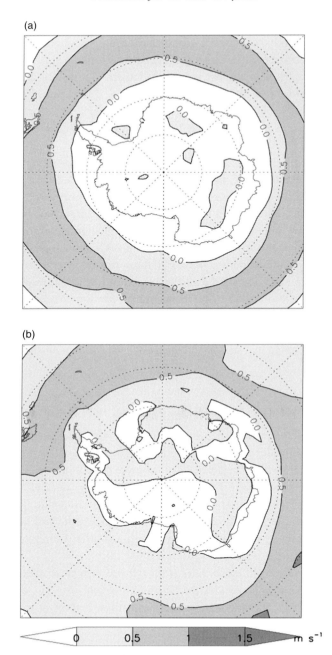

Fig. 7.7 Twenty-first-century near-surface (10 m) mean wind vector magnitude change for (a) DJF and (b) JJA. Difference between 2080–99 mean and 2004–23 mean. Calculated from monthly mean wind vectors. (From Bracegirdle et al., 2008.)

The poleward shift and increase of the westerly wind component results in a decrease of the coastal easterlies around Antarctica in summer and autumn of about 10–15%, but these do not penetrate sufficiently far south to influence coastal Antarctica in winter and spring

The average wind vector magnitude trend over the interior of Antarctica is close to zero, with cancellation between regions of increase and decrease. The increases (up to 10% in autumn) consistently occur in a region that extends over Queen Mary Land and Wilkes Land.

Cloud cover

As with the Arctic, over the next century there is expected to be an increase in cloud cover over the Southern Ocean and the Antarctic continent. However, the projected increases as produced by the AR4 models with the SRESA1B emission scenario are smaller than in the north and expected to be less than about 2–3%. Greater cloud cover is expected south of about 50° S, since there will be less sea ice and greater fluxes of heat and moisture into the atmosphere. However, north of this latitude the models suggest a reduction in cloud amount.

7.5 The ocean circulation and water masses

7.5.1 Introduction

As with estimating how the atmospheric conditions will evolve over the next century, the only tool that we have available to consider future oceanic conditions are climate models. However, it is much more difficult to validate the ocean models when run through the twentieth century because of the lack of in-situ data. Climate model projections of oceanic conditions are therefore more unreliable than those for the atmosphere.

Climate models suggest that over the coming century atmospheric temperatures will increase over the high latitude areas, where there will also be an increase in precipitation. These factors will result in the high latitude surface waters becoming less dense, which will increase their stability resulting in less vertical convection. But as discussed in the following sections, the model projections are quite different for the two polar regions.

As atmospheric CO_2 concentrations increase this will lead to greater acidification of the ocean. Projections based on the SRES scenarios give reductions in average global surface ocean pH[16] of between 0.14 and 0.35 units by the end of the twenty-first century.

7.5.2 The Arctic

Palaeoclimate records indicate the importance of the Atlantic MOC in controlling rapid climatic fluctuations. Over the coming century climate models project that there will be an increase in precipitation at high northern latitudes and a freshening of the Arctic Ocean, with a reduction of North Atlantic Deep Water (NADW) formation. Experiments in freshening the ocean with a range of ocean and climate models have suggested that between 0.06 and 0.15 Sv of additional fresh water entering the northern Atlantic could stop the formation of NADW, with consequent changes to the MOC. The export of fresh water is closely linked to the phase of the NAO and it has been suggested that if the cycle returns to its positive phase there could be a large transfer of fresh water out of the Arctic Ocean.

These factors, coupled with the input of fresh water from melt at the edges of Greenland, could have an impact on the strength of the Atlantic MOC. Modelling studies predict that over the next century there could be a weakening of the MOC, although the models project a wide range of changes, with some estimates indicating a weakening from zero to 30–50% (Cubasch *et al.*, 2001).

With a weaker MOC there will be less transport of warm water from subtropical latitudes to the northern North Atlantic. To date, no coupled climate model has projected a shutdown of the Atlantic MOC by 2100; however, the current generation of models do not include freshwater runoff from the ice sheets and glaciers so the impact of future climatic changes on the MOC may be greater than projected at present. However, model runs forced with large injections of fresh water corresponding to several times the present flux are needed to significantly alter the strength of the MOC.

Schmittner *et al.* (2005) applied a weighting scheme to the AR4 projects to try and derive improved estimates of the MOC changes over the next century. The weights were computed based on the skill of each model in simulating hydrographic properties. They considered 28 projections from nine coupled global climate models that used the SRESA1B emission scenario. The models predicted a weakening of the North Atlantic THC by 25% (± 25%) over the coming century.

Major freshwater injection events, such as occurred at 8.2 kyr BP in the North Atlantic are very unlikely over the next 100 years since the current ice masses are now much smaller or are more stable than the Laurentide Ice Sheet at the time of the release of water. However, it is important to understand the possible implications of even a small increase in freshwater flux into the ocean. For example, if there is an increase in global near-surface temperature of 3 °C in response to a doubling of atmospheric CO_2 with a twofold amplification in the polar regions, the total freshwater flux from the Greenland Ice Sheet has been predicted to increase by about 0.02 Sv (Warrick *et al.*, 1995). In addition, greater precipitation and melting of sea ice in response to anthropogenically induced temperature rise might cause an increase in freshwater flux of similar magnitude into the North Atlantic (Manabe and Stouffer, 1994). But this may result in a negative feedback on the climate of the North Atlantic region.

As the climate changes during the twenty-first century we can also expect changes in the discharge of fresh water into the Arctic Ocean from the surrounding land areas. A relationship has developed between the increase in global surface air temperature and river discharge into the Arctic Ocean (Peterson *et al.*, 2002). With estimates of a warming of 1.4–5.8 °C by 2100, these translate into discharge increases of 315–1260 km^3 (0.01 to 0.04 Sv) per year for the six largest Eurasian rivers. This would represent an 18–70% increase in Eurasian Arctic river discharge. An early assessment of the change in discharge based on a doubling CO_2 experiment suggested that it could increase from the large Eurasian and North American rivers by 10–45% (Miller and Russell, 1992). Arnell (1999) projected discharge increases of 3–10% to 30–40% for the largest rivers in the Arctic based on the output of six GCMs that gave data for 2050. ACIA projected that if CO_2 concentration doubles over the next century then the total annual discharge into the Arctic Ocean from Arctic land areas would increase by 10–20%. The change in discharge over the year is expected to vary, with winter seeing an increase as high as 50–80%. Of the order of 55–60% of the annual discharge is expected to enter the ocean during the peak runoff season of April–July.

7.5.3 The Antarctic

There have been fewer studies of the projected changes in the Southern Ocean circulation than for the high latitude ocean areas of the Northern Hemisphere. However, the studies carried out have suggested with the projected strengthening and poleward shift of the circumpolar westerlies there will be a consequent strengthening, poleward shift and narrowing of the Antarctic Circumpolar Current (Fyfe and Saenko, 2006). There will also be an enhanced Equatorward surface Ekman transport as a result of the stronger westerlies, balanced by an enhanced deep geostrophic poleward return flow below 2000 m.

Antarctic Bottom Water (AABW) is the densest water in the world's oceans and it plays a crucial role in the THC circulating heat, nutrients and gases throughout the oceans worldwide. The water mass forms as salt is expelling as water freezes around the Antarctic. As discussed in Section 7.6.3, if greenhouse gas levels continue to rise we expect a marked reduction in the amount of sea ice around the continent. We may therefore see a reduction in the amount of AABW formed, although it is difficult to estimate the magnitude of any such change since the current generation of climate models do not have realistic representations of this water mass.

Over recent decades near-surface temperatures across most of the Antarctic have changed much less than in other parts of the world, leading to discussion about a delayed response in the Antarctic to global climate change. Models predict that the atmosphere will warm markedly over the next century, but several studies have suggested that the response of the Southern Ocean to an increase in atmospheric greenhouse gas concentrations may be delayed compared with other regions (Bi *et al.*, 2001; Goosse and Renssen 2001, 2005), mainly because of the large thermal inertia of the Southern Ocean.

7.5.4 The carbon cycle

There are still many questions regarding how the different elements of the carbon cycle will change in the future. The ocean is a very important sink, currently sequestering about half of

356   *Predictions for the next 100 years*

the anthropogenic greenhouse gas emissions. However, the future is not clear because of possible changes in ocean circulation, biogeochemical cycling, and ecosystem dynamics.

The Southern Ocean is presently a very effective CO_2 sink since the cold ocean temperatures favour the uptake of carbon dioxide from the atmosphere. However, warmer temperatures can cause the ocean surface to release CO_2, but over the next 100 years the Southern Ocean is expected to experience some of the smallest temperature increases on Earth. By contrast, it has recently been suggested that the efficiency of the Southern Ocean in absorbing CO_2 may be becoming reduced as the winds around the continent increase as the SAM shifts predominantly into its positive phase. With the exception of summer, we expect the SAM to become more positive as greenhouse gas concentrations rise, so there may be a further reduction in CO_2 uptake by the Southern Ocean. In addition, greater stratification of the Southern Ocean could also limit its effectiveness as a carbon sink.

7.6   Sea ice

7.6.1   Introduction

Sea ice is extremely sensitive to both atmospheric and oceanic temperature changes and is often held up as a prime indicator of anthropogenic climate change. In recent decades there has been a spectacular reduction in Arctic sea ice extent, especially during the late summer and autumn, and, as discussed in earlier chapters, this is often used as evidence of humans' influence on the climate system. However, over the same period Antarctic sea ice extent has increased slightly. Nevertheless, over the next 100 years sea ice in both polar regions is expected to decrease under all SRES scenarios. Arzel *et al.* (2006) examined the projections of sea ice change in both polar regions in 13 of the models from the IPCC AR4 using the SRESA1B scenario. In the following we will examine the sea ice projections from this study, as well as that from Bracegirdle *et al.* (2008), who weighted the output from the AR4 models according to their performance in reproducing sea ice variability over the late twentieth century.

7.6.2   The Arctic

Under the SRESA1B greenhouse gas emission scenario large reductions in sea ice are projected for the Arctic, although there are substantial variations in the estimates of the amount and timing of the ice loss. The projections from 13 of the AR4 models are summarised in Table 7.1. For the Northern Hemisphere as a whole a 27.2% decrease of annual mean sea ice extent is expected by 2100. The loss is much larger during the late summer (61.7%) compared with the spring maximum (15.4%), so that the annual cycle of extent is enhanced.

In terms of ice volume, the annual mean decrease is 58.8%, with 78.9% in September and 47.8% in March. The sea ice extent and volume changes given here are taken as the difference between the averages over the period 2081–2100 from the SRESA1B scenario

Table 7.1 *Multimodel average relative changes (%) of sea ice extent and volume in the Arctic and Antarctic between the periods 2081–2100 and 1981–2000 for March, September and the annual mean in both hemispheres from 12 of the IPCC AR4 models*

(a)   Arctic

	March	September	Annual mean
Sea ice extent	−15.4	−61.7	−27.2
Sea ice volume	−47.8	−78.9	−58.8

(b)   Antarctic

	March	September	Annual mean
Sea ice extent	−49.0	−19.1	−24.0
Sea ice volume	−58.1	−27.4	−33.7

From Arzel *et al.* (2006).

runs and the averages over the period 1981–2000 from the 20C3M climate of the twentieth century experiment.

There has been much speculation in the press about the possibility that all the Arctic multi-year sea ice may disappear within a few decades, leaving an ice-free Arctic during the late summer period. Most models do suggest large reductions (to complete losses) in Arctic summer sea ice area accompanied by an increase in the duration of the open-water season by 2100. Indeed, more than half of the IPCC AR4 models predict that there will be an ice-free Arctic Ocean in late summer by the end of the twenty-first century (Arzel *et al.*, 2006). This is in agreement with the forecasts of the US National Snow and Ice Data Center that if current rates of decline in perennial Arctic sea ice continue, the summertime Arctic could be completely ice free well before the end of the twenty-first century.

Although we can have some confidence in the broadscale predictions of sea ice extent and volume for the next century, it is much more difficult to predict how sea ice will develop on a regional scale. Nevertheless, it is useful to examine how the AR4 models predict the regional sea ice changes, especially considering how consistent the models are in estimating changes in particular areas. Figure 7.8 shows the multimodel average sea ice cover changes by 2081–2100 from Arzel *et al.* (2006). The plots show the fraction of the model population that have ice at a given grid point, with the thick black line indicating the average observed sea ice edge for 1981–2000. As noted by Arzel *et al.*, the most remarkable difference from the recent ice climatology is that the area where more than 90% of the model population have ice has totally disappeared in summer in both hemispheres. They also note that during the winter the area has decreased much less in the north (−17.1%) compared with the

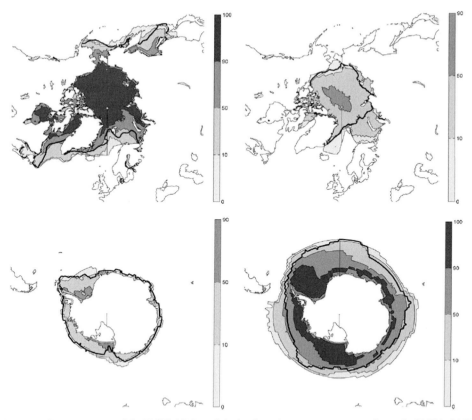

Fig 7.8 The percentage of the IPCC AR4 models that have ice on average over the period 2081–2100 (experiment A1B) in March (left) and September (right) for both hemispheres. The analysis is based on 10 models. For comparison, the thick black line represents the observed sea ice edge averaged over 1981–2000. (From Arzel et al., (2006).)

south (−38.5%), principally as a result of temperatures still being below freezing at that time of year. Figure 7.8 indicates that the greatest retreat in the Arctic will be in the North Pacific sector.

Arzel et al. (2006) considered the multimodel standard deviation of absolute sea ice concentration changes, taken as the difference between 2081–2100 and 1981–2000. They found that in the Arctic the largest differences were found in the northern parts of the Barents, Labrador, Bering and Okhotsk Seas with values of about 50% in late winter. In late summer, the multimodel differences are in excess of 65% from the north of the Canadian Archipelago and north of the Greenland coast to the central Arctic. It should be noted that the multimodel standard deviation is generally larger than the sea ice concentration change in many of these specific regions, demonstrating that there is large uncertainty in simulating the sea ice concentration changes at the regional scale.

7.6 Sea ice

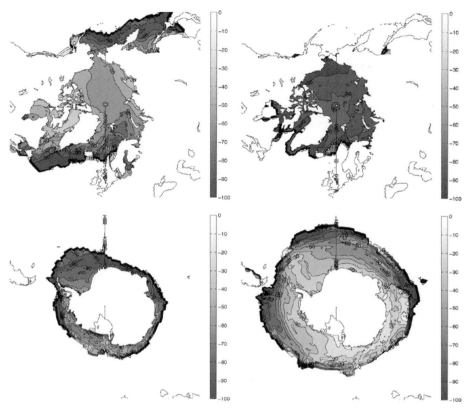

Fig 7.9 Multimodel average relative change (%) of sea ice thickness between the period 2081–2100 (experiment A1B) and 1981–2000 (experiment 20C3M) for March (left) and September (right) for both hemispheres. (From Arzel et al., 2006.)

As discussed in Section 6.5.3, there has been a marked decrease in mean sea ice thickness at the end of the melt season in the central Arctic over the last 50 years (Rothrock et al., 1999). Arzel et al. (2006) found that the AR4 models suggested that this thinning would continue throughout the twenty-first century. They determined the relative monthly mean ice thickness difference between 2081–2100 and 1981–2000 (Fig. 7.9). They identified two areas where large winter reductions in sea ice thickness were predicted. The first was in the Barents Sea, just east of Svalbard, with a thinning of about 70–80% with respect to 1981–2000. This might be a result of an enhanced oceanic poleward heat transport of waters from the North Atlantic into the Arctic, as proposed by Holland and Bitz (2003). The second was in the central Arctic where the models gave a thinning of 50–60%. In September the ice thickness changes in the central Arctic are larger at 70–80% and more homogeneous than for March. In this month the largest thinning of 90% is found north of the Bering Strait in the Chukchi Sea. However, it should be noted that the multimodel standard deviations of the sea

360   *Predictions for the next 100 years*

ice thickness changes are very large in some areas, such as the central Arctic Ocean where the standard deviation is about 100%.

7.6.3 *The Antarctic*

Although the extent of Southern Hemisphere sea ice has increased slightly in recent decades, it is expected to decrease markedly during the coming century. The multimodel trends from Arzel *et al.* (2006) are shown in Table 7.1b and suggest that the ice extent will decrease by 24.0% for the year as a whole, with a loss of 49.0% in March and 19.1% in September. The annual mean changes are therefore of comparable magnitude in both polar regions, and, as in the north, the amplitude of the seasonal cycle is expected to increase, although the magnitude of the change will be larger in the Northern Hemisphere.

The study of Bracegirdle *et al.* (2008) used a different set of IPCC AR4 sea ice projections and found that over the twenty-first century the annual average total sea ice area was projected to decrease by $2.6 \pm 0.73 \times 10^6$ km^2, or 33% with the SRESA1B scenario. Arzel *et al.* (2006) assessed different measures of sea ice amount using a different subset of 15 of the CMIP3 models to give projected twenty-first century deceases of 34% for sea ice volume and 24% for sea ice extent. Most of the ice retreat was expected to occur during winter and spring when the sea ice extent is largest. In the summer and autumn the main region of decrease was found to be in the Weddell Sea.

The annual mean ice volume is expected to decrease by 33.7%, with losses of 58.1% in March and 27.4% in September. The difference in the rate of change of sea ice volume compared with sea ice extent between the northern and southern hemispheres can be attributed to the different initial ice pack thicknesses. Sea ice experiences a growth-thickness negative feedback that is much more pronounced for thinner ice (Bitz and Roe, 2004). This feedback mechanism is such that thinner ice does not need to thin as much compared with thicker ice to increase its growth rate. We would therefore expect changes in sea ice thickness in the Southern Hemisphere to be smaller than those in the north since sea ice is usually thinner around the Antarctic compared with the Arctic. Consequently ice volume changes would be smaller in the south than in the north since the rates of changes of sea ice extent are observed to be comparable in both hemispheres.

Considering the expected regional changes, Fig. 7.8 suggests that the greatest ice retreat, taken as locations where 90% of the models have ice currently, will be in the Atlantic sector, and particularly north of the eastern part of the Weddell Sea. In late summer, the multimodel standard deviation of absolute sea ice concentration changes are largest in the western part of the Weddell Sea (45%). In late winter, the largest differences are found in the western part of the Pacific and Atlantic sectors, and in the eastern part of the Indian sector (50%). In terms of relative sea ice thickness, the changes over the next century are expected to be smaller around the Antarctic than in the north. But in March, the changes are predicted to be largest in the eastern part of the Weddell Sea at about 75% and there is also a loss of about 60% of

7.7 Snow cover and the terrestrial environment 361

the ice thickness in the central Weddell Sea. In September there are significant thickness reductions of 30–50% predicted in the Atlantic and Pacific sectors.

7.7 Seasonal snow cover and the terrestrial environment

7.7.1 Introduction

The extent and thickness of snow cover is dependent on many aspects of the atmospheric flow and condition of the surface. With an increase in temperature predicted for both polar regions over the next century we can expect a poleward migration of the snowline and a reduction in the thickness of the snow pack. This will have the greatest impact in the Arctic where many of the large centres of population are affected by winter season snow cover. In the Antarctic changes of snow cover will only be a factor in the coastal areas of the continent and on the sub-Antarctic islands; however, in these regions the changes will have a major impact on the environment and ecosystems.

7.7.2 The Arctic

The five models used by the ACIA suggested that there would only be a modest change in snow extent, even by the end of the twenty-first century. For the periods 2011–30, 2041–60 and 2071–90 compared with a 1981–2000 baseline the mean percentage changes in extent for the five models was projected to be -5, -9 and -13% respectively. However, there is a strong seasonal variability in the changes, with the mean changes in extent from the 1981–2000 baseline to 2071–90 for the four seasons being winter (-3.8×10^6 km^2), spring (-4.9×10^6 km^2), summer (-1.1×10^6 km^2) and autumn (-3.3×10^6 km^2). The percentage change is greatest in summer when the areal coverage is very small, but the actual decrease in area is largest in April and May during the spring and November in late autumn (Fig. 7.10). These changes demonstrate a reduction in the length of the snow season.

The changes in precipitation and P–E over the next century discussed above, which can be summarised as a projected increase in P–E during the winter and possibly a decrease over the major river basins during the summer, could have implications for river flow. Greater winter season P–E and early melt of the snow cover in the spring (see Fig. 7.10) suggest more river flow during these seasons, giving greater input of fresh water into the Arctic Ocean. However, river flow during the summer could be reduced.

If temperatures rise over the next century we can also expect changes in freshwater ice via reductions in ice cover on Arctic rivers and lakes. The warming is expected to be largest in the northernmost land areas, compared with the subpolar land areas. However, changes in lake and river ice may also be affected by changes in precipitation. In particular, greater winter snowfall could delay the break-up date. In some areas less snowfall will result in earlier break-up since there will be a lower spring albedo, but less insulation could also give

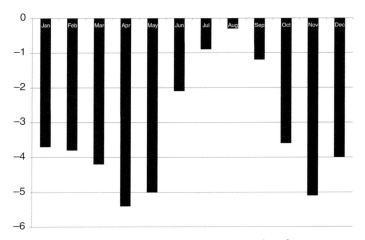

Fig. 7.10 Five-model mean change in snow cover extent (10^6 km^2) between 1981–2000 and 2071–90. (From ACIA, 2005.)

enhanced ice growth. As a general point, freeze-up and break-up timing respond more strongly to warming than cooling due to albedo-radiation feedbacks.

Changes to the extent of river ice are particularly difficult to estimate since changes across the catchments will affect conditions far downstream. In general terms, spring break-up timing is expected to advance by 15–35 days by 2100. There will also be changes in break-up severity (for example causing flooding) with changes in northward flowing rivers due to different rates of warming in downstream (higher latitude) areas versus upstream (lower latitude) driving zones.

7.7.3 The Antarctic

There is presently only about 0.33% of the Antarctic continent that is ice free, and most of this area is on the Antarctic Peninsula or the sub-Antarctic islands. Some parts of the Antarctic coastal zone are also snow free during the summer months and with the modest warming that we expect in these areas we may expect a small increase in extent of the snow-free area, although it is impossible to quantify this change at present since climate models do not have the spatial resolution necessary to discriminate ice-free areas of the continent.

7.8 Permafrost

7.8.1 Introduction

The stability of permafrost and seasonally frozen ground is very sensitive to near-surface air temperature so if temperatures rise through the twenty-first century there could be a widespread increase in thaw depth over most permafrost regions. Due to the poor

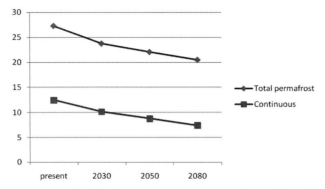

Fig. 7.11 The current area (10^6 km^2) of total (upper line) and continuous (lower line) permafrost and the areas predicted in 2030, 2050 and 2080 based on the mean of the projections from the five models used by the ACIA.

representation of soil in the current generation of GCMs they are not able to directly give projections of how permafrost will evolve over the next century. Currently the output of such models is used to drive stand-alone soil models that can give guidance on how permafrost extent will change.

7.8.2 The Arctic

If temperatures rise across the Northern Hemisphere over the next century we can expect to see a decrease in the area occupied by frozen ground and poleward shifts in the boundaries between continuous, discontinuous and sporadic permafrost. The ACIA used a simple frost-index-based model of permafrost forced by different scenarios of climate change to predict changes in permafrost area. Figure 7.11 shows the present and projected areas of total and continuous permafrost for the Northern Hemisphere based on the ACIA B2 scenario. There was a large variability in the projections between the models, but the mean of the projections suggests that by 2030, 2050 and 2080 the total area of permafrost will have dropped to 87%, 81% and 75% of the current value respectively. The comparable values for continuous permafrost are 81%, 70% and 59%.

The output of the five ACIA models suggest that the seasonal thaw depths are likely to increase by more than 50% in the northernmost permafrost locations, which include northern Canada, Siberia and the northern part of Alaska. At more southerly latitudes the thaw depth may increase by 30–50%.

Permafrost in Alaska has been warming for more than a century, with continuous permafrost on the North Slope region having warmed by 2–3 °C over this period. As temperatures at the upper surface of continuous permafrost are still low, typically below −5 °C, no significant loss of continuous permafrost is predicted over the twenty-first century. However, across northern Alaska the active layer thickness is projected to increase by up to

364   *Predictions for the next 100 years*

1 m by 2100 where there is coarse-grained material in the soil and by up to 0.5 m across the rest of the region.

With climate change, warming ranging from 1.5 to 10.0 °C over the next 100 years is possible, which will be accompanied by more precipitation in the summer and winter. Thawing of Alaska's discontinuous permafrost will likely result from continued warming and predicted increases in snow depth. This thawing, resulting in warmer soils, will speed decomposition reactions and release CO_2 and methane, both greenhouse gases, into the atmosphere. Thawing of any permafrost increases ground-water mobility and susceptibility to erosion and landslides, and can affect soil storage of CO_2. If the thawed soil drains and dries it could release CO_2 to the atmosphere, or storage of CO_2 in the soil could increase if the soil remains flooded.

Over the twenty-first century the top 10 m is likely to thaw throughout much of the ice-rich discontinuous permafrost zone, although complete thawing is likely to take centuries. Canadian studies have estimated that even present surface temperatures will cause an eventual further retreat of the southernmost permafrost fringe in the Canadian sub-Arctic by 100 to 160 km. Further retreat of 300 to 500 km could be anticipated under doubled-CO_2 equilibrium. The actual pattern of loss of permafrost will, of course, be more complex than a simple uniform retreat.

7.8.3   *The Antarctic*

With increasing temperatures we can expect permafrost degradation within the zones of discontinuous and sporadic permafrost, which include the northern Antarctic Peninsula and the South Shetland and South Orkney Islands. Some coastal areas around the coast of East Antarctica may also experience permafrost degradation. However, the total ice-free area of these regions is probably less than 5000 km^2.

Although we do not anticipate a large reduction in the permafrost area, melting of ice wedges in polygons may lead to thermokarst, which is subsidence from melting of ground ice. Areas where thermokarsts could occur include Casey Bay (70.5° S, 12° E), North Victoria land (70.5–73° S, 165–171° E), Scott Coast (74–78° S, 165° E), and throughout the Antarctic Peninsula and its offshore islands (55.5–72° S, 45–70° W). The total ice-free area that may be impacted is approximately 15 000 km^2.

7.9   Atmospheric gases and aerosols

7.9.1   *Introduction*

The possible changes in concentration of the main greenhouse gases have been discussed in Section 7.2 above. Clearly what happens over the next century will depend on many economic, political and social factors, and especially how successful humankind is in

7.9 Atmospheric gases and aerosols

moving away from a dependence on fossil fuels. However, it should be noted that a near-surface warming tends to reduce land and ocean uptake of CO_2, thus increasing the fraction of anthropogenic emissions that remains in the atmosphere.

7.9.2 Stratospheric ozone

Introduction

When the Montreal Protocol and its amendments came into force it provided a means for reducing the levels of chlorofluorocarbons (CFCs) in the atmosphere, which are the chemicals primarily responsible for stratospheric ozone depletion. Our understanding of all the factors that influence ozone loss is still incomplete; however, it is generally expected that as the levels of CFCs decrease, so ozone levels will recover during the twenty-first century, although natural variability in the atmospheric circulation of the high latitude regions may make it difficult to detect an upward trend in stratospheric ozone for some time.

The Arctic

Various assessments using a range of two- and three-dimensional models have been carried out of how stratospheric ozone levels in the Arctic will recover over the next century. The World Meteorological Organization conducted the Scientific Assessment of Ozone Depletion (WMO, 2003), which considered Arctic ozone recovery using several models. In this study about half the two-dimensional models predicted a recovery to 1980 levels by 2050. The models projected that Arctic ozone recovery would be slowest in the spring and earliest in the autumn. The two-dimensional models projected a range of upturn in Arctic ozone concentration, from about 0.5–2.0% per decade.

The three-dimensional models used in the WMO study included atmospheric dynamics, so giving greater insight into how levels of stratospheric ozone will evolve. The three-dimensional models projected a wide range of ozone recovery. The only model with projections beyond 2020 had ozone depletion occurring between 1995 and 2015, with only modest recovery by 2045. Overall, the two- and three-dimensional models had large differences in the projected total column ozone amounts, but all estimated that ozone levels would remain substantially below pre-depletion levels for at least the first two decades of the twenty-first century.

The three-dimensional models can give some insight into the locations where the greatest changes in total column ozone will occur over the coming decades, suggesting that over 2010–19 the largest changes will be near Greenland, although the models disagree as to the rate of recovery.

Clearly there are still many uncertainties in estimating the recovery of Arctic ozone and further work is required on the chemistry–climate models that are used to estimate how the ozone will change over the coming decades. Overall, there are large uncertainties in the estimates of the timing and magnitude of the recovery of the Arctic ozone layer (ACIA, 2005).

The Antarctic

Observations now show that levels of CFCs are declining; however, in the first decade of the twenty-first century there has been no clear indication of a reduction in ozone depletion. Nevertheless, during the next 100 years, the ozone levels are expected to stabilise then recover, but the exact rate of this is difficult to estimate since it is dependent on future levels of Antarctic chlorine and bromine. Newman *et al.* (2006) carried out modelling studies attempting to replicate recent changes in stratospheric ozone and predicting when the recovery would take place. They were able to explain 95% of the ozone hole area's variance in recent decades, indicating the skill of their model. Using estimates of future levels of CFCs, they suggested that the area of the ozone hole will very slowly decline between 2001 and 2017, and that it will not be possible to detect a statistically significant decrease in the area of the ozone hole until about 2024. They also predicted that the ozone hole will not fully recover to 1980 levels until around 2068. This is in line with other studies that have predicted a recovery by 2050 to 2070. As a consequence of this recovery, coupled chemistry–climate models suggest that the summer SAM will return to more neutral values over this period such that the circumpolar westerlies will reduce in strength from speeds observed at the beginning of the twenty-first century.

The recovery of the ozone hole will not take place in isolation from other changes occurring across the Antarctic, such as the increasing levels of greenhouse gases. However, Newman *et al.* (2006) showed that nominal Antarctic stratospheric greenhouse gas forced temperature change should have a small impact on the ozone hole.

7.10 Terrestrial ice, the ice shelves and sea level

7.10.1 Introduction

The state of the cryosphere and sea level change are inextricably linked and how they will change over the next century is one of the most important questions facing Earth scientists today. With regard to the prediction of the main meteorological parameters, it is possible to employ GCMs that can be tested by simulating climate change over recent decades. However, many of the changes that we have seen in the main ice sheets in recent decades, such as the rapid loss of ice from the Pine Island Glacier, were not predicted, and are still not fully understood. Even with the major changes taking place across the Greenland Ice Sheet (see Section 6.9.2) we are very much in the position of observing change in action and trying to interpret the mechanisms at work rather than being in a position to understand fully if and how present changes relate to anthropogenic activity and how the ice sheet will change in the future.

In this section we will review the current predictions of how terrestrial ice, consisting of the major ice sheets and their surrounding ice shelves, along with glaciers, will change over the twenty-first century and consider the impact on sea level. We include the ice shelves here since they are inextricably linked to the ice sheets and ice streams. The ice shelves are floating and their removal will not affect sea level directly. However, there is increasing

7.10 Terrestrial ice, the ice shelves and sea level 367

evidence that they buttress and provide a break on the ice streams that flow down from the interior of the ice sheets.

The range of projections for how the ice sheets will evolve is invariably broader than for many other aspects of the high latitude environment, reflecting our current poor understanding of the mechanisms behind variability and changes in the ice sheets.

7.10.2 *Terrestrial ice and sea level*

Introduction

It is essential to have reliable estimates of how the major ice masses will change over the coming century since it has been estimated that they could contribute a further 15–37 cm of sea level rise (IPCC, 2007). But this is very difficult since we still only have an incomplete understanding of the reasons for the changes that have been observed over recent decades. However, model predictions of surface temperature changes, along with changes in precipitation, can be used to estimate the input and loss of mass from the ice sheets and glaciers.

Predictions of global-scale sea level rise, as a result of greenhouse gas emissions, are based on models of the ocean's steric response to temperature increases and changes in the THC, in addition to eustatic effects contributed by potential ice volume changes in the Greenland and the Antarctic Ice Sheets, along with the smaller glaciers. In the following sections we will consider the projections of mass balance changes of Greenland and the Antarctic Ice Sheet, together with other ice masses. The methods used to predict changes in the ice sheets do not presently include glacier or ice sheet dynamics, calving or refreezing of meltwater that has percolated into the glacier, so there is still a large degree of uncertainty in these projections.

The Arctic

The IPCC AR4 models predict that the greatest warming over the next century will be at high northern latitudes, and will be accompanied by greater precipitation across much of the Arctic. More than half of the Greenland Ice Sheet is at elevations where temperatures never rise above the freezing point, so this part of the ice sheet is relatively insensitive to temperature changes. Here we expect that there will be greater accumulation during the twenty-first century. The IPCC AR4 report noted that for an average temperature change of 3 °C over the Greenland Ice Sheet, a combination of four high-resolution AGCM simulations and 18 AR4 AOGCMs gave a surface mass balance change of 0.3 ± 0.3 mm yr^{-1} for Greenland.

In contrast, at lower elevations there is expected to be greater melt. Since increased summer melting dominates over increased winter precipitation in model projections of future climate it is anticipated that there will be a continuation of the melting of the Greenland Ice Sheet that has been observed in recent decades. Loss of ice in the coastal region may also take place because there may be greater basal sliding as a result of surface meltwater reaching the base of the ice sheet (Zwally *et al.*, 2002b).

Quantifying the rate at which the Greenland Ice Sheet will melt is extremely difficult. Gregory *et al.* (2004) suggested that the ice sheet will melt at a faster rate in the future and will be essentially eliminated if the annual average temperature over the areas increases by more than about 3 °C. Of course a complete melting of the Greenland Ice Sheet will take many centuries, but Gregory *et al.* showed that greenhouse gas concentrations could reach levels that are sufficient to raise the temperature past this warming threshold before 2100. So over the next 1000 years such a loss of the whole ice sheet could raise the global average sea level by ~6 m.

A further study examining the fate of the ice sheet was carried out by Ridley *et al.* (2005) who used the UK Meteorological Office (UKMO)-HadCM3 model coupled to an ice sheet model. They considered the changes to the ice sheet under four times pre-industrial levels of atmospheric CO_2 and found that the complete ice sheet melted after 3000 years. The peak rate of melting was 0.06 Sv corresponding to about 5.5 mm yr^{-1} global sea level rise. In addition, Toniazzo *et al.* (2004) showed that in the UKMO-HadCM3, the complete melting of the Greenland Ice sheet is an irreversible process even if pre-industrial levels of atmospheric CO_2 are re-established after it melts.

Across the rest of the Arctic, most of the small icecaps and glaciers are at relatively low elevations, so that they are very susceptible to loss of ice as a result of temperature increases. However, estimating how the Arctic glaciers will change over the next century is quite difficult since we still have a relatively poor understanding of the relationship between glacier mass balance change and recent climatic changes. The ACIA therefore used a mass-balance sensitivity approach (Oerlemans and Reichert, 2000) to project future changes in ice sheets and glaciers. Monthly anomalies of surface temperature and precipitation from climate models were used to derive projected changes in mass balance. The projected contributions to sea level from Arctic land ice over 2000–2100 using the temperature and precipitation data from the five ACIA models and the B2 scenario is shown in Fig. 7.12. It can be seen that the models gave a very wide range of sea level contributions, although only one suggested an almost zero contribution, with the remainder having a range of 3 to 5.5 cm and a mean of approximately 4 cm by 2100. This is less than the estimate of 5.7 cm increase over 2000–70 produced by van de Wal and Wild (2001), although they forced their model with double CO_2 throughout the full 70-year simulation. The projections had very large variability on a regional scale, with no clear indication of the sign of the contribution to sea level in some areas, such as Svalbard. ACIA focused particularly on the output of the ECHAM4/OPYC3 model since it projected the largest temperature effects on the mass balance of glaciers and icecaps. Regional contributions to sea level rise from the temperature projections from this model (Fig. 7.13) suggest that the Greenland Ice Sheet will make the largest contribution of almost 4 cm. Although much smaller than Greenland, the Alaskan glaciers are projected to contribute over 2 cm to sea level rise as a result of their greater sensitivity to temperature increases.

Some of the smaller Arctic icecaps and glaciers, such as those in Svalbard and on the Canadian Arctic islands (e.g. the Meighen and Melville icecaps), may not survive the expected warming over the next 100 years.

7.10 *Terrestrial ice, the ice shelves and sea level* 369

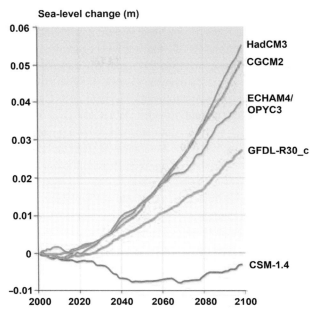

Fig. 7.12 Projected contribution of Arctic land ice to sea level change between 2000 and 2100 calculated using output from the five ACIA-designated models. (From ACIA, 2005.)

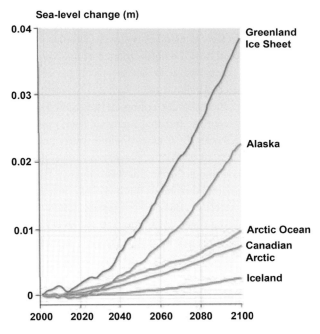

Fig 7.13 Projected contribution of Arctic land ice to sea level change between 2000 and 2100 for various regions, calculated from ECHAM4/OPYC3 model output using temperature effects only. (From ACIA, 2005.)

370   *Predictions for the next 100 years*

The Antarctic

The surface mass balance of the Antarctic Ice Sheet is extremely important in terms of sea level rise since it has an annual water turnover of 6 mm sea level equivalent (Budd, 1991). Over the next century we expect an increase in precipitation across Antarctica as temperatures rise, with there being perhaps an increase of 20% over the current snowfall (Bracegirdle *et al.*, 2008). The AR4 estimated that for an average temperature change of 3 °C over the Antarctic, four high-resolution AGCM simulations and 18 AR4 AOGCMs gave a surface mass balance change of -0.9 ± 0.5 mm yr^{-1} for the continent (sea level equivalent), in other words a sensitivity of -0.29 ± 0.18 mm yr^{-1} °C^{-1} for Antarctica. Most of the additional precipitation is expected to fall in the Antarctic coastal region, with there still being very small amounts of accumulation in the interior.

The greater precipitation and higher air temperatures will lead to a thickening of the inland areas and thinning at the margins, as a result of greater surface melting. Dynamic ice thinning may also take place close to the margins as a result of the removal of adjacent ice shelves. Warming can induce the break-up of ice shelves, which has been observed to coincide with the acceleration of nearby glacier export into the ocean (Rignot *et al.*, 2005). This is currently a key subject of research since ice sheet dynamics are not included in the AR4 models.

The West Antarctic Ice Sheet has received great attention recently because of the considerable changes observed in the elevation of the glaciers that terminate in the Amundsen Sea Embayment. Here the glaciers that flow down to the coast are the fastest in Antarctica, with the catchment holding enough ice to raise sea level by 1.3 m. It is estimated that once the ice shelves are removed and glacier retreat proceeds into the more inland, deeper parts of the glacier basins, they are likely to flow considerably faster (Thomas *et al.*, 2004).

Several groups have considered how the mass balance of the Antarctic may change over the next century and what its contribution to sea level will be. Huybrechts *et al.* (2004) reported that, based on a suite of climate model scenario experiments conducted for the IPCC Third Assessment Report, the Antarctic continent is expected to contribute negatively to sea level rise in the coming century, but that the associated error is the same order of magnitude as the change. In the fourth IPCC assessment it was concluded that the Antarctic ice sheet could contribute -0.12 to -0.02 m of sea level rise between 1980–99 and 2090–99 in terms of surface mass balance under the SRESA1B scenario. However, there are great uncertainties in terms of the dynamical changes that may take place within the ice sheet.

7.10.3   *Ice shelves*

The Arctic

Although the loss of the ice shelves around the Antarctic has received the most publicity, ice shelves on the north coast of Canada and around Greenland have also been disintegrating in recent decades. However, the difficulties in predicting how the Antarctic ice shelves will change over the next century also apply in the north. Events such as the break-up of the Ward

7.10 Terrestrial ice, the ice shelves and sea level

Hunt Ice Shelf on the north coast of Ellesmere Island, northern Canada during the first decade of the twenty-first century had not been predicted, so it is not possible to estimate when specific ice shelves will disintegrate. However, with the largest temperature increases over the next century predicted to be in the high northern latitudes, we can expect continued loss of the ice shelves in these regions.

The Antarctic

The ice shelves represent only about 13% of the area of the Antarctic, but they discharge much of the ice into the Southern Ocean. Melt/freeze under the shelves is particularly important as it varies from -1 to $10 \, \text{m yr}^{-1}$ compared with an accumulation rate of 0 to $0.1 \, \text{m yr}^{-1}$. Intrusion of water masses under the ice shelf is thought to be responsible for much of the change noted in the Amundsen Sea Embayment, where there has been melt of $20 \, \text{m yr}^{-1}$. The waters under the ice shelves are sensitive to the sea ice extent off the shelves, which is important when considering how they may change over the next century.

Ice sheet models do not currently have a good simulation of ice shelf processes. The AR4 made no projections about the state of the ice shelves, and to make progress on this we need higher resolution atmospheric and oceanic projections, improved sea ice projections and better bathymetry.

To make projections we need data to validate model physics, especially basal melt rates. We also need to know shelf water variability and to better resolve the key processes that take place within the shelf edge frontal system, polynyas and the ice–ocean boundary layer.

There has been a well-publicised retreat of many of the ice shelves around the Antarctic Peninsula and an acceleration of many of the glaciers that feed the shelves. However, the changes that will take place over the next century will be dictated to a large extent by the changes in near-surface air temperatures, although oceanographic conditions may also play a role. As we expect temperatures to continue to rise on the eastern side of the peninsula, it is reasonable to assume that there will be further retreat of the ice shelves in this area. On the western side of the peninsula it is not fully understood why the surface temperatures have risen so rapidly over the last 50 years, so it is not possible to estimate with a high degree of confidence how the remaining ice shelves will change in the future.

7.10.4 A summary of projected global sea level changes

The AR4 projected that global sea level will rise by 0.18–0.59 m over the period 1980–99 to 2090–99, depending on the greenhouse gas emission scenario that is considered. The SRESA1B mid-range scenario has a range of 0.21–0.48 m (Table 7.2). However, it should be noted that these figures do not include the possible impact on sea level of rapid dynamical changes in the flow of the ice streams of the Greenland and Antarctic Ice Sheets. They do include a contribution due to increased ice flow from Greenland and Antarctica at the rates observed for 1993 to 2003, but not any increased flow in the future. However, if the contributions from the icecaps were to increase linearly with global average temperature

Table 7.2 *Projected global average sea level rise over the period 1980–99 to 2090–99. Excludes future rapid dynamical changes in ice flow.*

Emission scenario	Expected sea level rise (m)
B1	0.18–0.38
A1T	0.20–0.45
B2	0.20–0.43
A1B	0.21–0.48
A2	0.23–0.51
A1F1	0.26–0.59

From IPCC (2007).

rise, the upper ranges of sea level rise for the greenhouse gas emission scenarios in Table 7.2 would increase by 0.1 to 0.2 m.

In reality the typical regional and decadal variability in sea level is of the same order of magnitude. Consequently, some oceanic regions where the decadal variability is high will experience an enhanced impact as a result of the overall rise in sea level. Conversely, some regions will experience little impact.

A different approach to estimating future sea level rise was taken by Rahmstorf (2007) and Vermeer and Rahmstorf (2009). They developed a simple relationship between global mean near-surface temperature and global sea level variations based on data that covered past decades. This was then applied to the next century using temperature predictions from the IPCC AR4 that spanned a range of greenhouse gas emission scenarios. In the initial study (Rahmstorf, 2007) the approach suggested that sea level could rise by 0.5–1.4 m above 1990 levels for the various emission scenarios. The second paper presented results from an improved and revised version of the method and this extended the range of possible sea level increase to 0.75–1.9 m by 2100. This work has resulted in considerable debate about how a technique that is based on past change is applicable to the future when increasing temperatures may result in melting of the major ice sheets rather than the mid and high latitude glaciers.

7.11   Concluding remarks

There are still many uncertainties regarding how the climates of the two polar regions will change over the coming century, but the coupled climate models, which are the only tools that we have to produce climate projections, are in agreement that the polar regions, and particularly the Arctic will experience some of the largest changes. During the first decade of the twenty-first century we are already seeing a marked loss of Arctic sea ice and increase in surface temperature. The persistence trends in many Arctic climate parameters support their

7.11 Concluding remarks

continuation over the next decade. One can expect large spatial and temporal differences, however, from the relative contributions of intrinsic variability, external forcing and internal feedbacks/amplifications. It is particularly important to resolve regional feedback processes in future projections based on modelling scenarios.

In the longer term (50–100 years) the robustness of the greenhouse signal in climate model projections shifts the balance of the likely changes toward the greenhouse-driven patterns, with major changes in autumn and winter in the Arctic arising from increasingly ice-free conditions. In such scenarios, the changes are expected to be larger in high latitudes than elsewhere. An improved understanding of the relevant feedbacks, and the incorporation of these feedbacks into models in a realistic manner, will be essential for projecting spatial and seasonal patterns of change. Even in the absence of a complete understanding of ongoing and projected environmental changes, it will be necessary to address the limits of adaptive capacity of local systems.

Beyond the Antarctic Peninsula the continent has not experienced any marked widespread warming in recent decades, we expect this to change in the first half of the next century. Detection of rising temperatures, greater precipitation and a reduction in sea ice extent will provide confirmation of the model projections, which all indicate large changes.

Perhaps the greatest uncertainties in the future evolution of the polar climates is in how the Greenland and Antarctic Ice Sheets will change. Presently we do not have the models needed to give reliable estimates of ice loss, but this has to be one of the greatest challenges facing scientists concerned with the polar regions because of the possible impact on sea level.

8

Summary and future research needs

8.1   Introduction

In earlier chapters we have given a summary of our current understanding of past climate change at high latitudes and provided scenarios of how the climates of the two polar regions may evolve over the next century under different levels of greenhouse gas emissions. We are at a critical period in the study of the Earth's climate. Improved data sets providing insight into past climate variability are constantly appearing from new ice and ocean coring initiatives. In addition, we are getting superior observations from satellite systems that give us an unprecedented coverage of the atmospheric, oceanic and cryospheric conditions at high latitudes, which allow us to better understand the mechanisms that are important there. In recent decades our ability to model the environment of the polar regions has also improved. However, there are still many questions unanswered regarding the reasons for past and current climate change and significant doubt about what may happen over the next century.

Regarding future climate changes, perhaps the largest uncertainty is over how greenhouse gas emissions will change over the coming decades. With the release of the IPCC's Fourth Assessment Report, there has been a greater willingness to accept that human activities since the start of the Industrial Revolution, and particularly over the last 100 years, have altered the climate of the Earth. However, it remains to be seen whether there is the political will to make the huge changes in society, industry and agriculture that are needed to reduce or even stabilise greenhouse gas emissions.

In this final chapter we will try and draw together many of the ideas presented earlier into a brief summary of our present understanding of past climate change at high latitudes. We will also consider the observations necessary to gain greater understanding of the processes of change in the polar regions and to be able to better represent them in models.

8.2   Gaining improved understanding of past climate change

8.2.1   The instrumental period

Changes observed in the in-situ meteorological record

We have reasonably good in-situ meteorological observations for approximately the last 100 years from the Arctic and about 50 years for the Antarctic. These show contrasting

8.2 Improving understanding of past climate change

pictures of change for the two polar regions, which are not fully explained at the moment. In the Arctic there has been a widespread increase in near-surface air temperature, with peaks over Alaska and Siberia. The Alaska warming appears to be associated with an upward jump in temperature around the mid 1970s, which has been linked to a change in the Pacific Decadal Oscillation at that time. The warming has resulted in extensive melting of permafrost, a retreat of glaciers across the region and other environmental changes. The loss of glacier ice appears to have started around the beginning of the twentieth century and the changes over the last 100 years may well be a combination of natural climate variability and anthropogenic forcing.

The Siberian warming is perhaps the least understood 'hot spot' on Earth. Concern has been expressed about the homogeneity of some of the station data from the region and further work is certainly needed to quality control the available observations, perhaps even creating a new central, high-quality data base of Arctic mean climate data, such as the one created for the Antarctic. However, other evidence of major change in the region is available, such as the large-scale loss of permafrost and the increase in freshwater input to the Arctic Ocean. Collectively such data all point to a major warming across the region. The rise in temperature has been linked to the shift of the NAM to more positive values in recent decades, and there were certainly more frequent westerly winds across Europe and into Russia during the 1990s. But in the early years of the twenty-first century the Northern Annular Mode (NAM) has become more neutral, yet the warming across Siberia has continued. More attribution studies are needed to understand the rapid changes observed in this region.

Great advances have been made in recent decades in understanding the climatic changes that have taken place across the Antarctic. The large warming on the eastern side of the Antarctic Peninsula has been shown to have occurred largely as a result of the more positive Southern Annular Mode (SAM) and the stronger westerly winds leading to more warmer, maritime air masses crossing this substantial orographic barrier. However, there is more work to be done on the detailed air flow across the peninsula and the relationship of the warming to the synoptic environment of the region.

On the western side of the peninsula the picture is much less clear. There is strong evidence that the rapid winter season warming is linked to a reduction in sea ice and that this in turn is driven by changes in the atmospheric flow. However, at the moment it is unclear whether this is a result of anthropogenic factors or natural climate variability, or a combination of these factors. Most atmosphere–ocean general circulation models (AOGCMs) do not simulate this warming during the late part of the twentieth century so such models at present cannot help in investigating the causes of this major change. Since there is very little in-situ data for the Bellingshausen Sea available, the best way to investigate atmospheric conditions in the middle of the twentieth century is probably via proxy data, such as ice cores from across the peninsula. Further insight might also be obtained via investigations of natural climate variability and decadal timescale circulation changes as simulated in climate models, provided that the next generation of models can correctly reproduce the recent climate of the region.

Across the rest of the Antarctic there has been little temperature change at the surface over the last 50 years. In recent decades there has been a small cooling around the coast of East

376   *Summary and future research needs*

Antarctica as a result of changes in the SAM, but no significant changes at Vostok station on the plateau of East Antarctica.

The winter season warming identified above Antarctica centred 5 km above sea level is not fully understood at present. The increase in polar stratospheric clouds (PSCs) that is believed to have occurred as stratospheric temperatures have dropped may well have played a role in the warming. However, much more research is needed on this possible mechanism. PSCs are not included in climate models at present and fully coupled stratosphere–troposphere climate models may well shed light on this important change that has been observed.

One of the most significant changes observed in the polar atmosphere in recent decades was the development of the Antarctic ozone hole. Much of the chemistry responsible for the springtime destruction of ozone is now understood, but there is still important work to be done on understanding the influence of stratospheric ozone depletion on features of the tropospheric circulation, such as the SAM. It is also very important to be able to produce better predictions of the recovery of the ozone hole over the coming decades.

The major modes of climate variability

Although we have gained considerable insight into the factors that influence the annular modes, we are far from having a complete picture of their natural variability and the influence of anthropogenic factors. In the Southern Hemisphere, the presence of the Antarctic ozone hole has clearly been a major driving force in moving the SAM predominantly into its positive phase during summer and autumn, but much further work is needed on the reasons for such changes in these modes and their natural variability. In the Arctic, the positive nature of the North Atlantic Oscillation (NAO)/NAM in the early 1990s was thought to be behind the large warming of the high northern latitudes. A positive NAO certainly promoted enhanced advection of warm air masses and surface waters into the Arctic Ocean via the Norwegian Sea. However, since about 2000 the NAO/NAM has been neutral, yet the atmospheric warming and loss of sea ice in the Arctic Ocean has been even more rapid. So, while the annular modes are clearly very important in influencing polar climate variability, other factors can be significant. Further research is needed into how other components of the climate system interact with the annular modes.

Research over the last couple of decades has shown that the high latitude climates are strongly influenced by atmospheric and oceanic conditions at lower latitudes. Features of the high latitude atmospheric circulation, such as the annular modes, are affected by sea surface temperatures in mid latitudes, as well as many other factors, such as greenhouse gas concentrations. We now know that such complex linkages are non-linear and, while we can see some consistent patterns of low latitude climate variability such as the El Niño–Southern Oscillation, there is a large degree of variability in the high latitude response. Research is needed to understand the high–low latitude teleconnections, but recent studies have indicated that there may be an inherent limit in the degree to which we can quantify this linkage, especially in the Southern Hemisphere.

The period for which we have reliable reanalysis fields indicates that there have been a number of rapid changes in high latitude climate parameters or indices of the atmospheric

8.2 Improving understanding of past climate change 377

circulation. These may be indications of decadal timescale variability or mode changes in the circulation. The meteorological records at high latitudes are rather short and therefore of limited value in investigating decadal timescale variability. For such analyses long runs of climate models are more appropriate provided that the models have realistic representations of polar processes. In addition, the data derived from ice cores and some of the high temporal resolution ocean cores are of value for looking at decadal timescale variability, provided that useful proxies of the atmospheric and oceanic conditions can be derived from them.

The terrestrial cryosphere

We have seen many large changes in the terrestrial cryosphere over the last few decades. There has been a melting of permafrost across Alaska and Siberia, and a reduction in the length of the season of lake and river ice. How features such as freeze-up and break-up of lake ice relate to temperature is poorly understood and further research is needed if we are to be able to produce reliable predictions of how the ice cover will change in the future.

Permafrost is extremely important to the infrastructure of many high latitude communities and environments and, with the rapid changes taking place in many areas, it is more important than ever to understand the processes within the permafrost and active layer, especially phase changes. Spatial complexity in ground surface response is best approached through remotely sensed data and field sampling within a geographic information system framework. We also need integration of high resolution climate modelling with substrate characterisation/modelling.

Sea ice

The pattern of sea ice change in the two polar regions in the period of satellite observations (the 1970s onwards) could not be more different. Late summer ice in the Arctic is decreasing rapidly, with there having been a series of record minima in the first decade of the twenty-first century. We are beginning to understand the reasons for the loss of ice, and the important pre-conditioning of the Arctic Ocean in the early 1990s via the flushing out of multi-year ice that took place. However, there is still a great deal to learn about the relative roles of the atmosphere and the ocean in the ice loss and how the ice will change in the future.

In the Antarctic we have seen a small but significant increase in the ice extent over the last 30 years, but with contrasting trends around the continent. There has been a large increase of ice in the Ross Sea sector, but extensive loss of ice to the west of the Antarctic Peninsula. Recently this pattern of change has been linked to the development of the Antarctic ozone hole and the stronger winds around the Antarctic continent, but more research is needed on why we have such contrasting pictures of change in the two polar regions.

The current generation of AOGCMs have great difficulty in reproducing the sea ice changes observed during recent decades. At present the models have larger errors in summer than winter, and also have less skill in the Southern Hemisphere than in the Northern Hemisphere. This last characteristic is attributed to problems in simulating realistically the Southern Ocean. Interestingly, very few models are able to reproduce the sharp decline in

Arctic September sea ice extent observed during the last decades, even though this is such a pronounced signal of climate change. Errors in the sea ice can also be attributed to biases in the wind forcing and to the representation of atmospheric boundary layer processes, cloudiness and oceanic mixing processes in models. Thus, to realistically simulate sea ice it is necessary to have many different elements of the Earth system reproduced correctly in climate models. Hence the sophistication of the sea ice models is often not the principal factor in the failings of current AOGCMs.

The ice sheets and sea level

The loss of ice from the Greenland and Antarctic Ice Sheets has been widely reported by the media and received a great deal of attention from polar scientists. However, investigating the reasons for recent changes is extremely difficult, since we have poor data on conditions within the ice sheets and models that still need much development before they are able to realistically simulate recent change.

The Greenland Ice Sheet is losing mass from its margins and inputting fresh water into the North Atlantic. This is not happening on the scale that occurred at 8.2 kyr BP, but is nevertheless of great concern considering that we expect further warming of the climate over the next century. Recent research has suggested that water from lakes on the surface of the ice sheet is penetrating to its base and possibly lubricating the bed of the ice and so promoting more rapid flow. Since it is so difficult to collect widespread observations below ice streams it is not presently clear whether this is indeed taking place, but understanding such key processes is essential if we are to produce realistic predictions of how the ice sheet may evolve in the future.

In recent decades the Antarctic Ice Sheet has shown much less change than Greenland since temperatures across the continent are mostly well below freezing. However, the West Antarctic Ice Sheet is grounded below sea level and susceptible to oceanographic changes. A rapid retreat of the glaciers in the Amundsen Sea Embayment has been observed over the last decades and this has been strongly linked to changes in water masses affecting the area. Airborne and ship-based investigations have started to reveal the factors responsible for this large loss of ice, but further work is needed to completely understand recent changes. It is not known if such changes will take place along other parts of the West Antarctic coast, but it is important to monitor these regions because of the possible impact on sea level. How the ice sheets will contribute to sea level rise over the next century is one of the most important questions concerning polar scientists, but it will be a difficult question to answer quantitatively since we currently have a relatively poor understanding of the magnitudes of the processes that affect ice sheet mass balance.

The ice shelves are now known to be important in buttressing the ice streams that flow down from the interior of the ice sheets, with the rate of ice discharge being at least partially controlled by the thickness and extent of the ice shelves. To understand how the ice shelves might respond to climate change we need to consider the role of surface melt in destabilising the ice shelves, the sensitivity of subsurface ice shelf melt to water temperature and the sensitivity of these shelf water temperatures to climate forcing. Observations suggest that the sensitivity of individual ice shelves to changes in water temperature varies over an order of magnitude. We also need to know more about shelf water variability and to understand the processes that drive it on seasonal and interannual timescales. We therefore need much more observational data on the ice shelf environment in order to understand how they might change in the future. Such observations can now be collected using sophisticated underwater autonomous vehicles.

In order to predict changes in the ice sheets we need reliable models. However, at the moment, there is no ice sheet model that predicts both the retreat of the Antarctic Ice Sheet since the end of the last ice age and correctly represents current variability. There are also major gaps in the physics included in the ice sheet models, including the interaction between ocean circulation in coastal waters and floating ice shelves, the coupling between this floating ice and the main grounded ice mass and the role of fracture and shear margins in the ice shelves. The development of realistic ice sheet models will be essential over the coming years for accurate prediction of sea level rise.

Glaciers

Glaciers are retreating in all parts of the world and contributing to increasing sea level. Yet only about 300 out of approximately 200 000 glaciers across the Earth have been sampled via mass balance. Global glacier change during the twenty-first century has been estimated by determining global mass balance sensitivity based on observed mass balance in combination with area–volume scaling, using different initial area and volume estimates. To produce better predictions we need more mass balance observations and knowledge of global glacier area and volume, along with improved mass balance models. We also need to know how glacier area and volume has changed in the past. To get changes in climate at the glacier surface we need to develop improved means of downscaling fields from global and regional climate models.

The oceans

The Arctic and Southern Oceans are two of the biggest unknowns in the high latitude climate systems. While we have reasonable reanalysis data sets for the atmosphere with which to investigate circulation changes, no such comparable fields exist for the ocean. To examine past changes we have to use spatially and temporally sparse observations from research cruises and moorings. These have provided some insight into recent changes, such as the freshening of the Ross Sea, but investigating changes of water masses is extremely difficult. Even determining the magnitude of the Meridional Overturning Circulation (MOC) in the Southern Ocean is not possible at this stage, yet we need to have reliable estimates of this quantity to validate climate models. It has been possible to determine some major changes taking place, such as the warming of the Southern Ocean over recent decades and the freshening of the Arctic Ocean. Future ocean observing systems will provide many more detailed observations of the world's oceans that will allow detailed monitoring of change at high latitudes.

380 *Summary and future research needs*

Early experiments are being carried out to develop ocean reanalyses that are comparable to those for the atmosphere. However, this is a much more difficult task because of the lack of ocean observations. Nevertheless, they still have the potential to greatly aid investigations into ocean change.

8.2.2   The Holocene

In recent years great advances have been made in understanding high latitude climate change during the Holocene as improved data have become available from ice and sediment cores and other proxy data. These have shown that the period was far from stable climatically, as was once thought, but that millennial-scale climatic cycles can be found throughout this period. The data have revealed periodicities of about 2500 and 950 years, possibly as a result of changes in solar flux. They also showed a 1500-year cycle, which may be a result of internal oscillations of the climate system.

We need to understand the suborbital timescale variability within the Holocene, such as the rapid change events and the proposed 1500-year variability in climate. Such questions can only be answered via combined observational and modelling studies.

8.2.3   The last one million years

There are many questions remaining about the Earth's climate history during the Pleistocene, particularly with regard to understanding the exact mechanisms by which changes in the planet's orbital characteristics, in particular obliquity and precession, modulate the climate and lead to the observed pattern of glacials and interglacials. The role of the North American ice sheets in the switch from a dominant 43-kyr climate cycle to a 100-kyr cycle after the Mid Pleistocene transition is one area of study that requires further analysis, as is the suitability of Marine Isotope Stage 11 as an analogue of present and future climate without anthropogenic forcing.

At millennial timescales we also need to increase our understanding of the interaction and variability of key parameters through time. For example, the millennial-scale changes in methane associated with Dansgaard–Oeschger events have a different pattern of relative amplitudes to the associated temperature changes: to understand why will require improved knowledge of how land–biosphere emission varies through stadial–interstadial transitions. Furthermore, more work is required to determine the exact origin of Heinrich events and how they relate to climate variability and the internal dynamics of the Laurentide Ice Sheet.

Finally, our knowledge about the means by which the so-called 'bipolar seesaw' operates at millennial timescales and longer needs to be enhanced, in particular regarding the transmission of signals between the North Atlantic MOC and heat storage within the Southern Ocean.

Future analysis of Pleistocene climate will be aided by new high-resolution cores, such as the North Greenland Eemian (NEEM) ice core drilled in North Greenland and completed in

2010, together with improved laboratory instrumentation and techniques to analyse and understand the chemical signals of past climate locked within the cores.

8.3 Modelling the high latitude climate system

The simulations of recent high latitude climate change by the current generation of AOGCMs are still relatively poor and unable to realistically reproduce some of the major observed changes, such as the large warming on the western side of the Antarctic Peninsula and the decrease of Arctic sea ice. Several factors are probably responsible for this. There needs to be better representation of some high latitude climate processes, such as clouds, clear sky precipitation, the stable boundary layer, improved representation of sea ice and the accumulation of snow on sea ice. We have poor observational data on some of these phenomena and the knowledge gained from field campaigns has to be used to improve the parameterisations.

In recent years it has become clear that in order to correctly simulate the high latitude climate, conditions in the tropics and mid latitudes must also be realistically simulated in models. There are physical links via both the ocean and atmosphere between high and low latitudes that transfer signals of events such as El Niño, and these must be included in models if past variability is to be simulated and reliable projections of future conditions produced.

Some terrestrial climate processes, such as permafrost, cannot be included realistically in climate models at present and the output of the global models has to be used to drive separate permafrost models. These are dependent on the quality of the forcing data from the climate models and good snow cover and temperature predictions are needed. The inclusion of blowing snow in the GCMs would also greatly aid the permafrost models.

While it has been possible to verify the atmospheric elements of these models in recent years, the ocean component is generally extremely difficult to assess, especially the time variability, because of the lack of validation data. One exception is the Antarctic Circumpolar Current, which can be assessed in models since we have sufficient observations to determine its speed and location and variability. However, there is relatively little information on the spatial variability of high latitude water masses and their penetration onto the continental shelves. Observational studies have shown that the flow of different water masses under the ice shelves is very important in determining ice shelf stability. Yet the current generation of global models do not have water flowing under the ice shelves, and they are represented in the same way as the ice sheets. Moreover, the ice shelves are important areas for bottom water formation and therefore it is essential that future global models have a realistic representation of these key regions.

In the Antarctic, oceanographic changes are thought to be largely responsible for driving the loss of ice from Pine Island glacier on the coast of West Antarctica and we need to be able to simulate such changes realistically in models. It is thought that difficulties in simulating the ocean are to a large extent responsible for the failings in reproducing sea ice changes in recent decades and it is important in the future to be able to have a much better representation of the oceans in climate models.

The oceanic MOC is an extremely important part of the global ocean system, yet important elements cannot be simulated at present. For example, Antarctic Bottom Water (AABW) is not present in most AOGCMs, yet it is a primary link from the Southern Ocean to the ocean basins in the Northern Hemisphere. In the future a high priority must be to include AABW in climate models so that the links between the oceans in the north and south are better simulated.

Global climate models have relatively coarse horizontal resolution at present and cannot simulate realistically important areas of rapidly changing orography, such as the Antarctic Peninsula, the Antarctic coastal region and the islands of the Arctic. Some experiments are being carried out with higher horizontal resolution global climate models, but a more realistic approach to obtain higher resolution in the polar regions is to use regional climate models nested within global models. Such regional climate models have often been run as atmosphere-only simulations, with the lateral boundary conditions being obtained from reanalyses and climatological ocean and ice conditions being applied. However, the high latitude atmosphere, ocean and ice are closely linked and these elements need to be fully coupled within future regional climate models. This will require considerable computer power, but this will become available with time.

There have already been a number of intercomparisons of the polar components of climate models. These have produced advances in the simulation of sea ice and other parts of the polar environment. However, there is an ongoing need for further comprehensive comparison of AOGCMs in the polar regions.

In relatively short runs of climate models out to about 100 years in the future, the extent and elevation of the major ice sheets have been kept fixed. This is a reasonable assumption on this timescale, although important processes, such as freshwater runoff, are not included. For longer runs of several millennia, often carried out with models of intermediate complexity, interactive ice sheet models have been incorporated in order to answer questions such as how the Greenland Ice Sheet will evolve in a warming world. These models consider changes in the accumulation on the ice sheet and melting at lower elevations, but not dynamical flow of the ice. Models that incorporate this latter feature are still under active development and cannot currently simulate realistically the recent changes observed in the Greenland and Antarctic Ice Sheets. However, in the future it is anticipated that there will be improved coupling between the ice sheets and ice shelves, the ocean and the atmosphere.

8.4   Data required

The polar regions are experiencing some of the largest climatic changes on Earth and it is essential that they be monitored intensively over the coming years and decades. This will not only allow us to monitor the state of important elements of the climate system, but will also provide essential information on the mechanisms of climate change, which need to be

8.4 Data required 383

included in models. The International Polar Year of 2007–08 established new high latitude observing systems and a primary goal was to leave a legacy of an improved means of monitoring the polar regions. It is to be hoped that this vision is realised and that we have enhanced observations of the polar regions over the coming decades in what may well be a crucial period of change.

8.4.1 In-situ data

In-situ observations are the most accurate form of data for monitoring most aspects of the polar regions and they have provided much of the evidence of high latitude climate change over recent decades. It is essential that these often expensive observing networks are maintained over the coming decades. Across the Arctic the number of synoptic reporting stations is falling and it is important that this trend is reversed. However, we require much more than just the meteorological data, and it is essential to have observations on permafrost state, river/lake ice and other parameters that it is difficult to obtain with satellite systems.

In the Antarctic there are a reasonable number of stations around most of the coastal regions, although it is very desirable to have a year-round station in the data void of West Antarctica, although access here would be a problem. However, the major difficulty is in the interior of the continent. There have always been few staffed stations here, but it is important to maintain the long observing programmes so that trends can be detected.

The automatic weather stations (AWSs) have been a remarkable source of data in recent decades from remote locations. It is hoped that these systems can be continued, with particular emphasis on maintaining systems at locations where we already have long time series. While these systems were originally established to make meteorological observations, an increasing variety of other observations are being made at isolated locations, such as aerosol concentration and surface ozone measurements. With the AWS networks we have a reasonable amount of surface meteorological data from the interior of Greenland and the Antarctic, but more upper air observations are needed. For example, Amundsen–Scott Station at the South Pole is the only Antarctic Plateau station that has a continuous radiosonde record that extends back to the IGY.

While the historical record of ocean data is sparse both temporally and spatially, observing systems, such as Argo floats and gliders, are starting to provide the in-situ data that are so urgently needed for monitoring and detecting change in the ocean. A combination of measurements made from ships, such as XBT probes, and autonomous observations will together provide a synoptic view of the high latitude oceans.

8.4.2 Proxy data

The ice cores collected in Greenland and the Antarctic have provided a rich source of data for investigating past climate change. The deep cores extending back through several ice ages or more have received the most publicity and have certainly increased our knowledge

384   *Summary and future research needs*

about the different ice ages and interglacials that have occurred. These are very expensive undertakings and can only be carried out at high domes. In the Antarctic it is likely that a deep core can be collected at Dome A. This core could provide proxy climate data extending back more than a million years, the longest ice core record that can be collected on Earth. However, much can also be learnt from shorter cores. These are much less expensive to collect, but can provide data extending back through the Holocene and earlier. In addition, it is important that many short cores, extending back for 100–200 years, are collected. The large number of such cores that can be collected means that the spatial variability of climate change can be investigated across large areas of the polar regions.

In parallel with such drilling initiatives, we also need better techniques for extracting environmental data from the ice cores. In the Antarctic, it has been possible to find proxies for sea ice extent in cores from some locations, although the relationships vary around the continent. We need greater understanding of how signals such as atmospheric circulation and sea ice extent reach the ice core sites and are stored there, in order to develop optimal retrieval algorithms to extract such data for the past.

In recent years remarkable advances have also been made in obtaining climate data from ocean cores. Here the temporal resolution is often less than that obtained from ice cores, but in recent years advances in core analysis have allowed decadal climate variability through the Holocene, providing fresh insight into climate processes. It is important that many further cores can be collected in climatically sensitive areas, such as around the Antarctic Peninsula and in the North Atlantic.

8.4.3   *Satellite data*

Although satellite records are much shorter than the in-situ climate data, they are fundamental in providing high spatial coverage in the remote polar regions. Without such data we would have no realistic way of determining the broadscale coverage of sea ice or the elevations of the ice sheets.

In recent years there have been major advances in satellite remote sensing techniques, with, for example, a quantity such as ice sheet mass balance, now being investigated with a range of sensors from laser and microwave altimeters, to gravity missions. It is essential that the observing programmes are continued with replacement satellites being planned for and new, improved sensors being flown.

8.4.4   *Reanalysis fields*

Atmospheric reanalysis fields have been extremely important in many high latitude climate investigations. A priority for the future must be the production of better reanalysis data sets, since these are the key tool for investigating the large-scale modes of atmospheric variability. Although small amounts of additional data may become available for future reanalysis exercises as old records are digitised, the analyses will be most improved by the use of better assimilation techniques, particularly in dealing more appropriately with the relatively sparse data network of the polar regions. To date the reanalyses that have been completed have been global in extent, but at the time of writing a high horizontal resolution, limited area reanalysis of the Arctic in being undertaken and a similar project for southern high latitudes may follow. The improved resolution of these reanalyses will allow better representation of the complex orography that is found in parts of both polar regions, along with improved handling of the atmosphere–ocean–cryosphere linkages.

8.5 Concluding remarks

The polar regions have experienced some of the largest climatic changes on Earth in recent decades and such events have been reported far beyond the scientific literature, often featuring in the press and on television broadcasts. It is often taken for granted by the general public, and often reported in the press, that the changes are caused by increasing greenhouse gas emissions. However, as discussed throughout this book, the changes that have taken place at high latitudes have multiple and complex causes and there is still much to be learnt about attributing the observed climatic changes to natural climate variability and anthropogenic activity. However, as noted by the IPCC Fourth Assessment Report, it is 90% certain that humankind is now affecting the climate of the Earth. In the future we expect to see the greatest changes taking place in the two polar regions, especially the Arctic. It remains to be seen how their climates evolve in the coming decades and how humankind adapts to these changes.

References

Abdalati, W., Krabill, W., Frederick, E., *et al.* (2001). Outlet glacier and margin elevation changes: near-coastal thinning of the Greenland ice sheet. *J. Geophys. Res.*, **106**, 33 729–41.

Abram, N. J., Mulvaney, R., Wolff, E. W. and Mudelsee, M. (2007). Ice core records as sea ice proxies: an evaluation from the Weddell Sea region of Antarctica. *J. Geophys. Res.*, **112**, D15101, doi:10.1029/2006JD008139.

ACIA (2005). *Arctic Climate Impact Assessment – Scientific Report*, Cambridge, UK: Cambridge University Press.

Adhémar, J. A. (1842). *Revolutions de la mer*, Paris: privately published.

Adler, R. F., Huffman, G. J., Chang, A., *et al.* (2003). The version-2 global precipitation climatology project (GPCP) monthly precipitation analysis (1979–present). *J. Hydromet.*, **4**, 1147–67.

Agassiz, L. (1840). *Etudes sur les glaciers*, Neuchâtel, Switzerland: privately published.

Ahn, J. and Brook, E. J. (2008). Atmospheric CO_2 and climate on millennial time scales during the last glacial period. *Science*, **322**, 83–5.

Ahn, J., Wahlen, M., Deck, B. L., *et al.* (2004). A record of atmospheric CO_2 during the last 40,000 years from the Siple Dome, Antarctica ice core. *J. Geophys. Res.*, **109**, D13305, doi:10.1029/2003JD064415.

Alexanderson, H., Hjort, C., Möller, P., Antonov, O., and Pavlov, M. (2001). The North Taymyr ice-marginal zone, Arctic Siberia: a preliminary overview and dating. *Global Planet. Change*, **31**, 427–45.

Alexeev, V. A. (2003). Sensitivity to CO_2 doubling of an atmospheric GCM coupled to an oceanic mixed layer: a linear analysis. *Clim. Dyn.*, **20**, 775–87.

Alexeev, V. A., Langen, P. L. and Bates, J. R. (2005). Polar amplification of surface warming on an aquaplanet in 'ghost forcing' experiments without sea ice feedbacks. *Clim. Dyn.*, **24**, 655–66.

Allen, C. S., Pike, J., Pudsey, C. J. and Leventer, A. (2005). Submillennial variations in ocean conditions during deglaciation based on diatom assemblages from the southwest Atlantic. *Paleoceanography*, **20**, PA2012, doi:10.1029/2004PA001055.

Alley, R. B., Meese, D. A., Shuman, C. A., *et al.* (1993). Abrupt increase in Greenland snow accumulation at the end of the Younger Dryas event. *Nature*, **362**, 527–9.

Alley, R. B., Mayewski, P. A., Sowers, T., *et al.* (1997). Holocene climatic instability: a prominent, widespread event 8200 yr ago. *Geology*, **25**, 483–6.

Ambaum, M. H. P., Hoskins, B. J. and Stephenson, D. B. (2001). Arctic Oscillation or North Atlantic Oscillation? *J. Clim.*, **14**, 3495–507.

Andersen, K. K., Azuma, N., Barnola, J. M., *et al.* (2004). High-resolution record of Northern Hemisphere climate extending into the last interglacial period. *Nature*, **431**, 147–51.

Anderson, J. B., Shipp, S. S., Lowe, A. L., Wellner, J. S., and Mosola, A. B. (2001a). The Antarctic Ice Sheet during the Last Glacial Maximum and its subsequent retreat history: a review. *Quat. Sci. Rev.*, **21**, 49–70.

Anderson, L., Abbott, M. B. and Finney, B. P. (2001b). Holocene climate inferred from oxygen isotope ratios in lake sediments, central Brooks Range, Alaska. *Quat. Res.*, **55**, 313–21.

Anderson, T. L., Charlson, R. J., Schwartz, S. E., *et al.* (2003). Climate forcing by aerosols: a hazy picture. *Science*, **300**, 1103–4.

Andrews, J. T., Helgadottir, G., Geirsdottir, A. and Jennings, A. E. (2001). Multicentury-scale records of carbonate (hydrographic ?) variability on the northern Iceland margin over the last 5000 years. *Quat. Res.*, **56**, 199–206.

Angell, J. K. (1997). Estimated impact of Agung, El Chichon and Pinatubo volcanic eruptions on global and regional total ozone after adjustment for the QBO. *Geophys. Res. Lett.*, **24**, 647–50.

Aoki, S. (2002). Coherent sea level response to the Antarctic Oscillation. *Geophys. Res. Lett.*, **29**, 1950, doi:10.1029/2002GL015733.

Aoki, S., Bindoff, N. L. and Church, J. (2005). Interdecadal water mass changes in the Southern Ocean between 30°E and 160°E. *Geophys. Res. Lett.*, **32**, L07607, doi:10.1029/2004GL022220.

Arbic, B. K., MacAyeal, D. R., Mitrovica, J. X., and Milne, G. A. (2004). Ocean tides and Heinrich events. *Nature*, **432**, 460.

Arblaster, J. and Meehl, G. A. (2006). Contributions of external forcings to Southern Annular Mode trends. *J. Clim.*, **19**, 2896–905.

Arguez, A., Waple, A. M. and Sanchez-Lugo, A. M. (2007). State of the climate in 2006. *Bull. Amer. Meteorol. Soc.*, **88**, 933–5.

Armstrong, R. L. and Brodzik, M. J. (2001). Recent northern hemisphere snow extent: a comparison of data derived from visible and microwave satellite sensors. *Geophys. Res. Lett.*, **28**, 3673–6.

Arnell, N. W. (1999). Climate change and global water resources. *Global Environ. Change*, **9**, S31–S49.

Arrhenius, S. (1896) On the influence of carbonic acid in the air upon the temperature of the ground. *Philos. Mag.*, **41**, 237–76.

Arzel, O., Fichefet, T. and Goosse, H. (2006). Sea ice evolution over the 20th and 21st centuries as simulated by current AOGCMs. *Ocean Model.*, **12**, 401–15.

Azumaya, T. and Ohtani, K. (1995). Effect of winter meteorological conditions on the formation of the cold bottom water in the eastern Bering Sea shelf. *J. Oceanogr.*, **51**, 665–80.

Baldwin, M. P. (2001). Annular modes in global daily surface pressure. *Geophys. Res. Lett.*, **28**, 4115–18.

Baldwin, M. P. and Dunkerton, T. J. (2001). Stratospheric harbingers of anomalous weather regimes. *Science*, **294**, 581–4.

Ballantyne, J. and Long, D. G. (2002). A multidecadal study of the number of Antarctic icebergs using scatterometer data. In *Geoscience and Remote Sensing Symposium 2002. IGARSS '02*, Washington, DC: IEEE International, pp. 3029–31.

Bals-Elsholz, T. M., Atallah, E. H., Bosart, L. F., *et al.* (2001). The wintertime Southern Hemisphere split jet: structure, variability, and evolution. *J. Clim.*, **14**, 4191–215.

Barbante, C., Turetta, C., Gambaro, A., Capodaglio, G. and Scarponi, G. (1998). Sources and origins of aerosols reaching Antarctica as revealed by lead concentration profiles in shallow snow. *Ann. Glaciol.*, **27**, 674–8.

Barber, D. C., Dyke, A., Hillaire-Marcel, C., *et al.* (1999). Forcing of the cold event of 8,200 years ago by catastrophic drainage of Laurentide lakes. *Nature*, **400**, 344–8.

Barker, S. P., Diz, M. J., Vautravers, J., *et al.* (2009). Interhemispheric Atlantic seesaw response during the last glaciations. *Nature*, **457**, 1097–103.

Bard, E., Raidbeck, G., Yiou, F. and Jouzel, J. (2000). Solar irradiance during the last 1200 years based on cosmogenic nuclides. *Tellus*, **52B**, 985–92.

Barnola, J. M., Pimienta, P., Raynaud, D. and Korotkevich, Y. S. (1991). CO_2 climate relationship as deduced from the Vostok ice core: a re-examination based on new measurements and on a re-evaluation of the air dating. *Tellus*, **43B**, 83–91.

Baroni, C. and Orombelli, G. (1994). Abandoned penguin rookeries as Holocene paleoclimatic indicators in Antarctica. *Geology*, **22**, 23–6.

Barrows, T. T., Lehman, S. J., Fifield, L. K., and De Deckker, P. (2007). Absence of cooling in New Zealand and the adjacent ocean during the Younger Dryas chronozone. *Science*, **318**, 86–8.

Barsch, D. and Mäusbacher, R. (1986). New data on the relief development of the South Shetland Islands, Antarctica. *Interdiscipl Sci. Rev.*, **11**, 211–18.

Batifol, F., Boutron, C. and Deangelis, M. (1989). Changes in copper, zinc and cadmium concentration in Antarctic ice during the past 40,000 years. *Nature*, **337**, 544–6.

Bauch, H. A., Erlenkeuser, H., Spielhagen, R. F., *et al.* (2001). A multiproxy reconstruction of the evolution of deep and surface waters in the subarctic Nordic seas over the last 30,000 years. *Quat. Sci. Rev.*, **20**, 659–78.

Baumann, K.-H., Lackschweitz, S., Mangerud, J., *et al.* (1995). Reflection of Scandinavian Ice Sheet fluctuations in Norwegian Sea sediments during the past 150,000 years. *Quat. Res.*, **43**, 185–97.

Beckley, B. D., Lemoine, F. G., Luthcke, S. B., Ray, R. D. and Zelensky, N. P. (2007). A reassessment of global and regional mean sea level trends from TOPEX and Jason-1 altimetry based on revised reference frame and orbits. *Geophys. Res. Lett.*, **34**, L14608, doi:10.1029/2007GL030002.

Becquey, S. and Gersonde, R. (2002). Past hydrographic and climatic changes in the Subantarctic zone of the South Atlantic: the Pleistocene record from ODP Site 1090. *Palaeogeogr. Palaeoclimatol. Palaeoecol.*, **182**, 221–39.

Beer, J., Mende, W., Stellmacher, R. and White, O. R. (1996). Intercomparisons of proxies for past solar variability. In *Climatic Variations and Forcing Mechanisms of the Last 2000 Years*, ed. P. D. Jones, R. S. Bradley, and J. Jouzel, Berlin: Springer, pp. 501–18.

Beilman, D. and Robinson, S. D. (2003). Peatland permafrost thaw and landform type along a climatic gradient. In *Proceedings of the 8th International Conference on Permafrost, Zurich, Switzerland, 21–25 July 2003*, Lisse, Netherlands: A.A. Balkema Publishers, pp. 61–5.

Bellucci, A. and Richards, K. J. (2006). Effects of NAO variability on the North Atlantic Ocean circulation. *Geophys. Res. Lett.*, **33**, L02612, doi:10.1029/2005GL024890.

Beltrami, H. and Taylor, A. E. (1995). Records of climatic change in the Canadian Arctic: towards calibrating oxygen isotope data with geothermal data. *Global Planet. Change*, **11**, 127–38.

Bengtsson, L., Semenov, V. A. and Johannessen, O. M. (2004). The early twentieth-century warming in the Arctic: a possible mechanism. *J. Clim.*, **17**, 4045–57.

Bentley, M. J. (1999). Volume of Antarctic ice at the Last Glacial Maximum, and its impact on global sea level change. *Quat. Sci. Rev.*, **18**, 1569–95.

Bentley, M. J., Hodgson, D. A., Sugden, D. E., *et al.* (2005). Early Holocene retreat of the George VI Ice Shelf, Antarctic Peninsula. *Geology*, **33**, 173–6.

Bentley, M. J., Fogwill, C. J., Kubik, P. W. and Sugden, D. E., (2006). Geomorphological evidence and cosmogenic ^{10}Be/^{26}Al exposure ages for the Last Glacial Maximum and deglaciation of the Antarctic Peninsula Ice Sheet. *Geol. Soc Amer. Bull.*, **118**, 1149–59.

Bentley, M. J., Hodgson, D. A., Smith, J. A., *et al.* (2009). Mechanisms of Holocene palaeoenvironmental change in the Antarctic Peninsula region. *Holocene*, **19**, 51–69.

Berger, A. (2001). The role of CO_2, sea-level and vegetation during the Milankovitch-forced glacial-interglacial cycles. In *Geosphere-Biosphere Interactions and Climate*, ed. L. Bengtsson and C. U. Hammer, New York: Cambridge University Press, pp. 119–46.

Berger, A., Li, X. S., and Loutre, M. F., (1999). Modelling northern hemisphere ice volume over the past 3 Ma. *Quat. Sci. Rev.*, **18**, 1–11.

Berger, W. H. and Wefer, G. (2003). On the dynamics of the ice ages: Stage-11 Paradox, Mid-Brunhes Climate Shift, and 100-ky cycle. In *Earth's Climate and Orbital Eccentricity: The Marine Isotope Stage 11 Question.*, ed. A. W. Droxler, R. Z. Poore and L. H. Burckle, Washington DC: AGU, pp. 41–59.

Berger, W. H., Yasuda, M. K., Bickert, T., Wefer, G. and Takayama, T. (1994). Quaternary time scale for the Ontong Java Plateau: Milankovitch template for ocean drilling program. *Geology*, **22**, 463–7.

Berresheim, H. (1987). Biogenic sulfur emissions from the subantarctic and Antarctic oceans. *J. Geophys. Res.*, **92**, 13 245–62.

Bi, D., Budd, W. F., Hirst, A. C. and Wu, X. (2001). Collapse and reorganisation of the Southern Ocean overturning under global warming in a coupled model. *Geophys. Res. Lett.*, **28**, 3927–30.

Bianchi, C. and Gersonde, R. (2002). The Southern Ocean surface between Marine Isotope Stages 6 and 5d: shape and timing of climate changes. *Palaeogeogr. Palaeoclimatol. Palaeoecol.*, **187**, 151–77.

Bigg, E. K. and Leck, C. (2001). Properties of the aerosol over the central Arctic Ocean. *J. Geophys. Res.*, **106**, 32 101–9.

Bindoff, N., Willebrand, J., Artale, V., *et al.* (2007). Observations: oceanic climate change and sea level. In *Climate Change 2007: The Physical Science Basis. Contribution of Working Group 1 to the Fourth Assessment Report of the Intergovernmental Panel on Climate Change*, ed. S. Solomon, D. Qin and M. Manning, Cambridge, UK: Cambridge University Press, pp. 385–432.

Bintanja, R. and van de Wal, R. S. W. (2008). North American ice-sheet dynamics and the onset of 100,000-year glacial cycles. *Nature*, **454**, 869–72.

Birks, C. J. A. and Koc, N. (2002). A high-resolution diatom record of late-Quaternary sea-surface temperatures and oceanographic conditions from the eastern Norwegian Sea. *Boreas*, **31**, 323–44.

Birks, H. H. and Birks, H. J. B. (2003). Reconstructing Holocene climates from pollen and plant macrofossils. In *Global Change in the Holocene*, ed. A. Mackay, R. Battarbee, J. Birks, and F. Oldfield, London: Arnold, pp. 342–357.

Bitz, C. M. and Roe, G. H. (2004). A mechanism for the high rate of sea ice thinning in the Arctic Ocean. *J. Clim.*, **17**, 3623–32.

Björck, S., Håkansson, H., Zale, R., Karlén, W. and Jönsson, B. L. (1991). A late Holocene lake sediment sequence from Livingston Island, South Shetland Islands, with palaeoclimatic implications. *Antarct. Sci.*, **3**, 61–72.

Björck, S., Håkansson, H., Olsson, S., Barnekow, L. and Janssens, J. (1993). Paleoclimatic studies in South Shetland Islands, Antarctica, based on numerous stratigraphic variables in lake sediments. *J. Paleolimnol.*, **8**, 233–72.

Björck, S., Olsson, S., Ellisevans, C., *et al.* (1996). Late Holocene palaeoclimatic records from lake sediments on James Ross Island, Antarctica. *Palaeogeog. Palaeoclimatol. Palaeoecol.*, **121**, 195–220.

Blanchon, P. and Shaw, J. (1995). Reef drowning during the last deglaciation: evidence for catastrophic sea-level rise and ice-sheet collapse. *Geology*, **23**, 4–8.

Blunier, T. and Brook, E. J. (2001). Timing of millenial-scale climate change in Antarctica and Greenland during the last glacial period. *Science*, **291**, 109–12.

Bockheim, J. and Hall, K. (2002). Periglacial processes and landforms of the Antarctic continent: a review. *South African J. Sci.*, **98**, 82–101.

Bodhaine, B. A., Deluisi, J. J., Harris, J. M., Houmere, P. and Bauman, S. (1986). Aerosol measurements at the South Pole. *Tellus*, **38B**, 223–35.

Bond, G., Heinrich, H., Broecker, W., *et al.* (1992). Evidence for massive discharges of icebergs into the North-Atlantic Ocean during the last glacial period. *Nature*, **360**, 245–49.

Bond, G., Showers, W., Cheseby, M., *et al.* (1997). A pervasive millennial-scale cycle in North Atlantic Holocene and glacial climates. *Science*, **278**, 1257–66.

Bond, G., Kromer, B., Beer, J., *et al.* (2001). Persistent solar influence on North Atlantic climate during the Holocene. *Science*, **294**, 2130–6.

Bormann, N. and Thépaut, J. N. (2004). Impact of MODIS polar winds in ECMWF's 4DVAR data assimilation system. *Mon. Wea. Rev.*, **132**, 929–40.

Boutron, C. F. and Wolff, E. W. (1989). Heavy-metal and sulfur emissions to the atmosphere from human activities in Antarctica. *Atmos. Environ.*, **23**, 1669–75.

Bowen, D. Q. (2010). Sea level ~400 000 years ago (MIS 11): analogue for present and future sea-level? *Clim. Past*, **6**, 19–29.

Boyer, T. P., Levitus, S., Antonov, J. I., Locarnini, R. A. and Garcia, H. E. (2005). Linear trends in salinity for the World Ocean, 1955–1998. *Geophys. Res. Lett.*, **32**, L01604, doi:10.1029/2004GL021791.

Bracegirdle, T. J., Connolley, W. M. and Turner, J. (2008). Antarctic climate change over the twenty first century. *J. Geophys. Res.*, **113**, D03103, doi:10.1029/2007JD008933.

Brachfeld, S., Domack, E., Kissel, C., *et al.* (2003). Holocene history of the Larsen-A Ice Shelf constrained by geomagnetic paleointensity dating. *Geology*, **31**, 749–52.

Bradley, R. S. (1988). The explosive volcano eruption signal in Northern Hemisphere continental temperature records. *Climatic Change*, **12**, 221–43.

Bradley, R. S. (2003). Climate forcing during the Holocene. In *Global Change in the Holocene*, ed. A. Mackay, R. Battarbee, J. Birks and F. Oldfield, London: Arnold, pp. 10–19.

Braithwaite, R. J. and Olesen, O. B. (1990). A simple energy-balance model to calculate ice ablation at the margin of the Greenland Ice-Sheet. *J. Glaciol.*, **36**, 222–8.

Briffa, K. R. (2000). Annual variability in the Holocene: interpreting the message of ancient trees. *Quat. Sci. Rev.*, **19**, 87–105.

Briffa, K. R., Bartholin, T. S., Eckstein, D., *et al.* (1990). A 1,400-year tree-ring record of summer temperatures in Fennoscandia. *Nature*, **346**, 434–9.

Broecker, W. S. (2000). Was a change in thermohaline circulation responsible for the Little Ice Age? *Proc. Natl. Acad. Sci.*, **97**, 1339–42.

Broecker, W. S. (2001). Paleoclimate: was the medieval warm period global? *Science*, **291**, 1497–9.

Broecker, W. S. (2006). Was the Younger Dryas triggered by a flood? *Science*, **312**, 1146–8.

Broecker, W. S., Bond, G., Klas, M., Bonani, G. and Wolfi, W. (1990). A salt oscillator in the glacial Atlantic? 1. The concept. *Paleoceanography*, **5**, 469–77.

Bromwich, D. H. (1988). Snowfall in high southern latitudes. *Rev. Geophys.*, **26**, 149–68.

Bromwich, D. H. and Fogt, R. L. (2004). Strong trends in the skill of the ERA-40 and NCEP/NCAR reanalyses in the high and middle latitudes of the Southern Hemisphere, 1958–2001. *J. Clim.*, **17**, 4603–19.

Bromwich, D. H. and Kurtz, D. D. (1984). Katabatic wind forcing of the Terra Nova Bay polynya. *J. Geophys. Res.*, **89**, 3561–72.

Bromwich, D. H., Carrasco, J. F. and Stearns, C. R. (1992). Satellite observations of katabatic-wind propagation for great distances across the Ross Ice Shelf. *Mon. Wea. Rev.*, **120**, 1940–9.

Bromwich, D. H., Rogers, A. N., Kållberg, P., *et al.* (2000). ECMWF analyses and reanalyses depiction of ENSO signal in Antarctic precipitation. *J. Clim.*, **13**, 1406–20.

Bromwich, D. H., Cassano, J. J., Klein, T., *et al.* (2001). Mesoscale modeling of katabatic winds over Greenland with the Polar MM5. *Mon. Wea. Rev.*, **129**, 2290–309.

Bromwich, D. H., Toracinta, E. R., Wei, H., *et al.* (2004). Polar MM5 simulations of the winter climate of the Laurentide Ice Sheet at the LGM. *J. Clim.*, **17**, 3415–33.

Bromwich, D. H., Monaghan, A. J., Manning, K. W. and Powers, J. G. (2005). Real-time forecasting for the Antarctic: an evaluation of the Antarctic Mesoscale Prediction System (AMPS). *Mon. Wea. Rev.*, **133**, 579–603.

Bromwich, D. H., Fogt, R. L., Hodges, K. I. and Walsh, J. E. (2007). A tropospheric assessment of the ERA-40, NCEP, and JRA-25 global reanalyses in the polar regions. *J. Geophys. Res.*, **112**, D10111, doi:10.1029/2006JD007859.

Brook, E. J., Harder, S., Severinghaus, J., Steig, E. J. and Sucher, C. M. (2000). On the origin and timing of rapid changes in atmospheric methane during the last glacial period. *Global Biogeochem. Cyc.*, **14**, 559–72.

Brown, J., Hinkel, K. M. and Nelson, F. E. (2000). The circumpolar active layer monitoring (calm) program: research designs and initial results. *Polar Geogr.*, **24**, 165–258.

Budd, W. F. (1991). Antarctica and global change. *Climatic Change*, **18**, 271–99.

Bugnion, V. (2000). Reducing the uncertainty in the contribution of Greenland to sea-level rise in the 20th and 21st centuries. *Ann. Glaciol.*, **31**, 121–5.

Cai, M. (2005). Dynamical amplification of polar warming. *Geophys. Res. Lett.*, **32**, L22710, doi:10.1029/2005GL 024481.

Cai, W. J., Baines, P. G. and Gordon, H. B. (1999). Southern mid- to high-latitude variability, a zonal wavenumber-3 pattern, and the Antarctic circumpolar wave in the CSIRO coupled model. *J. Clim.*, **12**, 3087–104.

Caillon, N., Severinghaus, J. P., Jouzel, J., *et al.* (2003). Timing of atmospheric CO_2 and Antarctic temperature changes across Termination III. *Science*, **299**, 1728–31.

Calkin, P. E., Wiles, G. C. and Barclay, D. J. (2001). Holocene coastal glaciation of Alaska. *Quat. Sci. Rev.*, **20**, 449–61.

Campbell, I. D., Campbell, C., Apps, M. J., Rutter, N. W. and Bush, A. B. G. (1998). Late Holocene ca.1500 yr climatic periodicities and their implications. *Geology*, **26**, 471–3.

Carril, A. F., Menendez, C. G. and Navarra, A. (2005). Climate response associated with the Southern Annular Mode in the surroundings of Antarctic Peninsula: a multimodel ensemble analysis. *Geophys. Res. Lett.*, **32**, L16713, doi:10.1029/GL023581.

Carsey, F. D. (1980). Microwave observations of the Weddell Polynya. *Mon. Wea. Rev.*, **108**, 2032–44.

Cash, B. A., Kushner, P. J. and Vallis, G. K. (2002). The structure and composition of the annular modes in an aquaplanet general circulation model. *J. Atmos. Sci.*, **59**, 3399–414.

Cassano, J. J., Uotila, P. and Lynch, A. H. (2006). Changes in synoptic weather patterns in the polar regions in the 20th and 21st centuries. Part 1. *Int. J. Climatol.*, **26**, 1027–49.

Cavalieri, D. J., Parkinson, C. L. and Vinnikov, K. Y. (2003). 30-year satellite record reveals contrasting Arctic and Antarctic decadal sea ice variability. *Geophys. Res. Lett.*, **30**, 1970, doi:10.1029/2003GL018031.

Chambers, M. J. G. (1966). Investigations of patterned ground at Signy Island, South Orkney Islands: II. Temperature regimes in the active layer. *BAS Bull.*, **10**, 71–83.

Chapman, W. L. and Walsh, J. E. (1993). Recent variations in sea ice and air temperature at high latitudes. *Bull. Amer. Meteorol. Soc.*, **74**, 33–48.

Chapman, W. L. and Walsh, J. E. (2007). A synthesis of Antarctic temperatures. *J. Clim.*, **20**, 4096–117.

Chappellaz, J., Barnola, J. M., Raynaud, D., Korotkevich, Y. S. and Lorius, C. (1990). Ice-core record of atmospheric methane over the past 160,000 years. *Nature*, **345**, 127–31.

Chappellaz, J., Blunier, T., Raynaud, D., *et al.* (1993). Synchronous changes in atmospheric CH_4 and Greenland climate between 40 and 8 kyr BP. *Nature*, **366**, 443–5.

Chappellaz, J., Brook, E., Blunier, T. and Malaiz, B. (1997). CH_4 and $\delta^{18}O$ of O_2 records from Antarctic and Greenland ice: a clue for stratigraphic disturbance in the bottom part of the Greenland Ice Core Project and the Greenland Ice Sheet Project 2 ice cores. *J. Geophys. Res.*, **102**, 26 547–57.

Charles, C. D., Rind, D., Jouzel, J., Koster, R. D., and Fairbanks, R. G. (1994). Glacial-interglacial changes in moisture sources for Greenland: influences on the ice core record of climate. *Science*, **263**, 508–11.

Charlson, R. J., Lovelock, J. E., Meinrat, O. A. and Warren, S. G. (1987). Oceanic phytoplankton, atmosphric sulphur, cloud albedo and climate. *Nature*, **326**, 655–61.

Chase, T. N., Herman, B., Pielke, R. A., Zeng, X. and Leuthold, M. (2002). A proposed mechanism for the regulation of minimum midtropospheric temperatures in the Arctic. *J. Geophys. Res.*, **107**, 4193, doi:10.1029/2001JD001425.

Cheddadi, R., Yu, G., Guiot, J., Harrison, S. P. and Prentice, I. C. (1996). The climate of Europe 6000 years ago. *Clim. Dyn.*, **13**, 1–9.

Chen, B., Smith, S. R. and Bromwich, D. H. (1996). Evolution of the tropospheric split jet over the South Pacific Ocean during the 1986–89 ENSO cycle. *Mon. Wea. Rev.*, **124**, 1711–31.

Chen, J. L., Wilson, C. R., Blankenship, D. D. and Tapley, B. D. (2006). Antarctic mass rates from GRACE. *Geophys. Res. Lett.*, **33**, L11502, doi:10.1029/2006GL026369.

Cheng, H., Edwards, R. L., Briecker, W. S., *et al.* (2009). Ice age terminations. *Science*, **326**, 248–52.

Christoph, M., Barnett, T. P. and Roeckner, E. (1998). The Antarctic Circumpolar Wave an a coupled ocean – atmosphere GCM. *J. Clim.*, **11**, 1659–72.

Church, J. A. and White, N. J. (2006). A 20th century acceleration in global sea-level rise. *Geophys. Res. Lett.*, **33**, L01602, doi:10.1029/2005GL024826.

Ciais, P., Petit, J. R., Jouzel, J., *et al.* (1992). Evidence for an Early Holocene climatic optimum in the Antarctic deep ice core record. *Clim. Dyn.*, **6**, 169–77.

Ciais, P., Jouzel, J., Petit, J. R., Lipenkov, V. and White, J. W. C. (1994). Holocene temperature variations inferred from six Antarctic ice cores. *Ann. Glaciol.*, **20**, 427–39.

Clague, J. J. and James, T. S. (2002). History and isostatic effects of the last ice sheet in southern British Columbia. *Quat. Sci. Rev.*, **21**, 71–87.

Clapperton, C. M. and Sugden, D. E. (1988). Holocene glacier fluctuations in South America and Antarctica. *Quat. Sci. Rev.*, **7**, 185–98.

Clapperton, C. M., Sugden, D. E., Birnie, J. and Wilson, M. J. (1989). Late-glacial and Holocene glacier fluctuations and environmental-change on South Georgia, Southern-Ocean. *Quat. Res.*, **31**, 210–28.

Clark, C. D., Knight, J. K., and Gray, J. T. (2000). Geomorphological reconstruction of the Labrador Sector of the Laurentide Ice Sheet. *Quat. Sci. Rev.*, **19**, 1343–66.

Clark, P. U., Alley, R. B. and Pollard, D. (1999). Climatology: Northern hemisphere ice-sheet influences on global climate change. *Science*, **286**, 1104–11.

Clark, P. U., Mitrovica, J. X., Milne, G. A. and Tamisiea, M. E. (2002). Sea-level finger-printing as a direct test for the source of global meltwater pulse 1A. *Science*, **295**, 2438–41.

Clark, P. U., Archer, D., Pollard, D., *et al.* (2006). The middle Pleistocene transition: characteristics, mechanisms, and implications for long-term changes in amospheric pCO_2. *Quat. Sci. Rev.*, **25**, 3150–84.

Clarke, G. K. C. (2005). Subglacial processes. *Annu. Rev. Earth Planet. Sci.*, **33**, 247–76.

Clarke, G. K. C., Leverington, D. W., Teller, J. T. and Dyke, A. S. (2004). Paleohydraulics of the last outburst flood from glacial Lake Agassiz and the 8200 BP cold event. *Quat. Sci. Rev.*, **23**, 389–407.

Claussen, M., Mysak, L. A., Weaver, A. J., *et al.* (2002). Earth system models of intermediate complexity: closing the gap in the spectrum of climate system models. *Clim. Dyn.*, **18**, 579–86.

COHMAP members (1988). Climatic changes of the last 18,000 years: observations and model simulations. *Science*, **241**, 1043–51.

Comiso, J. C. (2000). Variability and trends in Antarctic surface temperatures from in situ and satellite infrared measurements. *J. Clim.*, **13**, 1674–96.

Comiso, J. C. (2003). Warming trends in the Arctic from clear sky satellite observations. *J. Clim.*, **16**, 3498–510.

Comiso, J. C. and Nishio, F. (2008). Trends in the sea ice cover using enhanced and compatible AMSR-E, SSM/I and SMMR data. *J. Geophys. Res.*, **113**, C02S07, doi:10.1029/2007JC004257.

Comiso, J. C., Wadhams, P., Pedersen, L. T. and Gersten, R. A. (2001). Seasonal and interannual variability of the Odden ice tongue and a study of environmental effects, *J. Geophys. Res.*, **106**, 9093–116.

Compo, G. P., Whitaker, J. S. and Sardeshmukh, P. D. (2006). Feasibility of a 100 year reanalysis using only surface pressure data. *Bull. Amer. Meteorol Soc.*, **87**, 175–90.

Connolley, W. M. (1997). Variability in annual mean circulation in southern high latitudes. *Clim. Dyn.*, **13**, 745–56.

Connolley, W. M. (2002). Long-term variation of the Antarctic Circumpolar Wave. *J. Geophys. Res.*, **107**, 8076, doi:10.1029/2000JC000380.

Connolley, W. M. and Bracegirdle, T. J. (2007). An Antarctic assessment of IPCC AR4 coupled models. *Geophys. Res. Lett.*, **34**, L22505, doi:10.1029/2007GL031648.

Connolley, W. M. and King, J. C. (1993). Atmospheric water vapour transport to Antarctica inferred from radiosonde data. *Quart. J. Roy. Meteor Soc.*, **119**, 325–42.

Conway, H., Hall, B. L., Denton, G. H., Gades, A. M. and Waddington, E. D. (1999). Past and future grounding-line retreat of the West Antarctic Ice Sheet. *Science*, **286**, 280–3.

Cook, A. J., Fox, A. J., Vaughan, D. G. and Ferrigno, J. G. (2005). Retreating glacier fronts on the Antarctic Peninsula over the past half-century. *Science*, **308**, 541–4.

Cooper, A. P. R. (1997). Historical observations of Prince Gustav Ice Shelf. *Polar Rec.*, **33**, 285–94.

Cortijo, E., Lehman, S., Keigwin, L., *et al.* (1999). Changes in meridional temperature and salinity gradients in the North Atlantic Ocean (30°–72°N) during the last interglacial period. *Paleoceanography*, **14**, 23–33.

Croll, J. (1875). *Climate and Time*, New York: Appleton & Co.

Crosta, X., Debret, M., Denis, D., Courty, M.-A. and Ther, O. (2007). Holocene long- and short-term climate changes off Adelie Land, East Antarctica. *Geochem. Geophys. Geosys.*, **8**, Q11009, doi:10.1029/2007GC001718.

Crowley, T. J. (2000). Causes of climate change over the past 1000 years. *Science*, **289**, 270–7.

Cubasch, U., Meehl, G. A., Boer, G. J., *et al.* (2001). Projections of future climate change. In *Climate Change 2001: The Scientific Basis-Contribution of Working Group I to the Third Assessment Report of the Intergovernment Panel on Climate Change*, ed. J. T. Houghton, Y. Ding, D. J. Griggs, *et al.*, Cambridge, UK: Cambridge University Press, pp. 525–82.

Cuffey, K. M. and Clow, G. D. (1997). Temperature, accumulation, and ice sheet elevation in central Greenland through the last deglacial transition. *J. Geophys. Res.*, **102**, 26 383–96.

Cuffey, K. M., Clow, G. D., Alley, R. B., *et al.* (1995). Large Arctic temperature change at the Wisconsin-Holocene glacial transition. *Science*, **270**, 455–8.

Cullather, R. I. and Bromwich, D. H. (2000). The atmospheric hydrologic cycle over the Arctic basin from reanalyses. Part I: Comparison with observations and previous studies. *J. Clim.*, **13**, 923–37.

Cullather, R. I. and Lynch, A. H. (2003). The annual cycle and interannual variability of atmospheric pressure in the vicinity of the North Pole. *Int. J. Climatol.*, **23**, 1161–83.

Cullather, R. I., Bromwich, D. H. and Van Woert, M. L. (1996). Interannual variations in Antarctic precipitation related to El Niño–Southern Oscillation. *J. Geophys. Res.*, **101**, 19 109–18.

Cullather, R. I., Bromwich, D. H. and Van Woert, M. L. (1998). Spatial and temporal variability of Antarctic precipitation from atmospheric methods. *J. Clim.*, **11**, 334–67.

Curran, M. A. J., Vanommen, T. D., Morgan, V. I., Phillips, K. L. and Palmer, A. S. (2003). Ice core evidence for Antarctic sea ice decline since the 1950s. *Science*, **302**, 1203–6.

Curry, J. A., Rossow, W. B., Randall, D. and Schramm, J. L. (1996). Overview of Arctic cloud and radiation characteristics. *J. Clim.*, **9**, 1731–64.

Czaja, A. and Marshall, J. C. (2006). The partitioning of poleward heat transport between the atmosphere and ocean, *J. Atmos. Sci.*, **63**, 1498–511.

Dahl-Jensen, D., Mosegaard, K., Gundestrup, N., *et al.* (1998). Past temperatures directly from the Greenland Ice Sheet. *Science*, **282**, 268–71.

Dällenbach, A., Blunier, T., Flückiger, J., Stauffer, B., Chappellaz, J. and Raynaud, D. (2000). Changes in the atmospheric CH_4 gradient between Greenland and Antarctica during the last glacial and the transition to the Holocene. *Geophys. Res. Lett.*, **27**, 1005–8.

Dansgaard, W. (1985). Greenland ice core studies. *Palaeogeog. Palaeoclimatol. Palaeoecol.*, **50**, 185–7.

Dansgaard, W., Johnsen, S. J., Reeh, N., *et al.* (1975). Climatic changes, Norsemen and modern man. *Nature*, **255**, 24–8.

Dansgaard, W., White, J. W. C. and Johnsen, S. J. (1989). The abrupt termination of the Younger Dryas climate event. *Nature*, **339**, 532–4.

Dansgaard, W., Johnsen, S. J., Clausen, H. B., *et al.* (1993). Evidence for general instability of past climate from a 250-kyr ice-core record. *Nature*, **364**, 218–20.

Davis, C. H., Li, Y. H., McConnell, J. R., Frey, M. M. and Hanna, E. (2005). Snowfall-driven growth in East Antarctic ice sheet mitigates recent sea-level rise. *Science*, **308**, 1898–901.

de Angelis, H. and Kleman, J. (2005). Palaeo-ice streams in the northern Keewatin sector of the Laurentide ice sheet. *Ann. Glaciol.*, **42**, 135–44.

de la Mare, W. (1997). Abrupt mid-twentieth-century decline in Antarctic sea-ice extent from whaling records. *Nature*, **389**, 57–60.

de Vernal, A. and Hillaire-Marcel, C. (2000). Sea-ice cover, sea-surface salinity and halo-/thermocline structure of the northwest North Atlantic: modern versus full glacial conditions. *Quat. Sci. Rev.*, **19**, 65–85.

de Vernal, A. and Hillaire-Marcel, C. (2008). Natural variability of Greenland climate, vegetation, and ice volume during the past million years. *Science*, **320**, 1622–5.

de Vernal, A., Miller, G. H., and Hillaire-Marcel, C. (1991). Paleoenvironments of the last interglacial in northwest North Atlantic region and adjacent mainland Canada. *Quat. Int.*, **10–12**, 95–106.

de Vernal, A., Eynaud, F., Hillaire-Marcel, C., *et al.* (2005). Reconstruction of sea-surface conditions at middle to high latitudes of the Northern Hemisphere during the Last Glacial Maximum (LGM) based on dinoflagellate cyst assemblages. *Quat. Sci. Rev.*, **24**, 897–924.

de Vernal, A., Rosell-Melé, A., Kucera, M., *et al.* (2006). Comparing proxies for the reconstruction of LGM sea-surface conditions in the northern North Atlantic. *Quat. Sci. Rev.*, **25**, 2820–34.

Delmotte, M., Chappellaz, J., Brook, E., *et al.* (2004). Atmospheric methane during the last four glacial-interglacial cycles: rapid changes and their link with Antarctic temperature. *J. Geophys. Res.*, **109**, D12104, doi:10.1029/2003JD004417.

deMenocal, P., Ortiz, J., Guilderson, T. and Samthein, M. (2000). Coherent high- and low-latitude climate variability during the Holocene warm period. *Science*, **288**, 2198–202.

Denton, G. H. and Karlén, W. (1973). Holocene climatic variations: their pattern and possible cause. *Quat. Res.*, **3**, 155–205.

Deser, C. (2000). On the teleconnectivity of the 'Arctic Oscillation'. *Geophys. Res. Lett.*, **27**, 779–82.

Deser, C., Walsh, J. E. and Timlin, M. S. (2000). Arctic sea ice variability in the context of recent atmospheric circulation trends. *J. Clim.*, **13**, 617–33.

Dickson, R. R. (1999). All change in the Arctic. *Nature*, **397**, 389–91.

Dickson, R. R., Meincke, J., Malmberg, S.-A. and Lee, A. J. (1988). The 'Great Salinity Anomaly' in the Northern North Atlantic 1968–1982. *Prog. Oceanogr.*, **20**, 103–51.

Dickson, R. R., Lazier, J., Meincke, J., Rhines, P. and Swift, J. (1996). Long-term co-ordinated changes in the convective activity of the North Atlantic. *Prog. Oceanogr.*, **38**, 241–95.

Dickson, R. R., Osborn, T. J., Hurrell, J. W., *et al.* (2000). The Arctic Ocean response to the North Atlantic Oscillation. *J. Clim.*, **13**, 2671–96.

Dickson, R. R., Yashayaev, I., Meincke, J., *et al.* (2002). Rapid freshening of the deep North Atlantic Ocean over the past four decades. *Nature*, **416**, 832–7.

Divine, D. V. and Dick, C. (2006). Historical variability of sea ice edge position in the Nordic Seas. *J. Geophys. Res.*, **111**, C01001, doi:10.1029/2004JC002851.

Doake, C. S. and Vaughan, D. G. (1991). Rapid disintegration of the Wordie Ice Shelf in response to atmospheric warming. *Nature*, **350**, 328–30.

Domack, E., Duran, D., Leventer, A., *et al.* (2005). Stability of the Larsen B ice shelf on the Antarctic Peninsula during the Holocene epoch. *Nature*, **436**, 681–5.

Dorale, J. A., Onac, B. P., Fornós, J. J., *et al.* (2010). Sea-level highstand 81,000 years ago in Mallorca. *Science*, **327**, 860–3.

Doran, P. T., Wharton, R. A. and Lyons, W. B. (1994). Paleolimnology of the McMurdo Dry Valleys, Antarctica. *J. Paleolimnol.*, **19**, 85–114.

Doran, P. T., Priscu, J. C., Lyons, W. B., *et al.* (2002). Antarctic climate cooling and terrestrial ecosystem response. *Nature*, **415**, 517–20.

Douglas, M. S. V., Smol, J. P. and Blake, W. (1994). Marked post-18th century environmental-change in high-Arctic ecosystems. *Science*, **266**, 416–19.

Dowdeswell, J. A., Hagen, J. O., Bjornsson, H., *et al.* (1997). The mass balance of circum-Arctic glaciers and recent climate change. *Quat. Res.*, **48**. 1–14.

Drinkwater, M., Long, M. D. and Bingham, A. (2001). Greenland snow accumulation estimates from satellite radar scatterometer data. *J. Geophys. Res.*, **106**, 33 935–50.

Dukhovskoy, D. S., Johnson, M. A. and Proshutinsky, A. (2004). Arctic decadal variability: an auto-oscillatory system of heat and fresh water exchange. *Geophys. Res. Lett.*, **31**, L03302, doi:10.1029/2003GL019023.

Dyke, A. S. and Prest, V. K. (1987). Late Wisconsinan and Holocene history of the Laurentide Ice Sheet. *Geogr. Phys. Quatern.*, **41**, 237–63.

Dyke, A. S., Hooper, J. and Savelle, J. M. (1996). A history of sea ice in the Canadian Arctic Archipelago based on postglacial remains of the bowhead whale (*Balaena mysticetus*). *Arctic*, **49**, 235–55.

Dyke, A. S., England, J., Reimnitz, E. and Jette, H. (1997). Changes in driftwood delivery to the Canadian Arctic archipelago: the hypothesis of postglacial oscillations of the transpolar drift. *Arctic*, **50**, 1–8.

Dyke, A. S., Andrews, J. T., Clark, P. U., *et al.* (2002). The Laurentide and Innuitian ice sheets during the Last Glacial Maximum. *Quat. Sci. Rev.*, **21**, 9–31.

Dyurgerov, M. B. (2001). Mountain glaciers at the end of the twentieth century: global analysis in relation to climate and water cycle. *Polar Geogr.*, **25**, 241–336.

Dyurgerov, M. B. and Meier, M. F. (2000). Twentieth century climate change: evidence from small glaciers. *Proc. Natl. Acad. Sci.*, **97**, 1406–11.

Eddy, J. A. (1976). Maunder minimum. *Science*, **192**, 1189–202.

Elkibbi, M. and Rial, J. A. (2001). An outsider's review of the astronomical theory of the climate: is the eccentricity-driven insolation the main driver of the ice ages? *Earth-Sci. Rev.*, **56**, 161–77.

Ellis, J. M. and Calkin, P. E. (1984). Chronology of Holocene glaciation, central Brooks Range, Alaska. *Bull. Geol. Soc. Amer.*, **95**, 897–912.

Emiliani, C. (1955). Pleistocene temperatures. *J. Geol.*, **63**, 538–78.

Emslie, S. D., Coats, L. and Licht, K. (2007). A 45,000 yr record of Adélie penguins and climate change in the Ross Sea, Antarctica. *Geology*, **35**, 61–4.

England, J., Atkinson, N., Bednarski, J., *et al.* (2006). The Innuitian Ice Sheet: configuration, dynamics and chronology. *Quat. Sci. Rev.*, **25**, 689–703.

EPICA Community Members (2004). Eight glacial cycles from an Antarctic ice core. *Nature*, **429**, 623–8.

EPICA Community Members (2006). One-to-one coupling of glacial climate variability in Greenland and Antarctica. *Nature*, **444**, 195–8.

Etheridge, D. M., Steele, L. P., Langenfelds, R. L., *et al.* (1996). Natural and anthropogenic changes in atmospheric CO_2 over the last 1000 years from air in Antarctic ice and firn, *J. Geophys. Res.*, **101**, 4115–28.

Evans, J., Pudsey, C. J., Ó Cofaigh, C., Morris, P. and Domack, E. (2005). Late Quaternary glacial history, flow dynamics and sedimentation along the eastern margin of the Antarctic Peninsula Ice Sheet. *Quat. Sci. Rev.*, **24**, 741–74.

Fahrbach, E., Hoppema, M., Rohardt, G., Schröder, M. and Wisotzki, A. (2004). Decadal-scale variations of water mass properties in the deep Weddell Sea. *Ocean Dyn.*, **54**, 77–91.

Farley, K. A. and Patterson, D. B. (1995). A 100-kyr periodicity in the flux of extra-terrestrial ^{3}He to the sea floor. *Nature*, **378**, 600–3.

Farman, J. C., Gardiner, B. G. and Shanklin, J. D. (1985). Large losses of total ozone in Antarctica reveal seasonal $CClOx/NOx$ interaction. *Nature*, **315**, 207–10.

Feldstein, S. B. (2000). Teleconnections and ENSO: the timescale, power spectra, and climate noise properties. *J. Clim.*, **13**, 4430–40.

Feldstein, S. B. (2002). The recent trend and variance increase of the annular mode. *J. Clim.*, **15**, 88–94.

Fischer, H., Wahlen, M., Smith, J., Mastroianni, D., and Deck, B. (1999). Ice core records of atmospheric CO_2 around the last three glacial terminations. *Science*, **283**, 1712–14.

Fischer, H., Traufetter, F., Oerter, H., Weller, R. and Miller, H. (2004). Prevalence of the Antarctic Circumpolar Wave over the last two millenia recorded in Dronning Maud Land ice. *Geophys. Res. Lett.*, **31**, L08202, doi:10.1029/2003GL019186.

Fischer, H., Siggaard-Andersen, M. L., Ruth, U., Rothlisberger, R. and Wolff, E. (2007). Glacial/interglacial changes in mineral dust and sea-salt records in polar ice cores: sources, transport, and deposition. *Rev. Geophys.*, **45**, RG1002, doi:10.1029/2005RG000192.

Fisher, D. A., and Koerner, R. M. (2003). Holocene ice-core climate history: a multi-variable approach. In *Global Change in the Holocene*, ed. A. Mackay, R. Battarbee, J. Birks, and F. Oldfield, London: Arnold, pp. 281–93.

Fogt, R. L. and Bromwich, D. H. (2006). Decadal variability of the ENSO teleconnection to the high latitude South Pacific governed by coupling with the Southern Annular Mode. *J. Clim.*, **19**, 979–97.

Foldvik, A., Gammelsrod, T., Osterhus, S., *et al.* (2004). Ice shelf water overflow and bottom water formation in the southern Weddell Sea. *J. Geophys. Res.*, **109**, C02015, doi:10.1029/2003JC002008.

Fox, A. J. and Cooper, A. P. R. (1998). Climate change indicators from archival aerial photography of the Antarctic Peninsula. *Ann. Glaciol.*, **27**, 636–42.

Francis, J. A. (1994). Improvements to TOVS retrievals over sea ice and applications to estimating Arctic energy fluxes. *J. Geophys. Res.*, **99**, 10 395–408.

Frauenfeld, O. W., Zhang, T., Barry, R. G. and Gilichinsky, D. (2004). Interdecadal changes in seasonal freeze and thaw depths in Russia. *J. Geophys. Res.*, **109**, D05101, doi:10.1029/2003JD004245.

Fréchette, B., Wolfe, A. P., Miller, G. H., Richard, P. J. H., and de Vernal, A. (2006). Vegetation and climate of the last interglacial on Baffin Island, Arctic Canada. *Palaeogeogr. Palaeoclimatol. Palaeoecol.*, **236**, 91–106.

French, H. M. (1996). *The Periglacial Environment*, 2nd edn., Harlow, UK: Longmans.

Fronval, T. and Jansen, E. (1996). Rapid changes in ocean circulation and heat flux in the Nordic seas during the last interglacial period. *Nature*, **383**, 806–10.

Fyfe, J. C. (2006). Southern Ocean warming due to human influence. *Geophys. Res. Lett.*, **33**, L19701, doi:10.1029/2006GL027247.

Fyfe, J. C. and Lorenz, D. J. (2005). Characterizing midlatitude jet variability: lessons from a simple GCM. *J. Clim.*, **18**, 3400–4.

Fyfe, J. C. and Saenko, O. A. (2006). Simulated changes in the extratropical Southern Hemisphere winds and currents. *Geophys. Res. Lett.*, **33**, L06701, doi:10.1029/2005GL025332.

Gallée, H., Ypersele, J. P., Fichefet, T., Tricot, C. and Berger, A. (1991). Simulation of the last glacial cycle by a coupled, sectorially averaged climate-ice sheet model, 1. The climate model. *J. Geophys. Res.*, **96**, 13 139–61.

Ganachaud, A. and Wunsch, C. (2000). Improved estimates of global ocean circulation, heat transport and mixing from hydrographic data. *Nature*, **408**, 453–7.

Ganopolski, A. and Rahmstorf, S. (2001). Rapid changes of glacial climate simulated in a coupled climate model. *Nature*, **409**, 153–8.

Ganopolski, A., Rahmstorf, S., Petoukhov, V. and Claussen, M. (1998). Simulation of modern and glacial climates with a coupled global model of intermediate complexity. *Nature*, **391**, 351–6.

Garrett, J. F. (1980). Availability of the FGGE drifting buoy system data set. *Deep-Sea Res.*, **27A**, 1083–6.

Garrett, T. J., and Zhao, C. F. (2006). Increased Arctic cloud longwave emissivity associated with pollution from mid-latitudes. *Nature*, **440**, 787–9.

Genthon, C. and Krinner, G. (1998). Convergence and disposal of energy and moisture on the Antarctic polar cap from ECMWF reanalyses and forecasts. *J. Clim.*, **11**, 1703–16.

Genthon, C., Krinner, G. and Sacchettini, M. (2003). Interannual Antarctic tropospheric circulation and precipitation variability. *Clim. Dyn.*, **21**, 289–307.

Genthon, C., Kaspari, S. and Mayewski, P. A. (2005). Interannual variability of the surface mass balance of West Antarctica from ITASE cores and ERA40 reanalyses, 1958–2000. *Clim. Dyn.*, **24**, 759–70.

Genty, D., Blamart, D., Ouahdi, R., *et al.* (2003). Precise dating of Dansgaard-Oeschger climate oscillations in western Europe from stalagmite data. *Nature*, **421**, 833–7.

Gersonde, R. and Zielinski, U. (2000). The reconstruction of late Quaternary Antarctic sea-ice distribution: the use of diatoms as a proxy for sea-ice. *Palaeogeog. Palaeoclimatol. Palaeoecol.*, **162**, 263–86.

Gersonde, R., Crosta, X., Abelmann, A. and Armand, L. (2005). Sea-surface temperature and sea ice distribution of the Southern Ocean at the EPILOG Last Glacial Maximum: a circum-Antarctic view based on siliceous microfossil records. *Quat. Sci. Rev.*, **24**, 869–96.

Gibb, J. G. (1986). A New Zealand regional Holocene eustatic sea level curve and its application for determination of vertical tectonic movements. *Bull. Roy. Soc. New Zealand*, **24**, 377–95.

Gibson, J. K., Kållberg, P. and Uppala, S. (1996). The ECMWF re-analysis (ERA) project. *ECMWF Newsletter*, 7–17.

Gille, S. T. (2002). Warming of the Southern Ocean since the 1950s. *Science*, **295**, 1275–7.

Gille, S. T. (2008). Decadal-scale temperature trends in the Southern Hemisphere ocean. *J. Clim.*, **21**, 4749–65.

Gillett, N. P., Allen, M. R. and Williams, K. D. (2003). Modelling the atmospheric response to doubled CO_2 and depleted stratospheric ozone using a stratosphere-resolving coupled GCM. *Quart. J. Roy. Meteor. Soc.*, **129**, 947–66.

Gillett, N. P., Stone, D. A., Stott, P. A., *et al.* (2009). Attribution of polar warming to human influence. *Nature Geosci.*, **1**, 750–4.

Gloersen, P. (1995). Modulation of hemispheric sea-ice cover by ENSO events. *Nature*, **373**, 503–6.

Gloersen, P. and White, W. B. (2001). Reestablishing the circumpolar wave in sea ice around Antarctica from one winter to the next. *J. Geophys. Res.*, **106**, 4391–5.

Gong, D. and Wang, S. (1999). Definition of Antarctic oscillation index. *Geophys. Res. Lett.*, **26**, 459–62.

Goodwin, I. D. (1998). Did changes in Antarctic ice volume influence late Holocene sea-level lowering? *Quat. Sci. Rev.*, **17**, 319–32.

Goodwin, I. D., van Ommen, T. D., Curran, M. A. J. and Mayewski, P. A. (2004). Mid latitude winter climate variability in the South Indian and southwest Pacific regions since 1300 AD. *Clim. Dyn.*, **22**, 783–94.

Goosse, H. and Renssen, H. (2001). A two-phase response of the Southern Ocean to an increase in greenhouse gas concentrations. *Geophys. Res. Lett.*, **28**, 3469–72.

Goosse, H. and Renssen, H. (2005). A simulated reduction in antarctic sea-ice area since 1750: implications of the long memory of the ocean. *Int. J. of Climatol.*, **25**, 569–79.

Grant, A. N., Brönnimann, S. and Haimberger, L. (2008). Recent Arctic warming vertical structure contested. *Nature*, **455**, E2–E3.

Grachev, A. M. and Severinghaus, J. P. (2005) A revised $+10 \pm 4\,°C$ magnitude of the abrupt change in Greenland temperature at the Younger Dryas termination using published GISP2 gas isotope data and air thermal diffusion constants. *Quat. Sci Rev.* **24**, 513–19.

Graversen, R. G., Mauritsen, T., Tjernstrom, M., Kallen, E. and Svensson, G. (2008). Vertical structure of recent Arctic warming. *Nature*, **451**, 53–6.

Gregory, J. M., Huybrechts, P. and Raper, S. C. B. (2004). Climatology: threatened loss of the Greenland ice-sheet. *Nature*, **428**, 616.

Groisman, P. Y., Karl, T. R., Knight, R. W. and Stenchikov, G. L. (1994). Changes of snow cover, temperature, and radiative heat balance over the northern hemisphere. *J. Clim.*, **7**, 1633–56.

Grootes, P. M., Stuiver, M., White, J. W. C., Johnsen, S. and Jouzel, J. (1993). Comparison of oxygen isotope records from the GISP2 and GRIP Greenland ice cores. *Nature*, **366**, 552–4.

Grove, J. M. (1990). *The Little Ice Age*, London: Routledge.

Grudd, H., Briffa, K. R., Karlen, W., *et al.* (2002). A 7400-year tree-ring chronology in northern Swedish Lapland: natural climatic variability expressed on annual to millennial timescales. *Holocene*, **12**, 657–65.

Grumet, N. S., Wake, C. P., Mayewski, P. A., *et al.* (2001). Variability of sea-ice extent in Baffin Bay over the last millennium. *Climatic Change*, **49**, 129–45.

Guo, Z. C., Bromwich, D. H. and Cassano, J. J. (2003). Evaluation of Polar MM5 simulations of Antarctic atmospheric circulation. *Mon. Wea. Rev.*, **131**, 384–411.

Haas, C., Nicolaus, M., Willmes, S., Worby, A. and Flinspach, D. (2008). Sea ice and snow thickness and physical properties of an ice floe in the western Weddell Sea and their changes during spring warming. *Deep-Sea Res. II*, **55**, 963–74.

Haigh, J. D. (1996). The impact of solar variability on climate. *Science*, **272**, 981–4.

Hall, A. and Visbeck, M. (2002). Synchronous variability in the Southern Hemisphere atmosphere, sea ice, and ocean resulting from the Annular Mode. *J. Clim.*, **15**, 3043–57.

Hallberg, R. and Gnanadesikan, A. (2006). The role of eddies in determining the structure and response of the wind-driven Southern Hemisphere overturning: results from the modeling eddies in the Southern Ocean (MESO) project. *J. Phys. Oceanogr.*, **36**, 2232–52.

Hanna, E. and Cappelen, J. (2003). Recent cooling in coastal southern Greenland and relation with the North Atlantic Oscillation. *Geophys. Res. Lett.*, **30**, 1132, doi:10.1029/2002GL015797.

Hanna, E., Huybrechts, P. and Mote, T. L. (2002). Surface mass balance of the Greenland ice sheet from climate-analysis data and accumulation/runoff models. *Ann. Glaciol.*, **35**, 67–72.

Hanna, E., Jónsson, T. and Box, J. E. (2004). An analysis of the Icelandic climate since the nineteenth century. *Int. J. Climatol.*, **24**, 1193–210.

Harris, J. M. (1992). An analysis of 5-day midtropospheric flow patterns for the south pole: 1985–1989. *Tellus*, **44B**, 409–21.

Hartmann, D. L. and Lo, F. (1998). Wave-driven zonal flow vacillation on the Southern Hemisphere. *J. Atmos. Sci.*, **55**, 1303–15.

Hartmann, D. L., Wallace, J. M., Limpasuvan, V., Thompson, D. W. J. and Holton, J. R. (2000). Can ozone depletion and global warming interact to produce rapid climate change? *Proc. Natl. Acad. Sci.*, **97**, 1412–17.

Harvey, L. D. D. (1980). Solar variability as a contributing factor to Holocene climatic change. *Prog. Phys. Geogr.*, **4**, 487–530.

Hays, J. D., Imbrie, J. and Shackleton, N. J. (1976). Variations in the Earth's orbit: pacemaker of the ice ages. *Science*, **194**, 1121–32.

Hearty, P. J., Kindler, P., Cheng, H., and Edwards, R. L. (1999). A +20 m middle Pleistocene sea-level highstand (Bermuda and the Bahamas) due to partial collapse of Antarctic ice. *Geology*, **27**, 375–8.

Heinrich, H. (1988). Origin and consequences of cyclic ice rafting in the Northeast Atlantic-Ocean during the past 130,000 years. *Quat. Res.*, **29**, 142–52.

Hemming, S. R. (2004). Heinrich events: massive late Pleistocene detritus layers of the North Atlantic and their global climate imprint. *Rev. Geophys.*, **42**, RG1005, doi:10.1029/2003RG000128.

Herber, A., Thomason, L. W., Dethloff, K., *et al.* (1996). Volcanic perturbation of the atmosphere in both polar regions: 1991–1994. *J. Geophys. Res.*, **101**, 3921–8.

Herber, A., Thomason, L. W., Gernandt, H., *et al.* (2002). Continuous day and night aerosol optical depth observations in the Arctic between 1991 and 1999. *J. Geophys. Res.*, **107**, 4097, doi:10.1029/2001JD000536.

Herterich, K. (1988). A three-dimensional model of the Antarctic ice sheet. *Ann. Glaciol.*, **11**, 32–5.

Heusser, C. J. (1989). Southern westerlies during the last glacial maximum. *Quat. Res.*, **31**, 423–5.

Hewitt, C. D., Stouffer, R. J., Broccoli, A. J., Mitchell, J. F. B. and Valdes, P. J. (2003). The effect of ocean dynamics in a coupled GCM simulation of the Last Glacial Maximum. *Clim. Dyn.*, **20**, 203–18.

Hillaire-Marcel, C., de Vernal, A., Bilodeau, G. and Weaver, A. J. (2001). Absence of deep-water formation in the Labrador Sea during the last interglacial period. *Nature*, **410**, 1073–7.

Hines, K. M., Bromwich, D. H. and Marshall, G. J. (2000). Artificial surface pressure trends in the NCEP/NCAR reanalysis over the Southern Ocean and Antarctica. *J. Clim.*, **12**, 3940–52.

Hines, K. M. and D. H. Bromwich, (2008). Development and testing of Polar WRF. Part I. Greenland ice sheet meteorology, *Mon. Wea. Rev.*, **136**, 1971–89.

Hjort, C., Ingólfsson, O., Moller, P. and Lirio, J. M. (1997). Holocene glacial history and sea-level changes on James Ross Island, Antarctic Peninsula. *J. Quat. Sci.*, **12**, 259–73.

Hjort, C., Ingólfsson, O., Bentley, M. J. and Björck, S. (2003). Late Pleistocene and Holocene glacial and climate history of the Antarctic Peninsula region: a brief overview of the land and late sediment records. In *Antarctic Peninsula Climate Variability: A Historical and Paeoenvironmental Perspective*. Antarctic Research Series 79, ed. E. Domack, A. Burnett, P. Convey, M. Kirby, and R. Bindschadler, Washington, DC: American Geophysical Union, pp. 95–101.

Hodgson, D. and Vincent, J. S. (1984). A 10,000 year B.P. extensive ice shelf over Viscount Melville Sound, Arctic Canada. *Quat. Res.*, **22**, 18–30.

Hodgson, D. A. and Convey, P. (2005). A 7000-year record of oribatid mite communities on a maritime-Antarctic island: responses to climate change. *Arct. Antarct. Alp. Res.*, **37**, 239–45.

Hodgson, D. A., Verleyen, E., Sabbe, K., *et al.* (2005). Late Quaternary climate-driven environmental change in the Larsemann Hills, East Antarctica, multi-proxy evidence from a lake sediment core. *Quat. Res.*, **64**, 83–99.

Hogg, A. M., Meredith, M. P., Blundell, J. R. and Wilson, C. (2008). Eddy heat flux in the Southern Ocean: response to variable wind forcing. *J. Clim.*, **21**, 608–20.

Holland, D. M., Jacobs, S. S. and Jenkins, A. (2003). Modeling Ross Sea ice shelf – ocean interaction. *Antarct. Sci.*, **15**, 13–23.

Holland, M. M. and Bitz, C. M. (2003). Polar amplification of climate change in coupled models. *Clim. Dyn.*, **21**, 221–32.

Hoskins, B. J. and Hodges, K. I. (2002). New perspectives on the Northern Hemisphere winter storm tracks. *J. Atmos. Sci.*, **59**, 1041–61.

Hoskins, B. J. and Hodges, K. I. (2005). A new perspective on Southern Hemisphere storm tracks. *J. Clim.*, **18**, 4108–29.

Hoskins, B. J. and Karoly, D. J. (1981). The steady linear response of a spherical atmosphere to thermal and orographic forcing. *J. Atmos. Sci.*, **38**, 1179–96.

Hu, R. M., Blanchet, J. P. and Girard, E. (2005). Evaluation of the direct and indirect radiative and climate effects of aerosols over the western Arctic. *J. Geophys. Res.*, **110**, D11213, doi:10.1029/2004JD005043.

Huber, C., Leuenberger, M., Spahni, R., *et al.* (2006). Isotope calibrated Greenland temperature record over Marine Isotope Stage 3 and its relation to CH_4. *Earth Planet. Sci. Lett.*, **243**, 504–19.

Hughes, C. W., Woodworth, P. L., Meredith, M. P., *et al.* (2003). Coherence of Antarctic sea levels, Southern Hemisphere Annular Mode, amd flow through Drake Passage. *Geophys. Res. Lett.*, **30**, 1464, doi:10.1029/2003GL017240.

Hulbe, C. L., MacAyeal, D. R., Denton, G. H., Kleman, J. and Lowell, T. V. (2004). Catastrophic ice shelf breakup as the source of Heinrich event icebergs. *Paleoceanography*, **19**, PA1004, doi:10.1029/2003PA000890.

Hurrell, J. W. (1995). Decadal trends in the North-Atlantic oscillation: regional temperatures and precipitation. *Science*, **269**, 676–9.

402 *References*

Hurrell, J. W. (1996). Influence of variations in extratropical wintertime teleconnections on Northern Hemisphere temperature. *Geophys. Res. Lett.*, **23**, 665–8.

Hurrell, J. W., Kushnir, Y., Ottersen, G. and Visbeck, M. (2003). An overview of the North Atlantic Oscillation. In *The North Atlantic Oscillation*, ed. J. W. Hurrell, Y. Kushnir, G. Ottersen and M. Visbeck, Washington, DC: American Geophysical Union, pp. 1–35.

Huybers, P. and Denton, G. (2008). Antarctic temperature at orbital timescales controlled by local summer duration. *Nat. Geosci.*, **1**, 787–92.

Huybers, P. and Wunsch, C. (2005). Obliquity pacing of the late Pleistocene glacial terminations. *Nature*, **434**, 491–4.

Huybrechts, P. and de Wolde, J. (1999). The dynamic response of the Greenland and Antarctic ice sheets to multiple-century climatic warming. *J. Clim.*, **12**, 2169–88.

Huybrechts, P., Gregory, J., Janssens, I. and Wild, M. (2004). Modelling Antarctic and Greenland volume changes during the 20th and 21st centuries forced by GCM time slice integrations. *Global Planet. Change*, **42**, 83–105.

Ichiyanagi, K., Numaguti, A. and Kato, K. (2002). Interannual variation of stable isotopes in Antarctic precipitation in response to El Niño-Southern Oscillation. *Geophys. Res. Lett.*, **29**, 1001, doi:10.1029/2000GL012815.

Ikehara, M., Kawamura, K., Ohkouchi, N., *et al.* (1997). Alkenone sea surface temperature in the Southern Ocean for the last two deglaciations. *Geophys. Res. Lett.*, **24**, 679–82.

Imbrie, J., Berger, A., Boyle, E. A., *et al.* (1993). On the structure and origin of major glaciation cycles. 2. The 100,000-year cycle. *Paleoceanography*, **8**, 699–735.

Indermuhle, A., Stocker, T. F., Joos, F., *et al.* (1999). Holocene carbon-cycle dynamics based on CO_2 trapped in ice at Taylor Dome, Antarctica. *Nature*, **398**, 121–6.

Intrieri, J. M. and Shupe, M. D. (2004). Characteristics and radiative effects of diamond dust over the western Arctic Ocean region. *J. Clim.*, **17**, 2953–60.

IPCC (2007). *Climate Change 2007: The Physical Science Basis. Contribution of Working Group I to the Fourth Assessment Report of the Intergovernmental Panel on Climate Change*, Cambridge, UK: Cambridge University Press.

Jacobs, G. A. and Mitchell, J. L. (1996). Ocean circumpolar variations associated with the Antarctic Circumpolar Wave. *Geophys. Res. Lett.*, **23**, 2947–50.

Jacobs, S. S., Giulivi, C. F. and Mele, P. A. (2002). Freshening of the Ross Sea during the late 20th century. *Science*, **297**, 386–9.

Jakobsson, M., Løvlie, R., Arnold, E. M., *et al.* (2001). Pleistocene stratigraphy and paleoenvironmental variation from Lomonosov Ridge sediments, central Arctic Ocean. *Global Planet. Change*, **31**, 1–22.

Jansen, J. H. F., Kuijpers, A. and Troelstra, S. R. (1986). A mid-Brunhes climatic event: long-term changes in global atmosphere and ocean circulation. *Science*, **232**, 619–22.

Jennings, A. E. and Weiner, N. W. (1996). Environmental change in eastern Greenland during the last 1300 years: evidence from foraminifera and lithofacies in Nansen Fjord, 68° N. *Holocene*, **6**, 179–91.

Jennings, A. E., Knudsen, K. L., Hald, M., Hansen, C. V. and Andrews, J. T. (2002). A mid-Holocene shift in Arctic sea-ice variability on the East Greenland Shelf. *Holocene*, **12**, 49–58.

Jiang, H., Seidenkrantz, M.-S., Knudsen, K. L. and Eriksson, J. (2002). Late-Holocene summer sea-surface temperatures based on a diatom record from the north Icelandic shelf. *Holocene*, **12**, 137–47.

Jiang, X., Pawson, S., Camp, C. D., *et al.* (2008). Interannual variability and trends of extratropical ozone. Part I: Northern Hemisphere. *J. Atmos. Sci.*, **65**, 3013–29.

Johannessen, O. M., Bengtsson, L., Miles, M. W., *et al.* (2004). Arctic climate change: observed and modelled temperature and sea-ice variability. *Tellus*, **56A**, 328–41.

Johnsen, K. P., Miao, J. and Kidder, S. Q. (2004). Comparison of atmospheric water vapor over Antarctica derived from CHAMP/GPS and AMSU-B data. *Phys. Chem. Earth*, **29**, 251–5.

Johnsen, S. J., Clausen, H. B., Dansgaard, W., *et al.* (1992). Irregular glacial interstadials recorded in a new Greenland ice core. *Nature*, **359**, 311–13.

Johnson, R. G. and Lauritzen, S.-E. (1995). Hudson Bay–Hudson Strait Jökulhlaups and Heinrich events: a hypothesis. *Palaeogeog. Palaeoclimatol. Palaeoecol.*, **117**, 123–37.

Jones, A., Bowden, T. and Turner, J. (1998). Predicting total ozone based on GTS data: applications for Southern Hemisphere high-latitude populations. *J. Appl. Meteorol.*, **37**, 477–85.

Jones, D. A. and Simmonds, I. (1993). A climatology of Southern Hemisphere extratropical cyclones. *Clim. Dyn.*, **9**, 131–45.

Jones, J. M. and Widmann, M. (2003). Instrument- and tree-ring-based estimates of the Antarctic oscillation. *J. Clim.*, **16**, 3511–24.

Jones, J. M., Fogt, R. L., Widmann, M., *et al.* (2009) Historical Southern Hemisphere Annular Mode variability. Part I: Century length seasonal reconstructions of the Southern Hemisphere Annular Mode. *J. Clim.*, **22**, 5319–45.

Jones, P. D. (1990). Antarctic temperatures over the present century: a study of the early expedition record. *J. Clim.*, **3**, 1193–203.

Jones, P. D., Osborn, T. J. and Briffa, K. R. (2003). Pressure-based measures of the North Atlantic Oscillation (NAO): a comparison and an assessment of changes in the strength of the NAO and in its influence on surface climate parameters. In *The North Atlantic Oscillation*, ed. J. W. Hurrell, Y. Kushnir, G. Ottersen and M. Visbeck, Washington, DC: American Geophysical Union, pp. 51–62.

Jones, V. J., Hodgson, D. A. and Chepstow-Lusty, A. P. (2000). Palaeolimnological evidence for marked Holocene environmental changes on Signy Island, Antarctica. *Holocene*, **10**, 43–60.

Jorgenson, M. T., Racine, C. H. and Osterkamp, T. (2001). Permafrost degradation and ecological changes associated with a warming climate in central Alaska. *Climatic Change*, **48**, 551–79.

Jourdain, B. and Legrand, M. (2001). Seasonal variations of atmospheric dimethylsulfide, dimethylsulfoxide, sulfur dioxide, methanesulfonate, and non-sea-salt sulfate aerosols at Dumont d'Urville (coastal Antarctica) (December 1998 to July 1999). *J. Geophys. Res.*, **106**, 14 391–408.

Joussaume, S. and Taylor, K. E. (1995) Status of the Paleoclimate Modeling Intercomparison Project (PMIP). In *Proceedings of the First International AMIP Scientific Conference WCRP-92, Monterey, CA*. ed. L. W. Gates, Geneva: WMO, pp. 425–430.

Jouzel, J., Alley, R. B., Cuffey, K. M., *et al.* (1997). Validity of the temperature reconstruction from water isotopes in ice cores. *J. Geophys. Res.*, **102**, 26 471–87.

Jouzel, J., Masson-Delmotte, V., Cattani, O., *et al.* (2007). Orbital and millennial Antarctic climate variability over the past 800,000 years. *Science*, **317**, 793–6.

Kahl, J. D., Serreze, M. C., Shiotani, S., Skony, S. M. and Schnell, R. C. (1992). In-situ meteorological sounding archives for Arctic studies. *Bull. Amer. Meteorol. Soc.*, **73**, 1824–30.

Kahl, J. D. W., Serreze, M. C., Stone, R. S., Shiotani, S., Kisley, M. and Schnell, R. C. (1993). Tropospheric temperature trends in the Arctic: 1958–1986. *J. Geophys. Res.*, **98**, 12 825–38.

Kaleschke, L., Richter, A., Burrows, J., *et al.* (2004). Frost flowers on sea ice as a source of sea salt and their influence on tropospheric halogen chemistry. *Geophys. Res. Lett.*, **31**, L16114, doi:10.1029/2004GL020655.

Kalnay, E., Kanamitsu, M., Kistler, R., *et al.* (1996). The NCEP/NCAR 40-year reanalysis project. *Bull. Amer. Meteorol. Soc.*, **77**, 437–71.

Kanamitsu, M., Ebisuzaki, W., Woolen, J., *et al.* (2002). NCEP-DOE AMIP-II Reanalysis (R-2). *Bull. Amer. Meteorol. Soc.*, **83**, 1631–43.

Kapsner, W. R., Alley, R. B., Shuman, C. A., Anandakrishnan, S. and Grootes, P. M. (1995). Dominant influence of atmospheric circulation on snow accumulation in Greenland over the past 18,000 years. *Nature*, **373**, 52–4.

Kaspari, S., Mayewski, P. A., Dixon, D., *et al.* (2004). Climate variability in West Antarctica derived from annual accumulation rate records from ITASE firn/ice cores. *Ann. Glaciol.*, **39**, 585–94.

Kawamura, K., Parrenin, F., Lisiecki, L., *et al.* (2007). Northern Hemisphere forcing of climatic cycles in Antarctica over the past 360,000 years. *Nature*, **448**, 912–16.

Key, J. R., Santek, D., Velden, C. S., *et al.* (2003). Cloud-drift and water vapor winds in the polar regions from MODIS. *IEEE Trans. Geosci. Remote Sens.*, **41**, 482–92.

Khodri, M., Leclainche, Y., Ramstein, G., *et al.* (2001). Simulating the amplification of orbital forcing by ocean feedbacks in the last glaciation. *Nature*, **410**, 570–4.

Kidson, J. W. (1999). Principal modes of Southern Hemisphere low-frequency variability obtained from NCEP-NCAR reanalyses. *J. Clim.*, **12**, 2808–30.

King, J. C. and Harangozo, S. A. (1998). Climate change in the western Antarctic Peninsula since 1945: observations and possible causes. *Ann. Glaciol.*, **27**, 571–5.

King, J. C. and Turner, J. (1997). *Antarctic Meteorology and Climatology*, Cambridge, UK: Cambridge University Press.

Kistler, R., Kalnay, E., Collins, W., *et al.* (2001). The NCEP-NCAR 50-year reanalysis: monthly means CD-ROM and documentation. *Bull. Amer. Meteorol. Soc.*, **82**, 247–67.

Kleiber, H.-P., Knies, J., and Niessen, F. (2000). The Late Weichselian glaciation of the Franz Victoria Trough, northern Barents Sea: ice sheet extent and timing. *Mar. Geol.*, **168**, 25–44.

Kleman, J., Fastook, J., and Stroeven, A. P. (2002). Geologically and geomorphologically constrained numerical model of Laurentide Ice Sheet inception and build-up. *Quat. Int.*, **95–96**, 87–98.

Klitgaard-Kristensen, D., Rasmussen, T. L., Sejrup, H. P., Haflidason, H. and van Weering, T. C. E. (1998). Rapid changes in the oceanic fronts in the Norwegian Sea during the last deglaciation: implications for the Younger Dryas cooling event. *Mar. Geol.*, **152**, 177–88.

Knies, J., Kleiber, H.-P., Matthiessen, J., Müller, C. and Nowaczyk, N. (2001). Marine records indicate maximum extent of Saalian and Weichselian ice-sheets along the northern Eurasian margin. *Global Planet. Change*, **31**, 45–64.

Knorr, G. and Lohmann, G. (2003). Southern Ocean origin for the resumption of Atlantic thermohaline circulation during deglaciation. *Nature*, **424**, 532–6.

Knudsen, K.-L., Seidenkrantz, M.-S., and Kristensen, P. (2002). Last interglacial and early glacial circulation in the northern North Atlantic Ocean. *Quat. Res.*, **58**, 22–6.

Knutti, R., Flückiger, J., Stocker, T. F., and Timmermann, A. (2004). Strong hemispheric coupling of glacial climate through freshwater discharge and ocean circulation. *Nature*, **430**, 851–6.

Koch, L. (1945). The east Greenland ice. *Medd. Grønland*, **130**, 1–374.

Kohfeld, K. E., Le Quéré, C., Harrison, S. P. and Anderson, R. F. (2005). Role of marine biology in glacial-interglacial CO_2 cycles. *Science*, **308**, 74–8.

König-Langlo, G., King, J. C. and Pettré, P. (1998). Climatology of the three coastal Antarctic stations Dumont d'Urville, Neumayer and Halley. *J. Geophys. Res.*, **103**, 10 935–46.

Kopp, R. E., Simons, F. J., Mitrovica, J. X., Maloof, A. C. and Oppenheimer, M. (2009). Probabilistic assessment of sea level during the last interglacial stage. *Nature*, **462**, 863–7.

Krabill, W., Hanna, E., Huybrechts, P., *et al.* (2004). Greenland Ice Sheet: increased coastal thinning. *Geophys. Res. Lett.*, **31**, L24402, doi:10.1029/2004GL021533.

Krinner, G. and Genthon, C. (1998). GCM simulations of the Last Glacial Maximum surface climate of Greenland and Antarctica. *Clim. Dyn.*, **14**, 741–58.

Kuhlbrodt, T., Griesel, A., Montoya, M., *et al.* (2007). On the driving processes of the Atlantic meridional overturning circulation. *Rev. Geophys.*, **45**, RG2001, doi:10.1029/2004RG000166.

Kulbe, T., Melles, M., Verkulich, S. R. and Pushina, Z. V. (2001). East Antarctic climate and environmental variability over the last 9400 years inferred from marine sediments of the Bunger Oasis. *Arct. Antarct. Alpine Res.*, **33**, 223–30.

Kunz-Pirrung, M., Gersonde, R., and Hodell, D. A. (2002). Mid-Bruhnes century-scale diatom sea surface temperature and sea ice records from the Atlantic sector of the Southern Ocean (ODP Leg 177, sites 1093, 1094 and core PS2089–2). *Palaeogeogr. Palaeoclimatol. Palaeoecol.*, **182**, 305–28.

Kushner, P. J., Held, I. M. and Delworth, T. L. (2001). Southern Hemisphere atmospheric circulation response to global warming. *J. Clim.*, **14**, 2238–49.

Kushnir, Y., Robinson, W. A., Blade, I., (2002). Atmospheric GCM response to extratropical SST anomalies: synthesis and evaluation. *J. Clim.*, **15**, 2233–56.

Kutzbach, J. E. (1988). Climatic changes of the last 18,000 years: observations and model simulations. *Science*, **241**, 1043–52.

Kuzmina, S. I., Bengtsson, L., Johannessen, O. M., *et al.* (2005). The North Atlantic Oscillation and greenhouse gas forcing. *Geophys. Res. Lett.*, **32**, L04703, doi:10.1029/2004GL021064.

Kwok, R. and Comiso, J. C. (2002a). Spatial patterns of variability in antarctic surface temperature: connections to the Southern Hemisphere Annular Mode and the Southern Oscillation. *Geophys. Res. Lett.*, **29**, 1705, doi:10.1029GL015415.

Kwok, R. and Comiso, J. C. (2002b). Southern Ocean climate and sea ice anomalies associated with the Southern Oscillation. *J. Clim.*, **15**, 487–501.

Kwok, R., Cunningham, G. F. and Pang, S. S. (2004). Fram Strait sea ice outflow. *J. Geophys. Res.*, **109**, C01009, doi:10.1029/2003JC001785.

Labeyrie, L., Cole, J., Alverson, K., and Stocker, T. (2003). The history of climate dynamics in the Late Quaternary. In *Paleoclimate, Global Change and the Future*, ed. K. D. Alverson, R. Bradley, and T. Pedersen, Heidelberg, Germany: Springer, pp. 33–52.

Labracherie, M., Labeyrie, L. D., Duprat, J., *et al.* (1989). The last deglaciation in the southern ocean. *Paleoceanography*, **4**, 629–38.

Lachlan-Cope, A. (2010). Antarctic clouds. *Polar Res.*, **29**, 150–8.

Lachlan-Cope, T. A. and Connolley, W. M. (2006). Teleconnections between the tropical Pacific and the Amundsen-Bellingshausen Sea: role of the El Niño/Southern Oscillation. *J. Geophys. Res.*, **111**, D23101, doi:10.1029/2005JD006386.

Lachlan-Cope, T. A., Connolley, W. M., Turner, J., *et al.* (2009). Antarctic winter tropospheric warming: the potential role of polar stratospheric clouds, a sensitivity study. *Atmos. Sci. Lett.*, **10**, 262–6.

Lamb, H. H. (1977). *Climate: Present, Past and Future.* Vol. 2: *Climatic History and the Future*, London: Methuen.

Lamb, H. H. (1995). *Climate, History and the Modern World*, 2nd edn, London: Routledge.

Lambeck, K. and Chappell, J. (2001). Sea level change through the last glacial cycle. *Science*, **292**, 679–86.

Lambert, S. and Fyfe, J. C. (2006). Changes in winter cyclone frequencies and strengths simulated in enhanced greenhouse gas experiments: results from the models participating in the IPCC diagnostic exercize. *Clim. Dyn.*, **26**, 713–28.

Lammers, R. B., Shiklomanov, A. I., Vorosmarty, C. J., Fekete, B. M. and Peterson, B. J. (2001). Assessment of contemporary Arctic river runoff based on observational discharge records. *J. Geophys. Res.*, **106**, 3321–34.

Landais, A., Barnola, J.-M., Masson-Delmotte, V., *et al.* (2004). A continuous record of temperature evolution over a sequence of Dansgaard-Oeschger events during Marine Isotope Stage 4 (76 to 62 kyr BP). *Geophys. Res. Lett.*, **31**, L22211, doi:10.1029/2004GL021193.

Landvik, J. Y., Landvik, S., Bondevik, S., *et al.* (1998). The Last Glacial Maximum of Svalbard and the Barents Sea area: ice sheet extent and configuration. *Quat. Sci. Rev.*, **17**, 43–75.

Lang, C., Leuenberger, M., Schwander, J. and Johnsen, S. (1999). 16°C rapid temperature variation in central Greenland 70,000 years ago. *Science*, **286**, 934–7.

Lanzante, J. R., Klein, S. A. and Seidel, D. J. (2003). Temporal homogenization of monthly radiosonde temperature data. Part I: methodology. *J. Clim.*, **16**, 224–40.

Larsen, E., Lyså, A., Demidov, I., *et al.* (1999). Age and extent of the Scandinavian ice sheet in northwest Russia. *Boreas*, **28**, 115–32.

Larsen, H. C., Saunders, A. D., Clift, P. D., *et al.* and ODP Leg 152 Scientific Party (1994). Seven million years of glaciation in Greenland. *Science*, **264**, 952–55.

Laxon, S., Peacock, N. and Smith, D. (2003). High interannual variability of sea ice thickness in the Arctic region. *Nature*, **425**, 947–50.

Le Quéré, C., Rodenbeck, C., Buitenhuis, E. T., *et al.* (2007). Saturation of the Southern Ocean CO_2 sink due to recent climate change. *Science*, **316**, 1735–8.

Le Quéré, C., Raupach, M. R., Canadell, J. G., *et al.* (2009). Trends in the sources and sinks of carbon dioxide. *Nature Geosci.*, **2**(12), 831–6.

Lean, J. (2002). Solar forcing of climate change in recent millennia. In *Climate Development and History of the North Atlantic Realm*, ed. G. Wefer, W. H. Berger, K.-E. Behre and E. Jansen, Berlin: Springer-Verlag, pp. 75–88.

Lean, J., Skumanich, A. and White, O. (1992). Estimating the sun's radiative output during the Maunder Minimum. *Geophys. Res. Lett.*, **19**, 1591–4.

Lean, J., Beer, J. and Bradley, R. S. (1995). Reconstruction of solar irradiance since 1610: implications for climate change. *Geophys. Res. Lett.*, **22**, 3195–98.

Leck, C. and Bigg, E. K. (2005). Source and evolution of the marine aerosol: a new perspective. *Geophys. Res. Lett.*, **32**, L19803, doi:10.1029/GL023651.

Lee, A. M., Roscoe, H. K. and Oltmans, S. (2000). Model and measurements show Antarctic ozone loss follows edge of polar night. *Geophys. Res. Lett.*, **27**, 3845–8.

Lefebvre, W., Goosse, H., Timmermann, R. and Fichefet, T. (2004). Influence of the Southern Annular Mode on the sea ice-ocean system. *J. Geophys. Res.*, **109**, C09005, doi:10.1029/2004JC002403.

Legates, D. R. and Willmott, C. J. (1990). Mean seasonal and spatial variability in gauge-corrected, global precipitation. *Int. J. Climatol.*, **10**, 111–27.

Legrand, M. and Feniet-Saigne, C. (1991). Methanesulfonic acid in south polar snow layers: a record of strong El Nino? *Geophys. Res. Lett.*, **18**, 187–90.

Legrand, M. and Mayewski, P. (1997). Glaciochemistry of polar ice cores: a review. *Rev. Geophys.*, **35**, 219–43.

Legrand, M., Fenietsaigne, C., Saltzman, E. S., *et al.* (1991). Ice-core record of oceanic emissions of dimethylsulfide during the last climate cycle. *Nature*, **350**, 144–6.

Legrand, M., Hammer, C., Deangelis, M., *et al.* (1997). Sulfur-containing species (methanesulfonate and SO_4) over the last climatic cycle in the Greenland Ice Core Project (central Greenland) ice core. *J. Geophys. Res.*, **102**, 26 663–79.

Lehman, S. J. and Keigwin, L. D. (1992). Sudden changes in North Atlantic circulation during the last deglaciation. *Nature*, **356**, 757–62.

Leventer, A., Domack, E., Ishman, S., *et al.* (1996). Productivity cycles of 200–300 years in the Antarctic Peninsula region: understanding linkages among the sun, atmosphere, oceans, sea ice, and biota. *Geol. Soc. Amer. Bull.*, **108**, 1626–44.

Limpasuvan, V. and Hartmann, D. L. (1999). Eddies and the annular modes of climate variability. *Geophys. Res. Lett.*, **26**, 3133–6.

Limpasuvan, V. and Hartmann, D. L. (2000). Wave-maintained annular modes of climate variability. *J. Clim.*, **13**, 4414–29.

Lisiecki, L. E. and Raymo, M. E. (2005). A Pliocene-Pleistocene stack of 57 globally distributed benthic $\delta^{18}O$ records. *Paleoceanography*, **20**, PA1003, doi:10.1029/2004PA001071.

Liu, J., Yuan, X., Rind, D. and Martinson, D. G. (2002). Mechanism study of the ENSO and southern high latitude climate teleconnections. *Geophys. Res. Lett.*, **29**, 1679, doi:10.1029/2002GL015143.

Liu, J., Curry, J. A. and Martinson, D. G. (2004). Interpretation of recent Antarctic sea ice variability. *Geophys. Res. Lett.*, **31**, L02205, doi:10.1029/2003GL018732.

Loulergue, L., Schilt, A., Spahni, R., *et al.* (2008). Orbital and millennial-scale features of atmospheric CH_4 over the past 800,000 years. *Nature*, **453**, 383–6.

Lowe, A. L. and Anderson, J. B. (2002). Late Quaternary. advance and retreat of the West Antarctic Ice Sheet in Pine Island Bay, Antarctica. *Quat. Sci. Rev.*, **21**, 1879–97.

Lubin, D. and Massom, R. (2006). *Polar Remote Sensing.* Vol. I: *Atmosphere and Oceans*, Chichester, UK: Springer-Praxis.

Lubin, D. and Vogelmann, M. (2006). A climatologically significant aerosol longwave indirect effect in the Arctic. *Nature*, **439**, 453–6.

Lubin, D., Chen, B., Bromwich, D. H., *et al.* (1998). The impact of Antarctic cloud radiative properties on a GCM climate simulation. *J. Clim.*, **11**, 447–62.

Lucchitta, B. K., Mullins, K. F., Allison, A. L. and Ferrigno, J. G. (1993). Antarctic glacial tongue velocities from Landsat images: first results. *Ann. Glaciol.*, **17**, 356–66.

Lunkka, J. P., Saarnisto, M., Gey, V., Demidov, I., and Kiselova, V. (2001). Extent and age of the Last Glacial Maximum in the southeastern sector of the Scandinavian Ice Sheet. *Global Planet. Change*, **31**, 407–25.

Lüthi, D., Le Floch, M., Bereiter, B., *et al.* (2008). High-resolution carbon dioxide concentration record 650,000–800,000 years before present. *Nature*, **453**, 379–82.

Lynch, A. H., Uotila, P. and Cassano, J. J. (2006). Changes in synoptic weather patterns in the polar regions in the 20th and 21st centuries, Part 2: Antarctic. *Int. J. Climatol.*, **26**, 1181–99.

MacAyeal, D. R. (1993). Binge/purge oscillations of the Laurentide ice sheet as a cause of the North Atlantic's Heinrich events. *Paleoceanography*, **8**, 775–84.

Macdonald, G. M., Velichko, A. A., Kremenetski, C. V., *et al.* (2000). Holocene treeline history and climate change across northern Eurasia. *Quat. Res.*, **53**, 302–11.

Mahowald, N. M., Lamarque, J.-F., Tix, X. X., and Wolff, E. (2006). Sea-salt aerosol response to climate change: Last Glacial Maximum, preindustrial, and doubled dioxide climates. *J. Geophys. Res.*, **111**, D05303, doi:10.1029/2005JD006459.

Manabe, S. and Stouffer, R. J. (1994). Multiple-century response of a coupled ocean-atmosphere model to an increase of atmospheric carbon-dioxide. *J. Clim.*, **7**, 5–23.

Mangerud, J., Jakobsson, M., Alexanderson, H., *et al.* (2004). Ice-dammed lakes and rerouting of the drainage of Northern Eurasia during the last glaciation. *Quat. Sci. Rev.*, **23**, 1313–32.

Marshall, G. J. (2002). Trends in Antarctic geopotential height and temperature: a comparison between radiosonde and NCEP-NCAR reanalysis data. *J. Clim.*, **15**, 659–74.

Marshall, G. J. (2003). Trends in the Southern Annular Mode from observations and reanalyses. *J. Clim.*, **16**, 4134–43.

Marshall, G. J. (2007). Half-century seasonal relationships between the Southern Annular Mode and Antarctic temperatures. *Int. J. Climatol.*, **27**(3), 373–83.

Marshall, G. J. and Connolley, W. M. (2006). The effect of changing Southern Hemisphere winter sea surface temperatures on Southern Annular Mode strength. *Geophys. Res. Lett.*, **33**, L17717, doi:10.1029/2006GL026627.

Marshall, G. J. and Harangozo, S. A. (2000). An appraisal of NCEP/NCAR reanalysis MSLP viability for climate studies in the South Pacific. *Geophys. Res. Lett.*, **27**, 3057–60.

Marshall, G. J., and King, J. C. (1998). Southern Hemisphere circulation anomalies associated with extreme Antarctic Peninsula winter temperatures. *Geophys. Res. Lett.*, **25**, 2437–40.

Marshall, G. J., and Turner, J. (1997). Surface wind fields of Antarctic mesocyclones derived from ERS-1 scatterometer data. *J. Geophys. Res.*, **102**, 13 907–21.

Marshall, G. J., Turner, J. and Miners, W. D. (1998). Interpreting recent accumulation records through an understanding of the regional synoptic climatology: an example from the southern Antarctic Peninsula. *Ann. Glaciol.*, **27**, 610–16.

Marshall, G. J., Orr, A., van Lipzig, N. P. M. and King, J. C. (2006). The impact of a changing Southern Hemisphere Annular Mode on Antarctic Peninsula summer temperatures. *J. Clim.*, **19**, 5388–404.

Marshall, J., Johnson, H. and Goodman, J. (2001). A study of the interaction of the North Atlantic oscillation with ocean circulation. *J. Clim.*, **14**, 1399–421.

Martinez-Macchiavello, J. C., Tatur, A., Servant-Vildary, S. and del Valle, R. (1996). Holocene environmental change in a marine-estuarine-lacustrine sediment sequence, King George Island, South Shetland Islands. *Antarct. Sci.*, **8**, 313–22.

Martinson, D. G. and Iannuzzi, R. A. (2003). Spatial/temporal patterns in Weddell gyre characteristics and their relationship to global climate. *J. Geophys. Res.*, **108**, 8083, doi:10.1029/2000JC000538.

Maslin, M. A., Pike, J., Stickley, C., and Ettwein, V. (2003). Evidence of Holocene climate variability from marine sediments. In *Global Change in the Holocene*, ed. A. Mackay, R. Battarbee, J. Birks and F. Oldfield. London: Arnold, pp. 185–209.

Massom, R. and Lubin, D. (2006). *Polar Remote Sensing.* Vol. II: *Ice Sheets*, Chichester, UK: Springer-Praxis.

Masson, V., Vimeux, F., Jouzel, J., *et al.* (2000). Holocene climate variability in Antarctica based on 11 ice-core isotopic records. *Quat. Res.*, **54**, 348–58.

Masuda, K. (1990). Atmospheric heat and waterbudgets of polar regions: analysis of FGGE data. In *Proceedings of the Third NIPR Symposium on Polar Meteorology and Glaciology*, Tokyo: National Institute of Polar Research, pp. 79–88.

References

Mäusbacher, R., Müller, J. and Schmidt, R. (1989). Evolution of post glacial sedimentation in Antarctic lakes. *Z. Geomorphol.*, **33**, 219–34.

Mayewski, P. A., Meeker, L. D., Whitlow, S., *et al.* (1993). The atmosphere during the Younger Dryas. *Science*, **261**, 195–7.

Mayewski, P. A., Meeker, L. B., Whitlow, S., *et al.* (1994). Changes in atmospheric circulation and ocean ice cover over the North Atlantic during the last 41,000 years. *Science*, **263**, 1747–51.

Mayewski, P. A., Lyons, W. B., Zielinski, G. A., *et al.* (1995). An ice-core-based late Holocene history for the Transantarctic Mountains, Antarctica. *Antarct. Res. Ser.*, **67**, 33–45.

Mayewski, P. A., Meeker, L. D., Twickler, M., *et al.* (1997). Major features and forcing of high-latitude northern hemisphere atmospheric circulation using a 110,000-year-long glaciochemical series. *Science*, **102**, 26 345–66.

Mayewski, P. A. *et al.* (2006). The International Trans-Antarctic Scientific Expedition (ITASE): an overview. *Ann. Glaciol.*, **41**, 180–185.

McClelland, J. W., Dery, S. J., Peterson, B. J., Holmes, R. M. and Wood, E. F. (2006). A pan-arctic evaluation of changes in river discharge during the latter half of the 20th century. *Geophys. Res. Lett.*, **33**, L06715, doi:10.1029/2006GL025753.

McConnell, J. R., Lamorey, G. W., Hanna, E., *et al.* (2001). Annual net snow accumulation over southern Greenland from 1975 to 1998. *J. Geophys. Res.*, **106**, 33 827–37.

McConnell, J. R., Edwards, R., Kok, G. L., *et al.* (2007). 20th-century industrial black carbon emissions altered Arctic climate forcing. *Science*, **317**, 1381–4.

McLaren, A. (1989). The under-ice thickness distribution of the Arctic Basin as recorded in 1958 and 1970. *J. Geophys. Res.*, **94**, 4971–83.

McManus, J. F., Francois, R., Gherardi, J.-M., Keigwin, L. D. and Brown-Leger, S. (2004). Collapse and rapid resumption of Atlantic meridional circulation linked to deglacial climate changes. *Nature*, **428**, 834–7.

Meehl, G. A., Hurrell, J. W. and van Loon, H. (1998). A modulation of the mechanism of the semiannual oscillation in the Southern Hemisphere. *Tellus*, **50**A, 442–50.

Meese, D. A., Gow, A. J., Alley, R. B., *et al.* (1997). The Greenland Ice Sheet Project 2 depth-age scale: methods and results. *J. Geophys. Res., – Oceans*, **102**, 26 411–23.

Melles, M., Kulbe, T., Verkulich, S. R., Pushina, Z. V. and Hubberten, H. W. (1997). Late Pleistocene and Holocene environmental history of Bunger Hills, East Antarctica, as revealed by fresh-water and epishelf lake sediments. In *The Antarctic Region: Geological Evolution and Processes*, ed. C. A. Ricci, Siena, Italy: Università degli Studi di Siena, pp. 809–20.

Meredith, M. P. and Hogg, A. M. (2006). Circumpolar response of Southern Ocean eddy activity to a change in the Southern Annular Mode. *Geophys. Res. Lett.*, **33**, L16608, doi:10.1029/2006GL026499.

Meredith, M. P. and King, J. C. (2005). Climate change in the ocean to the west of the Antarctic Peninsula during the second half of the 20th century. *Geophys. Res. Lett.*, **32**, L19606, doi:10.1029/2005GL024042.

Meredith, M. P., Woodworth, P. L., Hughes, C. W. and Stepanov, V. (2004). Changes in the ocean transport through Drake Passage during the 1980s and 1990s, forced by changes in the Southern Annular Mode. *Geophys. Res. Lett.*, **31**, L21305, 10.1029/2004GL021169.

Meyerson, E. A., Mayewski, P. A., Kreutz, K. J., *et al.* (2002). The polar expression of ENSO and sea-ice variability as recorded in a South Pole ice core. *Ann. Glaciol.*, **35**, 430–6.

Milankovitch, M. (1930). Mathematische Klimalehre und Astronomische Theorie der Klimaschwankungen. In *Handbuch der Klimatologie I(A)*, ed. W. Köppen and R. Geiger, Berlin: Gebrüder Borntraeger, pp. 1–176.

Miller, J. R. and Russell, G. L. (1992). The impact of global warming on river runoff. *J. Geophys. Res.*, **97**, 2757–64.

Miller, R. L., Schmidt, G. A. and Shindell, D. T. (2006). Forced annular variations in the 20th century Intergovernmental Panel on Climate Change Fourth Assessment Report models. *J. Geophys. Res.*, **111**, D18101, doi:10.1029/2005JD006323.

Mitrovica, J. X., Tamisiea, M. E., Davis, J. L. and Milne, G. A. (2001). Recent mass balance of polar ice sheets inferred from patterns of global sea-level change. *Nature*, **409**, 1026–9.

Mo, K. C. and Ghil, M. (1987). Statistics and dynamics of persistent anomalies. *J. Atmos. Sci.*, **44**, 877–901.

Mo, K. C. and Higgins, R. W. (1998). The Pacific-South American modes and tropical convection during the Southern Hemisphere winter. *Mon. Weather Rev.*, **126**, 1581–96.

Molnia, B. F. (1986). Glacial history of the northeastern Gulf of Alaska: a synthesis. In *Glaciation in Alaska: The Geologic Record*, ed. T. D. Hamilton, K. M. Reed and R. M. Thorson, Anchorage, AL: Alaska Geological Society, pp. 219–35.

Monaghan, A. J., Bromwich, D. H., Wei, H. L., *et al.* (2003). Performance of weather forecast models in the rescue of Dr. Ronald Shemenski from South Pole in April 2001. *Weather Forecast.*, **18**, 142–60.

Monaghan, A. J., Bromwich, D. H., Fogt, R. L., *et al.* (2006a). Insignificant change in Antarctic snowfall since the International Geophysical Year. *Science*, **313**, 827–31.

Monaghan, A. J., Bromwich, D. H. and Wang, S.-H. (2006b). Recent trends in Antarctic snow accumulation from Polar MM5 simulations. *Phil. Trans. Roy. Soc. Lond. Ser. A*, **364**, 1683–708.

Monnin, E., Indermühle, A., Dällenbach, A., *et al.* (2001). Atmospheric CO_2 concentrations over the last glacial termination. *Science*, **291**, 112–14.

Moore, G. W. K., Alverson, K. and Renfrew, I. A. (2002). A reconstruction of the air-sea interaction associated with the Weddell polynya. *J. Phys. Oceanogr.*, **32**, 1685–98.

Moore, J. J., Hughen, K. A., Miller, G. H. and Overpeck, J. T. (2001). Little Ice Age recorded in summer temperature reconstruction from varved sediments of Donard Lake, Baffin Island, Canada. *J. Paleolimnol.*, **25**, 503–17.

Moran, K., Backman, J., Brinkhuis, H., *et al.* (2006). The Cenozoic palaeoenvironment of the Arctic Ocean. *Nature*, **441**, 601–5.

Mosley-Thompson, E., McConnell, J. R., Bales, R. C., *et al.* (2001). Local to regional-scale variability of annual net accumulation on the Greenland ice sheet from PARCA cores. *J. Geophys. Res.*, **106**, 33 839–51.

Mosley-Thompson, E. and Thompson, L. G. (2003). Ice core paleoclimate histories from the Antarctic Peninsula: where do we go from here? In *Antarctic Peninsula Climate Variability: A Historical and Paeoenvironmental Perspective*, ed. E. Domack, A. Burnett, P. Convey, M. Kirby and R. Bindschadler, Antarctic Research Series 79, Washington, DC, American Geophysical Union, pp. 115–27.

Motoi, T., Ono, N. and Wakatsuchi, M. (1987). A mechanism for the formation of the Weddell Polynya in 1974. *J. Phys. Oceanogr*, **17**, 2241–47.

Mueller, D. R., Vincent, W. F. and Jeffries, M. O. (2003). Break-up of the largest Arctic ice shelf and associated loss of an epishelf lake. *Geophys. Res. Lett.*, **30**, 2031, doi:10.1029/2003GL017931.

Muller, R. A., and MacDonald, G. J. (1997). Glacial cycles and astronomical forcing. *Science*, **277**, 215–18.

Mulvaney, R., Pasteur, E. C., Peel, D. A., Saltzman, E. S. and Whung, P. Y. (1992). The ratio of MSA to non-sea-salt sulphate in Antarctic Peninsular ice cores. *Tellus*, **44B**, 295–303.

Murphy, J. M., Sexton, D. M. H., Barnett, D. N., *et al.* (2004). Quantification of modelling uncertainties in a large ensemble of climate change simulations. *Nature*, **430**, 768–72.

Murton, J. B., Bateman, M. D., Dallimore S. R., Teller, J. T. and Yang, Z. (2010). Identification of Younger Dryas outburst flood path from Lake Agassiz to the Arctic Ocean. *Nature*, **464**, 740–3.

Muscheler, R., Beer, J., Wagner, G., and Finkel, R. C. (2000). Changes in deep-water formation during the Younger Dryas event inferred from ^{10}Be and ^{14}C records. *Nature*, **408**, 567–70.

Muscheler, R., Beer, J. and Vonmoos, M. (2004). Causes and timing of the 8200 yr BP event inferred from the comparison of the GRIP Be-10 and the tree ring delta C-14 record. *Quat. Sci. Rev.*, **23**, 2101–11.

Myneni, R. B., Keeling, C. D., Tucker, C. J., Asrar, G. and Nemani, R. R. (1997). Increased plant growth in the northern high latitudes from 1981 to 1991. *Nature*, **386**, 698–702.

Mysak, L. A. (1986). El Niño, interannual variability and fisheries in the northeast Pacific Ocean. *Can. J. Fish. Aquat. Sci.*, **43**, 464–97.

Mysak, L. A. (2001). Oceanography: patterns of Arctic circulation. *Science*, **293**, 1269–70.

Mysak, L. A., Manak, D. K. and Marsden, R. F. (1990). Sea-ice anomalies observed in the Greenland and Labrador Seas during 1901–1984 and their relation to an interdecadal Arctic climate cycle. *Clim. Dyn.*, **5**, 111–33.

Mysak, L. A., Ingram, R. G., Wang, J. and van der Baaren, A. (1996). The anomalous sea-ice extent in Hudson Bay, Baffin Bay and the Labrador Sea during three simultaneous NAO and ENSO episodes. *Atmos. Ocean*, **34**, 313–43.

Mysak, L. A., Wright, K. M., Sedlacek, J. and Eby, M. (2005). Simulation of sea ice and ocean variability in the Arctic during 1955–2002 with an intermediate complexity model. *Atmos. Ocean*, **43**, 101–18.

Nakada, M. and Lambeck, K. (1988). The melting history of the late Pleistocene Antarctic ice-sheet. *Nature*, **333**, 36–40.

Nakamura, N. and Oort, A. H. (1988). Atmospheric heat budgets of the polar regions. *J. Geophys. Res.*, **93**, 9510–24.

Nakicenovic, N., Alcamo, J., Davis, G., *et al.* (2000). *Special Report on Emissions Scenarios: A Special Report of Working Group III of the Intergovernmental Panel on Climate Change*, Cambridge, UK: Cambridge University Press.

Namias, J. (1950). The index cycle and its role in the general circulation. *J. Meteorol.*, **7**, 130–9.

Newman, P. A., Gleason, J. F., McPeters, R. D. and Stolarski, R. S. (1997). Anomalously low ozone over the Arctic. *Geophys. Res. Lett.*, **24**, 2689–92.

Newman, P. A., Nash, E. R., Kawa, S. R., Montzka, S. A. and Schauffler, S. M. (2006). When will the Antarctic ozone hole recover? *Geophys. Res. Lett.*, **33**, L12814, doi:10.1029/2005GL025232.

Nghiem, S. V., Chao, Y., Neumann, G., *et al.* (2006). Depletion of perennial sea ice in the East Arctic Ocean. *Geophys. Res. Lett.*, **33**, L17501, doi:10.1029/2006GL027198.

North Greenland Ice Core Project members (2004). High-resolution record of Northen Hemisphere climate extending into the last interglacial period. *Nature*, **431**, 147–51.

O'Brien, S. R., Mayewski, P. A., Meeker, L. D., *et al.* (1995). Complexity of Holocene climate as reconstructed from a Greenland ice core. *Science*, **270**, 1962–4.

Oechel, W., Hastings, S. J., Vourlitis, G., *et al.* (1993). Recent change of Arctic tundra ecosystems from a net carbon dioxide sink to a source. *Nature* **361**, 520–3.

Oerlemans, J. and Reichert, B. K. (2000). Relating glacier mass balance to meteorological data by using a seasonal sensitivity characteristic. *J. Glaciol.*, **46**, 1–6.

Ogilvie, A. E. J. (1984). The past climate and sea-ice record from Iceland. 1. Data to AD 1780. *Climatic Change*, **6**, 131–52.

Ogilvie, A. E. J. (1991). Climatic changes in Iceland A.D. c.865 to 1598. *Acta Archaeol.*, **61**, 233–51.

Ogilvie, A. E. J. (1992). Documentary evidence for changes in the climate of Iceland AD 1500 to 1800. In *Climate Since AD 1500*, ed. R. S. Bradley and P. D. Jones, London: Routledge, pp. 92–117.

Ogilvie, A. E. J. and Jónsdóttir, I. (2000). Sea ice, climate and Icelandic fisheries in historical times. *Arctic*, **53**, 383–394.

Ohmura, A., Dutton, E. G., Forgan, B., *et al.* (1998). Baseline Surface Radiation Network (BSRN/WCRP): new precision radiometry for climate research. *Bull. Amer. Meteorol. Soc.*, **79**, 2115–36.

Ohmura, A., Calanca, P., Wild, M. and Anklin, M. (1999). Precipitation, accumulation and mass balance of Greenland ice sheet. *Z. Gletscherkd. Glazialgeol.*, **35**, 1–20.

Oke, P. R. and England, M. H. (2004). Oceanic response to changes in the latitude of the Southern Hemisphere subpolar westerly winds. *J. Clim.*, **17**, 1040–54.

Orsi, A. H., Whitworth, T. and Nowlin, W. D. (1995). On the meridional extent and fronts of the Antarctic Circumpolar Current. *Deep-Sea Res. I*, **42**, 641–73.

Orsi, A. H., Johnson, G. C. and Bullister, J. L. (1999). Circulation, mixing and production of Antarctic Bottom Water. *Prog. Oceanogr.*, **43**, 55–109.

Otto-Bliesner, B. L., Marshall, S. J., Overpeck, J. T., *et al.* and CAPE Last Interglacial Project Members (2006). Simulating Arctic climate warmth and icefield retreat in the Last Interglaciation. *Science*, **311**, 1751–53.

Overland, J. E., and Wang, M. (2005). The Arctic climate paradox: the recent decrease of the Arctic Oscillation. *Geophys. Res. Lett.*, **32**, L06701, doi: 10.1029/2004GL021752.

Overland, J. E., McNutt, S. L., Groves, J., *et al.* (2000). Regional sensible and radiative heat flux estimates for the winter Arctic during the Surface Heat Budget of the Arctic Ocean (SHEBA) experiment. *J. Geophys. Res.*, **105**, 14 093–102.

Overland, J. E., Spillane, M. C., Percival, D. B., Wang, M. and Mofjeld, H. O. (2004). Seasonal and regional variation of pan-Arctic surface air temperature over the instrumental record. *J. Clim.*, **17**, 23 263–82.

Overpeck, J., Hughen, K., Hardy, D., *et al.* (1997). Arctic environmental change of the last four centuries. *Science*, **278**, 1251–6.

Overpeck, J. T., Otto-Bliesner, B. L., Miller, G. H., *et al.* (2006). Paleoclimatic evidence for future ice-sheet instability and rapid sea-level rise. *Science*, **311**, 1747–50.

Paillard, D. (1998). The timing of Pleistocene glaciations from a simple multiple-state climate model. *Nature*, **391**, 378–81.

Parish, T. R. and Bromwich, D. H. (1987). The surface windfield over the Antarctic ice sheets. *Nature*, **328**, 51–4.

Parish, T. R. and Cassano, J. J. (2003). The role of katabatic winds on the Antarctic surface wind regime. *Mon. Wea. Rev.*, **131**, 317–33.

Park, Y. H., Roquet, F. and Vivier, F. (2004). Quasi-stationary ENSO wave signals versus the Antarctic Circumpolar Wave scenario. *Geophys. Res. Lett.*, **31**, L09315, doi:10.1029/2004GL019806.

Parkinson, C. L. (1995). Recent sea-ice advances in Baffin Bay/Davis Strait and retreats in the Bellingshausen Sea. *Ann. Glaciol.*, **21**, 348–52.

Parkinson, C. L., Cavalieri, D. J., Gloersen, P., Zwally, H. J. and Comiso, J. C. (1999). Arctic sea ice extents, areas and trends 1978–1996. *J. Geophys. Res.*, **104**, 20 837–56.

Parrenin, F., Barnola, J.-M., Beer, J., *et al.* (2007). The EDC3 chronology for the EPICA Dome C ice core. *Clim. Past*, **3**, 485–97.

Peck, L. S. (2005). Prospects for survival in the Southern Ocean: vulnerability of benthic species to temperature change. *Antarct. Sci.*, **17**, 497–507.

Peltier, W. R. (1994). Ice-age paleotopography. *Science*, **265**, 195–201.

Peltier, W. R. (2002). Comments on the paper of Yokoyama *et al.* (2000), entitled 'Timing of the Last Glacial Maximum from observed sea level minima'. *Quat. Sci. Rev.*, **21**, 409–14.

Peltier, W. R. (2005). On the hemispheric origins of meltwater pulse 1a. *Quat. Sci. Rev.*, **24**, 1655–71.

Pereira, E. B., Evangelista, H., Pereira, K. C. D., Cavalcanti, I. F. A. and Setzer, A. W. (2006). Apportionment of black carbon in the South Shetland Islands, Antarctic Peninsula. *J. Geophys. Res.*, **111**, D03303, doi:10.1029/2005JD006086.

Perlwitz, J., Pawson, S., Fogt, R. L., Nielsen, J. E. and Neff, W. D. (2008). Impact of stratospheric ozone hole recovery on Antarctic climate. *Geophys. Res. Lett.*, **35**, L08714, doi:10.1029/2008GL033317.

Peterson, B. J., Holmes, R. M., McClelland, J. W., *et al.* (2002). Increasing river discharge to the Arctic Ocean. *Science*, **298**, 2171–73.

Peterson, B. J., McClelland, J. W., Curry, R., *et al.* (2006). Trajectory shifts in the Arctic and Subarctic freshwater cycle. *Science*, **313**, 1061–6.

Petit, J. R., Jouzel, J., Raynaud, D., *et al.* (1999). Climate and atmospheric history of the past 420,000 years from the Vostok ice core, Antarctica. *Nature*, **399**, 429–36.

Pilcher, J. R., Hall, V. A. and McCormac, F. G. (1995). Dates of Holocene Icelandic volcanic eruptions from tephra layers in Irish peats. *Holocene*, **5**, 103–10.

Pisias, N. G., Martinson, D. G., Moore, T. C., *et al.* (1984). High resolution stratigraphic correlation of benthic oxygen isotopic records spanning the last 300,000 years. *Mar. Geol.*, **56**, 119–36.

Polyakov, I. V., Bekryaev, R. V., Alekseev, G. V., *et al.* (2003). Variability and trends of air temperature and pressure in the maritime Arctic, 1875–2000. *J. Clim.*, **16**, 2067–77.

Porter, S. C. (1986). Late Holocene fluctuations of the fjord glacier variations during the last millennium. *Quat. Res.*, **26**, 27–48.

Portis, D. H., Walsh, J. E., El Hamly, M. and Lamb, P. J. (2001). Seasonality of the North Atlantic Oscillation. *J. Clim.*, **14**, 2069–78.

Przybylak, R. (2003). *The Climate of the Arctic*, Dordrecht, The Netherlands: Kluwer.

Pudsey, C. J. and Evans, J. (2001). First survey of Antarctic sub-ice shelf sediments reveals mid-Holocene ice shelf retreat. *Geology*, **29**, 787–90.

Quadrelli, R. and Wallace, J. M. (2004). A simplified linear framework for interpreting patterns of Northern Hemisphere wintertime climate variability. *J. Clim.*, **17**, 3728–44.

Raab, A., Melles, M., Berger, G. W., Hagedorn, B., and Hubberton, H.-W. (2003). Non-glacial paleoenvironments and the extent of Weichselian ice sheets on Severnaya Zemlya, Russian High Arctic. *Quat. Sci. Rev.*, **22**, 2267–83.

Radok, U. and Lile, R. C. (1977). A year of snow accumulation at plateau station. *Antarct. Res. Ser.*, **25**, 17–26.

Rahmstorf, S. (2002). Ocean circulation and climate during the past 120,000 years. *Nature*, **419**, 207–14.

Rahmstorf, S. (2007). A semi-empirical approach to projecting future sea-level rise. *Science*, **315**, 368–70.

Rahmstorf, S. and Ganapolski, A. (1999). Long-term global warming scenarios computed with an efficient coupled climate model. *Climatic Change*, **43**, 353–67.

Räisänen, J. (2001). CO_2-induced climate change in CMIP2 experiments: quantification of agreement and role of internal variability. *J. Clim.*, **14**, 2088–104.

Raisbeck, G. M., Yiou, F., Jouzel, J., and Stocker, T. F. (2007). Direct north-south synchronization of abrupt climate change record in ice cores using beryllium 10. *Clim. Past*, **3**, 755–69.

Randel, W. J. and Wu, F. (1999). Cooling of the Arctic and Antarctic polar stratospheres due to ozone depletion. *J. Clim.*, **12**, 1467–79.

Raphael, M. N. (2003). Impact of observed sea-ice concentration on the Southern Hemisphere extratropical atmospheric circulation in summer. *J. Geophys. Res.*, **108**, 4687, doi:10.1029/2002JD003308.

Raphael, M. N. (2004). A zonal wave 3 index for the Southern Hemisphere. *Geophys. Res. Lett.*, **31**, L23212, doi:10.1029/2004GL020365.

Raphael, M. N. (2007). The influence of atmospheric zonal wave three on Antarctic sea ice variability. *J. Geophys. Res.*, **112**, D12112, doi:10.1029/2006JD007852.

Rasmussen, E. A. and Turner, J. (2003). *Polar Lows: Mesoscale Weather Systems in the Polar Regions*, Cambridge, UK: Cambridge University Press.

Rasmussen, S. O., Andersen, K. K., Svensson, A. M., *et al.* (2006). A new Greenland ice core chronology for the last glacial termination. *J. Geophys. Res.*, **111**, D06102, doi:10.1029/2005JD006079.

Rauthe, M. A., Hense, A. and Paeth, H. (2004). A model intercomparison study of climate change. *Int. J. Climatol.*, **24**, 643–62.

Raymo, M. E. (1997). The timing of major climate terminations. *Paleoceanography*, **12**, 577–85.

Raymo, M. E., Lisiecki, L. E., and Nisancioglu, K. H. (2006). Plio-Pleistocene ice volume, Antarctic climate, and the global $\delta^{18}O$ record. *Science*, **313**, 492–5.

Raynaud, D., Barnola, J. M., Chappellaz, J., *et al.* (2000). The ice record of greenhouse gases: a view in the context of future changes. *Quat. Sci. Rev.*, **19**, 9–17.

Raynaud, D., Loutre, M. F., Ritz, C., *et al.* (2003). Marine Isotope Stage (MIS) 11 in the Vostok ice core: CO_2 forcing and stability of East Antarctica. In *Earth's Climate and Orbital Eccentricity: The Marine Isotope Stage 11 Question*, ed. A. W. Droxler, R. Z. Poore and L. H. Burckle, Washington, DC: American Geophysical Union, pp. 27–40.

Renfrew, I. A., King, J. C. and Markus, T. (2002). Coastal polynyas in the southern Weddell Sea: variability of the surface energy budget. *J. Geophys. Res.*, **107**, 3063, doi:10.1029/2000JC000720.

Renssen, H., van Geel, B., van der Plicht, J., and Magny, M. (2001a). Reduced solar activity as a trigger for the start of the Younger Dryas? *Quat. Int.*, **68–71**, 373–83.

Renssen, H., Goosse, H., Fichefet, T. and Campin, J. M. (2001b). The 8.2 kyr BP event simulated by a global atmosphere-sea-ice-ocean model. *Geophys. Res. Lett.*, **28**, 1567–70.

Renwick, J. A. (2004). Trends in the Southern Hemisphere polar vortex in NCEP and ECMWF reanalyses. *Geophys. Res. Lett.*, **31**, L07209, doi:10.1029/2003GL019302.

Renwick, J. A. (2005). Persistent positive anomalies in the Southern Hemisphere circulation. *Mon. Wea. Rev.*, **133**, 977–88.

Renwick, J. A. and Revell, M. J. (1999). Blocking over the South Pacific and Rossby wave propagation. *Mon. Wea. Rev.*, **127**, 2233–47.

Reusch, D. B., and Alley, R. B. (2002). Automatic weather stations and artificial neural networks: improving the instrumental record in West Antarctica. *Mon. Wea. Rev.*, **130**, 3037–53.

Reusch, D. B., Mayewski, P. A., Whitlow, S. I., Pittalwala, I. I. and Twickler, M. S. (1999). Spatial variability of climate and past atmospheric circulation patterns from central West Antarctic glaciochemistry. *J. Geophys. Res.*, **104**, 5985–6001.

Richter-Menge, J., Overland, J. E., Proshutinsky, A., *et al.* (2006). *State of the Arctic: NOAA OAR Special Report*, Seattle, WA: NOAA/OAR/PMEL.

Ridley, J. K., Huybrechts, P., Gregory, J. M. and Lowe, J. A. (2005). Elimination of the Greenland ice sheet in a high CO_2 climate. *J. Clim.*, **18**, 3409–27.

Rignot, E. J. (1998). Fast recession of a West Antarctic glacier. *Science*, **281**, 549–51.

Rignot, E. and Thomas, R. H. (2002). Mass balance of polar ice sheets. *Science*, **297**, 1502–6.

Rignot, E., Casassa, G., Gogineni, S., *et al.* (2005). Recent ice loss from the Fleming and other glaciers, Wordie Bay, West Antarctic Peninsula. *Geophys. Res. Lett.*, **32**, L07502, doi:10.1029/2004GL021947.

Rignot, E., Bamber, J. L., Van den Broeke, M. R., *et al.* (2008). Recent Antarctic ice mass loss from radar interferometry and regional climate modelling. *Nature Geosci.*, **1**(2), 106–110.

Rigor, I. G. and Wallace, J. M. (2004). Variations in the age of Arctic sea-ice and summer sea-ice extent. *Geophys. Res. Lett.*, **31**, L09401, doi:10.1029/2004GL019492.

Rigor, I. G., Colony, R. L. and Martin, S. (2000). Variations in surface air temperature observations in the Arctic, 1979–97. *J. Clim.*, **13**, 896–914.

Rigor, I. G., Wallace, J. M. and Colony, R. L. (2002). Response of sea ice to the Arctic Oscillation. *J. Clim.*, **15**, 2648–63.

Rintoul, S. R. (1998). On the origin and influence of Adélie Land Bottom Water. In *Ocean, Ice, and Atmosphere: Interactions at the Antarctic Continental Margin*, ed. S. S. Jacobs and R. F. Weiss, Washington DC: American Geophysical Union, pp. 151–71.

Rintoul, S. R. (2007). Rapid freshening of Antarctic Bottom Water formed in the Indian and Pacific Oceans. *Geophys. Res. Lett.*, **34**, L06606, doi:10.1029/2006GL028550.

Ritchie, J. C., Cwynar, L. C. and Spear, R. W. (1983). Evidence from North-West Canada for an early Holocene Milankovitch thermal maximum. *Nature*, **305**, 126–8.

Robertson, R., Visbeck, M., Gordon, A. L. and Fahrbach, E. (2002). Long-term temperature trends in the deep waters of the Weddell Sea. *Deep-Sea Res. II*, **49**, 4791–806.

Robinson, L. F., Adkins, J. F., Keigwin, L. D., *et al.* (2005). Radiocarbon variability in the western North Atlantic during the last deglaciation. *Science*, **310**, 1469–73.

Robock, A. and Free, M. P. (1995). Ice cores as an index of global volcanism from 1850 to the present. *J. Geophys. Res.*, **100**, 11 549–67.

Rogers, J. C. (1984). Association between the North Atlantic Oscillation and the Southern Oscillation in the Northern Hemisphere. *Mon. Wea. Rev.*, **112**, 1999–2015.

Rohling, E. J., Fenton, M., Jorissen, F. J., *et al.* (1998). Magnitudes of sea-level lowstands of the past 500,000 years. *Nature*, **394**, 162–5.

Rohling, E. J., Grant, K., Hemleben, C. H., *et al.* (2008). High rates of sea-level rise during the last interglacial period. *Nature Geosci.*, **1**, 38–42.

Romanovsky, V. E., Sazonova, T. S., Balobaev, V. T., Shender, N. I. and Sergueev, D. O. (2007). Past and recent changes in air and permafrost temperatures in eastern Siberia. *Global Planet. Change*, **56**, 399–413.

Roscoe, H. K., Colwell, S. R. and Shanklin, J. D. (2003). Stratospheric temperatures in Antarctic winter: does the 40-year record confirm trends in stratospheric water vapour? *Quart. J. Roy. Meteorol. Soc.*, **129**, 1745–59.

Roscoe, H. K., Carver, G. D. and Haynes, P. H. (2010). The 2002 split ozone hole: the wave of the century? *Weather*, **60**, 15–18.

Rosén, P., Segerström, U., Eriksson, L., Renberg, I. and Birks, H. J. B. (2001). Holocene climate change reconstructed from diatoms, chironomids, pollen and near-infrared spectroscopy at an alpine lake (Sjuodjijaure) in northern Sweden. *Holocene*, **11**, 551–62.

Röthlisberger, R., Bigler, M., Hutterli, M., *et al.* (2000). Technique for continuous high-resolution analysis of trace substances in firn and ice cores. *Environ. Sci. Technol.*, **34**, 338–42.

Röthlisberger, R., Mulvaney, R., Wolff, E. W., *et al.* (2002). Dust and sea salt variability in central East Antarctica (Dome C) over the last 45 kyrs and its implications for southern high-latitude climate. *Geophys. Res. Lett.*, **29**, 1963, doi:10.1029/2002GL015186.

Röthlisberger, R., Mudelsee, M., Bigler, M., *et al.* (2008). The Southern Hemisphere at glacial terminations: insights from the Dome C ice core. *Clim. Past*, **4**, 345–356.

Rothrock, D. A., Yu, Y. and Maykut, G. A. (1999). Thinning of the Arctic sea-ice cover. *Geophys. Res. Lett.*, **26**, 3469–72.

Rothrock, D. A. and Zhang, J. (2005). Arctic Ocean sea ice volume: what explains its recent depletion? *J. Geophys. Res.*, **110**, C01002, doi:10.1029/2004JC002282.

Rothrock, D. A., Zhang, J. and Yu, Y. (2003). The Arctic ice thickness anomaly of the 1990s: a consistent view from observations and models. *J. Geophys. Res.*, **108**, 3083, doi:10.1029/ 2001JC001208.

Rott, H., Skvarca, P. and Nagler, T. (1996). Rapid collapse of northern Larsen Ice Shelf, Antarctica. *Science*, **271**, 788–92.

Ruddiman, W. F. (2003). Orbital insolation, ice volume, and greenhouse gases. *Quat. Sci. Rev.*, **22**, 1597–629.

Rusin, N. P. (1961). *Meteorological and Radiational Regime of Antarctica*, Jerusalem: Israel Program for Scientific Translations.

Saltzman, E. S., Dioumaeva, I. and Finley, B. D. (2006). Glacial/interglacial variations in methanesulfonate (MSA) in the Siple Dome ice core, West Antarctica. *Geophys. Res. Lett.*, **33**, L11811, doi:10.1029/2005GL025629.

Sato, N., Kikuchi, K., Barnard, S. C. and Hogan, A. W. (1981). Some characteristic properties of ice crystal precipitation in the summer season at South Pole Station, Antarctica. *J. Meteorol. Soc. Jpn*, **59**, 772–80.

Schauer, U., Fahrbach, E., Osterhus, S. and Rohardt, G. (2004). Arctic warming through the Fram Strait: oceanic heat transport from 3 years of measurements. *J. Geophys. Res.*, **109**, C06026, doi:10.1029/2003JC001823.

Schmidt, R., Mäusbacher, R. and Müller, J. (1990). Holocene diatom flora and stratigraphy from sediment cores of two Antarctic lakes (King George Island). *J. Paleolimnol.*, **3**, 55–74.

Schmittner, A., Latif, M. and Schneider, B. (2005). Model projections of the North Atlantic thermohaline circulation for the 21st century assessed by observations. *Geophys. Res. Lett.*, **32**, L23710, doi:10.1029/2005GL024368.

Schneider, D. P., Steig, E. J. and Comiso, J. C. (2004). Recent climate variability in Antarctica from satellite-derived temperature data. *J. Clim.*, **17**, 1569–83.

Schneider, D. P., Steig, E. J., van Ommen, T. D., *et al.* (2006). Antarctic temperatures over the past two centuries from ice cores. *Geophys. Res. Lett.*, **33**, L16707, doi:10.1029/2006GL027057.

Schubert, S. D., Pfaendtner, J. and Rood, R. (1993). An assimilated data set for Earth Science applications. *Bull. Amer. Meteorol. Soc.*, **74**, 2331–42.

Schutz, B. E., Zwally, H. J., Shuman, C. A., Hancock, D. and DiMarzio, J. P. (2005). Overview of the ICESat Mission. *Geophys. Res. Lett.*, **32**, L21S01, doi:10.1029/2005GL024009.

Schwander, J., Jouzel, J., Hammer, C. U., *et al.* (2001). A tentative chronology for the EPICA Dome Concordia ice core. *Geophys. Res. Lett.*, **28**, 4243–46.

Screen, J. A., Gillett, N. P., Stevens, D. P., Marshall, G. J. and Roscoe, H. K. (2009). The role of eddies in the Southern Ocean temperature response to the Southern Annular Mode. *J. Clim.*, **22**, 806–18.

Sedwick, P. N., Harris, P. T., Robertson, L. G., *et al.* (2001). Holocene sediment records from the continental shelf of Mac. Robertson Land, East Antarctica. *Paleoceanography*, **16**, 212–25.

Serreze, M. C. and Barry, R. G. (2005). *The Arctic Climate System*, Cambridge, UK: Cambridge University Press.

Serreze, M. C. and Francis, J. A. (2006). The Arctic amplification debate. *Climatic Change*, **76**, 241–64.

Serreze, M. C., and Hurst, C. M. (2000). Representation of mean Arctic precipitation from NCEP-NCAR and ERA reanalyses. *J. Clim.*, **13**, 182–201.

Serreze, M. C., Kahl, J. and Schnell, R. (1992). Low-level temperature inversions of the Eurasian Arctic and comparisons with Soviet ice island data. *J. Clim.*, **5**, 599–613.

Serreze, M. C., Barry, R. G., Rehder, M. C. and Walsh, J. E. (1995). Variability in atmospheric circulation and moisture flux over the Arctic. *Phil. Trans. Roy. Soc. Lond. Ser. A*, **352**, 215–25.

Serreze, M. C., Rogers, J. C., Carsey, F. and Barry, R. G. (1997). Icelandic low cyclone activity: climatological features, linkages with the NAO and relationships with recent changes in the Northern Hemisphere circulation. *J. Clim.*, **10**, 453–64.

Serreze, M. C., Walsh, J. E., Chapin, F. S., *et al.* (2000). Observational evidence of recent change in the northern high-latitude environment. *Climatic Change*, **46**, 159–207.

Serreze, M. C., Barrett, A. P. and Lo, F. (2005). Northern high latitude precipitation as depicted by atmospheric reanalyses and satellite retrievals. *Mon. Wea. Rev.*, **133**, 3407–30.

Serreze, M. C., Barrett, A. P., Slater, A. G., *et al.* (2007a). The large-scale energy budget of the Arctic. *J. Geophys. Res.*, **112**, D11122, doi:10.1029/2006JD008230.

Serreze, M. C., Holland, M. M. and Stroeve, J. (2007b). Perspectives on the Arctic's shrinking sea-ice cover. *Science*, **315**, 1533–6.

Severinghaus, J. P. and Brook, E. J. (1999). Abrupt climate change at the end of the last glacial period inferred from trapped air in polar ice. *Science*, **286**, 930–4.

Severinghaus, J. P., Sowers, T., Brook, E. J., Alley, R. B. and Bender, M. L. (1998). Timing of abrupt climate change at the end of the Younger Dryas interval from thermally fractionated gases in polar ice. *Nature*, **391**, 141–6.

Severinghaus, J. P., Grachev, A., Luz, B., and Caillon, N. (2003). A method for precise measurement of argon 40/36 and krypton/argon ratios in trapped air in polar ice with applications to past firn thickness and abrupt climate change in Greenland and at Siple Some, Antarctica. *Geochim. Cosmochim. Acta*, **67**, 325–43.

Shackleton, N. J. (1969). The last interglacial in the marine and terrestrial record. *Proc. Roy. Soc. Lond., Ser. B*, **174**, 135–54.

Shackleton, N. J. (2000). The 100,000-year ice-age cycle identified and found to lag temperature, carbon dioxide, and orbital eccentricity. *Science*, **289**, 1897–902.

Shaffer, G., Olsen, S. M. and Bjerrum, C. J. (2004). Ocean subsurface warming as a mechanism for coupling Dansgaard-Oeschger climate cycles and ice-rafting events. *Geophys. Res. Lett.*, **31**, L24202, doi:10.1029/2004GL020968.

Shepherd, A. and Wingham, D. (2007). Recent sea-level contributions of the Antarctic and Greenland ice sheets. *Science*, **315**, 1529–32.

Shepherd, A., Wingham, D. J., Mansley, J. A. D. and Corr, H. F. J. (2001). Inland thinning of Pine Island Glacier, West Antarctica. *Science*, **291**, 862–4.

Shin, S.-I., Liu, Z., Otto-Bliesner, B., *et al.* (2003). A simulation of the Last Glacial Maximum climate using the NCAR-CCSM. *Clim. Dyn.*, **20**, 127–51.

Shindell, D. (2003). Whither Arctic climate? *Science*, **299**, 215–16.

Shindell, D. T. and Schmidt, G. A. (2004). Southern Hemisphere climate response to ozone changes and greenhouse gas increases. *Geophys. Res. Lett.*, **31**, L18209, doi:10.2029/2004GL020724.

Shuchman, R. A., Josberger, E. G., Russel, C. A., *et al.* (1998). Greenland Sea Odden sea ice feature: intra-annual and interannual variability. *J. Geophys. Res.*, **103**, 12 709–24.

Siddall, M., Rohling, E. J., Thompson, W. G., and Waelbroeck, C. (2008). Marine Isotope Stage 3 sea level fluctuatoins: data synthesis and new outlook. *Rev. Geophys.*, **46**, RG4003, doi:10.1029/2007RG000226.

Siegenthaler, U., Stocker, T. F., Monnin, E., *et al.* (2005). Stable carbon cycle-climate relationship during the late Pleistocene. *Science*, **310**, 1313–17.

Sigman, D. M. and Boyle, E. A. (2000). Glacial/interglacial variations in atmospheric carbon dioxide. *Nature*, **407**, 859–69.

Sime, L. C., Wolff, E. W., Oliver, K. I. C. and Tindall, J. C. (2009). Evidence for warmer interglacials in East Antarctic ice cores. *Nature*, **462**, 342–6.

Simmonds, I. and Keay, K. (2000a). Mean Southern Hemisphere extratropical cyclone behavior in the 40-year NCEP-NCAR reanalysis. *J. Clim.*, **13**, 873–85.

Simmonds, I. and Keay, K. (2000b). Variability of Southern Hemisphere extratropical cyclone behavior, 1958–97. *J. Clim.*, **13**, 550–561.

Simmonds, I., Keay, K. and Lim, E. P. (2003). Synoptic activity in the seas around Antarctica. *Mon. Wea. Rev.*, **131**, 272–88.

Sinclair, M. R. (1996). A climatology of anticyclones and blocking for the Southern Hemisphere. *Mon. Wea. Rev.*, **124**, 245–63.

Sinnhuber, B. M., Chipperfield, M. P., Davies, S., *et al.* (2000). Large loss of total ozone during the Arctic winter of 1999/2000. *Geophys. Res. Lett.*, **27**, 3473–6.

Skvarca, P., Rack, W., Rott, H. and Donángelo, T. I. (1999). Climatic trend, retreat and disintegration of ice shelves on the Antarctic Peninsula: an overview. *Polar Res.*, **18**, 151–7.

Smith, H. J., Wahlen, M., Mastroianni, D. and Taylor, K. C. (1997). The CO_2 concentration of air trapped in GISP2 ice from the Last Glacial Maximum-Holocene transition. *Geophys. Res. Lett.*, **24**, 1–4.

Smith, R. I. L. (1990). Signy island as a paradigm of biological and environmental change in Antarctic terrestrial ecosystems. In *Antarctic Ecosystems: Ecological Change and Conservation*, ed. K. R. Kerry and G. Hempel, Berlin: Springer-Verlag, pp. 32–50.

Son, S. W., Polvani, L. M., Waugh, D. W., *et al.* (2008). The impact of stratospheric ozone recovery on the Southern Hemisphere westerly jet. *Science*, **320**, 1486–9.

Sowers, T. (2001). N_2O record spanning the penultimate deglaciation from the Vostok ice core. *J. Geophys. Res.*, **106**, 31 903–14.

Sowers, T. (2006). Late Quaternary. atmospheric CH_4 isotope record suggests marine clathrates are stable. *Science*, **311**, 838–40.

Spahni, R., Chappellaz, J., Stocker, T. F., *et al.* (2005). Atmospheric methane and nitrous oxide of the Late Pleistocene from Antarctic ice cores. *Science*, **310**, 1317–21.

Spielhagen, R. F., Baumann, K.-H., Erlenkeuser, H., *et al.* (2004). Arctic Ocean deep-sea record of northern Eurasian ice sheet history. *Quat. Sci. Rev.*, **23**, 1455–83.

Stea, R. R., Piper, D. J. W., Fader, G. B. J. and Boyd, R. (1998). Wisconsinan glacial and sea-level history of Maritime Canada and the adjacent continental shelf: a correlation of land and sea events. *Geol. Surv. Amer. Bull.*, **110**, 821–45.

Steele, M., Morison, J., Ermond, W., *et al.* (2004). Circulation of summer Pacific halocline water in the Arctic Ocean. *J. Geophys. Res.*, **109**, C02027, 10.1029/2003JC002009.

Steffen, K. and Box, J. (2001). Surface climatology of the Greenland ice sheet: Greenland Climate Network 1995–1999. *J. Geophys. Res.*, **106**, 33 951–64.

Steig, E. J. (2006). The south-north connection. *Nature*, **444**, 152–3.

Steig, E. J. and Alley, R. B. (2002). Phase relationships between Antarctic and Greenland climate records. *Ann. Glaciol.*, **35**, 451–6.

Steig, E. J., Morse, D. L., Waddington, E. D., *et al.* (2000). Wisconsinan and Holocene climate history from an ice core at Taylor Dome, western Ross Embayment, Antarctica. *Geogr. Ann. A*, **82A**, 213–35.

Steig, E. J., Schneider, D. P., Rutherford, S. D., *et al.* (2009). Warming of the Antarctic ice-sheet surface since the 1957 International Geophysical Year. *Nature*, **457**, 459–62.

Stenchikov, G., Robock, A., Ramaswamy, V., *et al.* (2002). Arctic Oscillation response to the 1991 Mount Pinatubo eruption: effects of volcanic aerosols and ozone depletion. *J. Geophys. Res.*, **107**, 4803, doi:10.1029/2002JD002090.

Stenni, B., Masson-Delmotte, V., *et al.* (2001). An oceanic cold reversal during the last deglaciation. *Science*, **293**, 2074–77.

Stenni, B., Proposito, M., Gragnani, R., *et al.* (2002). Eight centuries of volcanic signal and climate change at Talos Dome (East Antarctica). *J. Geophys. Res.*, **107**, 4076, doi:10.1029/2000JD000317.

Stephens, B. B. and Keeling, R. (2000). The influence of Antarctic sea ice on glacial-interglacial CO_2 variations. *Nature*, **404**, 171–4.

Stocker, T. F. and Schmittner, A. (1997). Influence of CO_2 emission rates on the stability of the thermohaline circulation. *Nature*, **388**, 862–5.

Stocker, T. F., Wright, D. G. and Mysak, L. A. (1992). A zonally averaged, coupled ocean-atmosphere model for paleoclimate studies. *J. Clim.*, **5**, 773–97.

Stohl, A. (2006). Characteristics of atmospheric transport into the Arctic troposphere. *J. Geophys. Res.*, **111**, D11306, doi:10.1029/2005JD006888.

Stolarski, R. S., Krueger, A. J., Schoeberl, M. R., *et al.* (1986). Nimbus-7 satellite measurements of the springtime Antarctic ozone decrease. *Nature*, **322**, 808–11.

Stone, R. S., Dutton, E. G., Harris, J. M. and Longenecker, D. (2002). Earlier spring snowmelt in northern Alaska as an indicator of climate change. *J. Geophys. Res.*, **107**, 4089, doi:10.1029/2000JD000286.

Stotter, J., Wastl, M., Caseldine, C. and Haberle, T. (1999). Holocene palaeoclimatic reconstruction in northern Iceland: approaches and results. *Quat. Sci. Rev.*, **18**, 457–74.

Straus, D. M. and Shukla, J. (2002). Does ENSO force the PNA? *J. Clim.*, **15**, 2340–2358.

Stroeve, J., Serreze, M. C., Drobot, S., *et al.* (2008). Arctic sea ice extent plummets in 2007. *Eos*, **89**, 13.

Stuiver, M. and Braziunas, T. F. (1993). Sun, ocean, climate and atmospheric $^{14}CO_2$: an evaluation of causal and spectral relationships. *Holocene*, **3**, 289–305.

Stuiver, M. and Reimer, P. J. (1993). Extended ^{14}C data base and revised CALIB 3.0 ^{14}C age calibration program. *Radiocarbon*, **35**, 215–30.

Stuiver, M., Braziunas, T. F., Becker, B. and Kromer, B. (1991). Climatic, solar, oceanic, and geomagnetic influences on Late-Glacial and Holocene atmospheric ^{14}C/^{12}C change. *Quat. Res.*, **35**, 1–24.

Sturm, M., Schimel, J., Michaelson, G., *et al.* (2005). Winter biological processes could help convert arctic tundra to shrubland. *BioScience*, **55**, 17–26.

Su, F., Adam, J. C., Trenberth, K. E. and Lettenmaier, D. P. (2006). Evaluation of surface water fluxes of the pan-Arctic land region with a land surface model and ERA-40 reanalysis. *J. Geophys. Res.*, **111**, D05110, doi:10.1029/2005JD006387.

Suttie, T. K. (1970). Portrait of a polar low. *Weather*, **25**, 504–7.

Svendsen, J. I., Alexanderson, H., Astakhov, V. I., *et al.* (2004). Late Quaternary. ice sheet history of northern Eurasia. *Quat. Sci. Rev.*, **23**, 1229–71.

Svensson, A., Nielsen, S. W., Kipfstuhl, S., *et al.* (2005). Visual stratigraphy of the North Greenland Ice Core Project (NorthGRIP) ice core during the last glacial period. *J. Geophys. Res.*, **110**, C03012, D02108, doi.10.1029/2004JD005134.

Swift, J. H., Aagaard, K., Timokhov, L. and Nikiforov, E. G. (2005). Long-term variability of Arctic Ocean waters: evidence from a reanalysis of the EWG data set. *J. Geophys. Res.*, **110**, C03012, doi:10.1029/2004JC002312.

Talley, L. D. (2002). Ocean circulation. In *Encyclopedia of Global Environmental Change*, Vol. 1, ed. M. C. MacCracken and J. S. Perry, Chichester, UK: John Wiley and Sons, pp. 557–79.

Tangborn, W. (1980). 2 Models for Estimating Climate-Glacier Relationships in the North Cascades, Washington, USA. *J. Glaciol.*, **25**, 3–21.

Tarasov, L. and Peltier, W. R. (2004). A geophysically constrained large ensemble analysis of the deglacial history of the North American ice-sheet complex. *Quat. Sci. Rev.*, **23**, 359–88.

Tarasov, L. and Peltier, W. R. (2005). Arctic freshwater forcing of the Younger Dryas cold reversal. *Nature*, **435**, 662–5.

Tatur, A. and del Valle, R. (1986). Badania paleolimnologicne i geomorfologicne na wyspie krola jerzego-Antarktyka zachodnia (1984–1986). [Paleolimnological and geomorphological investigations on King George Island – West Antarctica]. (In Polish) *Przeglad Geologiczny*, **11**, 621–6.

Taylor, F. and McMinn, A. (2001). Evidence from diatoms for Holocene climate fluctuation along the East Antarctic margin. *Holocene*, **11**, 455–66.

Taylor, K. C., Hammer, C. U., Alley, R. B., *et al.* (1993). Electrical conductivity measurements from the GISP2 and GRIP Greenland ice cores. *Nature*, **366**, 549–52.

Teller, J. T., Boyd, M., Yang, Z., Kor, P. S. G. and Fard, A. M. (2005). Alternative routing of Lake Agassiz overflow during the Younger Dryas: new dates, paleotopography, and a re-evaluation. *Quat. Sci. Rev.*, **24**, 1890–905.

TEMPO (1996). Potential role of vegetation feedback in the climate sensitivity of high-latitude regions: a case study at 6000 years B.P. *Global Biogeochem. Cy.*, **10**, 727–36.

Thie, J. (1974). Distribution and thawing of permafrost in the southern part of the discontinuous permafrost zone in Manitoba. *Arctic*, **27**, 189–200.

Thoma, M., Jenkins, A., Holland, D. and Jacobs, S. (2008). Modelling Circumpolar Deep Water intrusions on the Amundsen Sea continental shelf, Antarctica. *Geophys. Res. Lett.*, **35**, L18602, doi:10.1029/2008GL034939.

Thomas, E. R., Wolff, E. W., Mulvaney, R., *et al.* (2007). The 8.2 ka event from Greenland ice cores. *Quat. Sci. Rev.*, **26**, 70–81.

Thomas, R., Rignot, E., Casassa, G., *et al.* (2004). Accelerated sea-level rise from West Antarctica. *Science*, **306**, 255–8.

Thompson, D. W. J. and Solomon, S. (2002). Interpretation of recent Southern Hemisphere climate change. *Science*, **296**, 895–9.

Thompson, D. W. J. and Wallace, J. M. (1998). The Arctic Oscillation signature in the wintertime geopotential height and temperature fields. *Geophys. Res. Lett.*, **25**, 1297–300.

Thompson, D. W. J. and Wallace, J. M. (2000). Annular modes in the extratropical circulation. Part I: Month-to-month variability. *J. Clim.*, **13**, 1000–16.

Thompson, D. W. J., Wallace, J. M. and Hegerl, G. C. (2000). Annular modes in the extratropical circulation. Part II: Trends. *J. Clim.*, **13**, 1018–36.

Thorarinsson, S. (1940). Present glacier shrinkage and eustatic change in sea level. *Geogr. Ann.*, **22**, 131–59.

Thorne, P. W. (2008). Arctic tropospheric warming amplification? *Nature*, **455**, E1–E2.

Thorne, P. W., Parker, D. E., Tett, S. F. B., *et al.* (2005). Revisiting radiosonde upper air temperatures from 1958 to 2002. *J. Geophys. Res.*, **110**, D18105, doi:10.1029/2004JD005753.

Tjernstrom, M., Zagar, M., Svensson, G., *et al.* (2005). Modelling the arctic boundary layer: an evaluation of six ARCMIP regional-scale models using data from the Sheba project. *Bound.-Lay. Meteorol.*, **117**, 337–81.

Toniazzo, T., Gregory, J. M. and Huybrechts, P. (2004). Climatic impact of a Greenland deglaciation and its possible irreversibility. *J. Clim.*, **17**, 21–33.

Tremblay, L.-B., Mysak, L. A. and Dyke, A. S. (1997). Evidence from driftwood records for century-to-millennial scale variations of the high latitude atmospheric circulation during the Holocene. *Geophys. Res. Lett.*, **24**, 2027–30.

Trenberth, K. E. and Hurrell, J. W. (1994). Decadal atmosphere-ocean variations in the Pacific. *Clim. Dyn.*, **9**, 303–19.

Tsukernik, M., Chase, T. N., Serreze, M. C., *et al.* (2004). On the regulation of minimum mid-tropospheric temperatures in the Arctic. *Geophys. Res. Lett.*, **31**, L06112, doi:10.1029/2003GL018831.

Tsukernik, M., Kindig, D. N. and Serreze, M. C. (2007). Characteristics of winter cyclone activity in the northern North Atlantic: insights from observations and regional modeling. *J. Geophys. Res.*, **112**, D03101, doi:10.1029/2006JD007184.

Turner, J. (2004). The El Niño-Southern Oscillation and Antarctica. *Int. J. Climatol.*, **24**, 1–31.

Turner, J., Lachlan-Cope, T. A. and Thomas, J. P. (1993). A comparison of Arctic and Antarctic mesoscale vortices. *J. Geophys. Res.*, **98** D7, 13 019–34.

Turner, J., Bromwich, D., Colwell, S., *et al.* (1996). The Antarctic First Regional Observing Study of the Troposphere (FROST) project. *Bull. Amer. Meteorol. Soc.*, **77**, 2007–32.

Turner, J., Marshall, G. J. and Lachlan-Cope, T. A. (1998). Analysis of synoptic-scale low pressure systems within the Antarctic Peninsula sector of the circumpolar trough. *Int. J. Climatol.*, **18**, 253–80.

Turner, J., Connolley, W. M., Leonard, S., Marshall, G. J. and Vaughan, D. G. (1999). Spatial and temporal variability of net snow accumulation over the Antarctic from ECMWF re-analysis project data. *Int. J. Climatol.*, **19**, 697–724.

Turner, J., Marshall, G. J. and Ladkin, R. (2001). An operational, real-time cloud detection scheme for use in the Antarctic based on AVHRR data. *Int. J. Rem. Sens.*, **22**, 3027–46.

Turner, J., Colwell, S. R., Marshall, G. J., *et al.* (2004). The SCAR READER project: towards a high-quality database of mean Antarctic meteorological observations. *J. Clim.*, **17**, 2890–8.

Turner, J., Colwell, S. R., Marshall, G. J., *et al.* (2005). Antarctic climate change during the last 50 years. *Int. J. Climatol.*, **25**, 279–94.

Turner, J., Lachlan-Cope, T. A., Colwell, S. R., Marshall, G. J. and Connolley, W. M. (2006a). Significant warming of the Antarctic winter troposphere. *Science*, **311**, 1914–17.

Turner, J., Connolley, W. M., Lachlan-Cope, T. A. and Marshall, G. J. (2006b). The performance of the Hadley Centre climate model (HadCM3) in high southern latitudes. *Int. J. Climatol.*, **26**, 91–112.

Turner, J., Overland, J. E. and Walsh, J. E. (2007). An Arctic and Antarctic perspective on recent climate change. *Int. J. Climatol.*, **27**, 277–93.

Turner, J., Anderson, P. S., Lachlan-Cope, T. A., *et al.* (2009a). Record low surface air temperature at Vostok station, Antarctica. *J. Geophys. Res.*, **114**, D24102, doi:10.1029/2009JD012104.

Turner, J., Comiso, J. C., Marshall, G. J., *et al.* (2009b). Non-annular atmospheric circulation change induced by stratospheric ozone depletion and its role in the recent increase of Antarctic sea ice extent. *Geophys. Res. Lett.*, **36**, L08502, doi:10.1029/2009GL037524.

Tziperman, E., Raymo, M. E., Huybers, P. and C. Wunsch, C. (2006). Consequences of pacing the Pleistocene 100 kyr ice ages by nonlinear phase locking to Milankovitch forcing. *Paleoceanography*, **21**, PA4206, doi:10.1029/2005PA001241.

Uppala, S. M., Kallberg, P. W., Simmons, A. J., *et al.* (2005). The ERA-40 re-analysis. *Quart. J. Roy. Meteorol. Soc.*, **131**, 2961–3012.

Uttal, T., Curry, J. A., McPhee, M. G., *et al.* (2002). Surface heat budget of the Arctic Ocean. *Bull. Amer. Meteorol. Soc.*, **83**, 255–75.

Valdes, P. J. (2003). An introduction to climate modelling of the Holocene. In *Global Change in the Holocene*, ed. A. Mackay, R. Battarbee, J. Birks and F. Oldfield, London: Arnold, pp. 20–35.

van de Berg, W. J., van den Broeke, M. R., Reijmer, C. H. and van Meijgaard, E. (2005). Characteristics of the Antarctic surface mass balance, 1958–2002, using a regional atmospheric climate model. *Ann. Glaciol.*, **41**, 97–104.

van de Berg, W. J., van den Broeke, M. R., Reijmer, C. H. and van Meijgaard, E. (2006). Reassessment of the Antarctic surface mass balance using calibrated output of a regional atmospheric climate model. *J. Geophys. Res.*, **111**, D11104, doi:10.1029/2005JD006495.

van de Wal, R. S. W. and Wild, M. (2001). Modelling the response of glaciers to climate change by applying volume-area scaling in combination with a high resolution GCM. *Clim. Dyn.*, **18**, 359–66.

van de Wal, R. S. W., Greuell, W., van den Broeke, M. R., Reijmer, C. H. and Oerlemans, J. (2006). Surface mass-balance observations and automatic weather station data along a transect near Kangerlussuaq, West Greenland. *Ann. Glaciol.*, **42**, 311–16.

van den Broeke, M. R. (1998). The semi-annual oscillation and Antarctic climate. Part 1: Influence on near surface temperatures (1957–79). *Antarct. Sci.*, **10**, 175–83.

van den Broeke, M. R., van de Wal, R. S. W. and Wind, M. (1997). Representation of Antarctic katabatic winds in a high-resolution GCM and a note on their climate sensitivity. *J. Clim.*, **10**, 3111–30.

van den Broeke, M., Reijmer, C. and van de Wal, R. (2004a). Surface radiation balance in Antarctica as measured with automatic weather stations. *J. Geophys. Res.*, **109**, D09103, doi:10.1029/2003JD004394.

van den Broeke, M. R., Reijmer, C. H. and van de Wal, R. S. W. (2004b). A study of the surface mass balance in Dronning Maud Land, Antarctica, using automatic weather stations. *J. Glaciol.*, **50**, 565–82.

van Lipzig, N. P. M., van Meijgaard, E. and Oerlemans, J. (1999). Evaluation of a regional atmospheric model using measurements of surface heat exchange processes from a site in Antarctica. *Mon. Wea. Rev.*, **127**, 1994–2011.

van Lipzig, N. P. M., Turner, J., Colwell, S. R. and van den Broeke, M. R. (2004). The near-surface wind field over the Antarctic continent. *Int. J. Climatol.*, **24**, 1973–82.

van Loon, H. (1972). Pressure in the Southern Hemisphere. In *Meteorology of the Southern Hemisphere*, ed. H. van Loon, J. J. Taljaard, T. Sasamori, *et al.*, Boston, MA: American Meteorological Society, pp. 59–86.

van Loon, H. and Jenne, R. L. (1972). Half-yearly oscillations in the Drake Passage. *Deep-Sea Res.*, **19**, 525–7.

van Loon, H. and Rogers, J. C. (1978). The seesaw in winter temperatures between Greenland and northern Europe. Part I: General description. *Mon. Wea. Rev.*, **106**, 296–310.

van Loon, H. and Rogers, J. C. (1984). Interannual variations in the half-yearly cycle of pressure gradients and zonal wind at sea level on the Southern Hemisphere. *Tellus*, **36A**, 76–86.

van Loon, H., Kidson, J. W. and Mullan, A. B. (1993). Decadal variation of the annual cycle in the Australian dataset. *J. Clim.*, **6**, 1227–31.

Vaughan, D. G. and Doake, C. S. (1996). Recent atmospheric warming and retreat of ice shelves on the Antarctic Peninsula. *Nature*, **379**, 328–31.

Veillette, J. (1994). Evolution and paleohydrology of glacial lakes Barlow and Ojibway. *Quat. Sci. Rev.*, **13**, 945–71.

Veillette, J. J., Dyke, A. S., and Roy, M. (1999). Ice-flow evolution of the Labrador Sector of the Laurentide Ice Sheet: a review, with new evidence from northern Quebec. *Quat. Sci. Rev.*, **18**, 993–1019.

Velicogna, I. and Wahr, J. (2005). Greenland mass balance from GRACE. *Geophys. Res. Lett.*, **32**, L18505, doi:10.1029/2005GL023955.

Venegas, S. A. (2003). The Antarctic Circumpolar Wave: a combination of two signals? *J. Clim.*, **16**, 2509–25.

Vermeer, M. and Rahmstorf, S. (2009). Global sea level linked to global temperature. *Proc. Natl. Acad. Sci.*, **106**, 21 527–32.

Vinther, B. M., Buchardt, S. L., Clausen, H. B., *et al.* (2009). Holocene thinning of the Greenland ice sheet. *Nature*, **461**, 385–8.

Visbeck, M., Chassignet, E. P., Curry, R. G., *et al.* (2003). The ocean's response to North Atlantic Oscillation variability. In *The North Atlantic Oscillation*, ed. J. W. Hurrell, Y. Kushnir, G. Ottersen and M. Visbeck, Washington, DC: American Geophysical Union, pp. 113–45.

Volodin, E. M. and Galin, V. Y. (1999). Interpretation of winter warming on Northern Hemisphere continents in 1977–94. *J. Clim.*, **12**, 2947–55.

Waddington, E. D., Conway, H., Steig, E. J., *et al.* (2005). Decoding the dipstick: thickness of Siple Dome, West Antarctica, at the Last Glacial Maximum. *Geology*, **33**, 281–4.

424 *References*

Wadhams, P. (1990). Evidence for thinning of the Arctic ice cover north of Greenland. *Nature*, **345**, 795–7.

Wadhams, P. (1992). Sea ice thickness distribution in the Greenland Sea and Eurasian Basin, May 1987. *J. Geophys. Res.*, **97**, doi:10.1029/91JC03137.

Wadhams, P. (1994). Sea ice thickness changes and their relation to climate. In *The Polar Oceans and Their Role in Shaping the Global Environment*, ed. O. M. Johannessen, R. D. Muench and J. E. Overland, Washington, DC: American Geophysical Union, pp. 337–61.

Walker, D. P., Brandon, M. A., Jenkins, A., *et al.* (2007). Oceanic heat transport onto the Amundsen Sea shelf through a submarine glacial trough. *Geophys. Res. Lett.*, **34**, L02602, doi:10.1029/2006GL028154.

Walker, G. T. and Bliss, E. W. (1932). World Weather V. *Mem. Roy. Meteorol. Soc.*, **4**, 53–84.

Wallace, J. M. (2000). North Atlantic Oscillation/annular mode: two paradigms-one phenomenon. *Quart. J. Roy. Meteorol. Soc.*, **126**, 791–805.

Wallace, J. M. and Gutzler, D. S. (1981). Teleconnections in the geopotential height field during the Northern Hemisphere winter. *Mon. Wea. Rev.*, **109**, 784–812.

Walsh, J. E. and Johnson, C. M. (1979). An analysis of Arctic sea ice. *J. Phys. Oceanogr.*, **9**, 580–91.

Walsh, J. E., Chapman, W. L. and Shy, T. L. (1996). Recent decrease of sea level pressure in the central Arctic. *J. Clim.*, **9**, 480–6.

Wang, M. and Overland, J. E. (2005). Detecting arctic climate change using Koppen classification. *Climatic Change*, **67**, 43–62.

Wang, M., Overland, J. E., Kattsov, V., *et al.* (2007). Intrinsic versus forced variation in coupled climate model simulations over the Arctic during the 20th century. *J. Clim.*, **20**, 1093–107.

Wang, P., Tian, J., Cheng, X., Liu, C., and Xu, J. (2003). Carbon reservoir changes precede major ice-sheet expansion at the mid-Brunhes event. *Geology*, **31**, 239–42.

Warren, S. G., Hahn, C. J., London, J., Chervin, R. M., and Jenne, R. L. (1986). *Global Distribution of Total Cloud and Cloud Type Amounts Over Land*. NCAR Technical Note TN-273+STR/DOE Technical Report ER/60085-H1 edition. Boulder, CO: NCAR.

Warrick, R. A., Le Provost, C., Meier, M. F., Oerlemans, J. and Woodworth, P. L. (1995). Changes in sea level. In *Climate Change 1995: The Science of Climate Change*, ed. J. T. Houghton, L. G. Meiro Filho, B. A. Callender, *et al.* Cambridge, UK: Cambridge University Press, pp. 359–405.

Watanabe, O., Jouzel, J., Johnsen, S., *et al.* (2003). Homogeneous climate variability across East Antarctica over the past three glacial cycles. *Nature*, **422**, 509–12.

Watkins, S. J., Maher, B. A. and Bigg, G. R. (2007). Ocean circulation at the Last Glacial Maximum: a combined modelling and magnetic proxy-based study. *Paleoceanography*, **22**, PA2204, doi:10.1029/2006PA001281.

Waugh, D. W. and Randel, W. J. (1999). Climatology of arctic and antarctic polar vortices using elliptical diagnostics. *J. Atmos. Sci.*, **56**, 1594–613.

Weaver, A. J., Eby, M., Wiebe, E. C., *et al.* (2001). The UVic Earth System Climate Model: model description, climatology and application to past, present and future climates. *Atmos. Ocean.*, **39**, 361–428.

Weaver, A. J., Saenko, O. A., Clark, P. U. and Mitrovica, J. X. (2003). Meltwater Pulse 1A from Antarctica as a trigger of the Bølling-Allerød warm interval. *Science*, **299**, 1709–13.

Webb, D. J., Killworth, P. D., Coward, A. C. and Thompson, S. R. (1991). *The FRAM Atlas of the Southern Ocean*, Swindon, UK: Natural Environment Research Council.

Webb III, T., Bartlein, P. J., Harrison, S. P. and Anderson, K. H. (1993). Vegetation, lake levels, and climate in eastern North America for the past 18,000 years. In *Global Climates since the Last Glacial Maximum*, ed. H. E. Wright, J. E. Kutzbach, T. Webb III, *et al.*, Minneapolis, MN: University of Minnesota Press, pp. 414–67.

Weisse, R., Mikolajewicz, U., Sterl, A. and Drijfhout, S. S. (1999). Stochastically forced variability in the Antarctic Circumpolar Current. *J. Geophys. Res.*, **104**, 11 049–64.

White, W. B. and Peterson, R. G. (1996). An Antarctic circumpolar wave in surface pressure, wind, temperature and sea-ice extent. *Nature*, **380**, 699–702.

White, W. B., Chen, S. C., Allan, R. J. and Stone, R. C. (2002). Positive feedbacks between the Antarctic Circumpolar Wave and the global El Niño-Southern Oscillation wave. *J. Geophys. Res.*, **107**, 3165, doi:10.1029/2000JC000581.

White, W. B., Gloersen, P. and Simmonds, I. (2004). Tropospheric response in the Antarctic Circumpolar Wave along the sea ice edge around Antarctica. *J. Clim.*, **17**, 2765–79.

Whitworth, T. and Peterson, R. G. (1985). Volume transport of the Antarctic Circumpolar Current from bottom pressure measurements. *J. Phys. Oceanogr.*, **15**, 810–16.

Williams, L. D. and Wigley, T. M. L. (1983). A comparison of evidence for late Holocene summer temperature-variations in the Northern Hemisphere. *Quat. Res.*, **20**, 286–307.

Winckler, G. and Fischer, H. (2006). 30,000 years of cosmic dust in Antarctic ice. *Science*, **313**, 491.

Wingham, D. J., Shepherd, A., Muir, A. and Marshall, G. J. (2006). Mass balance of the Antarctic ice sheet. *Phil. Trans. Roy. Soc. Lond. Ser. A*, **364**, 1627–35.

WMO (2003). *Scientific Assessment of Ozone Depletion: 2002*. Global Ozone Research and Monitoring Project, Report No. 47, Geneva: World Meteorological Organization.

Wolfe, B. B., Edwards, T. W. D., Aravena, R., *et al.* (2000). Holocene paleohydrology and paleoclimate at Treeline, north-central Russia, inferred from oxygen isotope records in lake sediment cellulose. *Quat. Res.*, **53**, 319–29.

Wolff, E. W. and Suttie, E. D. (1994). Antarctic snow record of Southern Hemisphere lead pollution. *Geophys. Res. Lett.*, **21**, 781–4.

Wolff, E. W., Rankin, A. M. and Röthlisberger, R. (2003). An ice core indicator of Antarctic sea ice production? *Geophys. Res. Lett.*, **30**, 2158, doi:10.1029/2003GL018454.

Wolff, E. W., Fischer, H., Fundel, F., *et al.* (2006). Southern Ocean sea-ice extent, productivity and iron flux over the past eight glacial cycles. *Nature*, **440**, 491–6.

Wolff, E. W., Chappellaz J., Blunier T., Rasmussen, S. O. and Svensson, A. (2010). Millenial-scale variability during the last glacial: the ice core record. *Quat. Sci. Rev.*, **29**, 2828–38.

Worby, A. P. and Allison, I. (1991). Ocean-atmosphere energy exchange over thin, variable concentration Antarctic pack ice. *Ann. Glaciol.*, **15**, 184–90.

Worby, A. P., Geiger, C. A., Paget, M. J., *et al.* (2008). The thickness distribution of Antarctic sea ice. *J. Geophys. Res.*, **113**, C05S92, doi:10.1029/2007JC004254.

426 *References*

Wu, B. Y., Wang, J. and Walsh, J. E. (2006). Dipole anomaly in the winter Arctic atmosphere and its association with sea ice motion. *J. Clim.*, **19**, 210–25.

Wu, Q. G. and Straus, D. M. (2004). AO, COWL, and observed climate trends. *J. Clim.*, **17**, 2139–56.

Wunsch, C. (1999). The interpretation of short climate records, with comments on the North Atlantic and Southern Oscillations. *Bull. Amer. Meteorol. Soc.*, **80**, 245–55.

Wunsch, C. (2003). The spectral description of climate change including the 100 ky energy. *Clim. Dyn.*, **20**, 353–63.

Yang, D. Q., Goodison, B., Metcalfe, J., *et al.* (2001). Compatibility evaluation of national precipitation gage measurements. *J. Geophys. Res.*, **106**, 1481–91.

Yang, F. and Schlesinger, M. E. (2002). On the surface and atmospheric temperature changes following the 1991 Pinatubo volcanic eruption: a GCM study. *J. Geophys. Res.*, **107**, 4073, doi:10.1029/2001JD000373.

Yokoyama, Y., Lambeck, K., De Deckker, P., Johnston, P. and Fifield, I. K. (2000). Timing of the Last Glacial Maximum from observed sea-level minima. *Nature*, **406**, 713.

Yuan, X. (2004). ENSO-related impacts on Antarctic sea ice: a synthesis of phenomenon and mechanisms. *Antarct. Sci.*, **16**, 415–25.

Yuan, X. and Martinson, D. G. (2000). Antarctic sea ice extent variability and its global connectivity. *J. Clim.*, **13**, 1697–717.

Yuan, X. J. and Martinson, D. G. (2001). The Antarctic Dipole and its predictability. *Geophys. Res. Lett.*, **28**, 3609–12.

Zazulie, N., Rusticucci, M. and Solomon, S. (2010). Changes in climate at high southern latitudes: a unique daily record at Orcadas spanning 1903–2008. *J. Clim.*, **23**, 189–96.

Zenk, W. and Morozov, E. (2007). Decadal warming of the coldest Antarctic Bottom Water flowing through the Vema Channel. *Geophys. Res. Lett.*, **34**, L14607, doi:10.1029/2007/GL030340.

Zhang, X. D., Walsh, J. E., Zhang, J., Bhatt, U. S. and Ikeda, M. (2004). Climatology and interannual variability of arctic cyclone activity: 1948–2002. *J. Clim.*, **17**, 2300–2317.

Zhang, Y., Wallace, J. M. and Battisti, D. S. (1997). ENSO-like interdecadal variability: 1900–93. *J. Clim.*, **10**, 1004–20.

Zielinski, G. A. (2000). Use of paleo-records in determining variability within the volcanism climate system. *Quat. Sci. Rev.*, **19**, 417–38.

Zielinski, G. A., Mayewski, P. A., Meeker, L. D., *et al.* (1994). Record of volcanism since 7000 B.C. from the GISP2 Greenland ice core and implications for the volcano-climate system. *Science*, **264**, 948–52.

Zoltai, S. C. (1995). Permafrost distribution in peatlands of west-central Canada during the Holocene warm period 6000 years ago. *Géogr. Phys Quatern.*, **49**, 45–54.

Zwally, H. J. (1989). Growth of Greenland ice sheet: interpretation. *Science*, **246**(4937), 1589–91.

Zwally, H. J., Comiso, J. C., Parkinson, C. L., *et al.* (1983a). *Antarctic Sea Ice, 1973–1976.* Report NASA SP-459, Washington, DC: NASA.

Zwally, H. J., Parkinson, C. L. and Comiso, J. C. (1983b). Variability of Antarctic sea ice and changes in carbon dioxide. *Science*, **220**, 1005–12.

Zwally, H. J., Comiso, J. C., Parkinson, C. L., Cavalieri, D. J. and Gloersen, P. (2002a). Variability of Antarctic sea ice 1979–1998. *J. Geophys. Res.*, **107**, 3041, 10.1029/2000JC000733.

Zwally, H. J., Abdalati, W., Herring, T., *et al.* (2002b). Surface melt-induced acceleration of Greenland ice-sheet flow. *Science*, **297**, 218–22.

Zwally, H. J., Giovinetto, M. B., Li, J., *et al.* (2005). Mass changes of the Greenland and Antarctic ice sheets and shelves and contributions to sea-level rise: 1992–2002. *J. Glaciol.*, **51**, 509–27.

Zweck, C. and Huybrechts, P. (2005). Modeling of the northern hemisphere ice sheets during the last glacial cycle and glaciological sensitivity. *J. Geophys. Res.*, **110**, D07103, doi:10.1029/2004JD005489.

Index

1500 year cycle, 200, 211, 240, 255
8.2 kyr event, 217, 235
active layer, 307
Advanced Very High Resolution Radiometer
  (AVHRR), 267
aerosols, 1, 3, 21, 62, 97, 99, 286
  volcanic, 99
air masses, 3, 79
Alaska, 5, 258, 274
  Barrow, 3, 32, 96, 304, 307
  Brooks Range, 248
  Fairbanks, 307
  glaciers, 228, 231, 326, 368
  Kenai Mountains, 232
  North Slope, 98
  permafrost, 310
  precipitation, 203, 347
albedo, 2, 9, 54, 68, 71, 90, 129, 133
Aleutian Low, 108, 114, 147, 258, 274, 346
altimeters, 144
Amery Oasis, 222, 227
Amundsen Sea, 120
Amundsen Sea Embayment, 15
Amundsen Sea Low, 120, 207, 297
Amundsen–Bellingshausen Sea, 110, 279, 296
Amundsen–Scott Station, 25, 96, 130, 265
Antarctic Bottom Water *see* water masses
Antarctic Circumpolar Current, 10, 106, 114, 152, 160,
  191, 285
  future, 355
Antarctic Circumpolar Wave, 112, 295
Antarctic Climate Change and the Environment
  report, 337
Antarctic climatic optimum, 199
Antarctic Cold Reversal, 185, 190
Antarctic Dipole, 110, 114
Antarctic Ice Sheet *see* ice sheet: Antarctic
Antarctic Isotope Maximum, 163, 181, 186
Antarctic Mesoscale Prediction System (AMPS), 30, 57
Antarctic Peninsula, 7, 8, 57, 140

  climate variability, 213
  deglaciation, 250
  glaciers, 252
  permafrost, 315
  warming, 15, 264
Antarctic Plateau, 2, 3, 122
Antarctic Sea Ice Processes and Climate (ASPeCt)
  project, 298
Antarctic warm events, 163
Antarctica
  area, 7
anticyclones, 125
Arctic Climate Impact Assessment, 337
Arctic Express, 109
Arctic front, 76, 140
Arctic haze, 3, 21, 95, 316
Arctic Ocean, 2, 3, 50, 118, 145, 154, 241, 282, 348
  sea ice, 148
Arctic Ocean Deep Water *see* water masses
Argo floats, 27, 280
Atlantic Water *see* water masses
Autonomous Lagrangian Circulation Explorer
  floats, 285

Baffin Bay, 118, 147, 238
Barents Sea, 78, 105, 115, 119, 155
Bear Island Current, 155
Beaufort Gyre, 4, 148, 155, 295
Beaufort Sea, 116
Bellingshausen Sea, 111, 125
Bellingshausen Station, 131
Bering Strait, 135
bipolar see-saw 89, 191
black carbon, 97, 317
blocking, 110, 120, 125, 177
blowing snow, 56, 70, 71
blue ice, 2
Bølling–Allerød warm period, 178, 187
Bond events, 201
boreal forest, 4, 13

Index

borehole measurements, 215, 224
boundary layer, 21, 27
Brewer–Dobson circulation, 92
brine rejection, 155, 156
Brunt Ice Shelf, 21
Bunger Hills, 250
Bunger Oasis, 227

Campbell Plateau, 160
Canadian Arctic, 116
carbon cycle, 11, 14, 53, 190, 288
  future, 355
carbon dioxide *see* greenhouse gases: carbon dioxide
chlorofluorocarbons, 75, 93, 365
Chukchi Sea, 126, 359
circumpolar trough, 97, 120, 125, 135, 137, 153, 279
clathrates, 92
CLIMBER model, 55
cloud, 3, 32, 57
  albedo, 133
  amount, 134
  condensation nuclei, 96
  convective, 75, 136
  feedbacks, 73
  future, 348, 353
  identification, 36
  measurement, 22
  optical depth, 73
  phase, 73
  properties, 98, 136
  streets, 75, 79
  thickness, 73
  tracking, 40
  type, 135
coastal erosion, 73
coccoliths, 46
Cockburn Stade, 204
cold halocline layer, 282
Cooperative Holocene Mapping Project, 238
coreless winter, 70, 130
Coupled Model Inter-comparison Project, 340
cyclogenesis, 115, 119, 125
cyclolysis, 119, 125

Dansgaard–Oeschger events, 88, 163, 168, 171, 211
Davis Strait, 146, 238, 261
Deep Western Boundary Current, 156
depressions, 10, 118
  tracking, 104, 115, 123, 135, 205, 275, 279
  trends, 277
diamond dust, 22, 24, 57, 137
diatoms, 46, 48, 183
dimethylsulphide, 44, 96, 111
Dome C Station, 25, 45
Drake Passage, 10, 106
drifting buoys, 27, 295
Dronning Maud Land, 2, 186

Dry Valley Drilling Project, 314
Dry Valleys, 9, 232, 266, 306, 314
Dumont d'Urville Station, 96
dust, 91, 97, 98, 180, 182, 190, 207, 246
Dye 3, 166

Early Holocene warm period, 213, 221
Early Weichselian, 169
Early Wisconsinan, 169
Earth Radiation Budget Experiment, 63
East Greenland Current, 148, 155, 237
East Spitsbergen Current, 155
Eemian interglacial, 165
El Niño–Southern Oscillation, 61, 107, 148, 161, 275
electrical conductivity measurement, 44
Elephant Island, 227, 252
Ellesmere Island, 126, 208
Erdalen event, 216
Esperanza Station, 265
European Centre for Medium-range Weather Forecasts
  (ECMWF) reanalysis, 31
European Project for Ice Coring in Antarctica
  (EPICA), 42
evaporation, 71, 348

Falklands Current, 160
Faraday/Vernadsky Station, 79, 130, 264
fast ice *see* sea ice: fast ice
feedback mechanisms, 2, 9, 54, 71, 198, 285, 346, 360
Ferrel cell, 109
Fine Resolution Antarctic Model (FRAM), 59
Finland, 173
firn, 42
foraminifera, 46
fossil tree data, 225
Fram Basin, 4
Fram Strait, 4, 105, 148, 155, 201
frost flowers, 44, 92, 97

glaciers, 36, 197
  Alaska, 251
  Antarctic Peninsula, 328
  Arctic, 326
  Canadian, 326
  future, 368
  Norway, 251
  outlet, 322
  Pine Island, 15, 39, 322
  Thwaites, 322
Global Precipitation Climatology Project, 32
Gravity Recovery and Climate Experiment (GRACE)
  mission, 39
Great Salinity Anomaly, 201, 282, 291
greenhouse gases, 12, 14, 17, 62, 74, 89, 102
  carbon dioxide, 43, 89, 181, 185, 245, 288, 311, 315,
    339, 364
  emission scenarios, 337

430   *Index*

greenhouse gases (cont.)
  methane, 43, 89, 181, 245, 309, 316
  nitrous oxide, 89, 316
  ozone, 89
  water vapour, 2, 89
Greenland, 3
  sea level contribution, 330
  snow cover, 300
  temperature, 177
Greenland Sea, 11, 105, 148, 282, 290
ground ice, 307
Gulf of Alaska, 78, 109, 120
Gulf of St. Lawrence, 293
Gulf Stream, 13
Gustav Channel, 232

Hadley circulation, 109, 201
Halley Station, 21, 277, 319
halogens, 62
halons, 93
heat flux, 75
Heinrich events, 86, 163, 172, 173, 176
Hudson Bay, 120, 147, 235, 248
Hudson Strait, 86
  ice stream, 173
humidity
  measurement, 20
hydrological cycle, 54, 305

ice ages, 47, 80, 90, 163, 164
icecaps, 320, 368
  Svalbard, 251
  Vavilov, Severnaya Zemlya, 251
ice cores
  Agassiz, 208, 231
  Byrd, 186, 228
  Camp Century, 231
  Devon, 231
  Dome C, 17, 96, 97, 181
  Dome Fuji, 184
  Dye 3, 204, 231
  EPICA Dronning Maud Land, 191
  GISP2, 45, 97, 191, 199, 201, 203, 207,
    240, 246
  GRIP, 45, 166, 168, 177,
    204, 209
  GRIP2, 212
  Law Dome, 106
  Nansen Fjord, 229
  North Greenland Eemian, 380
  North GRIP, 44, 177, 191, 231
  Renland, 218, 231
  Siple Dome, 207, 247
  Summit, 231
  Vostok, 43, 183, 245
ice-rafted debris, 86, 165, 201
ice rafting, 86

ice sheets, 56
  Antarctic, 7, 12, 59, 253
  Antarctic Peninsula, 187
  Barents, 248
  Barents–Kara, 165, 169, 172, 179
  Central Arctic, 170
  Cordilleran, 170, 175, 180
  East Antarctic, 85, 186
  Fennoscandian, 80, 204, 248
  future, 367
  Greenland, 12, 39, 59, 115, 165, 166, 321
  Innuitian, 174, 180
  Laurentide, 50, 80, 86, 97, 168, 170, 173, 177, 179,
    203, 219, 235, 248
  Quebec–Labrador, 170
  Ross, 250
  Scandinavian, 170, 173
  thickness, 8
  West Antarctic, 15, 186, 251, 323, 370
ice shelves, 8, 180, 320
  Antarctic Peninsula, 15, 37, 371
  Holocene, 252
  King George VI, 222, 252
  Larsen A, 228, 232, 329
  Larsen B, 48, 89, 222, 253, 330
  Prince Gustav, 329
  Ronne–Filchner, 8, 76, 140
  Ross, 8, 71
  Ward Hunt, 329, 371
ice streams, 8, 12
  velocity, 38
ice wedge *see* permafrost
icebergs, 8, 88, 89
  calving, 323
Iceland, 228, 231, 240, 252, 258
  sea ice, 50, 289
  volcanic eruptions, 202
Iceland basin, 237
Icelandic Low, 102, 114, 118, 119, 205, 261, 295,
  341, 347
indigenous people, 5, 13
interannual variability, 148
interglacials, 90
Intergovernmental Panel on Climate Change, 62, 336
International Arctic Buoy Programme, 126
International Geophysical Year, 18, 62, 257
International Polar Year, 17
International Trans-Antarctic Scientific Expedition
  (ITASE), 46

Jakobshavn Glacier, 322, 329
James Ross Island, 227, 250, 252
jets, atmospheric, 111
jökulhlaups, 86

Kangerdlugssuaq Glacier, 322
Kara Sea, 115

Index

Kerguelen Plateau, 160
King George Island, 22, 227, 250, 252
King Sejong Station, 317

Labrador Current, 155
Labrador Sea, 105, 136, 219, 282, 295
Lake Agassiz, 50, 179, 219, 248
lake ice, 361
Lake Ojibway, 219
Lapland, 50
Laptev Sea, 155
Larsemann Hills, 227, 250
Last Glacial Maximum, 52, 97, 186, 194, 207, 242, 247
Late Saalian glaciation, 165
Laurentide Ice Sheet *see* ice sheets: Laurentide
Laurie Island, 256
Law Dome, 296
Leads *see* sea ice
Lena River, 4
Little Ice Age, 205, 213, 229, 233, 251, 254, 289
Livingston Island, 227, 232, 250, 252
longwave radiation, 64, 133

Mackenzie River, 4
Madden–Julien Oscillation, 109
Marguerite Bay, 250
Marie Byrd Land, 267, 314
marine ecosystems, 14
Marine Isotope Stages, 162
mass balance, 35, 38, 71, 320
    Antarctic, 325
    future, 367, 370
    Greenland, 322
Maunder Minimum, 61, 199, 232
McMurdo Sound, 250
Medieval Cold Period, 229
Medieval Warm Period, 205, 213, 228, 233
Meltwater Pulse 1A, 187
Meridional Overturning Circulation, 176, 201, 346
    future, 353
mesocyclones, 37, 78, 136, 137
mesosphere, 129
methane *see* greenhouse gases: methane
methanesulphonic acid, 44, 96, 111, 244, 296
Miami Isopycnic Coordinate Ocean Model (MICOM), 59
Mid Holocene Hypsithermal, 226
mid-Brunhes event, 85, 181
Middle Weichselian, 170
Mid Holocene Climatic Optimum, 222, 244, 254
Mid Holocene Hypsithermal, 213
Mid Pleistocene transition, 84, 163
Milankovitch Theory, 80
Mirny Station, 130
moist static energy, 67
moisture flux, 75
Montreal Protocol, 95, 343
mother of pearl clouds *see* polar stratospheric cloud

Nansen Fjord, 229
Nansen Trough, 236
National Centers for Environmental Prediction (NCEP)
    reanalysis, 30
net radiation, 67
New Zealand, 79
Newfoundland, 238
nilas, 142
non sea-salt sulphate, 96
Normalised Difference Vegetation Index, 310
North Atlantic Current, 155, 168
North Atlantic Drift, 13
North Atlantic Oscillation, 101, 102, 150, 155, 156,
    180, 199, 205, 261, 272, 282, 294, 341
North Pole Environmental Observatory, 284
Northern Annular Mode, 99, 101, 116, 333
Norwegian Current, 146
Norwegian Sea, 78, 129, 135
Novolazarevskya Station, 266
nunataks, 306
Ny Ålesund Station, 128

ocean
    acidity, 353
    albedo, 2
    circulation, 61
    convection, 282
    deep convection, 78
    eddies, 58
    salinity, 159
    temperature, 155, 158
Ocean Circulation and Climate Advanced Modelling
    (OCCAM), 59
ocean fronts
    Antarctic polar front, 157
    Southern Antarctic Circumpolar Current Front, 157
    sub-Antarctic front, 157, 242
    subtropical front, 157
Odden, 105, 148
Oldest Dryas, 178, 192
Orcadas Station, 256
ozone
    measurement, 22
    trends, 318
ozone hole, 14, 15, 25, 54, 62, 75, 92, 137, 266, 269,
    277, 317, 334, 376
    area, 366
    future, 365

Pacific Decadal Oscillation, 259, 273, 375
Pacific North American Association, 108, 273
Pacific South American Association, 109, 120,
    131, 279
Palaeoclimate Model Intercomparison Project, 223
Palmer Deep drilling site, 250, 252
Patagonia, 45, 97
penguin optimum, 226

Penn State University/NCAR Mesoscale Model 5 (MM5), 57
permafrost, 4, 5, 12, 13, 57, 168, 225
  Antarctic, 8, 313
  extent, 5, 309
  future, 362
  Holocene, 254
  thickness, 5, 307
phytoplankton, 288
Pinatubo eruption, 99
Pine Island Bay, 186, 188
Pine Island Glacier *see* glaciers: Pine Island
pingos, 308
Piora Oscillation, 225
Plateau Station, 22
polar circulation index, 203
polar front jet, 110
polar front, atmospheric, 96, 97, 207
polar lows *see* mesocyclones
polar night jet, 92, 99
polar stratospheric cloud, 93, 137, 269
polar vortex, 92, 99, 102, 132, 203, 213, 224, 266, 272, 332
pole of variability, 120
poleward heat flux, 10
pollution, 3, 95, 317
polynya, 76, 123, 159
  coastal, 2
Pre-Boreal cold event, 216
preboreal interval, 204
precipitation, 3, 7, 15, 35, 70, 111, 204
  amounts, 138
  Antarctica, 322
  future, 346, 351
  Greenland, 321
  measurement, 22
  prediction, 344
  trends, 305
Prince Gustav Channel, 227, 253
Program for Arctic Regional Climate Assessment, 46

Queen Maud Land, 2, 186

radar altimeter, 297
radiocarbon analysis, 50
radiosonde, 25, 105, 129, 268
rain, 138
reflectivity *see* albedo
Regional Atmospheric Climate Model (RACMO), 57, 140
Regional Climate Model Intercomparison Project, 58
rime, 21
river discharge, 155, 281
river ice, 361
river runoff, 2
  future, 354
Ross Sea, 111, 151

Rossby wave propagation, 109
Rothera Station, 138, 264

salinity, 4, 105, 220, 236, 282
  future, 353
satellite imagery, 36
Scientific Ice Expeditions programme, 149, 284
Scotia Arc, 160
Scotia Sea, 287
sea ice, 2, 4
  albedo, 2
  classification, 37
  concentration, 37, 144
  decrease, 12, 356
  edge, 8
  extent, 4, 8, 48, 50, 63, 145, 182, 238, 265, 356
  fast ice, 73, 144
  first year, 8, 40, 73, 143, 153
  future, 356
  Holocene, 238
  Iceland, 240
  leads, 3, 70, 77, 92, 96, 135, 143
  melt ponds, 72
  motion, 27, 76, 148, 152
  multi-year, 2, 4, 8, 73, 143, 145, 357
  production, 76
  thickness, 4, 8, 40, 144, 149, 153, 297, 359, 360
  types, 142
sea level, 12, 59, 86, 165, 169, 171, 183, 187, 249, 251, 320, 323
  future, 371
  global rise, 330
  Holocene, 253
Sea of Japan, 78, 147
Sea of Okhotsk, 146
sea salt, 44
seasonally frozen ground *see* permafrost
semi-annual oscillation, 121, 130, 160, 343
Severnaya Zemlya, 170
Siberia, 5, 49, 258, 261, 375
Siberian High, 114
Siberian Low, 204
Signy Island, 222, 227, 232, 252, 262, 295, 315, 328
Siple Dome, 96, 186
snow, 71
  depth, 4
  extent, 299, 306
  future, 361
  reflectivity, 2
  trends, 301, 306
  variability, 301
snow ice, 71
snowfall, 138, 215
solar activity, 212
solar cycle, 199
solar radiation, 1, 9, 63, 197
South Georgia, 213, 227, 232, 252

South Pacific Convergence Zone, 138
South Pacific split jet, 110
South Shetland Islands, 227
Southern Annular Mode, 35, 101, 131, 265, 266, 277, 285, 333
  index, 106, 277
  prediction, 343
Southern Ocean, 11, 62, 157
  freshening, 286
  precipitation, 351
Southern Oscillation Index *see* El Niño-Southern Oscillation
storm surges, 13
stratospheric ozone depletion *see* ozone hole
stratospheric warmings, 116
sub-Antarctic islands, 305
Sub-Atlantic period, 228
Sub-Boreal period, 228
subpolar gyres, 286
subtropical jet, 109, 125
sulphates, 316
sulphuric acid, 45
surface energy balance, 71
Surface Heat Budget of the Arctic Ocean (SHEBA) experiment, 97
surface mass balance, 140
Svalbard, 95, 170, 238
Syowa Station, 266

taiga *see* boreal forest
taiga-tundra feedback, 217
talik, 14, 309
teleconnections, 101, 108, 110, 273
temperature, 7, 65
  future, 348
  highest, 7
  inversion, 3, 20, 41, 127, 129
  lowest, 2, 7, 126, 129
  measurement, 20
  Mid Holocene optimum, 208
  prediction, 344
terminations, 85, 91, 163, 165, 167
thermokarst lakes, 308
thermohaline circulation, 10, 13, 55, 61, 86, 89, 103, 106, 219, 321
thermokarst, 364
Thwaites Glacier *see* glaciers: Thwaites
tipping point, 294
tourism, 9
Transantarctic Mountains, 313
Transpolar Drift Stream, 4, 50, 148, 155, 205, 241, 295, 297
tree ring data, 50, 213, 231
treeline, 203
tropopause, 132
tundra, 4, 12, 13, 133, 165, 170, 224, 310

ultraviolet radiation, 92
ultraviolet-B (UV-B) radiation, 14, 92, 320
upward-looking sonar, 144, 297

visibility, 3
volcanic cooling, 202
volcanic eruptions, 45, 95, 202
volcanic forcing, 199
Vostok Station, 2, 7, 20, 35, 265

water masses, 4
  Antarctic Bottom Water, 8, 10, 14, 15, 77, 84, 159, 176, 286, 355
  Antarctic Intermediate Water, 157, 159
  Antarctic Surface Water, 157, 158
  Arctic Ocean Deep Water, 4
  Atlantic Intermediate Water, 236
  Atlantic Water, 105, 165, 284
  Circumpolar Deep Water, 158, 287
  Denmark Strait Overflow Water, 156
  High Salinity Shelf Water, 159
  Ice Shelf Water, 159
  Iceland-Scotland Overflow Water, 156, 237
  Labrador Sea Intermediate Water, 220, 235
  Labrador Sea Water, 156
  Lower Circumpolar Deep Water, 159
  North Atlantic Deep Water, 105, 156, 159, 176, 181, 201, 353
  Norwegian Atlantic Water, 283
  Pacific Water, 284
  Polar Water, 237, 282
  Sub-Antarctic Mode Water, 157, 159
  Upper Circumpolar Deep Water, 158, 159
  Warm Deep Water, 286
  Weddell Deep Water, 160
  Weddell Sea Bottom Water, 160, 286
  Weddell Sea Deep Water, 286
water vapour *see* greenhouse gases: water vapour
Weather Research and Forecasting (WRF) model, 57
weather systems
  mid-latitude, 3
  tracking, 37, 66
Weddell Gyre, 159, 161
Weddell Polynya, 77, 286
Weddell Sea, 2, 48, 153
  sea ice, 2
West Antarctica, 7
  warming, 266
West Greenland Current, 146
West Kamchatka Current, 146
West Spitsbergen Current, 155
westerlies, 207, 266, 272

westerlies, (cont.)
  future, 355
whaling records, 51, 295
wind
  future, 351
  katabatic, 10, 23, 37, 119, 122
  measurement, 20

speed, 40
trends, 271
Windmill Islands, 250

Yenisey River, 4
Younger Dryas, 43, 97, 173, 179, 180, 194, 201, 204,
  213, 215, 219, 235, 246

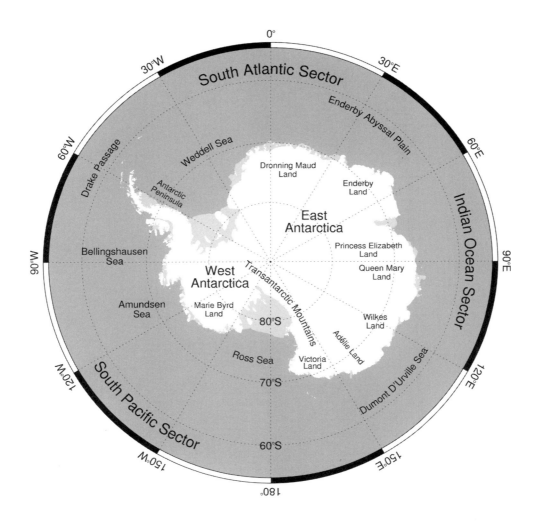